Fifth Edition

Partial Solutions Guide

Chemical Principles

Steven S. Zumdahl

Thomas J. Hummel
Steven S. Zumdahl

University of Illinois at Urbana-Champaign

HOUGHTON MIFFLIN COMPANY **Boston** **New York**

Vice President and Publisher: Charles Hartford
Executive Editor: Richard Stratton
Development Editor: Rebecca Berardy
Editorial Associate: Rosemary Mack
Senior Project Editor: Cathy Labresh Brooks
Senior Production/Design Coordinator: Jill Haber
Senior Manufacturing Coordinator: Marie Barnes
Senior Marketing Manager: Katherine Greig
Marketing Associate: Alexandra Shaw

Cover Image: Karl E. Deckart/Molecular Expressions

Printed in the U.S.A.

ISBN: **0-618-37208-3**

5 6 7 8 9-VHO-08 07

TABLE OF CONTENTS

TO THE STUDENT: HOW TO USE THIS GUIDE

Solutions to all of the odd-numbered end of chapter exercises are in this manual. This "Solutions Guide" can be very valuable if you use it properly. The way NOT to use it is to look at an exercise in the book and then immediately check the solution, often saying to yourself, "That's easy, I can do it." Developing problem solving skills takes practice. Don't look up a solution to a problem until you have tried to work it on your own. If you are completely stuck, see if you can find a similar problem in the Sample Exercises in the chapter. Only look up the solution as a last resort. If you do this for a problem, look for a similar problem in the end of chapter exercises and try working it. The more problems you do, the easier chemistry becomes. It is also in your self interest to try to work as many problems as possible. Most exams that you will take in chemistry will involve a lot of problem solving. If you have worked several problems similar to the ones on an exam, you will do much better than if the exam is the first time you try to solve a particular type of problem. No matter how much you read and study the text, or how well you think you understand the material, you don't really understand it until you have taken the information in the text and applied the principles to problem solving. You will make mistakes, but the good students learn from their mistakes.

In this manual we have worked problems as in the textbook. We have shown intermediate answers to the correct number of significant figures and used the rounded answer in later calculations. Thus, some of your answers may differ slightly from ours. When we have not followed this convention, we have usually noted this in the solution. The most common exception is when working with the natural logarithm (ln) function, where we usually carried extra significant figures in order to reduce round-off error. In addition, we tried to use constants and conversion factors reported to at least one more significant figure as compared to numbers given in the problem. For some problems, this required the use of more precise atomic masses for H, C, N and O as given in Chapter 3. This practice of carrying one extra significant figure in constants helps minimize round-off error.

We are grateful to Claire Szoke for her outstanding effort in preparing the manuscript of this manual. We also thank Linda C. Bush and Jon Booze for their careful and thorough accuracy review of the Solutions Guide. We also are grateful to Don DeCoste for his assistance in creating solutions to some of the problems in this Solutions Guide.

TJH
SSZ

CHAPTER TWO

ATOMS, MOLECULES, AND IONS

Development of the Atomic Theory

19. **From Avogadro's hypothesis, volume ratios are equal to molecule ratios at constant temperature and pressure. Therefore, we can write a balanced equation using the volume data, $Cl_2 + 3\ F_2 \rightarrow 2\ X$. Two molecules of X contain 6 atoms of F and two atoms of Cl. The formula of X is ClF_3 for a balanced reaction.**

21. **To get the atomic mass of H to be 1.00, we divide the mass that reacts with 1.00 g of oxygen by 0.126, i.e., $\frac{0.126}{0.126} = 1.00$. To get Na, Mg, and O on the same scale, we do the same division.**

Na: $\frac{2.875}{0.126} = 22.8$; Mg: $\frac{1.500}{0.126} = 11.9$; O: $\frac{1.00}{0.126} = 7.94$

	H	O	Na	Mg
Relative Value	**1.00**	**7.94**	**22.8**	**11.9**
Accepted Value	**1.0079**	**15.999**	**22.99**	**24.31**

The atomic masses of O and Mg are incorrect. The atomic masses of H and Na are close. Something must be wrong about the assumed formulas of the compounds. It turns out the correct formulas are H_2O, Na_2O, and MgO. The smaller discrepancies result from the error in the assumed atomic mass of H.

The Nature of the Atom

23. **β particles are electrons. A cathode ray is a stream of electrons (β particles).**

25. **First, divide all charges by the smallest quantity, 6.40×10^{-13}.**

$$\frac{2.56 \times 10^{-12}}{6.40 \times 10^{-13}} = 4.00;\quad \frac{7.68}{0.640} = 12.00;\quad \frac{3.84}{0.640} = 6.00$$

Since all charges are whole number multiples of 6.40×10^{-13} zirkombs then the charge on one electron could be 6.40×10^{-13} zirkombs. However, 6.40×10^{-13} zirkombs could be the charge of two electrons (or three electrons, etc.). All one can conclude is that the charge of an electron is 6.40×10^{-13} zirkombs or an integer fraction of 6.40×10^{-13}.

27. The proton and neutron have similar mass with the mass of the neutron slightly larger than that of the proton. Each of these particles has a mass approximately 1800 times greater than that of an electron. The combination of the protons and the neutrons in the nucleus makes up the bulk of the mass of an atom, but the electrons make the greatest contribution to the chemical properties of the atom.

Elements and the Periodic Table

29. The atomic number of an element is equal to the number of protons in the nucleus of an atom of that element. The mass number is the sum of the number of protons plus neutrons in the nucleus. The atomic mass is the actual mass of a particular isotope (including electrons). As we will see in chapter three, the average mass of an atom is taken from a measurement made on a large number of atoms. The average atomic mass value is listed in the periodic table.

31. a. Eight; Li to Ne
b. Eight; Na to Ar
c. Eighteen; K to Kr
d. Four; Fe, Ru, Os, Hs
e. Five; O, S, Se, Te, Po
f. Four; Ni, Pd, Pt, Uun (#110)

33. a. ${}^{24}_{12}Mg$: 12 protons, 12 neutrons, 12 electrons

b. ${}^{24}_{12}Mg^{2+}$: 12 p, 12 n, 10 e
c. ${}^{59}_{27}Co^{2+}$: 27 p, 32 n, 25 e

d. ${}^{59}_{27}Co^{3+}$: 27 p, 32 n, 24 e
e. ${}^{59}_{27}Co$: 27 p, 32 n, 27 e

f. ${}^{79}_{34}Se$: 34 p, 45 n, 34 e
g. ${}^{79}_{34}Se^{2-}$: 34 p, 45 n, 36 e

h. ${}^{63}_{28}Ni$: 28 p, 35 n, 28 e
i. ${}^{59}_{28}Ni^{2+}$: 28 p, 31 n, 26 e

35. Metals lose electrons to form cations in ionic compounds and nonmetals gain electrons to form anions. Group IA, IIA and IIIA metals form stable +1, +2 and +3 charged cations, respectively. Group VA, VIA and VIIA nonmetals form -3, -2 and -1 charged anions, respectively.

a. Lose one e^- to form Na^+.
b. Lose two e^- to form Sr^{2+}.

c. Lose two e^- to form Ba^{2+}.
d. Gain one e^- to form I^-.

e. Lose three e^- to form Al^{3+}.
f. Gain two e^- to form S^{2-}.

g. Gain three e^- to form N^{3-}.
h. Lose one e^- to form Cs^+.

i. Gain two e^- to form Se^{2-}.

Nomenclature

37. a. sulfur difluoride b. dinitrogen tetroxide
c. iodine trichloride d. tetraphosphorus hexoxide

39. a. copper(I) iodide b. copper(II) iodide c. cobalt(II) iodide
d. sodium carbonate e. sodium hydrogen carbonate or sodium bicarbonate
f. tetrasulfur tetranitride g. sulfur hexafluoride h. sodium hypochlorite
i. barium chromate j. ammonium nitrate

41. a. SO_2 b. SO_3 c. Na_2SO_3 d. $KHSO_3$
e. Li_3N f. $Cr_2(CO_3)_3$ g. $Cr(C_2H_3O_2)_2$ h. SnF_4
i. NH_4HSO_4: Composed of NH_4^+ and HSO_4^- ions.
j. $(NH_4)_2HPO_4$ k. $KClO_4$ l. NaH
m. HBrO n. HBr

43. a. $Pb(C_2H_3O_2)_2$: lead(II) acetate b. $CuSO_4$: copper(II) sulfate
c. CaO: calcium oxide d. $MgSO_4$: magnesium sulfate
e. $Mg(OH)_2$: magnesium hydroxide f. $CaSO_4$: calcium sulfate
g. N_2O: dinitrogen monoxide or nitrous oxide (common name)

Additional Exercises

45. a. nitric acid, HNO_3 b. perchloric acid, $HClO_4$ c. acetic acid, $HC_2H_3O_2$
d. sulfuric acid, H_2SO_4 e. phosphoric acid, H_3PO_4

47. From the XBr_2 formula, the charge on element X is +2. Therefore, the element has 88 protons, which identifies it as radium, Ra. 230 - 88 = 142 neutrons

49. In the case of sulfur, SO_4^{2-} is sulfate and SO_3^{2-} is sulfite. By analogy:
SeO_4^{2-}: selenate; SeO_3^{2-}: selenite; TeO_4^{2-} tellurate; TeO_3^{2-}: tellurite

51. If the formula is InO, then one atomic mass of In would combine with one atomic mass of O, or:

$$\frac{A}{16.00} = \frac{4.784 \text{ g In}}{1.000 \text{ g O}}, \text{ A = atomic mass of In} = 76.54$$

If the formula is In_2O_3, then two times the atomic mass of In will combine with three times the atomic mass of O, or:

$$\frac{2\text{A}}{(3)16.00} = \frac{4.784\text{ g In}}{1.000\text{ g O}},\ \text{A} = \text{atomic mass of In} = 114.8$$

The latter number is the atomic mass of In used in the modern periodic table.

53. Hydrazine: 1.44×10^{-1} g H/g N; Ammonia: 2.16×10^{-1} g H/g N

Hydrogen azide: 2.40×10^{-2} g H/g N

Let's try all of the ratios:

$$\frac{0.216}{0.144} = 1.50 = \frac{3}{2};\ \frac{0.144}{0.0240} = 6.00;\ \frac{0.216}{0.0240} = 9.00$$

All the masses of hydrogen in these three compounds can be expressed as simple whole number ratios of each other. The g H/g N in hydrazine, ammonia, and hydrogen azide are in the ratios 6:9:1.

55. Yes, 1.0 g H would react with 37.0 g ^{37}Cl and 1.0 g H would react with 35.0 g ^{35}Cl.

No, the mass ratio of H/Cl would always be 1 g H/37 g Cl for ^{37}Cl and 1 g H/35 g Cl for ^{35}Cl. As long as we had pure ^{35}Cl or pure ^{37}Cl, the above ratios will always hold. If we have a mixture (such as the natural abundance of chlorine), the ratio will also be constant as long as the composition of the mixture of the two isotopes does not change.

Challenge Problems

57. a. Both compounds have C_2H_6O as the formula. Since they have the same formula, their mass percent composition will be identical. However, these are different compounds with different properties since the atoms are bonded together differently. These compounds are called isomers of each other.

b. When wood burns, most of the solid material in wood is converted to gases, which escape. The gases produced are most likely CO_2 and H_2O.

c. The atom is not an indivisible particle, but is instead composed of other smaller particles, e.g., electrons, neutrons, protons.

d. The two hydride samples contain different isotopes of either hydrogen and/or lithium. Although the compounds are composed of different isotopes, their properties are similar because different isotopes of the same element have similar properties (except, of course, their mass).

59. Compound I: $\frac{14.0\text{ g R}}{3.00\text{ g Q}} = \frac{4.67\text{ g R}}{1.00\text{ g Q}}$; Compound II: $\frac{7.00\text{ g R}}{4.50\text{ g Q}} = \frac{1.56\text{ g R}}{1.00\text{ g Q}}$

The ratio of the masses of R that combines with 1.00 g Q is: $\frac{4.67}{1.56} = 2.99 \approx 3$

As expected from the law of multiple proportions, this ratio is a small whole number.

Since compound I contains three times the mass of R per gram of Q as compared to compound II (RQ), then the formula of compound I should be R_3Q.

61. Avogadro proposed that equal volumes of gases (at constant temperature and pressure) contain equal numbers of molecules. In terms of balanced equations, Avogadro's hypothesis implies that volume ratios will be identical to molecule ratios. Assuming one molecule of octane (C_xH_y) reacting, then 1 molecule of C_xH_y produces 8 molecules of CO_2 and 9 molecules of H_2O. $C_xH_y + n\ O_2 \rightarrow 8\ CO_2 + 9\ H_2O$. Since all the carbon in octane ends up as carbon in CO_2, then octane contains 8 atoms of C. Similarly, all hydrogen in octane ends up as hydrogen in H_2O, so one molecule of octane contains $9 \times 2 = 18$ atoms of H. Octane formula = C_8H_{18} and the ratio of C:H = 8:18 or 4:9.

CHAPTER THREE

STOICHIOMETRY

Atomic Masses and the Mass Spectrometer

23. A = atomic mass = 0.7899(23.9850 amu) + 0.1000(24.9858 amu) + 0.1101(25.9826 amu)

    A = 18.95 amu + 2.499 amu + 2.861 amu = 24.31 amu

25. 186.207 = 0.6260(186.956) + 0.3740(A), 186.207 - 117.0 = 0.3740(A)

    $A = \frac{69.2}{0.3740}$ = 185 amu (A = 184.95 amu without rounding to proper significant figures.)

27. There are three peaks in the mass spectrum, each 2 mass units apart. This is consistent with two isotopes, differing in mass by two mass units. The peak at 157.84 corresponds to a Br_2 molecule composed of two atoms of the lighter isotope. This isotope has mass equal to 157.84/2 or 78.92. This corresponds to ^{79}Br. The second isotope is ^{81}Br with mass equal to 161.84/2 = 80.92. The peaks in the mass spectrum correspond to $^{79}Br_2$, $^{79}Br^{81}Br$ and $^{81}Br_2$ in order of increasing mass. The intensities of the highest and lowest mass tell us the two isotopes are present at about equal abundance. The actual abundance is 50.68% ^{79}Br and 49.32% ^{81}Br.

29. GaAs can be either $^{69}GaAs$ or $^{71}GaAs$. The mass spectrum for GaAs will have 2 peaks at 144 (= 69 + 75) and 146 (= 71 + 75) with intensities in the ratio of 60:40 or 3:2.

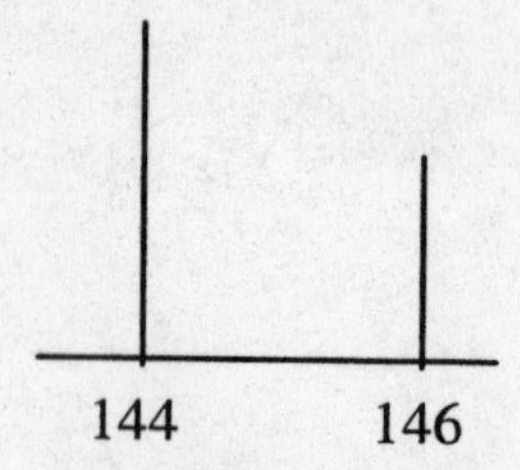

Ga_2As_2 can be $^{69}Ga_2As_2$, $^{69}Ga^{71}GaAs_2$ or $^{71}Ga_2As_2$. The mass spectrum will have 3 peaks at 288, 290, and 292 with intensities in the ratio of 36:48:16 or 9:12:4. We get this ratio from the following probability table:

	^{69}Ga (0.60)	^{71}Ga (0.40)
^{69}Ga (0.60)	0.36	0.24
^{71}Ga (0.40)	0.24	0.16

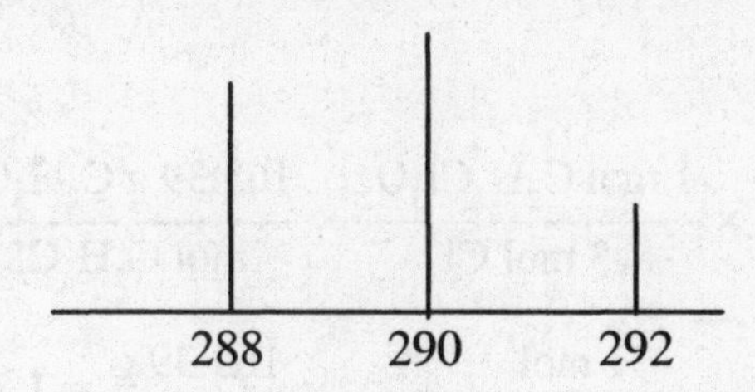

Moles and Molar Masses

31. a. 2.49×10^{20} molecules CO $\times \dfrac{1 \text{ mol CO}}{6.022 \times 10^{23} \text{ molecules CO}} = 4.13 \times 10^{-4}$ mol CO

b. $15.0 \text{ g } CuSO_4 \times \dfrac{1 \text{ mol } CuSO_4}{159.62 \text{ g } CuSO_4} = 9.40 \times 10^{-2} \text{ mol } CuSO_4$

c. $100 \text{ molecules } H_2SO_4 \times \dfrac{1 \text{ mol } H_2SO_4}{6.022 \times 10^{23} \text{ molecules } H_2SO_4} = 1.661 \times 10^{-22} \text{ mol } H_2SO_4$

d. $6.210 \text{ mg } K_2O \times \dfrac{1 \text{ g}}{1000 \text{ mg}} \times \dfrac{1 \text{ mol } K_2O}{94.20 \text{ g } K_2O} = 6.592 \times 10^{-5} \text{ mol } K_2O$

33. $4.0 \text{ g } H_2 \times \dfrac{1 \text{ mol } H_2}{2.016 \text{ g } H_2} \times \dfrac{2 \text{ mol H}}{1 \text{ mol } H_2} \times \dfrac{6.02 \times 10^{23} \text{ atoms H}}{1 \text{ mol H}} = 2.4 \times 10^{24} \text{ atoms}$

$4.0 \text{ g He} \times \dfrac{1 \text{ mol He}}{4.003 \text{ g He}} \times \dfrac{6.02 \times 10^{23} \text{ atoms He}}{1 \text{ mol He}} = 6.0 \times 10^{23} \text{ atoms}$

$1.0 \text{ mol } F_2 \times \dfrac{2 \text{ mol F}}{1 \text{ mol } F_2} \times \dfrac{6.02 \times 10^{23} \text{ atoms F}}{1 \text{ mol F}} = 1.2 \times 10^{24} \text{ atoms}$

$44.0 \text{ g } CO_2 \times \dfrac{1 \text{ mol } CO_2}{44.01 \text{ g } CO_2} \times \dfrac{3 \text{ mol atom}(1 \text{ C} + 2 \text{ O})}{1 \text{ mol } CO_2} \times \dfrac{6.022 \times 10^{23} \text{ atoms}}{1 \text{ mol atoms}} = 1.81 \times 10^{24} \text{ atoms}$

$146. \text{ g } SF_6 \times \dfrac{1 \text{ mol } SF_6}{146.07 \text{ g } SF_6} \times \dfrac{7 \text{ mol atoms } (1 \text{ S} + 6 \text{ F})}{1 \text{ mol } SF_6} \times \dfrac{6.022 \times 10^{23} \text{ atoms}}{1 \text{ mol atoms}} = 4.21 \times 10^{24} \text{ atoms}$

$4.0 \text{ g He} < 1.0 \text{ mol } F_2 < 44.0 \text{ g } CO_2 < 4.0 \text{ g } H_2 < 146 \text{ g } SF_6$

35. a. $2(12.01) + 3(1.008) + 3(35.45) + 2(16.00) = 165.39$ g/mol

b. $500.0 \text{ g} \times \frac{1 \text{ mol}}{165.39 \text{ g}} = 3.023 \text{ mol}$ c. $2.0 \times 10^{-2} \text{ mol} \times \frac{165.39 \text{ g}}{\text{mol}} = 3.3 \text{ g}$

d. $5.0 \text{ g } C_2H_3Cl_3O_2 \times \frac{1 \text{ mol}}{165.39 \text{ g}} \times \frac{6.02 \times 10^{23} \text{ molecules}}{\text{mol}} \times \frac{3 \text{ atoms Cl}}{\text{molecule}}$

$= 5.5 \times 10^{22}$ atoms of chlorine

e. $1.0 \text{ g Cl} \times \frac{1 \text{ mol Cl}}{35.45 \text{ g}} \times \frac{1 \text{ mol } C_2H_3Cl_3O_2}{3 \text{ mol Cl}} \times \frac{165.39 \text{ g } C_2H_3Cl_3O_2}{\text{mol } C_2H_3Cl_3O_2} = 1.6$ g chloral hydrate

f. $500 \text{ molecules} \times \frac{1 \text{ mol}}{6.022 \times 10^{23} \text{ molecules}} \times \frac{165.39 \text{ g}}{\text{mol}} = 1.373 \times 10^{-19} \text{ g}$

Percent Composition

37. In 1 mole of $YBa_2Cu_3O_7$ there are 1 mole of Y, 2 moles of Ba, 3 moles of Cu, and 7 moles of O.

$$\text{Molar mass} = 1 \text{ mol Y} \times \frac{88.91 \text{ g Y}}{\text{mol Y}} + 2 \text{ mol Ba} \times \frac{137.3 \text{ g Ba}}{\text{mol Ba}}$$

$$+ 3 \text{ mol Cu} \times \frac{63.55 \text{ g Cu}}{\text{mol Cu}} + 7 \text{ mol O} \times \frac{16.00 \text{ g O}}{\text{mol O}}$$

Molar mass = $88.91 + 274.6 + 190.65 + 112.00 = 666.2$ g/mol

$\% \text{ Y} = \frac{88.91 \text{ g}}{666.2 \text{ g}} \times 100 = 13.35\% \text{ Y}$; $\% \text{ Ba} = \frac{274.6 \text{ g}}{666.2 \text{ g}} \times 100 = 41.22\% \text{ Ba}$

$\% \text{ Cu} = \frac{190.65 \text{ g}}{666.2 \text{ g}} \times 100 = 28.62\% \text{ Cu}$; $\% \text{ O} = \frac{112.0 \text{ g}}{666.2 \text{ g}} \times 100 = 16.81\% \text{ O}$

or $\% \text{ O} = 100.00 - (13.35 + 41.22 + 28.62) = 100.00 - (83.19) = 16.81\% \text{ O}$

39. For each compound, determine the mass of one mol of compound then calculate the mass percentage of N in one mole of that compound.

a. NO: $\% \text{ N} = \frac{14.007 \text{ g N}}{30.006 \text{ g NO}} \times 100 = 46.681\% \text{ N}$

b. NO_2: $\% \text{ N} = \frac{14.007 \text{ g N}}{46.005 \text{ g } NO_2} \times 100 = 30.447\% \text{ N}$

c. N_2O_4: $\% \text{ N} = \frac{28.014 \text{ g N}}{92.010 \text{ g } N_2O_4} \times 100 = 30.447\% \text{ N}$

d. N_2O: $\%N = \frac{28.014 \text{ g N}}{44.013 \text{ g } N_2O} \times 100 = 63.649\% \text{ N}$

41. There are 0.390 g Cu for every 100.00 g of fungal laccase. Assuming 100.00 g fungal laccase:

$$\text{mol fungal laccase} = 0.390 \text{ g Cu} \times \frac{1 \text{ mol Cu}}{63.55 \text{ g Cu}} \times \frac{1 \text{mol fungal laccase}}{4 \text{ mol Cu}} = 1.53 \times 10^{-3} \text{ mol}$$

$$\frac{x \text{ g fungal laccase}}{1 \text{ mol fungal laccase}} = \frac{100.00 \text{ g}}{1.53 \times 10^{-3} \text{ mol}}, \; x = \text{molar mass} = 6.54 \times 10^4 \text{ g/mol}$$

Empirical and Molecular Formulas

43. a. Molar mass of CH_2O = $1 \text{ mol C}\left(\frac{12.011 \text{ g}}{\text{mol C}}\right) + 2 \text{ mol H}\left(\frac{1.0079 \text{ g H}}{\text{mol H}}\right)$

$$+ 1 \text{ mol O}\left(\frac{15.999 \text{ g}}{\text{mol O}}\right) = 30.026 \text{ g/mol}$$

$$\%C = \frac{12.011 \text{ g C}}{30.026 \text{ g } CH_2O} \times 100 = 40.002\% \text{ C}; \; \%H = \frac{2.0158 \text{ g H}}{30.026 \text{ g } CH_2O} \times 100 = 6.7135\% \text{ H}$$

$$\%O = \frac{15.999 \text{ g O}}{30.026 \text{ g } CH_2O} \times 100 = 53.284\% \text{ O or } \%O = 100.000 - (40.002 + 6.7135) = 53.285\%$$

b. Molar Mass of $C_6H_{12}O_6 = 6(12.011) + 12(1.0079) + 6(15.999) = 180.155$ g/mol

$$\%C = \frac{72.066 \text{ g C}}{180.155 \text{ g } C_6H_{12}O_6} \times 100 = 40.002\%; \quad \%H = \frac{12(1.0079) \text{ g}}{180.155 \text{ g}} \times 100 = 6.7136\%$$

$\%O = 100.00 - (40.002 + 6.7136) = 53.284\%$

c. Molar mass of $HC_2H_3O_2 = 2(12.011) + 4(1.0079) + 2(15.999) = 60.052$ g/mol

$$\%C = \frac{24.022 \text{ g}}{60.052 \text{ g}} \times 100 = 40.002\%; \quad \%H = \frac{4.0316 \text{ g}}{60.052 \text{ g}} \times 100 = 6.7135\%$$

$\%O = 100.00 - (40.002 + 6.7135) = 53.285\%$

All three compounds have the same empirical formula, CH_2O, and different molecular formulas. The composition of all three in mass percent is also the same (within rounding differences). Therefore, elemental analysis will give us only the empirical formula.

45. Compound I: mass O = 0.6498 g Hg_xO_y - 0.6018 g Hg = 0.0480 g O

$$0.6018 \text{ g Hg} \times \frac{1 \text{ mol Hg}}{200.6 \text{ g Hg}} = 3.000 \times 10^{-3} \text{ mol Hg}$$

$$0.0480 \text{ g O} \times \frac{1 \text{ mol O}}{16.00 \text{ g O}} = 3.00 \times 10^{-3} \text{ mol O}$$

The mol ratio between Hg and O is 1:1, so the empirical formula of compound I is HgO.

Compound II: mass Hg = 0.4172 g Hg_xO_y - 0.016 g O = 0.401 g Hg

$$0.401 \text{ g Hg} \times \frac{1 \text{ mol Hg}}{200.6 \text{ g Hg}} = 2.00 \times 10^{-3} \text{ mol Hg}; \quad 0.016 \text{ g O} \times \frac{1 \text{ mol O}}{16.00 \text{ g O}} = 1.0 \times 10^{-3} \text{ mol O}$$

The mol ratio between Hg and O is 2:1, so the empirical formula is Hg_2O.

47. First, we will determine composition in mass percent. We assume all of the carbon in 0.213 g CO_2 came from 0.157 g of the compound and that all of the hydrogen in the 0.0310 g H_2O came from the 0.157 g of the compound.

$$0.213 \text{ g } CO_2 \times \frac{12.01 \text{ g C}}{44.01 \text{ g } CO_2} = 0.0581 \text{ g C}; \quad \% \text{C} = \frac{0.0581 \text{ g C}}{0.157 \text{ g compound}} \times 100 = 37.0\% \text{ C}$$

$$0.0310 \text{ g } H_2O \times \frac{2.016 \text{ g H}}{18.02 \text{ g } H_2O} = 3.47 \times 10^{-3} \text{ g H}; \quad \% \text{H} = \frac{3.47 \times 10^{-3} \text{ g}}{0.157 \text{ g}} = 2.21\% \text{ H}$$

We get the mass % N from the second experiment:

$$0.0230 \text{ g } NH_3 \times \frac{14.01 \text{ g N}}{17.03 \text{ g } NH_3} = 1.89 \times 10^{-2} \text{ g N}$$

$$\% \text{N} = \frac{1.89 \times 10^{-2} \text{ g}}{0.103 \text{ g}} \times 100 = 18.3\% \text{ N}$$

The mass percent of oxygen is obtained by difference:

$$\% \text{O} = 100.00 - (37.0 + 2.21 + 18.3) = 42.5\%$$

So out of 100.00 g of compound, there are:

$$37.0 \text{ g C} \times \frac{1 \text{ mol C}}{12.01 \text{ g C}} = 3.08 \text{ mol C}; \quad 2.21 \text{ g H} \times \frac{1 \text{ mol H}}{1.008 \text{ g H}} = 2.19 \text{ mol H}$$

$$18.3 \text{ g N} \times \frac{1 \text{ mol N}}{14.01 \text{ g N}} = 1.31 \text{ mol N}; \quad 42.5 \text{ g O} \times \frac{1 \text{ mol O}}{16.00 \text{ g O}} = 2.66 \text{ mol O}$$

Lastly, and often the hardest part, we need to find simple whole number ratios. Divide all mole values by the smallest number:

$$\frac{3.08}{1.31} = 2.35; \quad \frac{2.19}{1.31} = 1.67; \quad \frac{1.31}{1.31} = 1.00; \quad \frac{2.66}{1.31} = 2.03$$

Multiplying all these ratios by 3 gives an empirical formula of $C_7H_5N_3O_6$.

49. **Assuming 100.00 g of compound (mass hydrogen = 100.00 g - 49.31 g C - 43.79 g O = 6.90 g H):**

$$49.31 \text{ g C} \times \frac{1 \text{ mol C}}{12.011 \text{ g C}} = 4.105 \text{ mol C}; \quad 6.90 \text{ g H} \times \frac{1 \text{ mol H}}{1.008 \text{ g H}} = 6.85 \text{ mol H}$$

$$43.79 \text{ g O} \times \frac{1 \text{ mol O}}{15.999 \text{ g O}} = 2.737 \text{ mol O}$$

Dividing all mole values by 2.737 gives:

$$\frac{4.105}{2.737} = 1.500; \quad \frac{6.85}{2.737} = 2.50; \quad \frac{2.737}{2.737} = 1.000$$

Since a whole number ratio is required, then the empirical formula is $C_3H_5O_2$.

Empirical formula mass ≈ 3(12.0) + 5(1.0) +2(16.0) = 73.0 g/mol

$$\frac{\text{molar mass}}{\text{empirical formula mass}} = \frac{146.1}{73.0} = 2.00; \quad \text{molecular formula} = (C_3H_5O_2)_2 = C_6H_{10}O_4$$

51. **The combustion data allows determination of the amount of hydrogen in cumene. One way to determine the amount of carbon in cumene is to determine the mass percent of hydrogen in the compound from the data in the problem, then determine the mass percent of carbon by difference (100.0 - mass % H = mass % C).**

$$42.8 \text{ mg } H_2O \times \frac{1 \text{ g}}{1000 \text{ mg}} \times \frac{2.016 \text{ g H}}{18.02 \text{ g } H_2O} \times \frac{1000 \text{ mg}}{\text{g}} = 4.79 \text{ mg H}$$

$$\% \text{ H} = \frac{4.79 \text{ mg H}}{47.6 \text{ mg cumene}} \times 100 = 10.1\% \text{ H}; \quad \% \text{ C} = 100.0 - 10.1 = 89.9\% \text{ C}$$

Now solve this empirical formula problem. Out of 100.0 g cumene, we have:

$$89.9 \text{ g C} \times \frac{1 \text{ mol C}}{12.01 \text{ g C}} = 7.49 \text{ mol C}; \quad 10.1 \text{ g H} \times \frac{1 \text{ mol H}}{1.008 \text{ g H}} = 10.0 \text{ mol H}$$

$\frac{10.0}{7.49} = 1.34 \approx \frac{4}{3}$, **i.e., mol H to mol C are in a 4:3 ratio. Empirical formula = C_3H_4**

Empirical formula mass ≈ 3(12.0) + 4(1.0) = 40.0 g/mol

The molecular formula is $(C_3H_4)_3$ or C_9H_{12} since the molar mass will be between 115 and 125 g/mol (molar mass ≈ 3 × 40.0 g/mol = 120. g/mol).

Balancing Chemical Equations

53. Unbalanced equation:

$$CaF_2{\bullet}3Ca_3(PO_4)_2(s) + H_2SO_4(aq) \rightarrow H_3PO_4(aq) + HF(aq) + CaSO_4{\bullet}2H_2O(s)$$

Balancing Ca^{2+}, F^-, and PO_4^{3-}:

$$CaF_2{\bullet}3Ca_3(PO_4)_2(s) + H_2SO_4(aq) \rightarrow 6\ H_3PO_4(aq) + 2\ HF(aq) + 10\ CaSO_4{\bullet}2H_2O(s)$$

On the right hand side there are 20 extra hydrogen atoms, 10 extra sulfates, and 20 extra water molecules. We can balance the hydrogen and sulfate with 10 sulfuric acid molecules. The extra waters came from the water in the sulfuric acid solution. The balanced equation is:

$$CaF_2{\bullet}3Ca_3(PO_4)_2(s) + 10\ H_2SO_4(aq) + 20\ H_2O(l) \rightarrow 6\ H_3PO_4(aq) + 2\ HF(aq) + 10\ CaSO_4{\bullet}2H_2O(s)$$

55. a. $16\ Cr(s) + 3\ S_8(s) \rightarrow 8\ Cr_2S_3(s)$

b. $2\ NaHCO_3(s) \rightarrow Na_2CO_3(s) + CO_2(g) + H_2O(g)$

c. $2\ KClO_3(s) \rightarrow 2\ KCl(s) + 3\ O_2(g)$

d. $2\ Eu(s) + 6\ HF(g) \rightarrow 2\ EuF_3(s) + 3\ H_2(g)$

e. $2\ C_6H_6(l) + 15\ O_2(g) \rightarrow 12\ CO_2(g) + 6\ H_2O(g)$

Reaction Stoichiometry

57. $1.000\text{ kg Al} \times \frac{1000\text{ g Al}}{\text{kg Al}} \times \frac{1\text{ mol Al}}{26.98\text{ g Al}} \times \frac{3\text{ mol } NH_4ClO_4}{3\text{ mol Al}} \times \frac{117.49\text{ g } NH_4ClO_4}{\text{mol } NH_4ClO_4} = 4355\text{ g } NH_4ClO_4$

59. $Fe_2O_3(s) + 2\ Al(s) \rightarrow 2\ Fe(l) + Al_2O_3(s)$

$15.0\text{ g Fe} \times \frac{1\text{ mol Fe}}{55.85\text{ g Fe}} = 0.269\text{ mol Fe}$; $0.269\text{ mol Fe} \times \frac{2\text{ mol Al}}{2\text{ mol Fe}} \times \frac{26.98\text{ g Al}}{\text{mol Al}} = 7.26\text{ g Al}$

$0.269\text{ mol Fe} \times \frac{1\text{ mol } Fe_2O_3}{2\text{ mol Fe}} \times \frac{159.70\text{ g } Fe_2O_3}{\text{mol } Fe_2O_3} = 21.5\text{ g } Fe_2O_3$

$0.269\text{ mol Fe} \times \frac{1\text{ mol } Al_2O_3}{2\text{ mol Fe}} \times \frac{101.96\text{ g } Al_2O_3}{\text{mol } Al_2O_3} = 13.7\text{ g } Al_2O_3$

61. $100.\text{ g } K_2PtCl_4 \times \frac{1\text{ mol } K_2PtCl_4}{415.1\text{ g } K_2PtCl_4} \times \frac{1\text{ mol } Pt(NH_3)_2Cl_2}{\text{mol } K_2PtCl_4} \times \frac{300.1\text{ g } Pt(NH_3)_2Cl_2}{\text{mol } Pt(NH_3)_2Cl_2}$

$= 72.3\text{ g } Pt(NH_3)_2Cl_2$

$$100.\ g\ K_2PtCl_4 \times \frac{1\ mol\ K_2PtCl_4}{415.1\ g\ K_2PtCl_4} \times \frac{2\ mol\ KCl}{mol\ K_2PtCl_4} \times \frac{74.55\ g\ KCl}{mol\ KCl} = 35.9\ g\ KCl$$

Limiting Reactants and Percent Yield

63. $$1.50\ g\ BaO_2 \times \frac{1\ mol\ BaO_2}{169.3\ g\ BaO_2} = 8.86 \times 10^{-3}\ mol\ BaO_2$$

$$25.0\ mL \times \frac{0.0272\ g\ HCl}{mL} \times \frac{1\ mol\ HCl}{36.46\ g\ HCl} = 1.87 \times 10^{-2}\ mol\ HCl$$

The required mol ratio from the balanced reaction is 2 mol HCl to 1 mol BaO_2. The actual ratio is:

$$\frac{1.87 \times 10^{-2}\ mol\ HCl}{8.86 \times 10^{-3}\ mol\ BaO_2} = 2.11$$

Since the actual mol ratio is larger than the required mol ratio, the denominator (BaO_2) is the limiting reagent.

$$8.86 \times 10^{-3}\ mol\ BaO_2 \times \frac{1\ mol\ H_2O_2}{mol\ BaO_2} \times \frac{34.02\ g\ H_2O_2}{mol\ H_2O_2} = 0.301\ g\ H_2O_2$$

The amount of HCl reacted is:

$$8.86 \times 10^{-3}\ mol\ BaO_2 \times \frac{2\ mol\ HCl}{mol\ BaO_2} = 1.77 \times 10^{-2}\ mol\ HCl$$

excess mol HCl = 1.87×10^{-2} mol - 1.77×10^{-2} mol = 1.0×10^{-3} mol HCl

$$\text{mass of excess HCl} = 1.0 \times 10^{-3}\ mol\ HCl \times \frac{36.46\ g\ HCl}{mol\ HCl} = 3.6 \times 10^{-2}\ g\ HCl$$

65. $$2.50\ \text{metric tons}\ Cu_3FeS_3 \times \frac{1000\ kg}{\text{metric ton}} \times \frac{1000\ g}{kg} \times \frac{1\ mol\ Cu_3FeS_3}{342.71\ g} \times \frac{3\ mol\ Cu}{1\ mol\ Cu_3FeS_3} \times \frac{63.55\ g}{mol\ Cu}$$

$$= 1.39 \times 10^6\ g\ Cu\ (\text{theoretical})$$

$$1.39 \times 10^6\ g\ Cu\ (\text{theoretical}) \times \frac{86.3\ g\ Cu\ (\text{actual})}{100.\ g\ Cu\ (\text{theoretical})} = 1.20 \times 10^6\ g\ Cu = 1.20 \times 10^3\ kg\ Cu$$

$$= 1.20\ \text{metric tons Cu (actual)}$$

67. An alternative method to solve limiting reagent problems is to assume each reactant is limiting then calculate how much product could be produced from each reactant. The reactant that produces the smallest amount of product will run out first and is the limiting reagent.

$$5.00 \times 10^6\ g\ NH_3 \times \frac{1\ mol\ NH_3}{17.03\ g\ NH_3} \times \frac{2\ mol\ HCN}{2\ mol\ NH_3} = 2.94 \times 10^5\ mol\ HCN$$

$$5.00 \times 10^6\ g\ O_2 \times \frac{1\ mol\ O_2}{32.00\ g\ O_2} \times \frac{2\ mol\ HCN}{3\ mol\ O_2} = 1.04 \times 10^5\ mol\ HCN$$

$$5.00 \times 10^6 \text{ g } CH_4 \times \frac{1 \text{ mol } CH_4}{16.04 \text{ g } CH_4} \times \frac{2 \text{ mol HCN}}{2 \text{ mol } CH_4} = 3.12 \times 10^5 \text{ mol HCN}$$

O_2 is limiting since it produces the smallest amount of HCN. Although more product could be produced from NH_3 and CH_4, only enough O_2 is present to produce 1.04×10^5 mol HCN. The mass of HCN produced is:

$$1.04 \times 10^5 \text{ mol HCN} \times \frac{27.03 \text{ g HCN}}{\text{mol HCN}} = 2.81 \times 10^6 \text{ g HCN}$$

$$5.00 \times 10^6 \text{ g } O_2 \times \frac{1 \text{ mol } O_2}{32.00 \text{ g } O_2} \times \frac{6 \text{ mol } H_2O}{3 \text{ mol } O_2} \times \frac{18.02 \text{ g } H_2O}{1 \text{ mol } H_2O} = 5.63 \times 10^6 \text{ g } H_2O$$

69. **$P_4(s) + 6\ F_2(g) \rightarrow 4\ PF_3(g)$; The theoretical yield of PF_3 is:**

$$120. \text{ g } PF_3 \text{ (actual)} \times \frac{100.0 \text{ g } PF_3 \text{ (theoretical)}}{78.1 \text{ g } PF_3 \text{ (actual)}} = 154 \text{ g } PF_3 \text{ (theoretical)}$$

$$154 \text{ g } PF_3 \times \frac{1 \text{ mol } PF_3}{87.97 \text{ g } PF_3} \times \frac{6 \text{ mol } F_2}{4 \text{ mol } PF_3} \times \frac{38.00 \text{ g } F_2}{\text{mol } F_2} = 99.8 \text{ g } F_2$$

99.8 g F_2 are needed to produce 120. g of PF_3 if the percent yield is 78.1%.

Additional Exercises

71. $$17.3 \text{ g H} \times \frac{1 \text{ mol H}}{1.008 \text{ g H}} = 17.2 \text{ mol H}; \quad 82.7 \text{ g C} \times \frac{1 \text{ mol C}}{12.01 \text{ g C}} = 6.89 \text{ mol C}$$

$\frac{17.2}{6.89} = 2.50$; **The empirical formula is C_2H_5.**

The empirical formula mass is ~29 g, so two times the empirical formula would put the compound in the correct range of the molar mass. Molecular formula = $(C_2H_5)_2 = C_4H_{10}$

$$2.59 \times 10^{23} \text{ atoms H} \times \frac{1 \text{ molecule } C_4H_{10}}{10 \text{ atoms H}} \times \frac{1 \text{ mol } C_4H_{10}}{6.022 \times 10^{23} \text{ molecules}} = 4.30 \times 10^{-2} \text{ mol } C_4H_{10}$$

$$4.30 \times 10^{-2} \text{ mol } C_4H_{10} \times \frac{58.12 \text{ g}}{\text{mol } C_4H_{10}} = 2.50 \text{ g } C_4H_{10}$$

73. **If the formula was Be_2O_3, then 2 times the atomic mass of Be would combine with three times the atomic mass of oxygen, or:**

$$\frac{2\text{ A}}{3(16.00)} = \frac{0.5633}{1.000} \quad \text{Solving, A = atomic mass Be = 13.52.}$$

The accepted value is 9.01 and the discrepancy is due to the assumed formula. The actual formula is BeO.

75. **Mass of H_2O = 0.755 g $CuSO_4 \cdot xH_2O$ - 0.483 g $CuSO_4$ = 0.272 g H_2O**

$$0.483 \text{ g } CuSO_4 \times \frac{1 \text{ mol } CuSO_4}{159.62 \text{ g } CuSO_4} = 0.00303 \text{ mol } CuSO_4$$

$$0.272 \text{ g } H_2O \times \frac{1 \text{ mol } H_2O}{18.02 \text{ g } H_2O} = 0.0151 \text{ mol } H_2O$$

$$\frac{0.0151 \text{ mol } H_2O}{0.00303 \text{ mol } CuSO_4} = \frac{4.98 \text{ mol } H_2O}{1 \text{ mol } CuSO_4}; \text{ Compound formula} = CuSO_4 \cdot 5H_2O,\ x = 5$$

77. **Consider the case of aluminum plus oxygen. Aluminum forms Al^{3+} ions; oxygen forms O^{2-} anions. The simplest compound of the two elements is Al_2O_3. Similarly we would expect the formula of any group 6A element with Al to be Al_2X_3. Assuming this, out of 100.00 g of compound there are 18.56 g Al and 81.44 g of the unknown element, X. Let's use this information to determine the molar mass of X which will allow us to identify X from the periodic table.**

$$18.56 \text{ g Al} \times \frac{1 \text{ mol Al}}{26.98 \text{ g Al}} \times \frac{3 \text{ mol X}}{2 \text{ mol Al}} = 1.032 \text{ mol X}$$

81.44 g of X must contain 1.032 mol of X.

$$\text{The molar mass of X} = \frac{81.44 \text{ g X}}{1.032 \text{ mol X}} = 78.91 \text{ g/mol}$$

From the periodic table, the unknown element is selenium and the formula is Al_2Se_3.

79. $$1.20 \text{ g } CO_2 \times \frac{1 \text{ mol } CO_2}{44.01 \text{ g}} \times \frac{1 \text{ mol C}}{\text{mol } CO_2} \times \frac{1 \text{ mol } C_{24}H_{30}N_3O}{24 \text{ mol C}} \times \frac{376.51 \text{ g}}{\text{mol } C_{24}H_{30}N_3O} = 0.428 \text{ g } C_{24}H_{30}N_3O$$

$$\frac{0.428 \text{ g } C_{24}H_{30}N_3O}{1.00 \text{ g sample}} \times 100 = 42.8\%\ C_{24}H_{30}N_3O$$

81. $$453 \text{ g Fe} \times \frac{1 \text{ mol Fe}}{55.85 \text{ g Fe}} \times \frac{1 \text{ mol } Fe_2O_3}{2 \text{ mol Fe}} \times \frac{159.70 \text{ g } Fe_2O_3}{\text{mol } Fe_2O_3} = 648 \text{ g } Fe_2O_3$$

$$\text{mass \% } Fe_2O_3 = \frac{648 \text{ g } Fe_2O_3}{752 \text{ g ore}} \times 100 = 86.2\%$$

83. $$\frac{^{85}\text{Rb atoms}}{^{87}\text{Rb atoms}} = 2.591;$$ **If we had exactly 100 atoms: x = number of ^{85}Rb atoms and 100 - x = number of ^{87}Rb atoms**

$$\frac{x}{100 - x} = 2.591,\ x = 259.1 - 2.591\,x,\ x = \frac{259.1}{3.591} = 72.15\%\ ^{85}\text{Rb}$$

$$0.7215\,(84.9117) + 0.2785\,(\text{A}) = 85.4678,\ \text{A} = \frac{85.4678 - 61.26}{0.2785} = 86.92 \text{ amu}$$

85. The volume of a gas is proportional to the number of molecules of gas. Thus the formulas are:

I: NH_3; II: N_2H_4; III: HN_3

The mass ratios are:

I: $\frac{4.634 \text{ g N}}{\text{g H}}$; II: $\frac{6.949 \text{ g N}}{\text{g H}}$; III: $\frac{41.7 \text{ g N}}{\text{g H}}$

If we set the atomic mass of H equal to 1.008, then the atomic mass for nitrogen is:

I: 14.01; II: 14.01; III. 14.0

For example for Compound I: $\frac{A}{3(1.008)} = \frac{4.634}{1}$, A = 14.01

87. $1.375 \text{ g AgI} \times \frac{1 \text{ mol AgI}}{234.8 \text{ g AgI}} = 5.856 \times 10^{-3} \text{ mol AgI} = 5.856 \times 10^{-3} \text{ mol I}$

$1.375 \text{ g AgI} \times \frac{126.9 \text{ g I}}{234.8 \text{ g AgI}} = 0.7431 \text{ g I}$; XI_2 contains 0.7431 g I and 0.257 g X.

$5.856 \times 10^{-3} \text{ mol I} \times \frac{1 \text{ mol X}}{2 \text{ mol I}} = 2.928 \times 10^{-3} \text{ mol X}$

Molar mass $= \frac{0.257 \text{ g X}}{2.928 \times 10^{-3} \text{ mol X}} = \frac{87.8 \text{ g}}{\text{mol}}$; atomic mass = 87.8 amu (X is Sr.)

89. X_2Z: 40.0% X and 60.0% Z by mass; $\frac{\text{mol X}}{\text{mol Z}} = 2 = \frac{40.0/A_x}{60.0/A_z} = \frac{40.0\ A_z}{60.0\ A_x}$ or $A_z = 3\ A_x$ where A = molar mass

For XZ_2, molar mass $= A_x + 2\ A_z = A_x + 2(3\ A_x) = 7\ A_x$

$\%\ X = \frac{A_x}{7A_x} \times 100 = 14.3\%$ X; % Z = 100.0 - 14.3 = 85.7% Z

91. Assuming one mol of vitamin A (286.4 g Vitamin A):

mol C $= 286.4 \text{ g Vitamin A} \times \frac{0.8386 \text{ g C}}{\text{g Vitamin A}} \times \frac{1 \text{ mol C}}{12.011 \text{ g C}} = 20.00 \text{ mol C}$

mol H $= 286.4 \text{ g Vitamin A} \times \frac{0.1056 \text{ g H}}{\text{g Vitamin A}} \times \frac{1 \text{ mol H}}{1.0079 \text{ g H}} = 30.01 \text{ mol H}$

Since one mol of Vitamin A contains 20 mol C and 30 mol H, then the molecular formula of Vitamin A is $C_{20}H_{30}E$. To determine E, lets calculate the molar mass of E.

286.4 g = 20(12.01) + 30(1.008) + molar mass E, molar mass E = 16.0 g/mol

From the periodic table, E = oxygen and the molecular formula of Vitamin A is $C_{20}H_{30}O$.

Challenge Problems

93. When the discharge voltage is low, the ions present are in the form of molecules. When the discharge voltage is increased, the bonds in the molecules are broken and the ions present are in the form of individual atoms. Therefore, the high discharge data indicates that the ions $^{16}O^+$, $^{18}O^+$ and $^{40}Ar^+$ are present. The only combination of these individual ions that can explain the mass data at low discharge is $^{16}O^{16}O^+$ (mass = 32), $^{16}O^{18}O^+$ (mass = 34) and $^{40}Ar^+$ (mass = 40). Therefore, the gas mixture contains $^{16}O^{16}O$, $^{16}O^{18}O$ and ^{40}Ar. To determine the percent composi-tion of each isotope, we use the relative intensity data from the high discharge data to determine the percentage that each isotope contributes to the total relative intensity. For ^{40}Ar:

$$\frac{1.0000}{0.7500 + 0.0015 + 1.0000} \times 100 = \frac{1.0000}{1.7515} \times 100 = 57.094\%\ ^{40}Ar$$

For ^{16}O: $\frac{0.7500}{1.7515} \times 100 = 42.82\%\ ^{16}O$; For ^{18}O: $\frac{0.0015}{1.7515} \times 100 = 8.6 \times 10^{-2}\%\ ^{18}O$

Note: ^{18}F instead of ^{18}O could also explain the data. However, OF(g) is not a stable compound. This is why ^{18}O is the best choice since $O_2(g)$ does form.

95. 10.00 g XCl_2 + excess Cl_2 → 12.55 g XCl_4; 2.55 g Cl reacted with XCl_2 to form XCl_4. XCl_4 contains 2.55 g Cl and 10.00 g XCl_2. From mol ratios, 10.00 g XCl_2 must also contain 2.55 g Cl with 10.00 - 2.55 = 7.45 g X.

$$2.55 \text{ g Cl} \times \frac{1 \text{ mol Cl}}{35.45 \text{ g Cl}} \times \frac{1 \text{ mol } XCl_2}{2 \text{ mol Cl}} \times \frac{1 \text{ mol X}}{\text{mol } XCl_2} = 3.60 \times 10^{-2} \text{ mol X}$$

So, 3.60×10^{-2} mol X must equal 7.45 g X. The molar mass of X is:

$$\frac{7.45 \text{ g X}}{3.60 \times 10^{-2} \text{ mol X}} = \frac{207 \text{ g}}{\text{mol X}};\quad \text{atomic mass of X} = 207 \text{ amu; X is Pb.}$$

97. For a gas, density and molar mass are proportional.

Molar mass $XH_n = 2.393(32.00) = \frac{76.58 \text{ g}}{\text{mol}}$; $0.803 \text{ g } H_2O \times \frac{2 \text{ mol H}}{18.02 \text{ g } H_2O} = 8.91 \times 10^{-2}$ mol H

$$\frac{8.91 \times 10^{-2} \text{ mol H}}{2.23 \times 10^{-2} \text{ mol } XH_n} = \frac{4 \text{ mol H}}{\text{mol } XH_n}$$

Molar mass X = 76.58 - 4(1.008 g) = 72.55 g/mol ; The element is Ge.

99. 4.000 g M_2S_3 → 3.723 g MO_2

There must be twice as many mol of MO_2 as mol of M_2S_3 in order to balance M in the reaction. Setting up an equation for 2 mol MO_2 = mol M_2S_3 where A = molar mass M:

$$2\left(\frac{4.000\text{ g}}{2\text{ A} + 3(32.07)}\right) = \frac{3.723\text{ g}}{\text{A} + 2(16.00)},\quad \frac{8.000}{2\text{ A} + 96.21} = \frac{3.723}{\text{A} + 32.00}$$

8.000 A + 256.0 = 7.446 A + 358.2, 0.554 A = 102.2, A = 184 g/mol; atomic mass = 184 amu

101. The balanced equations are:

$C(s) + 1/2\ O_2(g) \rightarrow CO(g)$ and $C(s) + O_2(g) \rightarrow CO_2(g)$

If we have 100.0 mol of products, we have 72.0 mol CO_2, 16.0 mol CO, and 12.0 mol O_2. The initial mixture contained 72.0 + 16.0 = 88.0 mol C and 72.0 (from CO_2) + $\frac{16.0}{2}$ (from CO) + 12.0 (unreacted) = 92.0 mol O_2. Initial reaction mixture contained:

$$\frac{92.0\text{ mol }O_2}{88.0\text{ mol C}} = 1.05\text{ mol }O_2\text{/mol C}$$

103. $LaH_{2.90}$ is the formula. If only La^{3+} is present, LaH_3 would be the formula. If only La^{2+} is present, LaH_2 would be the formula. Let x = mol La^{2+} and y = mol La^{3+}:

$(La^{2+})_x(La^{3+})_yH_{(2x+3y)}$ where x + y = 1.00 and 2x + 3y = 2.90

Solving by simultaneous equations:

$$\begin{array}{rl} 2x + 3y = & 2.90 \\ \underline{-2x - 2y =} & \underline{-2.00} \\ y = & 0.90 \text{ and } x = 0.10 \end{array}$$

$LaH_{2.90}$ contains $\frac{1}{10}$ La^{2+} or 10.% La^{2+} and $\frac{9}{10}$ La^{3+} or 90.% La^{3+}.

105. $NaCl(aq) + Ag^+(aq) \rightarrow AgCl(s)$; $KCl(aq) + Ag^+(aq) \rightarrow AgCl(s)$

$$8.5904\text{ g AgCl} \times \frac{1\text{ mol AgCl}}{143.4\text{ g AgCl}} \times \frac{1\text{ mol Cl}^-}{1\text{ mol AgCl}} = 5.991 \times 10^{-2}\text{ mol Cl}^-$$

Let x = g NaCl and y = g KCl:

$$x + y = 4.000\text{ g and } \frac{x}{58.44} + \frac{y}{74.55} = 5.991 \times 10^{-2}\text{ total mol Cl}^-\text{ or } 1.276\,x + y = 4.466$$

Solving using simultaneous equations:

$$\begin{array}{rl} 1.276\,x + y = & 4.466 \\ \underline{-x - y =} & \underline{-4.000} \\ 0.276\,x = & 0.466, \end{array} \quad x = 1.69\text{ g NaCl and } y = 2.31\text{ g KCl}$$

$$\%\text{ NaCl} = \frac{1.69\text{ g}}{4.000\text{ g}} \times 100 = 42.3\%\text{ NaCl};\ \%\text{ KCl} = 57.7\%$$

107. The balanced equations are:

$$4\ NH_3(g) + 5\ O_2(g) \rightarrow 4\ NO(g) + 6\ H_2O(g) \text{ and } 4\ NH_3(g) + 7\ O_2(g) \rightarrow 4\ NO_2(g) + 6\ H_2O(g)$$

Let 4x = number of mol of NO formed, and let 4y = number of mol of NO_2 formed. Then:

$$4x\ NH_3 + 5x\ O_2 \rightarrow 4x\ NO + 6x\ H_2O \text{ and } 4y\ NH_3 + 7y\ O_2 \rightarrow 4y\ NO_2 + 6y\ H_2O$$

All the NH_3 reacted, so 4x + 4y = 2.00. 10.00 - 6.75 = 3.25 mol O_2 reacted, so 5x + 7y = 3.25.

Solving by the method of simultaneous equations:

$$\begin{array}{r} 20x + 28y = 13.0 \\ \underline{-20x - 20y = -10.0} \\ 8y = 3.0, \end{array} \quad y = 0.38;\ \ 4x + 4 \times 0.38 = 2.00,\ \ x = 0.12$$

mol NO = 4x = 4 × 0.12 = 0.48 mol NO formed

Marathon Problems

109. To solve the limiting reagent problem, we must determine the formulas of all the compounds so we can get a balanced reaction.

a. 40 million trillion = $40 \times 10^6 \times 10^{12} = 4.000 \times 10^{19}$ (assuming 4 S.F.)

$$4.000 \times 10^{19} \text{ molecules A} \times \frac{1 \text{ mol A}}{6.022 \times 10^{23} \text{ molecules A}} = 6.642 \times 10^{-5} \text{ mol A}$$

$$\text{Molar mass of A} = \frac{4.26 \times 10^{-3} \text{ g A}}{6.642 \times 10^{-5} \text{ mol A}} = 64.1 \text{ g/mol}$$

Mass of carbon in one mol of A is:

$$64.1 \text{ g A} \times \frac{37.5 \text{ g C}}{100.0 \text{ g A}} = 24.0 \text{ g carbon} = 2 \text{ mol carbon in substance A}$$

The remainder of the molar mass (64.1 g - 24.0 g = 40.1 g) is due to the alkaline earth metal. From the periodic table, calcium has a molar mass of 40.08 g/mol. The formula of substance A is CaC_2.

b. 5.36 g H + 42.5 g O = 47.9 g; Substance B only contains H and O. Determining the empirical formula of B:

$$5.36 \text{ g H} \times \frac{1 \text{ mol H}}{1.008 \text{ g H}} = 5.32 \text{ mol H}; \quad \frac{5.32}{2.66} = 2.00$$

$$42.5 \text{ g O} \times \frac{1 \text{ mol O}}{16.00 \text{ g O}} = 2.66 \text{ mol O}; \quad \frac{2.66}{2.66} = 1.00$$

Empirical formula = H_2O; The molecular formula of substance B could be H_2O, H_4O_2, H_6O_3, etc. The most reasonable choice is water (H_2O) for substance B.

c. **Substance C + O_2 → CO_2 + H_2O; Substance C must contain carbon and hydrogen, and may contain oxygen. Determining the mass of carbon and hydrogen in substance C:**

$$33.8\ g\ CO_2 \times \frac{1\ mol\ CO_2}{44.01\ g\ CO_2} \times \frac{1\ mol\ C}{mol\ CO_2} \times \frac{12.01\ g\ C}{mol\ C} = 9.22\ g\ carbon$$

$$6.92\ g\ H_2O \times \frac{1\ mol\ H_2O}{18.02\ g\ H_2O} \times \frac{2\ mol\ H}{mol\ H_2O} \times \frac{1.008\ g\ H}{mol\ H} = 0.774\ g\ hydrogen$$

9.22 g carbon + 0.774 g hydrogen = 9.99 g; Since substance C initially weighed 10.0 g, then there is no oxygen present in substance C. Determining the empirical formula for substance C:

$$9.22\ g \times \frac{1\ mol\ C}{12.01\ g\ C} = 0.768\ mol\ carbon$$

$$0.774\ g\ H \times \frac{1\ mol\ H}{1.008\ g\ H} = 0.768\ mol\ hydrogen$$

mol C/mol H = 1.00; The empirical formula is CH which has an empirical formula mass ≈ 13. Since the mass spectrum data indicates a molar mass of 26 g/mol, then the molecular formula for substance C is C_2H_2.

d. **Substance D is $Ca(OH)_2$.**

Now we can answer the question. The balanced equation is:

$$CaC_2(s) + 2\ H_2O(l) \rightarrow C_2H_2(g) + Ca(OH)_2(aq)$$

$$45.0\ g\ CaC_2 \times \frac{1\ mol\ CaC_2}{64.10\ g\ CaC_2} = 0.702\ mol\ CaC_2$$

$$23.0\ g\ H_2O \times \frac{1\ mol\ H_2O}{18.02\ g\ H_2O} = 1.28\ mol\ H_2O$$

$$\frac{mol\ H_2O}{mol\ CaC_2} = \frac{1.28}{0.702} = 1.82$$

Since the actual mol ratio present is smaller than the required 2:1 mol ratio from the balanced equation, then H_2O is limiting.

$$1.28\ mol\ H_2O \times \frac{1\ mol\ C_2H_2}{2\ mol\ H_2O} \times \frac{26.04\ g\ C_2H_2}{mol\ C_2H_2} = 16.7\ g\ C_2H_2 = \text{mass of substance C produced}$$

CHAPTER FOUR

TYPES OF CHEMICAL REACTIONS AND SOLUTION STOICHIOMETRY

Aqueous Solutions: Strong and Weak Electrolytes

11. $MgSO_4(s) \rightarrow Mg^{2+}(aq) + SO_4^{2-}(aq)$; $NH_4NO_3(s) \rightarrow NH_4^+(aq) + NO_3^-(aq)$

Solution Concentration: Molarity

13. a. $2.00\text{ L} \times \dfrac{0.250\text{ mol NaOH}}{\text{L}} \times \dfrac{40.00\text{ g NaOH}}{\text{mol}} = 20.0\text{ g NaOH}$

Place 20.0 g NaOH in a 2 L volumetric flask; add water to dissolve the NaOH, and fill to the mark with water, mixing several times along the way.

b. $2.00\text{ L} \times \dfrac{0.250\text{ mol NaOH}}{\text{L}} \times \dfrac{1\text{ L stock}}{1.00\text{ mol NaOH}} = 0.500\text{ L}$

Add 500. mL of 1.00 M NaOH stock solution to a 2 L volumetric flask; fill to the mark with water, mixing several times along the way.

c. $2.00\text{ L} \times \dfrac{0.100\text{ mol } K_2CrO_4}{\text{L}} \times \dfrac{194.20\text{ g } K_2CrO_4}{\text{mol } K_2CrO_4} = 38.8\text{ g } K_2CrO_4$

Similar to the solution made in part a, instead using 38.8 g K_2CrO_4.

d. $2.00\text{ L} \times \dfrac{0.100\text{ mol } K_2CrO_4}{\text{L}} \times \dfrac{1\text{ L stock}}{1.75\text{ mol } K_2CrO_4} = 0.114\text{ L}$

Similar to the solution made in part b, instead using 114 mL of the 1.75 M K_2CrO_4 stock solution.

15. Molar mass of $NaHCO_3$ = 22.99 + 1.008 + 12.01 + 3(16.00) = 84.01 g/mol

$$\text{Volume} = 0.350\ \text{g NaHCO}_3 \times \frac{1\ \text{mol NaHCO}_3}{84.01\ \text{g NaHCO}_3} \times \frac{1\ \text{L}}{0.100\ \text{mol NaHCO}_3} = 0.0417\ \text{L} = 41.7\ \text{mL}$$

41.7 mL of 0.100 M $NaHCO_3$ contains 0.350 g $NaHCO_3$.

17. $$75.0\ \text{mL} \times \frac{0.79\ \text{g}}{\text{mL}} \times \frac{1\ \text{mol}}{46.1\ \text{g}} = 1.3\ \text{mol C}_2\text{H}_5\text{OH};\ \text{Molarity} = \frac{1.3\ \text{mol}}{0.250\ \text{L}} = 5.2\ M\ \text{C}_2\text{H}_5\text{OH}$$

19. Stock solution:

$$1.584\ \text{g Mn}^{2+} \times \frac{1\ \text{mol Mn}^{2+}}{54.94\ \text{g Mn}^{2+}} = 2.883 \times 10^{-2}\ \text{mol Mn}^{2+};\ M = \frac{2.883 \times 10^{-2}\ \text{mol}}{1.000\ \text{L}}$$

$$= 2.883 \times 10^{-2}\ M\ \text{Mn}^{2+}$$

Solution A contains:

$$50.00\ \text{mL} \times \frac{1\ \text{L}}{1000\ \text{mL}} \times \frac{2.883 \times 10^{-2}\ \text{mol}}{\text{L}} = 1.442 \times 10^{-3}\ \text{mol Mn}^{2+}$$

$$\text{Molarity} = \frac{1.442 \times 10^{-3}\ \text{mol}}{1000.0\ \text{mL}} \times \frac{1000\ \text{mL}}{\text{L}} = 1.442 \times 10^{-3}\ M\ \text{Mn}^{2+}$$

Solution B contains:

$$10.00\ \text{mL} \times \frac{1\ \text{L}}{1000\ \text{mL}} \times \frac{1.442 \times 10^{-3}\ \text{mol}}{\text{L}} = 1.442 \times 10^{-5}\ \text{mol Mn}^{2+}$$

$$\text{Molarity} = \frac{1.442 \times 10^{-5}\ \text{mol}}{0.2500\ \text{L}} = 5.768 \times 10^{-5}\ M\ \text{Mn}^{2+}$$

Solution C contains:

$$10.00 \times 10^{-3}\ \text{L} \times \frac{5.768 \times 10^{-5}\ \text{mol}}{\text{L}} = 5.768 \times 10^{-7}\ \text{mol Mn}^{2+}$$

$$\text{Molarity} = \frac{5.768 \times 10^{-7}\ \text{mol}}{0.5000\ \text{L}} = 1.154 \times 10^{-6}\ M\ \text{Mn}^{2+}$$

21. $$\text{mol Na}_2\text{CO}_3 = 0.0700\ \text{L} \times \frac{3.0\ \text{mol Na}_2\text{CO}_3}{\text{L}} = 0.21\ \text{mol Na}_2\text{CO}_3$$

$Na_2CO_3(s) \rightarrow 2\ Na^+(aq) + CO_3^{2-}(aq)$; mol Na^+ = 2(0.21) = 0.42 mol

$$\text{mol NaHCO}_3 = 0.0300\ \text{L} \times \frac{1.0\ \text{mol NaHCO}_3}{\text{L}} = 0.030\ \text{mol NaHCO}_3$$

$NaHCO_3(s) \rightarrow Na^+(aq) + HCO_3^-(aq)$; mol Na^+ = 0.030 mol

$$M_{Na^+} = \frac{\text{total mol Na}^+}{\text{total volume}} = \frac{0.42 \text{ mol} + 0.030 \text{ mol}}{0.0700 \text{ L} + 0.0300 \text{ L}} = \frac{0.45 \text{ mol}}{0.1000 \text{ L}} = 4.5\ M\ Na^+$$

23. $$1 \text{ ppm} = \frac{1\ \mu g}{mL} = \frac{1 \text{ mg}}{L};\quad \frac{1 \times 10^{-3} \text{ g F}^-}{L} \times \frac{1 \text{ mol F}^-}{19.0 \text{ g F}^-} = \frac{5 \times 10^{-5} \text{ mol}}{L} = 5 \times 10^{-5}\ M\ F^-$$

$$2 \text{ ppm F}^- = 1 \times 10^{-4}\ M\ F^-;\ 3 \text{ ppm F}^- = \frac{1.6 \times 10^{-4} \text{ mol}}{L} = 2 \times 10^{-4}\ M\ F^-$$

$$\frac{50. \times 10^{-3} \text{ g F}^-}{L} \times \frac{1 \text{ mol F}^-}{19.0 \text{ g F}^-} = 2.6 \times 10^{-3}\ M\ F^-$$

Precipitation Reactions

25. **For the following answers, the balanced molecular equation is first, followed by the complete ionic equation, then the net ionic equation.**

a. $(NH_4)_2SO_4(aq) + Ba(NO_3)_2(aq) \rightarrow 2\ NH_4NO_3(aq) + BaSO_4(s)$

$2\ NH_4^+(aq) + SO_4^{2-}(aq) + Ba^{2+}(aq) + 2\ NO_3^-(aq) \rightarrow 2\ NH_4^+(aq) + 2\ NO_3^-(aq) + BaSO_4(s)$

$Ba^{2+}(aq) + SO_4^{2-}(aq) \rightarrow BaSO_4(s)$ **is the net ionic equation (spectator ions omitted).**

b. $Pb(NO_3)_2(aq) + 2\ NaCl(aq) \rightarrow PbCl_2(s) + 2\ NaNO_3(aq)$

$Pb^{2+}(aq) + 2\ NO_3^-(aq) + 2\ Na^+(aq) + 2\ Cl^-(aq) \rightarrow PbCl_2(s) + 2\ Na^+(aq) + 2\ NO_3^-(aq)$

$Pb^{2+}(aq) + 2\ Cl^-(aq) \rightarrow PbCl_2(s)$

c. **The possible products, potassium phosphate and sodium nitrate, are both soluble in water. Therefore, no reaction occurs.**

d. **No reaction occurs since all possible products are soluble.**

e. $CuCl_2(aq) + 2\ NaOH(aq) \rightarrow Cu(OH)_2(s) + 2\ NaCl(aq)$

$Cu^{2+}(aq) + 2\ Cl^-(aq) + 2\ Na^+(aq) + 2\ OH^-(aq) \rightarrow Cu(OH)_2(s) + 2\ Na^+(aq) + 2\ Cl^-(aq)$

$Cu^{2+}(aq) + 2\ OH^-(aq) \rightarrow Cu(OH)_2(s)$

27. a. **When $CuSO_4(aq)$ is added to $Na_2S(aq)$, the precipitate that forms is $CuS(s)$. Therefore, Na^+ (the grey spheres) and SO_4^{2-} (the blueish-green spheres) are the spectator ions.**

$CuSO_4(aq) + Na_2S(aq) \rightarrow CuS(s) + Na_2SO_4(aq);\ Cu^{2+}(aq) + S^{2-}(aq) \rightarrow CuS(s)$

b. When $CoCl_2(aq)$ is added to NaOH(aq), the precipitate that forms is $Co(OH)_2(s)$. Therefore, Na^+ (the grey spheres) and Cl^- (the green spheres) are the spectator ions.

$$CoCl_2(aq) + 2\ NaOH(aq) \rightarrow Co(OH)_2(s) + 2\ NaCl(aq);\ \ Co^{2+}(aq) + 2\ OH^-(aq) \rightarrow Co(OH)_2(s)$$

c. When $AgNO_3(aq)$ is added to KI(aq), the precipitate that forms is AgI(s). Therefore, K^+ (the red spheres) and NO_3^- (the blue spheres) are the spectator ions.

$$AgNO_3(aq) + KI(aq) \rightarrow AgI(s) + KNO_3(aq);\ \ Ag^+(aq) + I^-(aq) \rightarrow AgI(s)$$

29. Since no precipitate formed upon addition of NaCl or Na_2SO_4, we can conclude that Hg_2^{2+} and Ba^{2+} are not present since Hg_2Cl_2 and $BaSO_4$ are insoluble salts. Since a precipitate formed with NaOH, then the solution must contain Mn^{2+} which forms $Mn(OH)_2(s)$.

31. $2\ AgNO_3(aq) + CaCl_2(aq) \rightarrow 2\ AgCl(s) + Ca(NO_3)_2(aq)$

$$\text{mol } AgNO_3 = 0.1000\ L \times \frac{0.20\ \text{mol } AgNO_3}{L} = 0.020\ \text{mol } AgNO_3$$

$$\text{mol } CaCl_2 = 0.1000\ L \times \frac{0.15\ \text{mol } CaCl_2}{L} = 0.015\ \text{mol } CaCl_2$$

The required mol $AgNO_3$ to mol $CaCl_2$ ratio is 2:1 (from the balanced equation). The actual mol ratio present is 0.020/0.015 = 1.3 (1.3:1). Therefore, $AgNO_3$ is the limiting reagent.

$$\text{mass AgCl} = 0.020\ \text{mol } AgNO_3 \times \frac{1\ \text{mol AgCl}}{1\ \text{mol } AgNO_3} \times \frac{143.4\ \text{g AgCl}}{\text{mol AgCl}} = 2.9\ \text{g AgCl}$$

The net ionic equation is: $Ag^+(aq) + Cl^-(aq) \rightarrow AgCl(s)$. The ions remaining in solution are the unreacted Cl^- ions and the spectator ions, NO_3^- and Ca^{2+} (all Ag^+ is used up in forming AgCl). The mol of each ion present initially (before reaction) can be easily determined from the mol of each reactant. 0.020 mol $AgNO_3$ dissolves to form 0.020 mol Ag^+ and 0.020 mol NO_3^-. 0.015 mol $CaCl_2$ dissolves to form 0.015 mol Ca^{2+} and 2(0.015) = 0.030 mol Cl^-.

mol unreacted Cl^- = 0.030 mol Cl^- initially - 0.020 mol Cl^- reacted = 0.010 mol Cl^- unreacted

$$M_{Cl^-} = \frac{0.010\ \text{mol } Cl^-}{\text{total volume}} = \frac{0.010\ \text{mol } Cl^-}{0.1000\ L + 0.1000\ L} = 0.050\ M\ Cl^-$$

The molarity of the spectator ions are:

$$M_{NO_3^-} = \frac{0.020\ \text{mol } NO_3^-}{0.2000\ L} = 0.10\ M\ NO_3^-;\ \ M_{Ca^{2+}} = \frac{0.015\ \text{mol } Ca^{2+}}{0.2000\ L} = 0.075\ M\ Ca^{2+}$$

33. $$1.00\ L \times \frac{0.200\ \text{mol } Na_2S_2O_3}{L} \times \frac{1\ \text{mol AgBr}}{2\ \text{mol } Na_2S_2O_3} \times \frac{187.8\ \text{g AgBr}}{\text{mol AgBr}} = 18.8\ \text{g AgBr}$$

35. All the sulfur in $BaSO_4$ came from the saccharin. The conversion from $BaSO_4$ to saccharin utilizes the molar masses of each compound.

$$0.5032 \text{ g } BaSO_4 \times \frac{32.07 \text{ g S}}{233.4 \text{ g } BaSO_4} \times \frac{183.19 \text{ g saccharin}}{32.07 \text{ g S}} = 0.3949 \text{ g saccharin}$$

$$\frac{\text{Avg. mass}}{\text{Tablet}} = \frac{0.3949 \text{ g}}{10 \text{ tablets}} = \frac{3.949 \times 10^{-2} \text{ g}}{\text{tablet}} = \frac{39.49 \text{ mg}}{\text{tablet}}$$

$$\text{Avg. mass \%} = \frac{0.3949 \text{ g saccharin}}{0.5894 \text{ g}} \times 100 = 67.00\% \text{ saccharin by mass}$$

37. $M_2SO_4(aq) + CaCl_2(aq) \rightarrow CaSO_4(s) + 2\ MCl(aq)$

$$1.36 \text{ g } CaSO_4 \times \frac{1 \text{ mol } CaSO_4}{136.15 \text{ g } CaSO_4} \times \frac{1 \text{ mol } M_2SO_4}{\text{mol } CaSO_4} = 9.99 \times 10^{-3} \text{ mol } M_2SO_4$$

From the problem, 1.42 g M_2SO_4 was reacted so:

$$1.42 \text{ g } M_2SO_4 = 9.99 \times 10^{-3} \text{ mol } M_2SO_4, \text{ molar mass} = \frac{1.42 \text{ g } M_2SO_4}{9.99 \times 10^{-3} \text{ mol } M_2SO_4} = 142 \text{ g/mol}$$

142 amu = 2(atomic mass M) + 32.07 + 4(16.00), atomic mass M = 23 amu

From periodic table, M = Na(sodium).

Acid-Base Reactions

39. a. Perchloric acid reacted with potassium hydroxide is a possibility.

$HClO_4(aq) + KOH(aq) \rightarrow H_2O(l) + KClO_4(aq)$

b. Nitric acid reacted with cesium hydroxide is a possibility.

$HNO_3(aq) + CsOH(aq) \rightarrow H_2O(l) + CsNO_3(aq)$

c. Hydroiodic acid reacted with calcium hydroxide is a possibility.

$2\ HI(aq) + Ca(OH)_2(aq) \rightarrow 2\ H_2O(l) + CaI_2(aq)$

41. If we begin with 50.00 mL of 0.100 *M* NaOH, then:

$$50.00 \times 10^{-3} \text{ L} \times \frac{0.100 \text{ mol}}{\text{L}} = 5.00 \times 10^{-3} \text{ mol NaOH to be neutralized.}$$

a. $NaOH(aq) + HCl(aq) \rightarrow NaCl(aq) + H_2O(l)$

$$5.00 \times 10^{-3} \text{ mol NaOH} \times \frac{1 \text{ mol HCl}}{\text{mol NaOH}} \times \frac{1 \text{ L soln}}{0.100 \text{ mol}} = 5.00 \times 10^{-2} \text{ L or } 50.0 \text{ mL}$$

b. $2\ NaOH(aq) + H_2SO_3(aq) \rightarrow 2\ H_2O(l) + Na_2SO_3(aq)$

$$5.00 \times 10^{-3} \text{ mol NaOH} \times \frac{1 \text{ mol } H_2SO_3}{2 \text{ mol NaOH}} \times \frac{1 \text{ L soln}}{0.100 \text{ mol } H_2SO_3} = 2.50 \times 10^{-2} \text{ L or } 25.0 \text{ mL}$$

c. $3\ NaOH(aq) + H_3PO_4(aq) \rightarrow Na_3PO_4(aq) + 3\ H_2O(l)$

$$5.00 \times 10^{-3} \text{ mol NaOH} \times \frac{1 \text{ mol } H_3PO_4}{3 \text{ mol NaOH}} \times \frac{1 \text{ L soln}}{0.200 \text{ mol } H_3PO_4} = 8.33 \times 10^{-3} \text{ L or } 8.33 \text{ mL}$$

d. $HNO_3(aq) + NaOH(aq) \rightarrow H_2O(l) + NaNO_3(aq)$

$$5.00 \times 10^{-3} \text{ mol NaOH} \times \frac{1 \text{ mol } HNO_3}{\text{mol NaOH}} \times \frac{1 \text{ L soln}}{0.150 \text{ mol } HNO_3} = 3.33 \times 10^{-2} \text{ L or } 33.3 \text{ mL}$$

e. $HC_2H_3O_2(aq) + NaOH(aq) \rightarrow H_2O(l) + NaC_2H_3O_2(aq)$

$$5.00 \times 10^{-3} \text{ mol NaOH} \times \frac{1 \text{ mol } HC_2H_3O_2}{\text{mol NaOH}} \times \frac{1 \text{ L soln}}{0.200 \text{ mol } HC_2H_3O_2} = 2.50 \times 10^{-2} \text{ L or } 25.0 \text{ mL}$$

f. $H_2SO_4(aq) + 2\ NaOH(aq) \rightarrow 2\ H_2O(l) + Na_2SO_4(aq)$

$$5.00 \times 10^{-3} \text{ mol NaOH} \times \frac{1 \text{ mol } H_2SO_4}{2 \text{ mol NaOH}} \times \frac{1 \text{ L soln}}{0.300 \text{ mol } H_2SO_4} = 8.33 \times 10^{-3} \text{ L or } 8.33 \text{ mL}$$

43. The pertinent reactions are:

$$2\ NaOH(aq) + H_2SO_4(aq) \rightarrow Na_2SO_4(aq) + 2\ H_2O(l)$$

$$HCl(aq) + NaOH(aq) \rightarrow NaCl(aq) + H_2O(l)$$

$$\text{Amount of NaOH added} = 0.0500 \text{ L} \times \frac{0.213 \text{ mol}}{\text{L}} = 1.07 \times 10^{-2} \text{ mol NaOH}$$

Amount of NaOH neutralized by HCl:

$$0.01321 \text{ L HCl} \times \frac{0.103 \text{ mol HCl}}{\text{L HCl}} \times \frac{1 \text{ mol NaOH}}{\text{mol HCl}} = 1.36 \times 10^{-3} \text{ mol NaOH}$$

The difference, 9.3×10^{-3} mol, is the amount of NaOH neutralized by the sulfuric acid.

$$9.3 \times 10^{-3} \text{ mol NaOH} \times \frac{1 \text{ mol } H_2SO_4}{2 \text{ mol NaOH}} = 4.7 \times 10^{-3} \text{ mol } H_2SO_4$$

$$\text{Concentration of } H_2SO_4 = \frac{4.7 \times 10^{-3} \text{ mol}}{0.1000 \text{ L}} = 4.7 \times 10^{-2}\ M\ H_2SO_4$$

45. $HC_2H_3O_2(aq) + NaOH(aq) \rightarrow H_2O(l) + NaC_2H_3O_2(aq)$

a. $16.58 \times 10^{-3}\ \text{L soln} \times \frac{0.5062\ \text{mol NaOH}}{\text{L soln}} \times \frac{1\ \text{mol acetic acid}}{\text{mol NaOH}} = 8.393 \times 10^{-3}\ \text{mol acetic acid}$

$\text{Concentration of acetic acid} = \frac{8.393 \times 10^{-3}\ \text{mol}}{0.01000\ \text{L}} = 0.8393\ M\ HC_2H_3O_2$

b. If we have 1.000 L of solution: $\text{total mass} = 1000.\ \text{mL} \times \frac{1.006\ \text{g}}{\text{mL}} = 1006\ \text{g solution}$

$\text{Mass of } HC_2H_3O_2 = 0.8393\ \text{mol} \times \frac{60.052\ \text{g}}{\text{mol}} = 50.40\ \text{g}\ HC_2H_3O_2$

$\text{Mass \% acetic acid} = \frac{50.40\ \text{g}}{1006\ \text{g}} \times 100 = 5.010\%$

47. $HNO_3(aq) + NaOH(aq) \rightarrow NaNO_3(aq) + H_2O(l)$

$15.0\ \text{g NaOH} \times \frac{1\ \text{mol NaOH}}{40.00\ \text{g}} = 0.375\ \text{mol NaOH}$

$0.1500\ \text{L} \times \frac{0.250\ \text{mol}\ HNO_3}{\text{L}} = 0.0375\ \text{mol}\ HNO_3$

We have added more moles of NaOH than mol of HNO_3 present. Since NaOH and HNO_3 react in a 1:1 mol ratio, then NaOH is in excess and the solution will be basic. The ions present after reaction will be the excess OH^- ions and the spectator ions, Na^+ and NO_3^-. The moles of ions present initially are:

$\text{mol NaOH} = \text{mol}\ Na^+ = \text{mol}\ OH^- = 0.375\ \text{mol}$

$\text{mol}\ HNO_3 = \text{mol}\ H^+ = \text{mol}\ NO_3^- = 0.0375\ \text{mol}$

The net ionic reaction occurring is: $H^+(aq) + OH^-(aq) \rightarrow H_2O(l)$

The mol of excess OH^- remaining after reaction will be the initial mol of OH^- minus the amount of OH^- neutralized by reaction with H^+:

$\text{mol excess}\ OH^- = 0.375\ \text{mol} - 0.0375\ \text{mol} = 0.338\ \text{mol}\ OH^-\ \text{excess}$

The concentration of ions present is:

$$M_{OH^-} = \frac{\text{mol}\ OH^-\ \text{excess}}{\text{volume}} = \frac{0.338\ \text{mol}\ OH^-}{0.1500\ \text{L}} = 2.25\ M\ OH^-$$

$$M_{NO_3^-} = \frac{0.0375\ \text{mol}\ NO_3^-}{0.1500\ \text{L}} = 0.250\ M\ NO_3^-;\quad M_{Na^+} = \frac{0.375\ \text{mol}}{0.1500\ \text{L}} = 2.50\ M\ Na^+$$

49. $Ba(OH)_2(aq) + 2\ HCl(aq) \rightarrow BaCl_2(aq) + 2\ H_2O(l);\ H^+(aq) + OH^-(aq) \rightarrow H_2O(l)$

$$75.0 \times 10^{-3}\ \text{L} \times \frac{0.250\ \text{mol HCl}}{\text{L}} = 1.88 \times 10^{-2}\ \text{mol HCl} = 1.88 \times 10^{-2}\ \text{mol H}^+ + 1.88 \times 10^{-2}\ \text{mol Cl}^-$$

$$225.0 \times 10^{-3}\ \text{L} \times \frac{0.0550\ \text{mol Ba(OH)}_2}{\text{L}} = 1.24 \times 10^{-2}\ \text{mol Ba(OH)}_2 = 1.24 \times 10^{-2}\ \text{mol Ba}^{2+} + 2.48 \times 10^{-2}\ \text{mol OH}^-$$

The net ionic equation requires a 1:1 mol ratio between OH^- and H^+. The actual mol OH^- to mol H^+ ratio is greater than 1:1 so OH^- is in excess.

Since 1.88×10^{-2} mol OH^- will be neutralized by the H^+, then we have $(2.48 - 1.88) \times 10^{-2} = 0.60 \times 10^{-2}$ mol OH^- remaining in excess.

$$M_{OH^-} = \frac{\text{mol OH}^-\ \text{excess}}{\text{total volume}} = \frac{6.0 \times 10^{-3}\ \text{mol OH}^-}{0.0750\ \text{L} + 0.2250\ \text{L}} = 2.0 \times 10^{-2}\ M\ \text{OH}^-$$

Oxidation-Reduction Reactions

51. Apply rules in Table 4.3.

a. $KMnO_4$ is composed of K^+ and MnO_4^- ions. Assign oxygen an oxidation state value of -2 which gives manganese a +7 oxidation state since the sum of oxidation states for all atoms in MnO_4^- must equal the -1 charge on MnO_4^-. K, +1; O, -2; Mn, +7.

b. Assign O a -2 oxidation state, which gives nickel a +4 oxidation state. Ni, +4; O, -2.

c. $K_4Fe(CN)_6$ is composed of K^+ cations and $Fe(CN)_6^{4-}$ anions. $Fe(CN)_6^{4-}$ is composed of iron and CN^- anions. For an overall anion charge of -4, iron must have a +2 oxidation state.

d. $(NH_4)_2HPO_4$ is made of NH_4^+ cations and HPO_4^{2-} anions. Assign +1 as oxidation state of H and -2 as the oxidation state of O. In NH_4^+, $x + 4(+1) = +1$, $x = -3$ = oxidation state of N. In HPO_4^{2-}, $+1 + y + 4(-2) = -2$, $y = +5$ = oxidation state of P.

e. O, -2; P, +3

f. O, -2; Fe, + 8/3

g. O, -2; F, -1; Xe, +6

h. F, -1; S, +4

i. O, -2; C, +2

j. Na, +1; O, -2; C, +3

53. a. -3 b. -3 c. $2(x) + 4(+1) = 0$, $x = -2$

d. +2 e. +1 f. +4

g. +3 h. +5 i. 0

55. To determine if the reaction is an oxidation-reduction reaction, assign oxidation states. If the oxidation states change for some elements, then the reaction is a redox reaction. If the oxidation states do not change, then the reaction is not a redox reaction. In redox reactions the species oxidized (called the reducing agent) shows an increase in the oxidation states and the species reduced (called the oxidizing agent) shows a decrease in oxidation states.

	Redox?	Oxidizing Agent	Reducing Agent	Substance Oxidized	Substance Reduced
a.	Yes	O_2	CH_4	CH_4 (C)	O_2 (O)
b.	Yes	HCl	Zn	Zn	HCl (H)
c.	No	-	-	-	-
d.	Yes	O_3	NO	NO (N)	O_3 (O)
e.	Yes	H_2O_2	H_2O_2	H_2O_2 (O)	H_2O_2 (O)
f.	Yes	CuCl	CuCl	CuCl (Cu)	CuCl (Cu)
g.	No	-	-	-	-
h.	No	-	-	-	-
i.	Yes	$SiCl_4$	Mg	Mg	$SiCl_4$ (Si)

In c, g, and h no oxidation states change from reactants to products.

57. a. Review section 4.11 of the text for rules on balancing by the half-reaction method. The first step is to separate the reaction into two half-reactions then balance each half-reaction separately.

$$(Cu \rightarrow Cu^{2+} + 2\,e^-) \times 3$$

$$NO_3^- \rightarrow NO + 2\,H_2O$$
$$(3\,e^- + 4\,H^+ + NO_3^- \rightarrow NO + 2\,H_2O) \times 2$$

Adding the two balanced half-reactions so electrons cancel:

$$3\,Cu \rightarrow 3\,Cu^{2+} + 6\,e^-$$
$$6\,e^- + 8\,H^+ + 2\,NO_3^- \rightarrow 2\,NO + 4\,H_2O$$

$$3\,Cu(s) + 8\,H^+(aq) + 2\,NO_3^-(aq) \rightarrow 3\,Cu^{2+}(aq) + 2\,NO(g) + 4\,H_2O(l)$$

b. $(2\,Cl^- \rightarrow Cl_2 + 2\,e^-) \times 3$

$$Cr_2O_7^{2-} \rightarrow 2\,Cr^{3+} + 7\,H_2O$$
$$6\,e^- + 14\,H^+ + Cr_2O_7^{2-} \rightarrow 2\,Cr^{3+} + 7\,H_2O$$

Add the two balanced half-reactions with six electrons transferred:

$$6\,Cl^- \rightarrow 3\,Cl_2 + 6\,e^-$$
$$6\,e^- + 14\,H^+ + Cr_2O_7^{2-} \rightarrow 2\,Cr^{3+} + 7\,H_2O$$

$$14\,H^+(aq) + Cr_2O_7^{2-}(aq) + 6\,Cl^-(aq) \rightarrow 3\,Cl_2(g) + 2\,Cr^{3+}(aq) + 7\,H_2O(l)$$

c.

$$Pb \rightarrow PbSO_4$$
$$Pb + H_2SO_4 \rightarrow PbSO_4 + 2\,H^+$$
$$Pb + H_2SO_4 \rightarrow PbSO_4 + 2\,H^+ + 2\,e^-$$

$$PbO_2 \rightarrow PbSO_4$$
$$PbO_2 + H_2SO_4 \rightarrow PbSO_4 + 2\,H_2O$$
$$2\,e^- + 2\,H^+ + PbO_2 + H_2SO_4 \rightarrow PbSO_4 + 2\,H_2O$$

Add the two half-reactions with two electrons transferred:

$$2\ e^- + 2\ H^+ + PbO_2 + H_2SO_4 \rightarrow PbSO_4 + 2\ H_2O$$
$$Pb + H_2SO_4 \rightarrow PbSO_4 + 2\ H^+ + 2\ e^-$$

$$Pb(s) + 2\ H_2SO_4(aq) + PbO_2(s) \rightarrow 2\ PbSO_4(s) + 2\ H_2O(l)$$

This is the reaction that occurs in an automobile lead storage battery.

d. $Mn^{2+} \rightarrow MnO_4^-$

$(4\ H_2O + Mn^{2+} \rightarrow MnO_4^- + 8\ H^+ + 5\ e^-) \times 2$

$NaBiO_3 \rightarrow Bi^{3+} + Na^+$

$6\ H^+ + NaBiO_3 \rightarrow Bi^{3+} + Na^+ + 3\ H_2O$

$(2\ e^- + 6\ H^+ + NaBiO_3 \rightarrow Bi^{3+} + Na^+ + 3\ H_2O) \times 5$

$$8\ H_2O + 2\ Mn^{2+} \rightarrow 2\ MnO_4^- + 16\ H^+ + 10\ e^-$$
$$10\ e^- + 30\ H^+ + 5\ NaBiO_3 \rightarrow 5\ Bi^{3+} + 5\ Na^+ + 15\ H_2O$$

$$8\ H_2O + 30\ H^+ + 2\ Mn^{2+} + 5\ NaBiO_3 \rightarrow 2\ MnO_4^- + 5\ Bi^{3+} + 5\ Na^+ + 15\ H_2O + 16\ H^+$$

Simplifying:

$$14\ H^+(aq) + 2\ Mn^{2+}(aq) + 5\ NaBiO_3(s) \rightarrow 2\ MnO_4^-(aq) + 5\ Bi^{3+}(aq) + 5\ Na^+(aq) + 7\ H_2O(l)$$

e. $H_3AsO_4 \rightarrow AsH_3$ $(Zn \rightarrow Zn^{2+} + 2\ e^-) \times 4$

$H_3AsO_4 \rightarrow AsH_3 + 4\ H_2O$

$8\ e^- + 8\ H^+ + H_3AsO_4 \rightarrow AsH_3 + 4\ H_2O$

$$8\ e^- + 8\ H^+ + H_3AsO_4 \rightarrow AsH_3 + 4\ H_2O$$
$$4\ Zn \rightarrow 4\ Zn^{2+} + 8\ e^-$$

$$8\ H^+(aq) + H_3AsO_4(aq) + 4\ Zn(s) \rightarrow 4\ Zn^{2+}(aq) + AsH_3(g) + 4\ H_2O(l)$$

f. $As_2O_3 \rightarrow H_3AsO_4$

$As_2O_3 \rightarrow 2\ H_3AsO_4$

$(5\ H_2O + As_2O_3 \rightarrow 2\ H_3AsO_4 + 4\ H^+ + 4\ e^-) \times 3$

$NO_3^- \rightarrow NO + 2\ H_2O$

$4\ H^+ + NO_3^- \rightarrow NO + 2\ H_2O$

$(3\ e^- + 4\ H^+ + NO_3^- \rightarrow NO + 2\ H_2O) \times 4$

$$12\ e^- + 16\ H^+ + 4\ NO_3^- \rightarrow 4\ NO + 8\ H_2O$$
$$15\ H_2O + 3\ As_2O_3 \rightarrow 6\ H_3AsO_4 + 12\ H^+ + 12\ e^-$$

$$7\ H_2O(l) + 4\ H^+(aq) + 3\ As_2O_3(s) + 4\ NO_3^-(aq) \rightarrow 4\ NO(g) + 6\ H_3AsO_4(aq)$$

g. $(2\ Br^- \rightarrow Br_2 + 2\ e^-) \times 5$

$MnO_4^- \rightarrow Mn^{2+} + 4\ H_2O$

$(5\ e^- + 8\ H^+ + MnO_4^- \rightarrow Mn^{2+} + 4\ H_2O) \times 2$

$10\ Br^- \rightarrow 5\ Br_2 + 10\ e^-$

$10\ e^- + 16\ H^+ + 2\ MnO_4^- \rightarrow 2\ Mn^{2+} + 8\ H_2O$

$16\ H^+(aq) + 2\ MnO_4^-(aq) + 10\ Br^-(aq) \rightarrow 5\ Br_2(l) + 2\ Mn^{2+}(aq) + 8\ H_2O(l)$

h. $CH_3OH \rightarrow CH_2O$

$(CH_3OH \rightarrow CH_2O + 2\ H^+ + 2\ e^-) \times 3$

$Cr_2O_7^{2-} \rightarrow Cr^{3+}$

$14\ H^+ + Cr_2O_7^{2-} \rightarrow 2\ Cr^{3+} + 7\ H_2O$

$6\ e^- + 14\ H^+ + Cr_2O_7^{2-} \rightarrow 2\ Cr^{3+} + 7\ H_2O$

$3\ CH_3OH \rightarrow 3\ CH_2O + 6\ H^+ + 6\ e^-$

$6\ e^- + 14\ H^+ + Cr_2O_7^{2-} \rightarrow 2\ Cr^{3+} + 7\ H_2O$

$8\ H^+(aq) + 3\ CH_3OH(aq) + Cr_2O_7^{2-}(aq) \rightarrow 2\ Cr^{3+}(aq) + 3\ CH_2O(aq) + 7\ H_2O(l)$

59. a. HCl(aq) dissociates to $H^+(aq) + Cl^-(aq)$. For simplicity let's use H^+ and Cl^- separately.

$H^+ \rightarrow H_2$

$(2\ H^+ + 2\ e^- \rightarrow H_2) \times 3$

$Fe \rightarrow HFeCl_4$

$(H^+ + 4\ Cl^- + Fe \rightarrow HFeCl_4 + 3\ e^-) \times 2$

$6\ H^+ + 6\ e^- \rightarrow 3\ H_2$

$2\ H^+ + 8\ Cl^- + 2\ Fe \rightarrow 2\ HFeCl_4 + 6\ e^-$

$8\ H^+ + 8\ Cl^- + 2\ Fe \rightarrow 2\ HFeCl_4 + 3\ H_2$

or $8\ HCl(aq) + 2\ Fe(s) \rightarrow 2\ HFeCl_4(aq) + 3\ H_2(g)$

b. $IO_3^- \rightarrow I_3^-$

$3\ IO_3^- \rightarrow I_3^-$

$3\ IO_3^- \rightarrow I_3^- + 9\ H_2O$

$16\ e^- + 18\ H^+ + 3\ IO_3^- \rightarrow I_3^- + 9\ H_2O$

$I^- \rightarrow I_3^-$

$(3\ I^- \rightarrow I_3^- + 2\ e^-) \times 8$

$16\ e^- + 18\ H^+ + 3\ IO_3^- \rightarrow I_3^- + 9\ H_2O$

$24\ I^- \rightarrow 8\ I_3^- + 16\ e^-$

$18\ H^+ + 24\ I^- + 3\ IO_3^- \rightarrow 9\ I_3^- + 9\ H_2O$

Reducing: $6\ H^+(aq) + 8\ I^-(aq) + IO_3^-(aq) \rightarrow 3\ I_3^-(aq) + 3\ H_2O(l)$

c. $(Ce^{4+} + e^- \rightarrow Ce^{3+}) \times 97$

$Cr(NCS)_6^{4-} \rightarrow Cr^{3+} + NO_3^- + CO_2 + SO_4^{2-}$

$54\ H_2O + Cr(NCS)_6^{4-} \rightarrow Cr^{3+} + 6\ NO_3^- + 6\ CO_2 + 6\ SO_4^{2-} + 108\ H^+$

Charge on left -4. Charge on right $= +3 + 6(-1) + 6(-2) + 108(+1) = +93$. Add 97 e^- to the right, then add the two balanced half-reactions with a common factor of 97 e^- transferred.

$$54\ H_2O + Cr(NCS)_6^{4-} \rightarrow Cr^{3+} + 6\ NO_3^- + 6\ CO_2 + 6\ SO_4^{2-} + 108\ H^+ + 97\ e^-$$
$$97\ e^- + 97\ Ce^{4+} \rightarrow 97\ Ce^{3+}$$

$$97\ Ce^{4+}(aq) + 54\ H_2O(l) + Cr(NCS)_6^{4-}(aq) \rightarrow 97\ Ce^{3+}(aq) + Cr^{3+}(aq) + 6\ NO_3^-(aq) + 6\ CO_2(g) + 6\ SO_4^{2-}(aq) + 108\ H^+(aq)$$

This is very complicated. A check of the net charge is a good check to see if the equation is balanced. Left: charge = 97(+4) - 4 = +384. Right: charge = 97(+3) + 3 + 6(-1) + 6(-2) + 108(+1) = +384.

d. $CrI_3 \rightarrow CrO_4^{2-} + IO_4^-$ $\qquad$ $Cl_2 \rightarrow Cl^-$

$(16\ H_2O + CrI_3 \rightarrow CrO_4^{2-} + 3\ IO_4^- + 32\ H^+ + 27\ e^-) \times 2$ $\qquad$ $(2\ e^- + Cl_2 \rightarrow 2\ Cl^-) \times 27$

Common factor is a transfer of 54 e^-.

$$54\ e^- + 27\ Cl_2 \rightarrow 54\ Cl^-$$
$$32\ H_2O + 2\ CrI_3 \rightarrow 2\ CrO_4^{2-} + 6\ IO_4^- + 64\ H^+ + 54\ e^-$$

$$32\ H_2O + 2\ CrI_3 + 27\ Cl_2 \rightarrow 54\ Cl^- + 2\ CrO_4^{2-} + 6\ IO_4^- + 64\ H^+$$

Add 64 OH^- to both sides and convert 64 H^+ into 64 H_2O.

$$64\ OH^- + 32\ H_2O + 2\ CrI_3 + 27\ Cl_2 \rightarrow 54\ Cl^- + 2\ CrO_4^{2-} + 6\ IO_4^- + 64\ H_2O$$

Reducing gives:

$$64\ OH^-(aq) + 2\ CrI_3(s) + 27\ Cl_2(g) \rightarrow 54\ Cl^-(aq) + 2\ CrO_4^{2-}(aq) + 6\ IO_4^-(aq) + 32\ H_2O(l)$$

e. $Ce^{4+} \rightarrow Ce(OH)_3$

$(e^- + 3\ H_2O + Ce^{4+} \rightarrow Ce(OH)_3 + 3\ H^+) \times 61$

$$Fe(CN)_6^{4-} \rightarrow Fe(OH)_3 + CO_3^{2-} + NO_3^-$$
$$Fe(CN)_6^{4-} \rightarrow Fe(OH)_3 + 6\ CO_3^{2-} + 6\ NO_3^-$$

There are 39 extra O atoms on right. Add 39 H_2O to left, then add 75 H^+ to right to balance H^+.

$39\ H_2O + Fe(CN)_6^{4-} \rightarrow Fe(OH)_3 + 6\ CO_3^{2-} + 6\ NO_3^- + 75\ H^+$

net charge = -4 $\qquad$ net charge = +57

Add 61 e^- to the right then add the two balanced half-reactions with a common factor of 61 e^- transferred.

$$39\ H_2O + Fe(CN)_6^{4-} \rightarrow Fe(OH)_3 + 6\ CO_3^{2-} + 6\ NO_3^- + 75\ H^+ + 61\ e^-$$
$$61\ e^- + 183\ H_2O + 61\ Ce^{4+} \rightarrow 61\ Ce(OH)_3 + 183\ H^+$$

$$222\ H_2O + Fe(CN)_6^{4-} + 61\ Ce^{4+} \rightarrow 61\ Ce(OH)_3 + Fe(OH)_3 + 6\ CO_3^{2-} + 6\ NO_3^- + 258\ H^+$$

Adding 258 OH^- to each side then reducing gives:

$$258\ OH^-(aq) + Fe(CN)_6^{4-}(aq) + 61\ Ce^{4+}(aq) \rightarrow 61\ Ce(OH)_3(s) + Fe(OH)_3(s) + 6\ CO_3^{2-}(aq) + 6\ NO_3^-(aq) + 36\ H_2O(l)$$

61. $Mn \rightarrow Mn^{2+} + 2\ e^-$ $HNO_3 \rightarrow NO_2$

$HNO_3 \rightarrow NO_2 + H_2O$

$(e^- + H^+ + HNO_3 \rightarrow NO_2 + H_2O) \times 2$

$$Mn \rightarrow Mn^{2+} + 2\ e^-$$
$$2\ e^- + 2\ H^+ + 2\ HNO_3 \rightarrow 2\ NO_2 + 2\ H_2O$$

$$2\ H^+(aq) + Mn(s) + 2\ HNO_3(aq) \rightarrow Mn^{2+}(aq) + 2\ NO_2(g) + 2\ H_2O(l) \text{ or}$$

$$4\ H^+(aq) + Mn(s) + 2\ NO_3^-(aq) \rightarrow Mn^{2+}(aq) + 2\ NO_2(g) + 2\ H_2O(l)$$ (HNO_3 is a strong acid.)

$(4\ H_2O + Mn^{2+} \rightarrow MnO_4^- + 8\ H^+ + 5\ e^-) \times 2$ $(2\ e^- + 2\ H^+ + IO_4^- \rightarrow IO_3^- + H_2O) \times 5$

$$8\ H_2O + 2\ Mn^{2+} \rightarrow 2\ MnO_4^- + 16\ H^+ + 10\ e^-$$
$$10\ e^- + 10\ H^+ + 5\ IO_4^- \rightarrow 5\ IO_3^- + 5\ H_2O$$

$$3\ H_2O(l) + 2\ Mn^{2+}(aq) + 5\ IO_4^-(aq) \rightarrow 2\ MnO_4^-(aq) + 5\ IO_3^-(aq) + 6\ H^+(aq)$$

63. $(H_2C_2O_4 \rightarrow 2\ CO_2 + 2\ H^+ + 2\ e^-) \times 5$ $(5\ e^- + 8\ H^+ + MnO_4^- \rightarrow Mn^{2+} + 4\ H_2O) \times 2$

$$5\ H_2C_2O_4 \rightarrow 10\ CO_2 + 10\ H^+ + 10\ e^-$$
$$10\ e^- + 16\ H^+ + 2\ MnO_4^- \rightarrow 2\ Mn^{2+} + 8\ H_2O$$

$$6\ H^+(aq) + 5\ H_2C_2O_4(aq) + 2\ MnO_4^-(aq) \rightarrow 10\ CO_2(g) + 2\ Mn^{2+}(aq) + 8\ H_2O(l)$$

$$0.1058\text{ g } H_2C_2O_4 \times \frac{1\text{ mol } H_2C_2O_4}{90.034\text{ g}} \times \frac{2\text{ mol } MnO_4^-}{5\text{ mol } H_2C_2O_4} = 4.700 \times 10^{-4}\text{ mol } MnO_4^-$$

$$\text{Molarity} = \frac{4.700 \times 10^{-4}\text{ mol } MnO_4^-}{28.97\text{ mL}} \times \frac{1000\text{ mL}}{\text{L}} = 1.622 \times 10^{-2}\ M\ MnO_4^-$$

65.

$$(Fe^{2+} \rightarrow Fe^{3+} + e^-) \times 5$$
$$5\ e^- + 8\ H^+ + MnO_4^- \rightarrow Mn^{2+} + 4\ H_2O$$

$$8\ H^+(aq) + MnO_4^-(aq) + 5\ Fe^{2+}(aq) \rightarrow 5\ Fe^{3+}(aq) + Mn^{2+}(aq) + 4\ H_2O(l)$$

From the titration data we can get the number of moles of Fe^{2+}. We then convert this to a mass of iron and calculate the mass percent of iron in the sample.

$$38.37 \times 10^{-3}\text{ L } MnO_4^- \times \frac{0.0198\text{ mol } MnO_4^-}{\text{L}} \times \frac{5\text{ mol } Fe^{2+}}{\text{mol } MnO_4^-} = 3.80 \times 10^{-3}\text{ mol } Fe^{2+}$$

$$= 3.80 \times 10^{-3}\text{ mol Fe present}$$

$$3.80 \times 10^{-3} \text{ mol Fe} \times \frac{55.85 \text{ g Fe}}{\text{mol Fe}} = 0.212 \text{ g Fe}$$

$$\text{Mass \% Fe} = \frac{0.212 \text{ g}}{0.6128 \text{ g}} \times 100 = 34.6\% \text{ Fe}$$

67. $Mg(s) + 2\ HCl(aq) \rightarrow MgCl_2(aq) + H_2(g)$

$$3.00 \text{ g Mg} \times \frac{1 \text{ mol Mg}}{24.31 \text{ g Mg}} \times \frac{2 \text{ mol HCl}}{\text{mol Mg}} \times \frac{1 \text{ L HCl}}{5.0 \text{ mol HCl}} = 0.0494 \text{ L} = 49.4 \text{ mL HCl}$$

Additional Exercises

69. $$\text{mol CaCl}_2 \text{ present} = 0.230 \text{ L CaCl}_2 \times \frac{0.275 \text{ mol CaCl}_2}{\text{L CaCl}_2} = 6.33 \times 10^{-2} \text{ mol CaCl}_2$$

The volume of $CaCl_2$ solution after evaporation is:

$$6.33 \times 10^{-2} \text{ mol CaCl}_2 \times \frac{1 \text{ L CaCl}_2}{1.10 \text{ mol CaCl}_2} = 5.75 \times 10^{-2} \text{ L} = 57.5 \text{ mL CaCl}_2$$

Volume H_2O evaporated = 230. mL - 57.5 mL = 173 mL H_2O evaporated

71. a. $MgCl_2(aq) + 2\ AgNO_3(aq) \rightarrow 2\ AgCl(s) + Mg(NO_3)_2(aq)$

$$0.641 \text{ g AgCl} \times \frac{1 \text{ mol AgCl}}{143.4 \text{ g AgCl}} \times \frac{1 \text{ mol MgCl}_2}{2 \text{ mol AgCl}} \times \frac{95.21 \text{ g}}{\text{mol MgCl}_2} = 0.213 \text{ g MgCl}_2$$

$$\frac{0.213 \text{ g MgCl}_2}{1.50 \text{ g mixture}} \times 100 = 14.2\% \text{ MgCl}_2$$

b. $$0.213 \text{ g MgCl}_2 \times \frac{1 \text{ mol MgCl}_2}{95.21 \text{ g}} \times \frac{2 \text{ mol AgNO}_3}{\text{mol MgCl}_2} \times \frac{1 \text{ L}}{0.500 \text{ mol AgNO}_3} \times \frac{1000 \text{ mL}}{1 \text{ L}}$$
$$= 8.95 \text{ mL AgNO}_3$$

73. a. $$0.308 \text{ g AgCl} \times \frac{35.45 \text{ g Cl}}{143.4 \text{ g AgCl}} = 0.0761 \text{ g Cl}; \ \%\text{Cl} = \frac{0.0761 \text{ g}}{0.256 \text{ g}} \times 100 = 29.7\% \text{ Cl}$$

Cobalt(III) oxide, Co_2O_3: 2(58.93) + 3(16.00) = 165.86 g/mol

$$0.145 \text{ g Co}_2\text{O}_3 \times \frac{117.86 \text{ g Co}}{165.86 \text{ g Co}_2\text{O}_3} = 0.103 \text{ g Co}; \ \%\text{Co} = \frac{0.103 \text{ g}}{0.416 \text{ g}} \times 100 = 24.8\% \text{ Co}$$

The remainder, 100.0 - (29.7 + 24.8) = 45.5%, is water.

Assuming 100.0 g of compound:

$$45.5 \text{ g H}_2\text{O} \times \frac{2.016 \text{ g H}}{18.02 \text{ g H}_2\text{O}} = 5.09 \text{ g H}; \ \%\text{H} = \frac{5.09 \text{ g H}}{100.0 \text{ g compound}} \times 100 = 5.09\% \text{ H}$$

$$45.5 \text{ g } H_2O \times \frac{16.00 \text{ g O}}{18.02 \text{ g } H_2O} = 40.4 \text{ g O}; \ \%O = \frac{40.4 \text{ g O}}{100.0 \text{ g compound}} \times 100 = 40.4\% \text{ O}$$

The mass percent composition is 24.8% Co, 29.7% Cl, 5.09% H and 40.4% O.

b. **Out of 100.0 g of compound, there are:**

$$24.8 \text{ g Co} \times \frac{1 \text{ mol}}{58.93 \text{ g Co}} = 0.421 \text{ mol Co}; \ 29.7 \text{ g Cl} \times \frac{1 \text{ mol}}{35.45 \text{ g Cl}} = 0.838 \text{ mol Cl}$$

$$5.09 \text{ g H} \times \frac{1 \text{ mol}}{1.008 \text{ g H}} = 5.05 \text{ mol H}; \ 40.4 \text{ g O} \times \frac{1 \text{ mol}}{16.00 \text{ g O}} = 2.53 \text{ mol O}$$

Dividing all results by 0.421, we get $CoCl_2 \bullet 6H_2O$.

c. $CoCl_2 \bullet 6H_2O(aq) + 2\ AgNO_3(aq) \rightarrow 2\ AgCl(s) + Co(NO_3)_2(aq) + 6\ H_2O(l)$

$CoCl_2 \bullet 6H_2O(aq) + 2\ NaOH(aq) \rightarrow Co(OH)_2(s) + 2\ NaCl(aq) + 6\ H_2O(l)$

$Co(OH)_2 \rightarrow Co_2O_3$ **This is an oxidation-reduction reaction. Thus, we also need to include an oxidizing agent. The obvious choice is O_2.**

$4\ Co(OH)_2(s) + O_2(g) \rightarrow 2\ Co_2O_3(s) + 4\ H_2O(l)$

75. $Ag^+(aq) + Cl^-(aq) \rightarrow AgCl(s)$; **Let x = mol NaCl and y = mol KCl.**

$22.90 \times 10^{-3} \text{ L} \times 0.1000 \text{ mol/L} = 2.290 \times 10^{-3} \text{ mol } Ag^+ = 2.290 \times 10^{-3} \text{ mol } Cl^- \text{ total}$

$x + y = 2.290 \times 10^{-3} \text{ mol } Cl^-, \ x = 2.290 \times 10^{-3} - y$

Since the molar mass of NaCl is 58.44 g/mol and the molar mass of KCl is 74.55 g/mol, then:

$58.44\ x + 74.55\ y = 0.1586 \text{ g}$

$58.44\ (2.290 \times 10^{-3} - y) + 74.55\ y = 0.1586, \ 16.11\ y = 0.0248, \ y = 1.54 \times 10^{-3} \text{ mol KCl}$

$$\text{Mass \% KCl} = \frac{1.54 \times 10^{-3} \text{ mol} \times 74.55 \text{ g/mol}}{0.1586 \text{ g}} \times 100 = 72.4\% \text{ KCl}$$

% NaCl = 100.0 - 72.4 = 27.6% NaCl

77. $Cr(NO_3)_3(aq) + 3\ NaOH(aq) \rightarrow Cr(OH)_3(s) + 3\ NaNO_3(aq)$

$$\text{mol NaOH used to form precipitate} = 2.06 \text{ g } Cr(OH)_3 \times \frac{1 \text{ mol } Cr(OH)_3}{103.02 \text{ g}} \times \frac{3 \text{ mol NaOH}}{1 \text{ mol } Cr(OH)_3} = 6.00 \times 10^{-2} \text{ mol NaOH}$$

$NaOH(aq) + HCl(aq) \rightarrow NaCl(aq) + H_2O(l)$

$$\text{mol NaOH used to react with HCl} = 0.1000 \text{ L} \times \frac{0.400 \text{ mol HCl}}{\text{L}} \times \frac{1 \text{ mol NaOH}}{\text{mol HCl}} = 4.00 \times 10^{-2} \text{ mol NaOH}$$

$$M_{NaOH} = \frac{\text{mol NaOH}}{\text{volume}} = \frac{6.00 \times 10^{-2}\text{ mol} + 4.00 \times 10^{-2}\text{ mol}}{0.0500\text{ L}} = 2.00\ M\text{ NaOH}$$

79. The amount of KHP used = $0.4016\text{ g} \times \frac{1\text{ mol}}{204.22\text{ g}} = 1.967 \times 10^{-3}$ mol KHP

Since one mole of NaOH reacts completely with one mole of KHP, then the NaOH solution contains 1.967×10^{-3} mol NaOH.

$$\text{Molarity of NaOH} = \frac{1.967 \times 10^{-3}\text{ mol}}{25.06 \times 10^{-3}\text{ L}} = \frac{7.849 \times 10^{-2}\text{ mol NaOH}}{\text{L}}$$

$$\text{Maximum molarity} = \frac{1.967 \times 10^{-3}\text{ mol}}{25.01 \times 10^{-3}\text{ L}} = \frac{7.865 \times 10^{-2}\text{ mol NaOH}}{\text{L}}$$

$$\text{Minimum molarity} = \frac{1.967 \times 10^{-3}\text{ mol}}{25.11 \times 10^{-3}\text{ L}} = \frac{7.834 \times 10^{-2}\text{ mol NaOH}}{\text{L}}$$

We can express this as $0.07849 \pm 0.00016\ M$. An alternate way is to express the molarity as $0.0785 \pm 0.0002\ M$. This second way shows the actual number of significant figures in the molarity. The advantage of the first method is that it shows that we made all of our individual measurements to four significant figures.

81. mol $C_6H_8O_7 = 0.250\text{ g } C_6H_8O_7 \times \frac{1\text{ mol } C_6H_8O_7}{192.1\text{ g } C_6H_8O_7} = 1.30 \times 10^{-3}\text{ mol } C_6H_8O_7$

Let H_xA represent citric acid where x is the number of acidic hydrogens. The balanced neutralization reaction is:

$$H_xA(aq) + x\ OH^-(aq) \rightarrow x\ H_2O(l) + A^{x-}(aq)$$

$$\text{mol OH}^- \text{ reacted} = 0.0372\text{ L} \times \frac{0.105\text{ mol OH}^-}{\text{L}} = 3.91 \times 10^{-3}\text{ mol OH}^-$$

$$x = \frac{\text{mol OH}^-}{\text{mol citric acid}} = \frac{3.91 \times 10^{-3}\text{ mol}}{1.30 \times 10^{-3}\text{ mol}} = 3.01$$

Therefore, the general acid formula for citric acid is H_3A meaning that citric acid has three acidic hydrogens per citric acid molecule (citric acid is a triprotic acid).

Challenge Problems

83. $Zn(s) + 2\ AgNO_2(aq) \rightarrow 2\ Ag(s) + Zn(NO_2)_2(aq)$

Let x = mass of Ag and y = mass of Zn after the reaction has stopped. Then x + y = 29.0 g. Since the mol of Ag produced will equal two times the mol of Zn reacted, then:

$$(19.0 - y)\text{ g Zn} \times \frac{1\text{ mol Zn}}{65.38\text{ g Zn}} \times \frac{2\text{ mol Ag}}{1\text{ mol Zn}} = x\text{ g Ag} \times \frac{1\text{ mol Ag}}{107.9\text{ g Ag}}$$

Simplifying:

$$3.059 \times 10^{-2}(19.0 - y) = 9.268 \times 10^{-3}\,x$$

Substituting $x = 29.0 - y$ into the equation gives:

$$3.059 \times 10^{-2}(19.0 - y) = 9.268 \times 10^{-3}(29.0 - y)$$

Solving:

$$0.581 - 3.059 \times 10^{-2}\,y = 0.269 - 9.268 \times 10^{-3}\,y,\ \ 2.132 \times 10^{-2}\,y = 0.312,\ y = 14.6\text{ g Zn}$$

14.6 g Zn are present and 29.0 - 14.6 = 14.4 g Ag are present after the reaction is stopped.

85. **Molar masses: KCl, 39.10 + 35.45 = 74.55 g/mol; KBr, 39.10 + 79.90 = 119.00 g/mol**

AgCl, 107.9 + 35.45 = 143.4 g/mol; AgBr, 107.9 + 79.90 = 187.8 g/mol

Let x = number of moles of KCl in mixture and y = number of moles of KBr in mixture. Since $Ag^+ + Cl^- \rightarrow AgCl$ and $Ag^+ + Br^- \rightarrow AgBr$, then x = moles AgCl and y = moles AgBr. Setting up simultaneous equations from the given information:

$$0.1024\text{ g} = 74.55\,x + 119.0\,y \text{ and } 0.1889\text{ g} = 143.4\,x + 187.8\,y$$

Multiply the first equation by $\frac{187.8}{119.0}$, then subtract from the second.

$$\begin{array}{rl} 0.1889 = & 143.4\,x + 187.8\,y \\ -0.1616 = & -117.7\,x - 187.8\,y \\ \hline 0.0273 = & 25.7\,x, \qquad x = 1.06 \times 10^{-3}\text{ mol KCl} \end{array}$$

$$1.06 \times 10^{-3}\text{ mol KCl} \times \frac{74.55\text{ g KCl}}{\text{mol KCl}} = 0.0790\text{ g KCl}$$

$$\text{Mass \% KCl} = \frac{0.0790\text{ g}}{0.1024\text{ g}} \times 100 = 77.1\%,\ \ \%\text{ KBr} = 100.0 - 77.1 = 22.9\%$$

87. a. $C_{12}H_{10-n}Cl_n + n\,Ag^+ \rightarrow n\,AgCl$; **molar mass (AgCl) = 143.4 g/mol**

molar mass (PCB) = 12(12.01) + (10-n) (1.008) + n(35.45) = 154.20 + 34.44 n

Since n mol AgCl are produced for every 1 mol PCB reacted, then n(143.4) grams of AgCl will be produced for every (154.20 + 34.44 n) grams of PCB reacted.

$$\frac{\text{mass of AgCl}}{\text{mass of PCB}} = \frac{143.4\,n}{154.20 + 34.44\,n} \text{ or } \text{mass}_{AgCl}\,(154.20 + 34.44\,n) = \text{mass}_{PCB}\,(143.4\,n)$$

b. $0.4971 (154.20 + 34.44\ n) = 0.1947 (143.4\ n),\ 76.65 + 17.12\ n = 27.92\ n$

$76.65 = 10.80\ n,\ n = 7.097$

89. a. Flow rate = 5.00×10^4 L/s + 3.50×10^3 L/s = 5.35×10^4 L/s

b. $C_{HCl} = \dfrac{3.50 \times 10^3 (65.0)}{5.35 \times 10^4} = 4.25$ ppm HCl

c. 1 ppm = 1 mg/kg H_2O = 1 mg/L (assuming density = 1.00 g/mL)

$$8.00\ \text{hr} \times \frac{60\ \text{min}}{\text{hr}} \times \frac{60\ \text{s}}{\text{min}} \times \frac{1.80 \times 10^4\ \text{L}}{\text{s}} \times \frac{4.25\ \text{mg HCl}}{\text{L}} \times \frac{1\ \text{g}}{1000\ \text{mg}} = 2.20 \times 10^6\ \text{g HCl}$$

$$2.20 \times 10^6\ \text{g HCl} \times \frac{1\ \text{mol HCl}}{36.46\ \text{g HCl}} \times \frac{1\ \text{mol CaO}}{2\ \text{mol HCl}} \times \frac{56.08\ \text{g CaO}}{\text{mol CaO}} = 1.69 \times 10^6\ \text{g CaO}$$

d. The concentration of Ca^{2+} going into the second plant was:

$$\frac{5.00 \times 10^4 (10.2)}{5.35 \times 10^4} = 9.53\ \text{ppm}$$

The second plant used: 1.80×10^4 L/s × (8.00 × 60 × 60) sec = 5.18×10^8 L of water.

$$1.69 \times 10^6\ \text{g CaO} \times \frac{40.08\ \text{g Ca}^{2+}}{56.08\ \text{g CaO}} = 1.21 \times 10^6\ \text{g Ca}^{2+}$$ was added to this water.

$$C_{Ca^{2+}}\ (\text{plant water}) = 9.53 + \frac{1.21 \times 10^9\ \text{mg}}{5.18 \times 10^8\ \text{L}} = 9.53 + 2.34 = 11.87\ \text{ppm}$$

Since 90.0% of this water is returned, $1.80 \times 10^4 \times 0.900 = 1.62 \times 10^4$ L/s of water with 11.87 ppm Ca^{2+} is mixed with $(5.35 - 1.80) \times 10^4 = 3.55 \times 10^4$ L/s of water containing 9.53 ppm Ca^{2+}.

$$C_{Ca^{2+}}\ (\text{final}) = \frac{(1.62 \times 10^4\ \text{L/s})(11.87\ \text{ppm}) + (3.55 \times 10^4\ \text{L/s})(9.53\ \text{ppm})}{1.62 \times 10^4\ \text{L/s} + 3.55 \times 10^4\ \text{L/s}} = 10.3\ \text{ppm}$$

91. a. $YBa_2Cu_3O_{6.5}$:

$+3 + 2(+2) + 3x + 6.5(-2) = 0$

$7 + 3x - 13 = 0,\ 3x = 6,\ x = +2$ Only Cu^{2+} present.

$YBa_2Cu_3O_7$:

$+3 + 2(+2) + 3x + 7(-2) = 0,\ x = +2\ 1/3$ or 2.33

This corresponds to two Cu^{2+} and one Cu^{3+} present.

$YBa_2Cu_3O_8$:

$+3 + 2(+2) + 3x + 8(-2) = 0,\ x = +3$ Only Cu^{3+} present.

b. $(e^- + Cu^{2+} + I^- \rightarrow CuI) \times 2$ $\quad$ $2\,e^- + Cu^{3+} + I^- \rightarrow CuI$

$3I^- \rightarrow I_3^- + 2\,e^-$ $\quad$ $3I^- \rightarrow I_3^- + 2\,e^-$

$2\,Cu^{2+}(aq) + 5\,I^-(aq) \rightarrow 2\,CuI(s) + I_3^-(aq)$ $\quad$ $Cu^{3+}(aq) + 4\,I^-(aq) \rightarrow CuI(s) + I_3^-(aq)$

$2\,S_2O_3^{2-} \rightarrow S_4O_6^{2-} + 2\,e^-$

$2\,e^- + I_3^- \rightarrow 3\,I^-$

$2\,S_2O_3^{2-}(aq) + I_3^-(aq) \rightarrow 3\,I^-(aq) + S_4O_6^{2-}(aq)$

c. Step II data: All Cu is converted to Cu^{2+}. Note: superconductor abbreviated as "123."

$$22.57 \times 10^{-3}\text{ L} \times \frac{0.1000\text{ mol } S_2O_3^{2-}}{\text{L}} \times \frac{1\text{ mol } I_3^-}{2\text{ mol } S_2O_3^{2-}} \times \frac{2\text{ mol } Cu^{2+}}{\text{mol } I_3^-} = 2.257 \times 10^{-3}\text{ mol } Cu^{2+}$$

$$2.257 \times 10^{-3}\text{ mol Cu} \times \frac{1\text{ mol "123"}}{3\text{ mol Cu}} = 7.523 \times 10^{-4}\text{ mol "123"}$$

$$\text{Molar mass of } YBa_2Cu_3O_x = \frac{0.5042\text{ g}}{7.523 \times 10^{-4}\text{ mol}} = 670.2\text{ g/mol}$$

$670.2 = 88.91 + 2(137.3) + 3(63.55) + x(16.00)$, $670.2 = 554.2 + x(16.00)$

$x = 7.250$; Formula is $YBa_2Cu_3O_{7.25}$.

Check with Step I data: Both Cu^{2+} and Cu^{3+} present.

$$37.77 \times 10^{-3}\text{ L} \times \frac{0.1000\text{ mol } S_2O_3^{2-}}{\text{L}} \times \frac{1\text{ mol } I_3^-}{2\text{ mol } S_2O_3^{2-}} = 1.889 \times 10^{-3}\text{ mol } I_3^-$$

We get 1 mol I_3^- per mol Cu^{3+} and 1 mol I_3^- per 2 mol Cu^{2+}. Let $n_{Cu^{3+}}$ = mol Cu^{3+} and $n_{Cu^{2+}}$ = mol Cu^{2+}, then:

$$n_{Cu^{3+}} + \frac{n_{Cu^{2+}}}{2} = 1.889 \times 10^{-3}\text{ mol}$$

In addition: $\dfrac{0.5625\text{ g}}{670.2\text{ g/mol}} = 8.393 \times 10^{-4}$ mol "123"

This amount of "123" contains: $3(8.393 \times 10^{-4}) = 2.518 \times 10^{-3}$ mol Cu total $= n_{Cu^{3+}} + n_{Cu^{2+}}$

Solving by simultaneous equations:

$$n_{Cu^{3+}} + n_{Cu^{2+}} = 2.518 \times 10^{-3}$$

$$-n_{Cu^{3+}} - \frac{n_{Cu^{2+}}}{2} = -1.889 \times 10^{-3}$$

$$\frac{n_{Cu^{2+}}}{2} = 6.29 \times 10^{-4}$$

$n_{Cu^{2+}} = 1.26 \times 10^{-3}$ mol Cu^{2+}; $n_{Cu^{3+}} = 2.518 \times 10^{-3} - 1.26 \times 10^{-3} = 1.26 \times 10^{-3}$ mol Cu^{3+}

This sample of superconductor contains equal moles of Cu^{2+} and Cu^{3+}. Therefore, 1 mol of $YBa_2Cu_3O_x$ contains 1.50 mol Cu^{2+} and 1.50 mol Cu^{3+}. Solving for x using oxidation states:

$$+3 + 2(+2) + 1.50(+2) + 1.50(+3) + x(-2) = 0, \quad 14.50 = 2x, \quad x = 7.25$$

The two experiments give the same result, $x = 7.25$ with formula $YBa_2Cu_3O_{7.25}$.

Average oxidation state of Cu:

$$+3 + 2(+2) + 3(x) + 7.25(-2) = 0, \quad 3x = 7.50, \quad x = +2.50$$

As determined from step I data, this superconductor sample contains equal moles of Cu^{2+} and Cu^{3+}, giving an average oxidation state of +2.50.

93. There are 3 unknowns so we need 3 equations to solve for the unknowns. Let x = mass $AgNO_3$, y = mass $CuCl_2$ and z = mass $FeCl_3$. Then x + y + z = 1.0000 g. The Cl^- in $CuCl_2$ and $FeCl_3$ will react with the excess $AgNO_3$ to form the precipitate AgCl(s). Assuming silver has an atomic mass of 107.90:

$$\text{Mass of Cl in mixture} = 1.7809 \text{ g AgCl} \times \frac{35.45 \text{ g Cl}}{143.35 \text{ g AgCl}} = 0.4404 \text{ g Cl}$$

$$\text{mass of Cl from CuCl}_2 = \text{y g CuCl}_2 \times \frac{2(35.45) \text{ g Cl}}{134.45 \text{ g CuCl}_2} = 0.5273 \text{ y}$$

$$\text{mass of Cl from FeCl}_3 = \text{z g FeCl}_3 \times \frac{3(35.45) \text{ g Cl}}{162.20 \text{ g FeCl}_3} = 0.6557 \text{ z}$$

The second equation is: 0.4404 g Cl = 0.5273 y + 0.6557 z

Similarly, let's calculate the mass of metals in each salt.

$$\text{mass of Ag in AgNO}_3 = \text{x g AgNO}_3 \times \frac{107.9 \text{ g Ag}}{169.91 \text{ g AgNO}_3} = 0.6350 \text{ x}$$

For $CuCl_2$ and $FeCl_3$, we already calculated the amount of Cl in each initial amount of salt; the remainder must be the mass of metal in each salt.

mass of Cu in $CuCl_2$ = y - 0.5273 y = 0.4727 y

mass of Fe in $FeCl_3$ = z - 0.6557 z = 0.3443 z

The third equation is: 0.4684 g metals = 0.6350 x + 0.4727 y + 0.3443 z

We now have three equations with three unknowns. Solving:

$$-0.6350\ (1.0000 = x + y + z)$$
$$0.4684 = 0.6350\ x + 0.4727\ y + 0.3443\ z$$

$$-0.1666 = -0.1623\ y - 0.2907\ z$$

$$\frac{0.5273}{0.1623}\ (-0.1666 = -0.1623\ y - 0.2907\ z)$$
$$0.4404 = 0.5273\ y + 0.6557\ z$$

$$-0.1009 = -0.2888\ z, \quad z = \frac{0.1009}{0.2888} = 0.3494\ g\ FeCl_3$$

$0.4404 = 0.5273\ y + 0.6557\ (0.3494),\ y = 0.4007\ g\ CuCl_2$

$x = 1.0000 - y - z = 1.0000 - 0.4007 - 0.3494 = 0.2499\ g\ AgNO_3$

$$\text{mass \% } AgNO_3 = \frac{0.2499\ g}{1.0000\ g} \times 100 = 24.99\%\ AgNO_3$$

$$\text{mass \% } CuCl_2 = \frac{0.4007\ g}{1.0000\ g} \times 100 = 40.07\%\ CuCl_2, \quad \text{mass \% } FeCl_3 = 34.94\%$$

Marathon Problems

95. a. Compound A = $M(NO_3)_x$; In 100.00 g compound: $8.246\ g\ N \times \frac{47.997\ g\ O}{14.007\ g\ N} = 28.26\ g\ O$

Thus, the mass of nitrate is: 100.0 g A - 36.51 g if x = 1.

If x = 1, mass of M = 100.00 - 36.51 g = 63.49 g

$$\text{mol M} = \text{mol N} = \frac{8.246\ g}{14.007\ g/mL} = 0.5887$$

$$\text{MM of metal M} = \frac{63.49\ g}{0.5887\ g/mL} = 107.8\ g/mol\ (Ag)$$

If x = 2, 0.5887 mass of M = 100.00 - 2(36.51) = 26.98 g

$$\text{mol M} = 1/2\ \text{mol N} = \frac{0.5887\ mol}{2} = 0.2944\ mol$$

$$\text{MM of metal M} = \frac{26.98\ g}{0.2944\ mol} = 91.64\ g/mol$$

This is close to Zr, but Zr does not form stable +2 ions in solution; it forms stable +4 ions. Thus, compound A is $AgNO_3$.

For compound B, the only chromium salt discussed in the solubility rules in Table 4.1 was CrO_4^{2-}. It has Cr in the +6 oxidation state, so let's assume compound B is K_2CrO_4.

b. The reaction is:

$$2\ AgNO_3(aq) + K_2CrO_4(aq) \rightarrow Ag_2CrO_4(s) + 2\ KNO_3(aq)$$

The blood red precipitate is $Ag_2CrO_4(s)$.

c. 331.8 g Ag_2CrO_4 is formed, which is 1 mol Ag_2CrO_4.

It is important to note that we begin with equal masses of $AgNO_3$ and K_2CrO_4. From the formula for Ag_2CrO_4, we must have at least 2 mol $AgNO_3$ and 1 mol K_2CrO_4.

$$2.000\ \text{mol}\ AgNO_3 = 339.8\ \text{g}\ AgNO_3$$

$$1.000\ \text{mol}\ K_2CrO_4 = 194.2\ \text{g}\ K_2CrO_4$$

$$\frac{2.000\ \text{mol}\ AgNO_3}{1.000\ \text{mol}\ K_2CrO_4} = \frac{339.8}{194.2} = 1.750;$$ The balanced equation requires a 2:1 mol ratio, so $AgNO_3$ is limiting and for equal masses we have 339.8 g $AgNO_3$ and 339.8 g K_2CrO_4.

Initially, solution A $= \frac{2.000\ \text{mol}\ Ag^+}{0.5000\ \text{L}} = 4.000\ M\ Ag^+$ and $\frac{2.000\ \text{mol}\ NO_3^-}{0.5000\ \text{L}} = 4.000\ M\ NO_3^-$.

Solution B $= 339.8\ \text{g}\ K_2CrO_4 \times \frac{1\ \text{mol}}{194.2\ \text{g}} = 1.750\ \text{mol}$

$$\frac{2 \times 1.750\ \text{mol}\ K^+}{0.5000\ \text{L}} = 7.000\ M\ K^+ \text{ and } \frac{1.750\ \text{mol}\ CrO_4^{2-}}{0.5000\ \text{L}} = 3.500\ M\ CrO_4^{2-}$$

d. After the reaction, moles of K^+ and moles of NO_3^- remain unchanged since they are spectator ions.

	$2\ Ag^+$ +	CrO_4^{2-}	$\rightarrow\ Ag_2CrO_4$
initial	2.000	1.750 mol	
after	0	0.750 mol	

$$[K^+] = \frac{2 \times 1.750\ \text{mol}}{1.000\ \text{L}} = 3.500\ M\ K^+$$

$$[NO_3^-] = \frac{2.000\ \text{mol}}{1.000\ \text{L}} = 2.000\ M\ NO_3^-$$

$$[CrO_4^{2-}] = \frac{0.750\ \text{mol}}{1.000\ \text{L}} = 0.750\ M\ CrO_4^{2-};\quad [Ag^+] = 0\ M \text{ (the limiting reagent)}$$

CHAPTER FIVE

GASES

Pressure

21. $4.75 \text{ cm} \times \frac{10 \text{ mm}}{\text{cm}} = 47.5$ mm Hg or 47.5 torr; $47.5 \text{ torr} \times \frac{1 \text{ atm}}{760 \text{ torr}} = 6.25 \times 10^{-2}$ atm

$$6.25 \times 10^{-2} \text{ atm} \times \frac{1.013 \times 10^5 \text{ Pa}}{\text{atm}} = 6.33 \times 10^3 \text{ Pa}$$

23. Suppose we have a column of Hg 1.00 cm × 1.00 cm × 76.0 cm = V = 76.0 cm^3:

$$\text{mass} = 76.0 \text{ cm}^3 \times 13.59 \text{ g/cm}^3 = 1.03 \times 10^3 \text{ g} \times \frac{1 \text{ kg}}{1000 \text{ g}} = 1.03 \text{ kg}$$

$$F = mg = 1.03 \text{ kg} \times 9.81 \text{ m/s}^2 = 10.1 \text{ kg m/s}^2 = 10.1 \text{ N}$$

$$\frac{\text{Force}}{\text{area}} = \frac{10.1 \text{ N}}{\text{cm}^2} \times \left(\frac{100 \text{ cm}}{\text{m}}\right)^2 = 1.01 \times 10^5 \frac{\text{N}}{\text{m}^2} \text{ or } 1.01 \times 10^5 \text{ Pa}$$

(Note: 76.0 cm Hg = 1 atm = 1.01×10^5 Pa.)

To exert the same pressure, a column of water will have to contain the same mass as the 76.0 cm column of Hg. Thus, the column of water will have to be 13.59 times taller or 76.0 cm × 13.59 = 1.03×10^3 cm = 10.3 m.

25. a. $4.8 \text{ atm} \times \frac{760 \text{ mm Hg}}{\text{atm}} = 3.6 \times 10^3$ mm Hg; b. $3.6 \times 10^3 \text{ mm Hg} \times \frac{1 \text{ torr}}{\text{mm Hg}} = 3.6 \times 10^3$ torr

c. $4.8 \text{ atm} \times \frac{1.013 \times 10^5 \text{ Pa}}{\text{atm}} = 4.9 \times 10^5$ Pa; d. $4.8 \text{ atm} \times \frac{14.7 \text{ psi}}{\text{atm}} = 71$ psi

Gas Laws

27. Treat each gas separately and use the relationship $P_1V_1 = P_2V_2$, since for each gas, n and T are constant.

For H_2: $P_2 = \frac{P_1V_1}{V_2} = 475 \text{ torr} \times \frac{2.00 \text{ L}}{3.00 \text{ L}} = 317 \text{ torr}$

For N_2: $P_2 = 0.200 \text{ atm} \times \frac{1.00 \text{ L}}{3.00 \text{ L}} = 0.0667 \text{ atm}$; $0.0667 \text{ atm} \times \frac{760 \text{ torr}}{\text{atm}} = 50.7 \text{ torr}$

$P_{total} = P_{H_2} + P_{N_2} = 317 + 50.7 = 368 \text{ torr}$

29. $PV = nRT$, $\frac{nT}{P} = \frac{V}{R} = \text{constant}$, $\frac{n_1T_1}{P_1} = \frac{n_2T_2}{P_2}$; mol × molar mass = mass

$$\frac{n_1 \text{ (molar mass) } T_1}{P_1} = \frac{n_2 \text{ (molar mass) } T_2}{P_2}, \quad \frac{\text{mass}_1 \times T_1}{P_1} = \frac{\text{mass}_2 \times T_2}{P_2}$$

$$\text{mass}_2 = \frac{\text{mass}_1 \times T_1P_2}{T_2P_1} = \frac{1.00 \times 10^3 \text{ g} \times 291 \text{ K} \times 650. \text{ psi}}{299 \text{ K} \times 2050. \text{ psi}} = 309 \text{ g}$$

31. $$P = P_{CO_2} = \frac{n_{CO_2}RT}{V} = \frac{\left(22.0 \text{ g} \times \frac{1 \text{ mol}}{44.01 \text{ g}}\right) \times \frac{0.08206 \text{ L atm}}{\text{mol K}} \times 300. \text{ K}}{4.00 \text{ L}} = 3.08 \text{ atm}$$

With air present, the partial pressure of CO_2 will still be 3.08 atm. The total pressure will be the sum of the partial pressures.

$$P_{total} = P_{CO_2} + P_{air} = 3.08 \text{ atm} + \left(740. \text{ torr} \times \frac{1 \text{ atm}}{760 \text{ torr}}\right) = 3.08 + 0.974 = 4.05 \text{ atm}$$

33. $$n = \frac{PV}{RT} = \frac{135 \text{ atm} \times 200.0 \text{ L}}{0.08206 \frac{\text{L atm}}{\text{mol K}} \times (273 + 24) \text{ K}} = 1.11 \times 10^3 \text{ mol}$$

For He: $1.11 \times 10^3 \text{ mol} \times \frac{4.003 \text{ g He}}{\text{mol}} = 4.44 \times 10^3 \text{ g He}$

For H_2: $1.11 \times 10^3 \text{ mol} \times \frac{2.016 \text{ g He}}{\text{mol}} = 2.24 \times 10^3 \text{ g } H_2$

35. As NO_2 is converted completely into N_2O_4, the moles of gas present will decrease by a factor of one-half (from the 2:1 mol ratio in the balanced equation). Using Avogadro's law,

$$\frac{V_1}{n_1} = \frac{V_2}{n_2}, \quad V_2 = V_1 \times \frac{n_2}{n_1} = 25.0 \text{ mL} \times \frac{1}{2} = 12.5 \text{ mL}$$

$N_2O_4(g)$ will occupy one-half the original volume of $NO_2(g)$.

37. PV = nRT, P is constant. $\frac{nT}{V} = \frac{P}{R}$ = constant, $\frac{n_1T_1}{V_1} = \frac{n_2T_2}{V_2}$

$$\frac{n_2}{n_1} = \frac{T_1V_2}{T_2V_1} = \frac{294\ K}{335\ K} \times \frac{4.20 \times 10^3\ m^3}{4.00 \times 10^3\ m^3} = 0.921$$

39. At constant T and P, Avogadro's law applies, that is, equal volumes contain equal moles of molecules. In terms of balanced equations, we can say that mol ratios and volume ratios between the various reactants and products will be equal to each other. $Br_2 + 3\ F_2 \rightarrow 2\ X$; Two moles of X must contain two moles of Br and 6 moles of F; X must have the formula BrF_3.

41. $P_{He} + P_{H_2O} = 1.00\ atm = 760.\ torr = P_{He} + 23.8\ torr,\ P_{He} = 736\ torr$

$$n_{He} = 0.586\ g \times \frac{1\ mol}{4.003\ g} = 0.146\ mol\ He$$

$$V = \frac{n_{He}RT}{P_{He}} = \frac{0.146\ mol \times \frac{0.08206\ L\ atm}{mol\ K} \times 298\ K}{736\ torr \times \frac{1\ atm}{760\ torr}} = 3.69\ L$$

43. a. mol fraction $CH_4 = \chi_{CH_4} = \frac{P_{CH_4}}{P_{total}} = \frac{0.175\ atm}{0.175\ atm + 0.250\ atm} = 0.412$; $\chi_{O_2} = 1.000 - 0.412 = 0.588$

b. PV = nRT, $n_{total} = \frac{P_{total} \times V}{RT} = \frac{0.425\ atm \times 10.5\ L}{\frac{0.08206\ L\ atm}{mol\ K} \times 338\ K} = 0.161\ mol$

c. $\chi_{CH_4} = \frac{n_{CH_4}}{n_{total}}$, $n_{CH_4} = \chi_{CH_4} \times n_{total} = 0.412 \times 0.161\ mol = 6.63 \times 10^{-2}\ mol\ CH_4$

$$6.63 \times 10^{-2}\ mol\ CH_4 \times \frac{16.04\ g\ CH_4}{mol\ CH_4} = 1.06\ g\ CH_4$$

$$n_{O_2} = 0.588 \times 0.161\ mol = 9.47 \times 10^{-2}\ mol\ O_2;\ 9.47 \times 10^{-2}\ mol\ O_2 \times \frac{32.00\ g\ O_2}{mol\ O_2} = 3.03\ g\ O_2$$

Gas Density, Molar Mass, and Reaction Stoichiometry

45. Out of 100.0 g of compound, there are:

$$87.4\ g\ N \times \frac{1\ mol\ N}{14.01\ g\ N} = 6.24\ mol\ N;\ \frac{6.24}{6.24} = 1.00$$

$$12.6\ g\ H \times \frac{1\ mol\ H}{1.008\ g\ H} = 12.5\ mol\ H;\ \frac{12.5}{6.24} = 2.00$$

Empirical formula is NH_2. P × (molar mass) = dRT where d = density.

$$\text{Molar mass} = \frac{dRT}{P} = \frac{\frac{0.977\text{ g}}{\text{L}} \times \frac{0.08206\text{ L atm}}{\text{mol K}} \times 373\text{ K}}{710.\text{ torr} \times \frac{1\text{ atm}}{760\text{ torr}}} = 32.0\text{ g/mol}$$

Empirical formula mass of NH_2 = 16.0 g. Therefore, molecular formula is N_2H_4.

47. If Be^{3+}, the formula is $Be(C_5H_7O_2)_3$ and molar mass ≈ 13.5 + 15(12) + 21(1) + 6(16) = 311 g/mol.

If Be^{2+}, the formula is $Be(C_5H_7O_2)_2$ and molar mass ≈ 9.0 + 10(12) + 14(1) + 4(16) = 207 g/mol.

Data Set I (molar mass = dRT/P and d = mass/V):

$$\text{molar mass} = \frac{\text{mass} \times RT}{PV} = \frac{0.2022\text{ g} \times \frac{0.08206\text{ L atm}}{\text{mol K}} \times 286\text{ K}}{(765.2\text{ torr} \times \frac{1\text{ atm}}{760\text{ torr}}) \times 22.6 \times 10^{-3}\text{ L}} = 209\text{ g/mol}$$

Data Set II:

$$\text{molar mass} = \frac{\text{mass} \times RT}{PV} = \frac{0.2224\text{ g} \times \frac{0.08206\text{ L atm}}{\text{mol K}} \times 290.\text{ K}}{(764.6\text{ torr} \times \frac{1\text{ atm}}{760\text{ torr}}) \times 26.0 \times 10^{-3}\text{ L}} = 202\text{ g/mol}$$

These results are close to the expected value of 207 g/mol for $Be(C_5H_7O_2)_2$. Thus, we conclude from these data that beryllium is a divalent element with an atomic weight (mass) of 9.0 g/mol.

49. $$\text{molar mass} = \frac{dRT}{P} = \frac{\frac{0.70902\text{ g}}{\text{L}} \times \frac{0.08206\text{ L atm}}{\text{mol K}} \times 273.2\text{ K}}{1.000\text{ atm}} = 15.90\text{ g/mol}$$

15.90 g/mol is the average molar mass of the mixture of methane and helium. Let x = mol % of CH_4. This is also the volume % of CH_4 since T and P are constant.

$$15.90 = \frac{x\,(16.04) + (100 - x)\,(4.003)}{100}, \quad 1590. = 16.04\,x + 400.3 - 4.003\,x, \quad 1190. = 12.04\,x$$

x = 98.84 % CH_4 by volume; % He = 100.00 - x = 1.16 % He by volume

51. $$n_{H_2} = \frac{PV}{RT} = \frac{1.0\text{ atm} \times \left[4800\text{ m}^3 \times \left(\frac{100\text{ cm}}{\text{m}}\right)^3 \times \frac{1\text{ L}}{1000\text{ cm}^3}\right]}{\frac{0.08206\text{ L atm}}{\text{mol K}} \times 273\text{ K}} = 2.1 \times 10^5\text{ mol}$$

2.1×10^5 mol H_2 are in the balloon. This is 80.% of the total amount of H_2 that had to be generated:

$$0.80\text{ (total mol } H_2) = 2.1 \times 10^5, \text{ total mol } H_2 = 2.6 \times 10^5 \text{ mol } H_2$$

$$2.6 \times 10^5 \text{ mol } H_2 \times \frac{1 \text{ mol Fe}}{\text{mol } H_2} \times \frac{55.85 \text{ g Fe}}{\text{mol Fe}} = 1.5 \times 10^7 \text{ g Fe}$$

$$2.6 \times 10^5 \text{ mol } H_2 \times \frac{1 \text{ mol } H_2SO_4}{\text{mol } H_2} \times \frac{98.09 \text{ g } H_2SO_4}{\text{mol } H_2SO_4} \times \frac{100 \text{ g reagent}}{98 \text{ g } H_2SO_4} = 2.6 \times 10^7 \text{ g of 98\% sulfuric acid}$$

53. $C_6H_{12}O_6(s) + 6\ O_2(g) \rightarrow 6\ CO_2(g) + 6\ H_2O(g)$

$$5.00 \text{ g } C_6H_{12}O_6 \times \frac{1 \text{ mol } C_6H_{12}O_6}{180.16 \text{ g}} \times \frac{6 \text{ mol } O_2}{\text{mol } C_6H_{12}O_6} = 0.167 \text{ mol } O_2$$

$$V = \frac{nRT}{P} = \frac{0.167 \text{ mol} \times 0.08206 \frac{\text{L atm}}{\text{mol K}} \times 301 \text{ K}}{0.976 \text{ atm}} = 4.23 \text{ L } O_2$$

Since T and P are constant, the volume of each gas will be directly proportional to the mol of gas present. The balanced equation says that equal mol of CO_2 and H_2O will be produced as mol of O_2 reacted. So the volumes of CO_2 and H_2O produced will equal the volume of O_2 reacted.

$$V_{CO_2} = V_{H_2O} = V_{O_2} = 4.23 \text{ L}$$

55. $2\ NaClO_3(s) \rightarrow 2\ NaCl(s) + 3\ O_2(g)$

$$P_{total} = P_{O_2} + P_{H_2O},\ P_{O_2} = P_{total} - P_{H_2O} = 734 \text{ torr} - 19.8 \text{ torr} = 714 \text{ torr}$$

$$n_{O_2} = \frac{P_{O_2} \times V}{RT} = \frac{\left(714 \text{ torr} \times \frac{1 \text{ atm}}{760 \text{ torr}}\right) \times 0.0572 \text{ L}}{\frac{0.08206 \text{ L atm}}{\text{mol K}} \times (273 + 22) \text{ K}} = 2.22 \times 10^{-3} \text{ mol } O_2$$

$$\text{Mass } NaClO_3 \text{ decomposed} = 2.22 \times 10^{-3} \text{ mol } O_2 \times \frac{2 \text{ mol } NaClO_3}{3 \text{ mol } O_2} \times \frac{106.44 \text{ g } NaClO_3}{\text{mol } NaClO_3} = 0.158 \text{ g } NaClO_3$$

$$\text{Mass \% } NaClO_3 = \frac{0.158 \text{ g}}{0.8765 \text{ g}} \times 100 = 18.0\%$$

57. $P_{total} = P_{N_2} + P_{H_2O},\ P_{N_2} = 726 \text{ torr} - 23.8 \text{ torr} = 702 \text{ torr} \times \frac{1 \text{ atm}}{760 \text{ torr}} = 0.924 \text{ atm}$

$$n_{N_2} = \frac{P_{N_2} \times V}{RT} = \frac{0.924 \text{ atm} \times 31.8 \times 10^{-3} \text{ L}}{\frac{0.08206 \text{ L atm}}{\text{mol K}} \times 298 \text{ K}} = 1.20 \times 10^{-3} \text{ mol } N_2$$

$$\text{Mass of N in compound} = 1.20 \times 10^{-3} \text{ mol } N_2 \times \frac{28.02 \text{ g } N_2}{\text{mol}} = 3.36 \times 10^{-2} \text{ g nitrogen}$$

$$\text{Mass \% N} = \frac{3.36 \times 10^{-2} \text{ g}}{0.253 \text{ g}} \times 100 = 13.3\% \text{ N}$$

59. For NH_3: $P_2 = \frac{P_1V_1}{V_2} = 0.500 \text{ atm} \times \frac{2.00 \text{ L}}{3.00 \text{ L}} = 0.333 \text{ atm}$

For O_2: $P_2 = \frac{P_1V_1}{V_2} = 1.50 \text{ atm} \times \frac{1.00 \text{ L}}{3.00 \text{ L}} = 0.500 \text{ atm}$

After the stopcock is opened, V and T will be constant, so $P \propto n$. The balanced equation requires:

$$\frac{n_{O_2}}{n_{NH_3}} = \frac{P_{O_2}}{P_{NH_3}} = \frac{5}{4} = 1.25$$

The actual ratio present is: $\frac{P_{O_2}}{P_{NH_3}} = \frac{0.500 \text{ atm}}{0.333 \text{ atm}} = 1.50$

The actual ratio is larger than the required ratio, so NH_3 in the denominator is limiting. Since equal mol of NO will be produced as NH_3 reacted, the partial pressure of NO produced is 0.333 atm (the same as P_{NH_3} reacted).

61. Rigid container (constant volume): As reactants are converted to products, the mol of gas particles present decrease by one-half. As n decreases, the pressure will decrease (by one-half). Density is the mass per unit volume. Mass is conserved in a chemical reaction, so the density of the gas will not change since mass and volume do not change.

Flexible container (constant pressure): Pressure is constant since the container changes volume in order to keep a constant pressure. As the mol of gas particles decrease by a factor of 2, the volume of the container will decrease (by one-half). We have the same mass of gas in a smaller volume, so the gas density will increase (is doubled).

Kinetic Molecular Theory and Real Gases

63. $(KE)_{avg} = 3/2\ RT$; KE depends only on temperature. At each temperature CH_4 and N_2 will have the same average KE. For energy units of joules (J), use $R = 8.3145 \text{ J mol}^{-1} \text{ K}^{-1}$. To determine average KE per molecule, divide by Avogadro's number, 6.022×10^{23} molecules/mol.

at 273 K: $(KE)_{avg} = \frac{3}{2} \times \frac{8.3145 \text{ J}}{\text{mol K}} \times 273 \text{ K} = 3.40 \times 10^3 \text{ J/mol} = 5.65 \times 10^{-21} \text{ J/molecule}$

at 546 K: $(KE)_{avg} = \frac{3}{2} \times \frac{8.3145 \text{ J}}{\text{mol K}} \times 546 \text{ K} = 6.81 \times 10^3 \text{ J/mol} = 1.13 \times 10^{-20} \text{ J/molecule}$

65. No; The numbers calculated in Exercise 5.63 are the average kinetic energies at the various temperatures. At each temperature, there is a distribution of energies. Similarly, the numbers calculated in Exercise 5.64 are a special kind of average velocity. There is a distribution of velocities as shown in Figures 5.15-5.17 of the text. Note that the major reason there is a distribution of kinetic energies is because there is a distribution of velocities for any gas sample at some temperature.

67. a. They will all have the same average kinetic energy since they are all at the same temperature. Average kinetic energy depends only on temperature.

b. Flask C; At constant T, $u_{rms} \propto (1/M)^{1/2}$. In general, the lighter the gas molecules the greater the root mean square velocity (at constant T).

c. Flask A: Collision frequency is proportional to average velocity × n/V (as the average velocity doubles, the number of collisions will double and as the number of molecules in the container doubles, the number of collisions again doubles). At constant T and V, n is proportional to P and average velocity is proportional to $(1/M)^{1/2}$. We use these relationships and the data in the problem to determine the following relative values.

	n (relative)	u_{avg} (relative)	Coll. Freq. (relative) = n × u_{avg}
A	1.0	1.0	1.0
B	0.33	1.0	0.33
C	0.13	3.7	0.48

69. $$\frac{\text{Rate}_1}{\text{Rate}_2} = \left(\frac{M_2}{M_1}\right)^{1/2}$$ where Rate = rate of effusion and M = molar mass.

$$\frac{\text{Rate}_1}{\text{Rate}_2} = \frac{31.50}{30.50} = \left(\frac{31.998}{M}\right)^{1/2} = 1.033, \quad \frac{31.998}{M} = 1.067, \quad M = 29.99$$

Of the choices, the gas with the molar mass closest to 29.99 is NO.

71. $$\frac{\text{Rate}_1}{\text{Rate}_2} = \left(\frac{M_2}{M_1}\right)^{1/2}; \ \text{Rate}_1 = \frac{24.0\ \text{mL}}{\text{min}}, \ \text{Rate}_2 = \frac{47.8\ \text{mL}}{\text{min}}, \ M_2 = \frac{16.04\ \text{g}}{\text{mol}} \ \text{and} \ M_1 = ?$$

$$\frac{24.0}{47.8} = \left(\frac{16.04}{M_1}\right)^{1/2} = 0.502, \quad 16.04 = (0.502)^2 \times M_1, \quad M_1 = \frac{16.04}{0.252} = \frac{63.7\ \text{g}}{\text{mol}}$$

73. a. PV = nRT

$$P = \frac{nRT}{V} = \frac{0.5000\ \text{mol} \times \frac{0.08206\ \text{L atm}}{\text{mol K}} \times (25.0 + 273.2)\ \text{K}}{1.0000\ \text{L}} = 12.24\ \text{atm}$$

b. $\left[P + a\left(\frac{n}{V}\right)^2\right](V - nb) = nRT$; For N_2: $a = 1.39$ atm L^2/mol^2 and $b = 0.0391$ L/mol

$$\left[P + 1.39\left(\frac{0.5000}{1.0000}\right)^2 \text{atm}\right](1.0000\text{ L} - 0.5000 \times 0.0391\text{ L}) = 12.24\text{ L atm}$$

$$(P + 0.348\text{ atm})(0.9805\text{ L}) = 12.24\text{ L atm}$$

$$P = \frac{12.24\text{ L atm}}{0.9805\text{ L}} - 0.348\text{ atm} = 12.48 - 0.348 = 12.13\text{ atm}$$

c. The ideal gas law is high by 0.11 atm or $\frac{0.11}{12.13} \times 100 = 0.91\%$.

75. The kinetic molecular theory assumes that gas particles do not exert forces on each other and that gas particles are volumeless. Real gas particles do exert attractive forces for each other, and real gas particles do have volumes. A gas behaves most ideally at low pressures and high temperatures. The effect of attractive forces is minimized at high temperatures since the gas particles are moving very rapidly. At low pressure, the container volume is relatively large (P and V are inversely related) so the volume of the container taken up by the gas particles is negligible.

77. Rearranging the Van der Waals' equation gives: $P = \frac{nRT}{V - nb} - a\left(\frac{n}{V}\right)^2$

P is the measured pressure and V is the volume of the container.

For NH_3: $a = 4.17$ atm L^2/mol^2 and $b = 0.0371$ L/mol

For the first experiment (assuming four significant figures in all values):

$$P = \frac{nRT}{V - nb} - a\left(\frac{n}{V}\right)^2 = \frac{1.000 \times 0.08206 \times 273.2}{172.1 - 0.0371} - 4.17\left(\frac{1.000}{172.1}\right)^2$$

$P = 0.1303 - 0.0001 = 0.1302$ atm. In Example 5.1, $P_{obs} = 0.1300$ atm. The difference is less than 0.1%. The ideal gas law also gives 0.1302 atm. At low pressures, the van der Waals equation agrees with the ideal gas law.

For experiment 6, the measured pressure is 1.000 atm. The ideal gas law gives $P = 1.015$ atm and the van der Waals equation gives $P = 1.017 - 0.086 = 1.008$ atm. The van der Waals equation accounts for approximately one-half of the deviation from ideal behavior. Note: As pressure increases, deviation from the ideal gas law increases.

79. The corrected (ideal) volume is the volume accessible to the gas molecules. For a real gas, this volume is less than the container volume because of the space occupied by the gas molecules.

81. The values of a are: H_2, $\frac{0.244\text{ atm L}^2}{\text{mol}^2}$; CO_2, 3.59; N_2, 1.39; CH_4, 2.25

Since a is a measure of intermolecular attractions, the attractions are greatest for CO_2.

83. $$u_{rms} = \left(\frac{3RT}{M}\right)^{1/2} = \left[\frac{3\left(\dfrac{8.3145\ \text{kg m}^2}{\text{s}^2\ \text{mol K}}\right)(227°\text{C} + 273)\text{K}}{28.02 \times 10^{-3}\ \text{kg/mol}}\right]^{1/2} = 667\ \text{m/s}$$

$$u_{mp} = \left(\frac{2RT}{M}\right)^{1/2} = \left[\frac{2\left(\dfrac{8.3145\ \text{kg m}^2}{\text{s}^2\ \text{mol K}}\right)(500.\ \text{K})}{28.02 \times 10^{-3}\ \text{kg/mol}}\right]^{1/2} = 545\ \text{m/s}$$

$$u_{avg} = \left(\frac{8RT}{\pi M}\right)^{1/2} = \left[\frac{8\left(\dfrac{8.3145\ \text{kg m}^2}{\text{s}^2\ \text{mol K}}\right)(500.\ \text{K})}{\pi(28.02 \times 10^{-3}\ \text{kg/mol})}\right]^{1/2} = 615\ \text{m/s}$$

85. **The force per impact is proportional to $\Delta(mu) = 2mu$. Since $m \propto M$, the molar mass, and $u \propto (1/M)^{1/2}$ at constant T, then the force per impact at constant T is proportional to $M \times (1/M)^{1/2} = \sqrt{M}$.**

$$\frac{\text{Impact Force (H}_2)}{\text{Impact Force (He)}} = \sqrt{\frac{M_{H_2}}{M_{He}}} = \sqrt{\frac{2.016}{4.003}} = 0.7097$$

87. **$\Delta(mu) = 2mu$ = change in momentum per impact. Since m is proportional to M, the molar mass, and u is proportional to $(T/M)^{1/2}$, then:**

$$\Delta(mu)_{O_2} \propto 2M_{O_2}\left(\frac{T}{M_{O_2}}\right)^{1/2} \quad \text{and} \quad \Delta(mu)_{He} \propto 2M_{He}\left(\frac{T}{M_{He}}\right)^{1/2}$$

$$\frac{\Delta(mu)_{O_2}}{\Delta(mu)_{He}} = \frac{2M_{O_2}\left(\dfrac{T}{M_{O_2}}\right)^{1/2}}{2M_{He}\left(\dfrac{T}{M_{He}}\right)^{1/2}} = \frac{M_{O_2}}{M_{He}}\left(\frac{M_{He}}{M_{O_2}}\right)^{1/2} = \frac{31.998}{4.003}\left(\frac{4.003}{31.998}\right)^{1/2} = 2.827$$

The change in momentum per impact is 2.827 times larger for O_2 molecules than for He atoms.

$$Z = A\frac{N}{V}\left(\frac{RT}{2\pi M}\right)^{1/2} = \text{collision rate}$$

$$\frac{Z_{O_2}}{Z_{He}} = \frac{A\left(\dfrac{N}{V}\right)\left(\dfrac{RT}{2\pi M_{O_2}}\right)^{1/2}}{A\left(\dfrac{N}{V}\right)\left(\dfrac{RT}{2\pi M_{He}}\right)^{1/2}} = \frac{\left(\dfrac{1}{M_{O_2}}\right)^{1/2}}{\left(\dfrac{1}{M_{He}}\right)^{1/2}} = \left(\frac{M_{He}}{M_{O_2}}\right)^{1/2} = 0.3537; \quad \frac{Z_{He}}{Z_{O_2}} = 2.827$$

There are 2.827 times as many impacts per second for He as compared to O_2.

89. Intermolecular collision frequency = $Z = 4\frac{N}{V}d^2\left(\frac{\pi RT}{M}\right)^{1/2}$ where d = diameter of He atom

$$\frac{n}{V} = \frac{P}{RT} = \frac{3.0\ \text{atm}}{\frac{0.08206\ \text{L atm}}{\text{mol K}} \times 300.\ \text{K}} = 0.12\ \text{mol/L}$$

$$\frac{N}{V} = \frac{0.12\ \text{mol}}{\text{L}} \times \frac{6.022 \times 10^{23}\ \text{molecules}}{\text{mol}} \times \frac{1000\ \text{L}}{\text{m}^3} = \frac{7.2 \times 10^{25}\ \text{molecules}}{\text{m}^3}$$

$$Z = 4 \times \frac{7.2 \times 10^{25}\ \text{molecules}}{\text{m}^3} \times (50. \times 10^{-12}\ \text{m})^2 \times \left(\frac{\pi(8.3145)(300.)}{4.00 \times 10^{-3}}\right)^{1/2} = 1.0 \times 10^9\ \text{collisions/s}$$

$$\text{mean free path} = \lambda = \frac{u_{avg}}{Z};\ u_{avg} = \left(\frac{8RT}{\pi M}\right)^{1/2} = 1260\ \text{m/s};\ \lambda = \frac{1260\ \text{m/s}}{1.0 \times 10^9\ \text{s}^{-1}} = 1.3 \times 10^{-6}\ \text{m}$$

Atmospheric Chemistry

91. a. If we have 1.0×10^6 L of air, then there are 3.0×10^2 L of CO.

$$P_{CO} = \chi_{CO}P_{total};\ \chi_{CO} = \frac{V_{CO}}{V_{total}}\ \text{since}\ V \propto n;\ P_{CO} = \frac{3.0 \times 10^2}{1.0 \times 10^6} \times 628\ \text{torr} = 0.19\ \text{torr}$$

b. $n_{CO} = \frac{P_{CO}V}{RT}$; Assuming 1.0 m³ air, 1 m³ = 1000 L:

$$n_{CO} = \frac{\frac{0.19}{760}\ \text{atm} \times (1.0 \times 10^3\ \text{L})}{\frac{0.08206\ \text{L atm}}{\text{mol K}} \times 273\ \text{K}} = 1.1 \times 10^{-2}\ \text{mol CO}$$

$$1.1 \times 10^{-2}\ \text{mol} \times \frac{6.02 \times 10^{23}\ \text{molecules}}{\text{mol}} = 6.6 \times 10^{21}\ \text{CO molecules in 1.0 m}^3\ \text{of air}$$

c. $$\frac{6.6 \times 10^{21}\ \text{molecules}}{\text{m}^3} \times \left(\frac{1\ \text{m}}{100\ \text{cm}}\right)^3 = \frac{6.6 \times 10^{15}\ \text{molecules CO}}{\text{cm}^3}$$

93. $N_2(g) + O_2(g) \rightarrow 2\ NO(g)$, automobile combustion or formed by lightning

$2\ NO(g) + O_2(g) \rightarrow 2\ NO_2(g)$, reaction with atmospheric O_2

$2\ NO_2(g) + H_2O(l) \rightarrow HNO_3(aq) + HNO_2(aq)$, reaction with atmospheric H_2O

$S(s) + O_2(g) \rightarrow SO_2(g)$, combustion of coal

$2\ SO_2(g) + O_2(g) \rightarrow 2SO_3(g)$, reaction with atmospheric O_2

$H_2O(l) + SO_3(g) \rightarrow H_2SO_4(aq)$, reaction with atmospheric H_2O

$2\ HNO_3(aq) + CaCO_3(s) \rightarrow Ca(NO_3)_2(aq) + H_2O(l) + CO_2(g)$

$H_2SO_4(aq) + CaCO_3(s) \rightarrow CaSO_4(aq) + H_2O(l) + CO_2(g)$

Additional Exercises

95. $0.050 \text{ mL} \times \frac{1.149 \text{ g}}{\text{mL}} \times \frac{1 \text{ mol } O_2}{32.0 \text{ g}} = 1.8 \times 10^{-3} \text{ mol } O_2$

$$V = \frac{nRT}{P} = \frac{1.8 \times 10^{-3} \text{ mol} \times \frac{0.08206 \text{ L atm}}{\text{mol K}} \times 310. \text{ K}}{1.0 \text{ atm}} = 4.6\ 10^{-2} \text{ L} = 46 \text{ mL}$$

97. $Mn(s) + x\ HCl(g) \rightarrow MnCl_x(s) + \frac{x}{2}\ H_2(g)$

$$n_{H_2} = \frac{PV}{RT} = \frac{0.951 \text{ atm} \times 3.22 \text{ L}}{\frac{0.08206 \text{ L atm}}{\text{mol K}} \times 373 \text{ K}} = 0.100 \text{ mol } H_2$$

$$\text{mol Cl in compound} = \text{mol HCl} = 0.100 \text{ mol } H_2 \times \frac{x \text{ mol Cl}}{\frac{x}{2} \text{ mol } H_2} = 0.200 \text{ mol Cl}$$

$$\frac{\text{mol Cl}}{\text{mol Mn}} = \frac{0.200 \text{ mol Cl}}{2.747 \text{ g Mn} \times \frac{1 \text{ mol Mn}}{54.94 \text{ g Mn}}} = \frac{0.200 \text{ mol Cl}}{0.05000 \text{ mol Mn}} = 4.00$$

The formula of compound is $MnCl_4$.

99. PV = nRT, V and T are constant. $\frac{P_1}{n_1} = \frac{P_2}{n_2}, \quad \frac{P_2}{P_1} = \frac{n_2}{n_1}$

We will do this limiting reagent problem using an alternative method than described in Chapter 3. Let's calculate the partial pressure of C_3H_3N that can be produced from each of the starting materials assuming each reactant is limiting. The reactant that produces the smallest amount of product will run out first and is the limiting reagent.

$$P_{C_3H_3N} = 0.500 \text{ MPa} \times \frac{2 \text{ MPa } C_3H_3N}{2 \text{ MPa } C_3H_6} = 0.500 \text{ MPa if } C_3H_6 \text{ is limiting.}$$

$$P_{C_3H_3N} = 0.800 \text{ MPa} \times \frac{2 \text{ MPa } C_3H_3N}{2 \text{ MPa } NH_3} = 0.800 \text{ MPa if } NH_3 \text{ is limiting.}$$

$$P_{C_3H_3N} = 1.500 \text{ MPa} \times \frac{2 \text{ MPa } C_3H_3N}{3 \text{ MPa } O_2} = 1.000 \text{ MPa if } O_2 \text{ is limiting.}$$

Thus, C_3H_6 is limiting. Although more product could be produced from NH_3 and O_2, there is only enough C_3H_6 to produce 0.500 MPa of C_3H_3N. The partial pressure of C_3H_3N in atm after the reaction is:

$$0.500 \times 10^6 \text{ Pa} \times \frac{1 \text{ atm}}{1.013 \times 10^5 \text{ Pa}} = 4.94 \text{ atm}$$

$$n = \frac{PV}{RT} = \frac{4.94 \text{ atm} \times 150. \text{ L}}{\frac{0.08206 \text{ L atm}}{\text{mol K}} \times 298 \text{ K}} = 30.3 \text{ mol } C_3H_3N$$

$$30.3 \text{ mol} \times \frac{53.06 \text{ g}}{\text{mol}} = 1.61 \times 10^3 \text{ g } C_3H_3N \text{ can be produced.}$$

101. $P_{TOT} = P_{H_2} + P_{H_2O}$, $1.032 \text{ atm} = P_{H_2} + 32 \text{ torr} \times \frac{1 \text{ atm}}{760 \text{ torr}}$, $P_{H_2} = 1.032 - 0.042 = 0.990 \text{ atm}$

$$n_{H_2} = \frac{P_{H_2}V}{RT} = \frac{0.990 \text{ atm} \times 0.240 \text{ L}}{0.08206 \frac{\text{L atm}}{\text{mol K}} \times 303 \text{ K}} = 9.56 \times 10^{-3} \text{ mol } H_2$$

$$9.56 \times 10^{-3} \text{ mol } H_2 \times \frac{1 \text{ mol Zn}}{\text{mol } H_2} \times \frac{65.38 \text{ g Zn}}{\text{mol Zn}} = 0.625 \text{ g Zn}$$

103. $P_1V_1 = P_2V_2$; The total volume is 1.00 L + 1.00 L + 2.00 L = 4.00 L.

For He: $P_2 = \frac{P_1V_1}{V_2} = 200. \text{ torr} \times \frac{1.00 \text{ L}}{4.00 \text{ L}} = 50.0 \text{ torr He}$

For Ne: $P_2 = 0.400 \text{ atm} \times \frac{1.00 \text{ L}}{4.00 \text{ L}} = 0.100 \text{ atm}$; $0.100 \text{ atm} \times \frac{760 \text{ torr}}{\text{atm}} = 76.0 \text{ torr Ne}$

For Ar: $P_2 = 24.0 \text{ kPa} \times \frac{2.00 \text{ L}}{4.00 \text{ L}} = 12.0 \text{ kPa}$; $12.0 \text{ kPa} \times \frac{1 \text{ atm}}{101.3 \text{ kPa}} \times \frac{760 \text{ torr}}{\text{atm}} = 90.0 \text{ torr Ar}$

$P_{total} = 50.0 + 76.0 + 90.0 = 216.0 \text{ torr}$

105. Out of 100.00 g compounds, there are:

$$58.51 \text{ g C} \times \frac{1 \text{ mol C}}{12.011 \text{ g C}} = 4.871 \text{ mol C}; \frac{4.871}{2.436} = 2.000$$

$$7.37 \text{ g H} \times \frac{1 \text{ mol H}}{1.008 \text{ g H}} = 7.31 \text{ mol H}; \frac{7.31}{2.436} = 3.00$$

$$34.12 \text{ g N} \times \frac{1 \text{ mol N}}{14.007 \text{ g N}} = 2.436 \text{ mol N}; \frac{2.436}{2.436} = 1.000$$

Empirical formula: C_2H_3N

$$\frac{\text{Rate}_1}{\text{Rate}_2} = \left(\frac{M_2}{M_1}\right)^{1/2}; \text{ Let gas (1) = He; } 3.20 = \left(\frac{M_2}{4.003}\right)^{1/2}, \ M_2 = 41.0 \text{ g/mol}$$

Empirical formula mass of $C_2H_3N \approx 2(12.0) + 3(1.0) + 1(14.0) = 41.0$. So molecular formula is also C_2H_3N.

107. We will apply Boyle's law to solve. $PV = nRT =$ constant, $P_1V_1 = P_2V_2$

Let condition (1) correspond to He from the tank that can be used to fill balloons. We must leave 1.0 atm of He in the tank, so P_1 = 200. atm - 1.00 = 199 atm and V_1 = 15.0 L. Condition (2) will correspond to the filled balloons with P_2= 1.00 atm and V_2 = N(2.00 L) where N is the number of filled balloons, each at a volume of 2.00 L.

199 atm × 15.0 L = 1.00 atm × N(2.00 L), N = 1492.5; We can't fill 0.5 of a balloon. So N = 1492 balloons or to 3 significant figures, 1490 balloons.

109. If we had 100.0 g of the gas, we would have 50.0 g He and 50.0 g Xe.

$$\chi_{He} = \frac{n_{He}}{n_{He} + n_{Xe}} = \frac{\frac{50.0 \text{ g}}{4.003 \text{ g/mol}}}{\frac{50.0 \text{ g}}{4.003 \text{ g/mol}} + \frac{50.0 \text{ g}}{131.3 \text{ g/mol}}} = \frac{12.5 \text{ mol He}}{12.5 \text{ mol He} + 0.381 \text{ mol Xe}} = 0.970$$

$P_{He} = \chi_{He}P_{total}$ = 0.970 × 600. torr = 582 torr; P_{Xe} = 600. - 582 = 18 torr

111. $$n_{Ar} = \frac{228 \text{ g}}{39.95 \text{ g/mol}} = 5.71 \text{ mol Ar}; \ \chi_{CH_4} = \frac{n_{CH_4}}{n_{CH_4} + n_{Ar}} = 0.650 = \frac{n_{CH_4}}{n_{CH_4} + (5.71)}$$

$$0.650(n_{CH_4} + 5.71) = n_{CH_4}, \ 3.71 = 0.350\ n_{CH_4}, \ n_{CH_4} = 10.6 \text{ mol } CH_4; \ KE_{avg} = \frac{3}{2}RT \text{ for 1 mol}$$

So KE_{total} = (10.6 + 5.71 mol) × 3/2 × 8.3145 J mol^{-1} K^{-1} × 298 K = 6.06 × 10^4 J = 60.6 kJ

113. $CH_3OH(l) + 3/2\ O_2(g) \rightarrow CO_2(g) + 2\ H_2O(g)$ or $2\ CH_3OH(l) + 3\ O_2(g) \rightarrow 2\ CO_2(g) + 4\ H_2O(g)$

$$50.0 \text{ mL} \times \frac{0.850 \text{ g}}{\text{mL}} \times \frac{1 \text{ mol}}{32.04 \text{ g}} = 1.33 \text{ mol } CH_3OH(l) \text{ available}$$

$$n_{O_2} = \frac{PV}{RT} = \frac{2.00 \text{ atm} \times 22.8 \text{ L}}{\frac{0.08206 \text{ L atm}}{\text{mol K}} \times 300. \text{ K}} = 1.85 \text{ mol } O_2 \text{ available}$$

$$1.33 \text{ mol } CH_3OH \times \frac{3 \text{ mol } O_2}{2 \text{ mol } CH_3OH} = 2.00 \text{ mol } O_2 \text{ required for complete reaction}$$

We only have 1.85 mol O_2, so O_2 is limiting. $1.85 \text{ mol } O_2 \times \frac{4 \text{ mol } H_2O}{3 \text{ mol } O_2} = 2.47 \text{ mol } H_2O$

115. For 1 mol of gas: $KE_{ave} = 3/2\ RT$

For 1 molecule of gas (N_A = Avogadro's number): $KE_{ave} = \frac{3RT}{2N_A} = \frac{3}{2}k_BT$, $k_B = 1.3807 \times 10^{-23}$ J/K

$$KE_{ave} = \frac{3}{2}(1.3807 \times 10^{-23} \text{ J/K})(400.\text{ K}) = 8.28 \times 10^{-21} \text{ J/molecule}$$

117. a. Out of 100.00 g of Z, we have:

$$34.38 \text{ g Ni} \times \frac{1 \text{ mol}}{58.69 \text{ g}} = 0.5858 \text{ mol Ni}$$

$$28.13 \text{ g C} \times \frac{1 \text{ mol}}{12.011 \text{ g}} = 2.342 \text{ mol C};\quad \frac{2.342}{0.5858} = 3.998$$

$$37.48 \text{ g O} \times \frac{1 \text{ mol}}{15.999 \text{ g}} = 2.343 \text{ mol O};\quad \frac{2.343}{0.5858} = 4.000$$

The empirical formula is NiC_4O_4.

b. $\frac{\text{rate Z}}{\text{rate Ar}} = \left(\frac{M_{Ar}}{M_Z}\right)^{1/2} = \left(\frac{39.95}{M_Z}\right)^{1/2}$; Since initial mol Ar = mol Z, then:

$$0.4837 = \left(\frac{39.95}{M_Z}\right)^{1/2},\quad M_z = 170.8 \text{ g/mol}$$

c. NiC_4O_4: $M = 58.69 + 4(12.01) + 4(16.00) = 170.73$ g/mol

Molecular formula is also NiC_4O_4.

d. Each effusion step changes the concentration of Z in the gas by a factor of 0.4837. The original concentration of Z molecules to Ar atoms is a 1:1 ratio. After 5 stages:

$$n_Z/n_{Ar} = (0.4837)^5 = 2.648 \times 10^{-2}$$

119. a. $2\ CH_4(g) + 2\ NH_3(g) + 3\ O_2(g) \rightarrow 2\ HCN(g) + 6\ H_2O(g)$

b. Volumes of gases are proportional to moles at constant T and P. Using the balanced equation, methane and ammonia are in stoichiometric amounts and oxygen is in excess. In 1 second:

$$n_{CH_4} = \frac{PV}{RT} = \frac{1.00 \text{ atm} \times 20.0 \text{ L}}{0.08206 \text{ L atm K}^{-1} \text{mol}^{-1} \times 423 \text{ K}} = 0.576 \text{ mol CH}_4$$

$$\frac{0.576 \text{ mol CH}_4}{\text{s}} \times \frac{2 \text{ mol HCN}}{2 \text{ mol CH}_4} \times \frac{27.03 \text{ g HCN}}{\text{mol HCN}} = 15.6 \text{ g HCN/s}$$

Challenge Problems

121. $Cr(s) + 3\ HCl(aq) \rightarrow CrCl_3(aq) + 3/2\ H_2(g)$; $Zn(s) + 2\ HCl(aq) \rightarrow ZnCl_2(aq) + H_2(g)$

$$\text{mol H}_2 \text{ produced} = n = \frac{PV}{RT} = \frac{\left(750.\text{ torr} \times \frac{1 \text{ atm}}{760 \text{ torr}}\right) \times 0.225 \text{ L}}{\frac{0.08206 \text{ L atm}}{\text{mol K}} \times (273 + 27) \text{ K}} = 9.02 \times 10^{-3} \text{ mol H}_2$$

9.02×10^{-3} mol H_2 = mol H_2 from Cr reaction + mol H_2 from Zn reaction

From the balanced equation: 9.02×10^{-3} mol H_2 = mol Cr × (3/2) + mol Zn × 1

Let x = mass of Cr and y = mass of Zn, then:

$$x + y = 0.362 \text{ g and } 9.02 \times 10^{-3} = \frac{1.5\,x}{52.00} + \frac{y}{65.38}$$

We have two equations and two unknowns. Solving by simultaneous equations:

$$9.02 \times 10^{-3} = 0.02885\,x + 0.01530\,y$$
$$\underline{-0.01530 \times 0.362 = -0.01530\,x - 0.01530\,y}$$
$$3.48 \times 10^{-3} = 0.01355\,x$$

$$x = \text{mass of Cr} = \frac{3.48 \times 10^{-3}}{0.01355} = 0.257 \text{ g}$$

$$y = \text{mass of Zn} = 0.362 \text{ g} - 0.257 \text{ g} = 0.105 \text{ g Zn; mass \% Zn} = \frac{0.105 \text{ g}}{0.362 \text{ g}} \times 100 = 29.0\% \text{ Zn}$$

123. $\text{molar mass} = \frac{dRT}{P}$, P and molar mass are constant; $dT = \frac{P \times \text{molar mass}}{R} = \text{constant}$

d = constant (1/T) or $d_1T_1 = d_2T_2$, where T is in kelvin (K).

T = x + °C; 1.2930(x + 0.0) = 0.9460(x + 100.0)

1.2930 x = 0.9460 x + 94.60, 0.3470 x = 94.60, x = 272.6

From these data absolute zero would be -272.6°C. Actual value is -273.15°C.

125. $\frac{PV}{nRT} = 1 + \beta P$; $\frac{n}{V} \times \text{molar mass} = d$

$$\frac{\text{molar mass}}{RT} \times \frac{P}{d} = 1 + \beta P, \quad \frac{P}{d} = \frac{RT}{\text{molar mass}} + \frac{\beta RTP}{\text{molar mass}}$$

This is in the equation for a straight line: y = b + mx. If we plot P/d vs P and extrapolate to P = 0, we get a y-intercept = b = 1.398 = RT/molar mass.

$$\text{At } 0.00°\text{C, molar mass} = \frac{0.08206 \times 273.15}{1.398} = 16.03 \text{ g/mol}$$

127. Figure 5.16 shows the effect of temperature on the Maxwell-Boltzmann distribution of velocities of molecules. Note that as temperature increases, the probability that a gas particle has the most probable velocity decreases. Thus, since the probability of the gas particle with the most probable velocity decreased by one-half, then the temperature must be higher than 300. K.

The equation that determines the probability that a gas molecule has a certain velocity is:

$$f(u) = 4\pi \left(\frac{m}{2\pi k_B T}\right)^{3/2} u^2 e^{-mu^2/2k_B T}$$

Let T_x = the unknown temperature, then:

$$\frac{f(u_{mp,\,x})}{f(u_{mp,\,300})} = \frac{1}{2} = \frac{4\pi\left(\frac{m}{2\pi k_B T_x}\right)^{3/2} u_{mp,\,x}^2 \, e^{-mu_{mp,\,x}^2/2k_B T_x}}{4\pi\left(\frac{m}{2\pi k_B T_{300}}\right)^{3/2} u_{mp,\,300}^2 \, e^{-mu_{mp,\,300}^2/2k_B T_{300}}}$$

Since $u_{mp} = \sqrt{\frac{2k_B T}{m}}$, then the equation reduces to:

$$\frac{1}{2} = \frac{\left(\frac{1}{T_x}\right)^{3/2}(T_x)}{\left(\frac{1}{T_{300}}\right)^{3/2}(T_{300})} = \left(\frac{T_{300}}{T_x}\right)^{1/2}$$

Note that the overall exponent term cancels from the expression when $2k_BT/m$ is substituted for u_{mp}^2 in the exponent term; the temperatures cancel. Solving for T_x:

$$\frac{1}{2} = \left(\frac{300.\text{ K}}{T_x}\right)^{1/2}, \quad T_x = 1.20 \times 10^3 \text{ K}$$

As expected, T_x is higher than 300. K.

129. $C_6H_{14}(l) + 19/2\ O_2(g) \rightarrow 6\ CO_2(g) + 7\ H_2O(l)$ or $2\ C_6H_{14}(l) + 19\ O_2(g) \rightarrow 12\ CO_2(g) + 14\ H_2O(l)$

$C_3H_8(l) + 5\ O_2(g) \rightarrow 3\ CO_2(g) + 4\ H_2O(l)$

$$0.8339\ g\ CO_2 \times \frac{1\ mol\ CO_2}{44.009\ g\ CO_2} = 1.895 \times 10^{-2}\ mol\ CO_2$$

Let x = mass of C_6H_{14} and y = mass of C_3H_8, so x + y = 0.2759 g.

1.895×10^{-2} mol CO_2 = 6 × mol C_6H_{14} + 3 × mol C_3H_8

$$1.895 \times 10^{-2} = \frac{6x}{86.177} + \frac{3y}{44.096};\ \text{Rearranging:}$$

$$x + \frac{3y}{44.096} \times \frac{86.177}{6} = 1.895 \times 10^{-2} \times \frac{86.177}{6},\ x + 0.9772\,y = 0.2722$$

Solving:

$$\begin{array}{r} x + \quad\quad\ \ y = \ 0.2759 \\ \underline{-x - 0.9772\,y = -0.2722} \\ 0.0288\,y = \ 0.0037,\ y = 0.16 = \text{mass of } C_3H_8 \end{array}$$

$$\text{mass \% } C_3H_8 = \frac{0.16\ g}{0.2759\ g} \times 100 = 58\%\ C_3H_8;\ \%\ C_6H_{14} = 42\%$$

131. The reactions are:

$C(s) + 1/2\ O_2(g) \rightarrow CO(g)$ and $C(s) + O_2(g) \rightarrow CO_2(g)$

$$PV = nRT,\ P = n\left(\frac{RT}{V}\right) = n\ (\text{constant})$$

Since the pressure has increased by 17.0%, then the number of moles of gas has also increased by 17.0%.

$n_{final} = 1.170\ n_{initial} = 1.170\ (5.00) = 5.85$ mol gas $= n_{O_2} + n_{CO} + n_{CO_2}$

$n_{CO} + n_{CO_2} = 5.00$ (balancing moles of C)

If all C was converted to CO_2, no O_2 would be left. If all C was converted to CO, we would get 5 mol CO and 2.5 mol excess O_2 in the reaction mixture. In the final mixture: $n_{CO} = 2n_{O_2}$

$$\begin{array}{rl} n_{O_2} + n_{CO} + n_{CO_2} & = 5.85 \\ -(n_{CO} + n_{CO_2} & = 5.00) \\ \hline n_{O_2} & = 0.85 \end{array}$$

$n_{CO} = 2n_{O_2} = 1.70$ mol CO; $1.70 + n_{CO_2} = 5.00$, $n_{CO_2} = 3.30$ mol CO_2

$$\chi_{CO} = \frac{1.70}{5.85} = 0.291; \quad \chi_{CO_2} = \frac{3.30}{5.85} = 0.564; \quad \chi_{O_2} = \frac{0.85}{5.85} = 0.145 \approx 0.15$$

133. a. The reaction is: $CH_4(g) + 2\ O_2(g) \rightarrow CO_2(g) + 2\ H_2O(g)$

$$PV = nRT, \quad \frac{PV}{n} = RT = \text{constant}, \quad \frac{P_{CH_4}V_{CH_4}}{n_{CH_4}} = \frac{P_{air}V_{air}}{n_{air}}$$

The balanced equation requires 2 mol O_2 for every mol of CH_4 that reacts. For three times as much oxygen, we would need 6 mol O_2 per mol of CH_4 reacted ($n_{O_2} = 6\ n_{CH_4}$). Air is 21% mol percent O_2, so $n_{O_2} = 0.21\ n_{air}$. Therefore, the mol of air we would need to delivery the excess O_2 is:

$$n_{O_2} = 0.21\ n_{air} = 6\ n_{CH_4}, \quad n_{air} = 29\ n_{CH_4}, \quad \frac{n_{air}}{n_{CH_4}} = 29$$

In one minute:

$$V_{air} = V_{CH_4} \times \frac{n_{air}}{n_{CH_4}} \times \frac{P_{CH_4}}{P_{air}} = 200.\ \text{L} \times 29 \times \frac{1.50\ \text{atm}}{1.00\ \text{atm}} = 8.7 \times 10^3\ \text{L air/min}$$

b. If x moles of CH_4 were reacted, then 6 x mol O_2 were added, producing 0.950 x mol CO_2 and 0.050 x mol of CO. In addition, 2 x mol H_2O must be produced to balance the hydrogens.

$CH_4(g) + 2\ O_2(g) \rightarrow CO_2(g) + 2\ H_2O(g)$; $CH_4(g) + 3/2\ O_2(g) \rightarrow CO(g) + 2\ H_2O(g)$

Amount O_2 reacted:

$$0.950\ \text{x mol } CO_2 \times \frac{2\ \text{mol } O_2}{\text{mol } CO_2} = 1.90\ \text{x mol } O_2$$

$$0.050\ \text{x mol CO} \times \frac{1.5\ \text{mol } O_2}{\text{mol CO}} = 0.075\ \text{x mol } O_2$$

Amount of O_2 left in reaction mixture = 6.00 x - 1.90 x - 0.075 x = 4.03 x mol O_2

$$\text{Amount of } N_2 = 6.00\ \text{x mol } O_2 \times \frac{79\ \text{mol } N_2}{21\ \text{mol } O_2} = 22.6\ \text{x} \approx 23\ \text{x mol } N_2$$

The reaction mixture contains:

0.950 x mol CO_2 + 0.050 x mol CO + 4.03 x mol O_2 + 2.00 x mol H_2
+ 23 x mol N_2 = 30. x mol of gas total

$$\chi_{CO} = \frac{0.050\ \text{x}}{30.\ \text{x}} = 0.0017; \quad \chi_{CO_2} = \frac{0.950\ \text{x}}{30.\ \text{x}} = 0.032; \quad \chi_{O_2} = \frac{4.03\ \text{x}}{30.\ \text{x}} = 0.13$$

$$\chi_{H_2O} = \frac{2.00\ \text{x}}{30.\ \text{x}} = 0.067; \quad \chi_{N_2} = \frac{23\ \text{x}}{30.\ \text{x}} = 0.77$$

c. The partial pressures are determined by $P = \chi P_{tot}$. Since $P_{tot} = 1.00$ atm, then $P_{CO} = 0.0017$ atm, $P_{CO_2} = 0.032$ atm, $P_{O_2} = 0.13$ atm, $P_{H_2O} = 0.067$ atm and $P_{N_2} = 0.77$ atm.

135. Each stage will give an enrichment of:

$$\frac{\text{Diff. Rate } ^{12}CO_2}{\text{Diff. Rate } ^{13}CO_2} = \left(\frac{M_{^{13}CO_2}}{M_{^{12}CO_2}}\right)^{1/2} = \left(\frac{45.001}{43.998}\right)^{1/2} = 1.0113$$

Since $^{12}CO_2$ moves slightly faster, each successive stage will have less $^{13}CO_2$.

$$\frac{99.90\ ^{12}CO_2}{0.10\ ^{13}CO_2} \times 1.0113^N = \frac{99.990\ ^{12}CO_2}{0.010\ ^{13}CO_2}$$

$$1.0113^N = \frac{9{,}999.0}{999.00} = 10.009 \quad \text{(carrying extra significant figures)}$$

$$N \log(1.0113) = \log(10.009),\ N = \frac{1.000391}{4.88 \times 10^{-3}} = 2.05 \times 10^2 \approx 2.1 \times 10^2 \text{ stages are needed.}$$

Marathon Problems

137. We must determine the identities of element A and compound B in order to answer the questions. Use the first set of data to determine the identity of element A.

Mass N_2 = 659.452 g - 658.572 g = 0.880 g N_2

$$0.880 \text{ g } N_2 \times \frac{1 \text{ mol } N_2}{28.02 \text{ g } N_2} = 0.0314 \text{ mol } N_2$$

$$V = \frac{nRT}{P} = \frac{0.0314 \text{ mol} \times \frac{0.08206 \text{ L atm}}{\text{mol K}} \times 288 \text{ K}}{790.\text{ torr} \times \frac{1 \text{ atm}}{760 \text{ torr}}} = 0.714 \text{ L}$$

$$\text{moles of A} = n = \frac{\left(745 \text{ torr} \times \frac{1 \text{ atm}}{760 \text{ torr}}\right) \times 0.714 \text{ L}}{0.08206 \text{ L atm K}^{-1} \text{mol}^{-1} \times (273 + 26) \text{ K}} = 0.0285 \text{ mol A}$$

Mass of A = 660.59 - 658.572 g = 2.02 g A

$$\text{Molar mass of A} = \frac{2.02 \text{ g A}}{0.0285 \text{ mol A}} = 70.9 \text{ g/mol}$$

The only element that is a gas at 26°C and 745 torr and has a molar mass close to 70.9 g/mol is chlorine = Cl_2 = element A.

The remainder of the information is used to determine the formula of compound B. Assuming 100.00 g of B:

$$85.6\text{ g C} \times \frac{1\text{ mol C}}{12.01\text{ g C}} = 7.13\text{ mol C}; \quad \frac{7.13}{7.13} = 1.00$$

$$14.4\text{ g H} \times \frac{1\text{ mol H}}{1.008\text{ g H}} = 14.3\text{ mol H}; \quad \frac{14.3}{7.13} = 2.01$$

Empirical formula of B = CH_2; Molecular formula = C_xH_{2x} where x is a whole number.

The balanced combustion reaction of C_xH_{2x} with O_2 is:

$$C_xH_{2x}(g) + 3x/2\ O_2(g) \rightarrow x\ CO_2(g) + x\ H_2O(l)$$

To determine the formula of C_xH_{2x}, we need to determine the actual moles of all species present.

Mass of $CO_2 + H_2O$ produced = 846.7 g - 765.3 g = 81.4 g

Since mol CO_2 = mol H_2O = x (see balanced equation), then:

$$81.4\text{ g} = x\text{ mol } CO_2 \times \frac{44.01\text{ g } CO_2}{\text{mol } CO_2} + x\text{ mol } H_2O \times \frac{18.02\text{ g } H_2O}{\text{mol } H_2O}, \quad x = 1.31\text{ mol}$$

$$\text{mol } O_2 \text{ reacted} = 1.31\text{ mol } CO_2 \times \frac{1.5\text{ mol } O_2}{\text{mol } CO_2} = 1.97\text{ mol } O_2$$

From the data, we can calculate moles excess O_2 since only $O_2(g)$ remains after the combustion reaction has gone to completion.

$$n_{O_2} = \frac{PV}{RT} = \frac{6.02\text{ atm} \times 10.68\text{ L}}{0.08206\text{ L atm K}^{-1}\text{mol}^{-1} \times (273 + 22)\text{ K}} = 2.66\text{ mol excess } O_2$$

mol O_2 present initially = 1.97 mol + 2.66 mol = 4.63 mol O_2

$$\text{Total mol gaseous reactants before reaction} = \frac{PV}{RT} = \frac{11.98\text{ atm} \times 10.68\text{ L}}{0.08206 \times 295\text{ K}} = 5.29\text{ mol total}$$

mol C_xH_{2x} = 5.29 mol total - 4.63 mol O_2 = 0.66 mol C_xH_{2x}

Summarizing:

$$0.66\text{ mol } C_xH_{2x} + 1.97\text{ mol } O_2 \rightarrow 1.31\text{ mol } CO_2 + 1.31\text{ mol } H_2O$$

Dividing all quantities by 0.66 gives:

$$C_xH_{2x} + 3\ O_2 \rightarrow 2\ CO_2 + 2\ H_2O$$

To balance the equation, C_xH_{2x} must be C_2H_4 = compound B.

a. Now we can answer the questions. The reaction is:

$$\begin{array}{ccccc} C_2H_4(g) & + & Cl_2(g) & \rightarrow & C_2H_4Cl_2(g) \\ B & + & A & & C \end{array}$$

$$\text{mol } Cl_2 = n = \frac{PV}{RT} = \frac{1.00 \text{ atm} \times 10.0 \text{ L}}{0.08206 \text{ L atm K}^{-1}\text{mol}^{-1} \times 273 \text{ K}} = 0.446 \text{ mol } Cl_2$$

$$\text{mol } C_2H_4 = n = \frac{PV}{RT} = \frac{1.00 \text{ atm} \times 8.60 \text{ L}}{0.08206 \text{ L atm K}^{-1}\text{mol}^{-1} \times 273 \text{ K}} = 0.384 \text{ mol } C_2H_4$$

Since a 1:1 mol ratio is required by the balanced reaction, then C_2H_4 is limiting.

$$\text{Mass } C_2H_4Cl_2 \text{ produced} = 0.384 \text{ mol } C_2H_4 \times \frac{1 \text{ mol } C_2H_4Cl_2}{\text{mol } C_2H_4} \times \frac{98.95 \text{ g}}{\text{mol } C_2H_4Cl_2} = 38.0 \text{ g } C_2H_4Cl_2$$

b. excess mol Cl_2 = 0.446 mol Cl_2 - 0.384 mol Cl_2 reacted = 0.062 mol Cl_2

$$P_{total} = \frac{n_{total}RT}{V};\ n_{total} = 0.384 \text{ mol } C_2H_4Cl_2 \text{ produced} + 0.062 \text{ mol } Cl_2 \text{ excess} = 0.446 \text{ mol}$$

$$V = 10.0 \text{ L} + 8.60 \text{ L} = 18.6 \text{ L}$$

$$P_{total} = \frac{0.446 \text{ mol} \times 0.08206 \text{ L atm K}^{-1}\text{mol}^{-1} \times 273 \text{ K}}{18.6 \text{ L}} = 0.537 \text{ atm}$$

CHAPTER SIX

CHEMICAL EQUILIBRIUM

Characteristics of Chemical Equilibrium

11. $2\ NOCl(g) \rightleftharpoons 2\ NO(g) + Cl_2(g)$ $K = 1.6 \times 10^{-5}$ mol/L

    The expression for K is the product concentrations divided by the reactant concentrations. When K has a value much less than one, the product concentrations are relatively small and the reactant concentrations are relatively large.

    $2\ NO(g) \rightleftharpoons N_2(g) + O_2(g)$ $K = 1 \times 10^{31}$

    When K has a value much greater than one, the product concentrations are relatively large and the reactant concentrations are relatively small. In both cases, however, the rate of the forward reaction equals the rate of the reverse reaction at equilibrium (this is a definition of equilibrium).

13. No, it doesn't matter which direction the equilibrium position is reached. Both experiments will give the same equilibrium position since both experiments started with stoichiometric amounts of reactants or products.

15. When equilibrium is reached, there is no net change in the amount of reactants and products present since the rates of the forward and reverse reactions are equal to each other. The first diagram has 4 A_2B molecules, 2 A_2 molecules and 1 B_2 molecule present. The second diagram has 2 A_2B molecules, 4 A_2 molecules, and 2 B_2 molecules. Therefore, the first diagram cannot represent equilibrium since there was a net change in reactants and products. Is the second diagram the equilibrium mixture? That depends on whether there is a net change between reactants and products when going from the second diagram to the third diagram. The third diagram contains the same number and type of molecules as the second diagram, so the second diagram is the first illustration that represents equilibrium.

    The reaction container initially contained only A_2B. From the first diagram, 2 A_2 molecules and 1 B_2 molecule are present (along with 4 A_2B molecules). From the balanced reaction, these 2 A_2 molecules and 1 B_2 molecule were formed when 2 A_2B molecules decomposed. Therefore, the initial number of A_2B molecules present equals 4 + 2 = 6 molecules A_2B.

The Equilibrium Constant

17. For the gas phase reaction $a\,A + b\,B \rightleftharpoons c\,C + d\,D$:

the equilibrium constant expression is: $K = \dfrac{[C]^c\,[D]^d}{[A]^a\,[B]^b}$

and the reaction quotient has the same form: $Q = \dfrac{[C]^c\,[D]^d}{[A]^a\,[B]^b}$

The difference is that in the expression for K we use equilibrium concentrations, i.e., [A], [B], [C] and [D] are all in equilibrium with each other. Any set of concentrations can be plugged into the reaction quotient expression. Typically, we plug in initial concentrations into the Q expression then compare the value of Q to K to see how far we are from equilibrium. If Q = K, then the reaction is at equilibrium with these concentrations. If Q ≠ K, then the reaction will have to shift either to products (Q < K) or to reactants (Q > K) to reach equilibrium.

19. $K_p = K(RT)^{\Delta n}$ where Δn = sum of gaseous product coefficients - sum of gaseous reactant coefficients. For this reaction, $\Delta n = 1 - 2 = -1$.

$$K_p = \frac{3.7 \times 10^9\ \text{L}}{\text{mol}} \times \left(\frac{0.08206\ \text{L atm}}{\text{mol K}} \times 298\ \text{K}\right)^{-1} = 1.5 \times 10^8\ \text{atm}^{-1}$$

21. $[NO] = \dfrac{4.5 \times 10^{-3}\ \text{mol}}{3.0\ \text{L}} = 1.5 \times 10^{-3}\ M$; $[Cl_2] = \dfrac{2.4\ \text{mol}}{3.0\ \text{L}} = 0.80\ M$

$[NOCl] = \dfrac{1.0\ \text{mol}}{3.0\ \text{L}} = 0.33\ M$; $K = \dfrac{[NO]^2[Cl_2]}{[NOCl]^2} = \dfrac{(1.5 \times 10^{-3})^2(0.80)}{(0.33)^2} = 1.7 \times 10^{-5}\ M$

23. $K_p = \dfrac{P_{H_2}^4}{P_{H_2O}^4}$; $P_{tot} = P_{H_2O} + P_{H_2}$, 36.3 torr = 15.0 torr + P_{H_2}, P_{H_2} = 21.3 torr

Since 1 atm = 760 torr, then: $K_P = \dfrac{\left(21.3\ \text{torr} \times \dfrac{1\ \text{atm}}{760\ \text{torr}}\right)^4}{\left(15.0\ \text{torr} \times \dfrac{1\ \text{atm}}{760\ \text{torr}}\right)^4} = 4.07$

Note: Solids and pure liquids are not included in K expressions.

25. $PCl_5(g) \rightleftharpoons PCl_3(g) + Cl_2(g)$ $K_p = \dfrac{P_{PCl_3} \times P_{Cl_2}}{P_{PCl_5}}$

To determine K_p, we must determine the equilibrium partial pressures of each gas. Initially, P_{PCl_5} = 0.50 atm and $P_{PCl_3} = P_{Cl_2} = 0$ atm. To reach equilibrium some of the PCl_5 reacts to produce some PCl_3 and Cl_2, all in a 1:1 mol ratio. We must determine the change in partial pressures necessary to

reach equilibrium. Since moles ∝ P at constant V and T, then if we let x = atm of PCl_5 that reacts to reach equilibrium, this will produce x atm of PCl_3 and x atm of Cl_2 at equilibrium. The equilibrium partial pressures of each gas will be the initial partial pressure of each gas plus the change necessary to reach equilibrium. The equilibrium partial pressures are:

$$P_{PCl_5} = 0.50 \text{ atm} - x, \quad P_{PCl_3} = P_{Cl_2} = x$$

Now we solve for x using the information in the problem:

$$P_{total} = P_{PCl_5} + P_{PCl_3} + P_{Cl_2}, \quad 0.84 \text{ atm} = 0.50 - x + x + x, \quad 0.84 \text{ atm} = 0.50 + x, \quad x = 0.34 \text{ atm}$$

The equilibrium partial pressures are:

$$P_{PCl_5} = 0.50 - 0.34 = 0.16 \text{ atm}, \quad P_{PCl_3} = P_{Cl_2} = 0.34 \text{ atm}$$

$$K_p = \frac{P_{PCl_3} \times P_{Cl_2}}{P_{PCl_5}} = \frac{(0.34)(0.34)}{(0.16)} = 0.72 \text{ atm}$$

$$K = \frac{K_p}{(RT)^{\Delta n}}, \quad \Delta n = 2 - 1 = 1; \quad K_p = \frac{0.72}{(0.08206)(523)} = 0.017 \text{ mol/L}$$

27. **When solving equilibrium problems, a common method to summarize all the information in the problem is to set up a table. We commonly call this table the ICE table since it summarizes initial concentrations, changes that must occur to reach equilibrium and equilibrium concentrations (the sum of the initial and change columns). For the change column, we will generally use the variable *x* which will be defined as the amount of reactant (or product) that must react to reach equilibrium. In this problem, the reaction must shift right to reach equilibrium since there are no products present initially. Therefore, *x* is defined as the amount of reactant SO_3 that reacts to reach equilibrium and we use the coefficients in the balanced equation to relate the net change in SO_3 to the net change in SO_2 and O_2. The general ICE table for this problem is:**

$$2\,SO_3(g) \rightleftharpoons 2\,SO_2(g) + O_2(g) \qquad K = \frac{[SO_2]^2[O_2]}{[SO_3]^2}$$

	$2\,SO_3(g)$		$2\,SO_2(g)$	$O_2(g)$
Initial	12.0 mol/3.0 L		0	0
	Let *x* mol/L of SO_3 react to reach equilibrium			
Change	$-x$	→	$+x$	$+x/2$
Equil.	$4.0 - x$		x	$x/2$

From the problem, we are told that the equilibrium SO_2 concentration is 3.0 mol/3.0 L = 1.0 *M* ($[SO_2]_e = 1.0\,M$). From the ICE table set-up, $[SO_2]_e = x$ so $x = 1.0$. Solving for the other equilibrium concentrations: $[SO_3]_e = 4.0 - x = 4.0 - 1.0 = 3.0\,M$; $[O_2] = x/2 = 1.0/2 = 0.50\,M$.

$$K = \frac{[SO_2]^2[O_2]}{[SO_3]^2} = \frac{(1.0\,M)^2(0.50\,M)}{(3.0\,M)^2} = 0.056 \text{ mol/L}$$

Alternate Method: Fractions in the change column can be avoided (if you want) be defining *x* differently. If we were to let 2*x* mol/L of SO_3 react to reach equilibrium then the ICE table set-up is:

	$2\ SO_3(g)$	$\rightleftharpoons$	$2\ SO_2(g)$	+	$O_2(g)$	$K = \frac{[SO_2]^2[O_2]}{[SO_3]^2}$
Initial	4.0 *M*		0		0	
	Let 2*x* mol/L of SO_3 react to reach equilibrium					
Change	$-2x$	→	$+2x$		$+x$	
Equil.	$4.0 - 2x$		$2x$		x	

Solving: $2x = [SO_2]_e = 1.0\ M$, $x = 0.50\ M$; $[SO_3]_e = 4.0 - 2(0.50) = 3.0\ M$; $[O_2]_e = x = 0.50\ M$

These are exactly the same equilibrium concentrations as solved for previously, thus K will be the same (as it must be). The moral of the story is define *x* in a manner that is most comfortable for you. Your final answer is independent of how you define *x* initially.

Equilibrium Calculations

29. $CH_3CO_2H + C_2H_5OH \rightleftharpoons CH_3CO_2C_2H_5 + H_2O \quad K = \frac{[CH_3CO_2C_2H_5][H_2O]}{[CH_3CO_2H][C_2H_5OH]} = 2.2$

a. $Q = \frac{[CH_3CO_2C_2H_5]_o[H_2O]_o}{[CH_3CO_2H]_o[C_2H_5OH]_o} = \frac{(0.22\ M)(0.10\ M)}{(0.010\ M)(0.010\ M)} = 220 > K;$

Reaction will shift left to reach equilibrium since Q > K, so the concentration of water will decrease.

b. $Q = \frac{(0.22)(0.0020)}{(0.0020)(0.10)} = 2.2 = K;$ **Reaction is at equilibrium since Q = K, so the concentration of water will remain the same.**

c. $Q = \frac{(0.88)(0.12)}{(0.044)(6.0)} = 0.40 < K;$ **Since Q < K, the concentration of water will increase since the reaction shifts right to reach equilibrium.**

d. $Q = \frac{(4.4)(4.4)}{(0.88)(10.0)} = 2.2 = K;$ **At equilibrium so the water concentration is unchanged.**

e. $K = 2.2 = \frac{(2.0\ M)[H_2O]}{(0.10\ M)(5.0\ M)}$, $[H_2O] = 0.55\ M$

f. **Water is a product of the reaction, but it is not the solvent. Thus, the concentration of water must be included in the equilibrium expression since it is a solute in the reaction. Only when water is the solvent do we not include it in the equilibrium expression.**

31. $H_2O(g) + Cl_2O(g) \rightleftharpoons 2\ HOCl(g) \qquad K = 0.090 = \frac{[HOCl]^2}{[H_2O][Cl_2O]}$

a. The initial concentrations of H_2O and Cl_2O are:

$$\frac{1.0\text{ g } H_2O}{1.0\text{ L}} \times \frac{1\text{ mol}}{18.0\text{ g}} = 5.6 \times 10^{-2}\text{ mol/L}; \quad \frac{2.0\text{ g } Cl_2O}{1.0\text{ L}} \times \frac{1\text{ mol}}{86.9\text{ g}} = 2.3 \times 10^{-2}\text{ mol/L}$$

Since only reactants are present initially, the reaction must proceed to the right to reach equilibrium. Summarizing the problem in a table:

	$H_2O(g)$ +	$Cl_2O(g)$	$\rightleftharpoons$	$2\ HOCl(g)$
Initial	$5.6 \times 10^{-2}\ M$	$2.3 \times 10^{-2}\ M$		0
	x mol/L of H_2O reacts to reach equilibrium			
Change	$-x$	$-x$	$\rightarrow$	$+2x$
Equil.	$5.6 \times 10^{-2} - x$	$2.3 \times 10^{-2} - x$		$2x$

$$K = 0.090 = \frac{(2x)^2}{(5.6 \times 10^{-2} - x)(2.3 \times 10^{-2} - x)}, \quad 1.16 \times 10^{-4} - 7.11 \times 10^{-3}\,x + 0.090\,x^2 = 4\,x^2$$

$3.91\,x^2 + 7.11 \times 10^{-3}\,x - 1.16 \times 10^{-4} = 0$ (We carried extra significant figures.)

Solving using the quadratic formula (see Appendix 1.4 of the text):

$$x = \frac{-7.11 \times 10^{-3} \pm (5.06 \times 10^{-5} + 1.81 \times 10^{-3})^{1/2}}{7.82} = 4.6 \times 10^{-3}\ M \text{ or } -6.4 \times 10^{-3}\ M$$

A negative answer makes no physical sense; we can't have less than nothing.
So $x = 4.6 \times 10^{-3}\ M$.

$[HOCl] = 2x = 9.2 \times 10^{-3}\ M$; $[Cl_2O] = 2.3 \times 10^{-2} - x = 0.023 - 0.0046 = 1.8 \times 10^{-2}\ M$

$[H_2O] = 5.6 \times 10^{-2} - x = 0.056 - 0.0046 = 5.1 \times 10^{-2}\ M$

b.

	$H_2O(g)$ +	$Cl_2O(g)$	$\rightleftharpoons$	$2\ HOCl(g)$
Initial	0	0		1.0 mol/2.0 L = 0.50 M
	$2x$ mol/L of HOCl reacts to reach equilibrium			
Change	$+x$	$+x$	$\leftarrow$	$-2x$
Equil.	x	x		$0.50 - 2x$

$$K = 0.090 = \frac{[HOCl]^2}{[H_2O][Cl_2O]} = \frac{(0.50 - 2x)^2}{x^2}$$

The expression is a perfect square, so we can take the square root of each side:

$$0.30 = \frac{0.50 - 2x}{x}, \quad 0.30\,x = 0.50 - 2x, \quad 2.30\,x = 0.50$$

$x = 0.217\ M$ (We carried extra significant figures.)

$x = [H_2O] = [Cl_2O] = 0.217 = 0.22\ M$; $[HOCl] = 0.50 - 2x = 0.50 - 0.434 = 0.07\ M$

33. $2\ SO_2(g) + O_2(g) \rightleftharpoons 2\ SO_3(g)\quad K_p = 0.25$

	$2\ SO_2(g)$	$O_2(g)$		$2\ SO_3(g)$
Initial	0.50 atm	0.50 atm		0
	$2x$ atm of SO_2 reacts to reach equilibrium			
Change	$-2x$	$-x$	→	$+2x$
Equil.	$0.50 - 2x$	$0.50 - x$		$2x$

$$K_p = 0.25 = \frac{P_{SO_3}^2}{P_{SO_2}^2 \times P_{O_2}} = \frac{(2x)^2}{(0.50 - 2x)^2(0.50 - x)}$$

This will give a cubic equation. Graphing calculators can be used to solve this expression. If you don't have a graphing calculator, an alternative method for solving a cubic equation is to use the method of successive approximations (see Appendix 1.4 of the text). The first step is to guess a value for x. Since the value of K is small (K < 1), then not much of the forward reaction will occur to reach equilibrium. This tells us that x is small. Lets guess that $x = 0.050$ atm. Now we take this estimated value for x and substitute it into the equation everywhere that x appears except for one. For equilibrium problems, we will substitute the estimated value for x into the denominator, then solve for the numerator value of x. We continue this process until the estimated value of x and the calculated value of x converge on the same number. This is the same answer we would get if we were to solve the cubic equation exactly. Applying the method of successive approximations and carrying extra significant figures:

$$\frac{4x^2}{[0.50 - 2(0.050)]^2\,[0.50 - (0.050)]} = \frac{4x^2}{(0.40)^2(0.45)} = 0.25,\ x = 0.067$$

$$\frac{4x^2}{[0.50 - 2(0.067)]^2\,[0.50 - (0.067)]} = \frac{4x^2}{(0.366)^2(0.433)} = 0.25,\ x = 0.060$$

$$\frac{4x^2}{(0.38)^2(0.44)} = 0.25,\ x = 0.063;\quad \frac{4x^2}{(0.374)^2(0.437)} = 0.25,\ x = 0.062$$

The next trial gives the same value for $x = 0.062$ atm. We are done except for determining the equilibrium concentrations. They are:

$$P_{SO_2} = 0.50 - 2x = 0.50 - 2(0.062) = 0.376 = 0.38 \text{ atm}$$

$$P_{O_2} = 0.50 - x = 0.438 = 0.44 \text{ atm};\quad P_{SO_3} = 2x = 0.124 = 0.12 \text{ atm}$$

35. a. **The reaction must proceed to products to reach equilibrium since only reactants are present initially. Summarizing the problem in a table:**

$2\ NOCl(g) \rightleftharpoons 2\ NO(g) + Cl_2(g)\quad K = 1.6 \times 10^{-5}$

	$2\ NOCl(g)$		$2\ NO(g)$	$Cl_2(g)$
Initial	$\frac{2.0 \text{ mol}}{2.0 \text{ L}} = 1.0\ M$		0	0
	$2x$ mol/L of NOCl reacts to reach equilibrium			
Change	$-2x$	→	$+2x$	$+x$
Equil.	$1.0 - 2x$		$2x$	x

$$K = 1.6 \times 10^{-5} = \frac{[NO]^2[Cl_2]}{[NOCl]^2} = \frac{(2x)^2\,(x)}{(1.0 - 2x)^2}$$

If we assume that $1.0 - 2x \approx 1.0$ (from the small size of K, we know that the product concentrations will be small so x will be small), then:

$$1.6 \times 10^{-5} = \frac{4x^3}{1.0^2}, \quad x = 1.6 \times 10^{-2}\ M; \text{ Now we must check the assumption.}$$

$$1.0 - 2x = 1.0 - 2(0.016) = 0.97 = 1.0 \text{ (to proper significant figures)}$$

Our error is about 3%, i.e., $2x$ is 3.2% of 1.0 M. Generally, if the error we introduce by making simplifying assumptions is less than 5%, we go no further, the assumption is said to be valid. We call this the 5% rule. Solving for the equilibrium concentrations:

$$[NO] = 2x = 0.032\ M; \quad [Cl_2] = x = 0.016\ M; \quad [NOCl] = 1.0 - 2x = 0.97\ M \approx 1.0\ M$$

Note: If we were to solve this cubic equation exactly (a longer process), we get $x = 0.016$. This is the exact same answer we determined by making a simplifying assumption. We saved time and energy. Whenever K is a very small value, always make the assumption that x is small. If the assumption introduces an error of less than 5%, then the answer you calculated making the assumption will be considered the correct answer.

b. There is a little trick we can use to solve this problem in order to avoid solving a cubic equation. Since K for this reaction is very small (K << 1), then the reaction will contain mostly reactants at equilibrium (the equilibrium position lies far to the left). We will let the products react to completion by the reverse reaction, then we will solve the forward equilibrium problem to determine the equilibrium concentrations. Summarizing these steps in a table:

	$2\ NOCl(g)$	$\rightleftharpoons$	$2\ NO(g)$	+ $Cl_2(g)$	$K = 1.6 \times 10^{-5}$
Before	0		2.0 M	1.0 M	
	Let 1.0 mol/L Cl_2 react completely.				(K is small, reactants dominate.)
Change	+2.0	←	−2.0	−1.0	React completely
After	2.0		0	0	New initial conditions
	$2x$ mol/L of NOCl reacts to reach equilibrium				
Change	$-2x$	→	$+2x$	$+x$	
Equil.	$2.0 - 2x$		$2x$	x	

$$K = 1.6 \times 10^{-5} = \frac{(2x)^2\,(x)}{(2.0 - 2x)^2} \approx \frac{4x^3}{2.0^2} \text{ (assuming } 2.0 - 2x \approx 2.0)$$

$x^3 = 1.6 \times 10^{-5}$, $x = 2.5 \times 10^{-2}\ M$; Assumption good by the 5% rule ($2x$ is 2.5% of 2.0).

$[NOCl] = 2.0 - 0.050 = 1.95\ M = 2.0\ M$; $[NO] = 0.050\ M$; $[Cl_2] = 0.025\ M$

Note: If we do not break this problem into two parts (a stoichiometric part and an equilibrium part), then we are faced with solving a cubic equation. The set-up would be:

	2 NOCl	⇌	2 NO	+	Cl_2
Initial	0		2.0 *M*		1.0 *M*
Change	+2*y*	←	-2*y*		-*y*
Equil.	2*y*		2.0 - 2*y*		1.0 - *y*

$1.6 \times 10^{-5} = \dfrac{(2.0 - 2y)^2(1.0 - y)}{(2y)^2}$; **If we say that y is small to simplify the problem, then:**

$1.6 \times 10^{-5} = \dfrac{2.0^2}{4y^2}$; **We get $y = 250$. This is impossible!**

To solve this equation, we cannot make any simplifying assumptions; we have to find a way to solve a cubic equation. Or, we can use some chemical common sense and solve the problem the easier way.

c. $2\ NOCl(g) \rightleftharpoons 2\ NO(g) + Cl_2(g)$

	2 NOCl(g)	⇌	2 NO(g)	+	Cl_2(g)
Initial	1.0 *M*		1.0 *M*		0
	2*x* mol/L NOCl reacts to reach equilibrium				
Change	-2*x*	→	+2*x*		+*x*
Equil.	1.0 - 2*x*		1.0 + 2*x*		*x*

$1.6 \times 10^{-5} = \dfrac{(1.0 + 2x)^2(x)}{(1.0 - 2x)^2} \approx \dfrac{(1.0)^2(x)}{(1.0)^2}$ **(Assuming $2x \ll 1.0$)**

$x = 1.6 \times 10^{-5}\ M$; Assumptions are great ($2x$ is 3.2×10^{-3} % of 1.0).

$[Cl_2] = 1.6 \times 10^{-5}\ M$ and $[NOCl] = [NO] = 1.0\ M$

d. $2\ NOCl(g) \rightleftharpoons 2\ NO(g) + Cl_2(g)$

	2 NOCl(g)	⇌	2 NO(g)	+	Cl_2(g)	
Before	0		3.0 *M*		1.0 *M*	
	Let 1.0 mol/L Cl_2 react completely.					
Change	+2.0	←	-2.0		-1.0	React completely
After	2.0		1.0		0	New Initial
	2*x* mol/L NOCl reacts to reach equilibrium					
Change	-2*x*	→	+2*x*		+*x*	
Equil.	2.0 - 2*x*		1.0 + 2*x*		*x*	

$1.6 \times 10^{-5} = \dfrac{(1.0 + 2x)^2(x)}{(2.0 - 2x)^2} \approx \dfrac{x}{4.0}$; **Solving: $x = 6.4 \times 10^{-5}\ M$**

Assumptions great ($2x$ is 1.3×10^{-2} % of 1.0).

$[Cl_2] = 6.4 \times 10^{-5}\ M$; $[NOCl] = 2.0$ M; $[NO] = 1.0\ M$

e. $2\,NOCl(g) \rightleftharpoons 2\,NO(g) + Cl_2(g)$

	$2\,NOCl(g)$		$2\,NO(g)$	$Cl_2(g)$	
Before	2.0 *M*		2.0 *M*	1.0 *M*	
	Let 1.0 mol/L Cl_2 react completely.				
Change	+2.0	←	-2.0	-1.0	React completely
After	4.0		0	0	New Initial
	$2x$ mol/L NOCl reacts to reach equilibrium				
Change	$-2x$	→	$+2x$	$+x$	
Equil.	$4.0 - 2x$		$2x$	x	

$$1.6 \times 10^{-5} = \frac{(2x)^2\,(x)}{(4.0-2x)^2} \approx \frac{4x^3}{16},\ x = 4.0 \times 10^{-2}\ M;\ \text{Assumption good (2\% error).}$$

$[Cl_2] = 0.040\ M$; $[NO] = 0.080\ M$; $[NOCl] = 4.0 - 2(0.040) = 3.92\ M \approx 3.9\ M$

f. $2\,NOCl(g) \rightleftharpoons 2\,NO(g) + Cl_2(g)$

	$2\,NOCl(g)$		$2\,NO(g)$	$Cl_2(g)$	
Before	1.00 *M*		1.00 *M*	1.00 *M*	
	Let 1.00 mol/L NO react completely (the limiting reagent).				
Change	+1.00	←	-1.00	-0.500	React completely
After	2.00		0	0.50	New Initial
	$2x$ mol/L NOCl reacts to reach equilibrium				
Change	$-2x$	→	$+2x$	$+x$	
Equil.	$2.00 - 2x$		$2x$	$0.50 + x$	

$$K = \frac{(2x)^2\,(0.50+x)}{(2.00-2x)^2} \approx \frac{4x^2\,(0.50)}{(2.00)^2} = 1.6 \times 10^{-5},\ x = 5.7 \times 10^{-3}\ M$$

Assumptions good (x is 1.1% of 0.50).

$[NO] = 2x = 1.1 \times 10^{-2}\ M$; $[Cl_2] = 0.50 + 0.0057 = 0.51\ M$; $[NOCl] = 2.00 - 2(0.0057) = 1.99\ M$

37. a. The reaction must proceed to products to reach equilibrium since no product is present initially. Summarizing the problem in a table where x atm of N_2O_4 reacts to reach equilibrium:

$N_2O_4(g) \rightleftharpoons 2\,NO_2(g)\quad K_p = 0.25$

	$N_2O_4(g)$		$2\,NO_2(g)$
Initial	4.5 atm		0
Change	$-x$	→	$+2x$
Equil.	$4.5 - x$		$2x$

$$K_p = \frac{P_{NO_2}^2}{P_{N_2O_4}} = \frac{(2x)^2}{4.5-x} = 0.25,\ 4x^2 = 1.125 - 0.25\,x,\ 4x^2 + 0.25\,x - 1.125 = 0$$

We carried extra significant figures in this expression (as will be typical when we solve an expression using the quadratic formula). Solving using the quadratic formula (see Appendix 1.4 of text):

$$x = \frac{-0.25 \pm [(0.25)^2 - 4(4)(-1.125)]^{1/2}}{2(4)} = \frac{-0.25 \pm 4.25}{8},\ x = 0.50\ \text{(Other value is negative.)}$$

$P_{NO_2} = 2x = 1.0$ atm; $P_{N_2O_4} = 4.5 - x = 4.0$ atm

b. **The reaction must shift to reactants (shift left) to reach equilibrium.**

	$N_2O_4(g)$	$\rightleftharpoons$	$2\,NO_2(g)$
Initial	0		9.0 atm
Change	$+x$	$\leftarrow$	$-2x$
Equil.	x		$9.0 - 2x$

$$K_p = \frac{(9.0 - 2x)^2}{x} = 0.25,\ 4x^2 - 36.25\,x + 81 = 0 \text{ (carrying extra sig. figs.)}$$

Solving using quadratic formula: $x = \dfrac{-(-36.25) \pm [(-36.25)^2 - 4(4)(81)]^{1/2}}{2(4)}$, $x = 4.0$ atm

The other value, 5.1, is impossible. $P_{N_2O_4} = x = 4.0$ atm; $P_{NO_2} = 9.0 - 2x = 1.0$ atm

c. **No, we get the same equilibrium position starting with either pure N_2O_4 or pure NO_2 in stoichiometric amounts.**

d. **From part a, the equilibrium partial pressures are $P_{NO_2} = 1.0$ atm and $P_{N_2O_4} = 4.0$ atm. Halving the container volume will increase each of these partial pressures by a factor of 2. $Q = (2.0)^2 / 8.0 = 0.50$. Since $Q > K_p$, then the reaction will shift left to reestablish equilibrium.**

	$N_2O_4(g)$	$\rightleftharpoons$	$2\,NO_2(g)$
Initial	4.0 atm		1.0 atm
New Initial	8.0		2.0
Change	$+x$	$\leftarrow$	$-2x$
Equil.	$8.0 + x$		$2.0 - 2x$

$$K_p = \frac{(2.0 - 2x)^2}{8.0 + x} = 0.25,\ 4\,x^2 - 8.25\,x + 2.0 = 0 \text{ (carrying extra sig. figs.)}$$

Solving using the quadratic formula: $x = 0.28$ atm

$P_{N_2O_4} = 8.0 + x = 8.3$ atm; $P_{NO_2} = 2.0 - 2x = 1.4$ atm

39. a. $K_p = K(RT)^{\Delta n} = \dfrac{4.5 \times 10^9\ \text{L}}{\text{mol}} \left(\dfrac{0.08206\ \text{L atm}}{\text{mol K}} \times 373\ \text{K} \right)^{-1}$ where $\Delta n = 1 - 2 = -1$

$K_p = 1.5 \times 10^8\ \text{atm}^{-1}$

b. **K_p is so large that at equilibrium we will have almost all $COCl_2$. Assume $P_{total} \approx P_{COCl_2} \approx 5.0$.**

$CO(g) + Cl_2(g) \rightleftharpoons COCl_2(g)$ $\quad K_p = 1.5 \times 10^8$

	$CO(g)$	$Cl_2(g)$		$COCl_2(g)$
Initial	0	0		5.0 atm
	x atm $COCl_2$ reacts to reach equilibrium			
Change	$+x$	$+x$	$\leftarrow$	$-x$
Equil.	x	x		$5.0 - x$

$$K_p = 1.5 \times 10^8 = \frac{5.0 - x}{x^2} \approx \frac{5.0}{x^2} \quad \text{(Assuming } 5.0 - x \approx 5.0\text{)}$$

Solving: $x = 1.8 \times 10^{-4}$ atm. Check assumptions: $5.0 - x = 5.0 - 1.8 \times 10^{-4} = 5.0$ atm. Assumptions are good (well within the 5% rule).

$P_{CO} = P_{Cl_2} = 1.8 \times 10^{-4}$ atm and $P_{COCl_2} = 5.0$ atm

Le Chatelier's Principle

41. For this reaction, we want to maximize the amount of ethyl butyrate produced and minimize the amount of butyric acid present at equilibrium. First, we should avoid water. Any extra water we add from the solvent tends to push the equilibrium to the left. This eliminates water and 95% ethanol as solvent choices. Of the remaining two solvents, acetonitrile will not take part in the reaction, whereas ethanol is a reactant. If we use ethanol as the solvent it will drive the equilibrium to the right, thereby reducing the concentration of the objectionable butyric acid to a minimum while maximizing the yield of ethyl butyrate. Thus, the best solvent is 100% ethanol.

43. a. No effect; Adding more of a pure solid or pure liquid has no effect on the equilibrium position.

    b. Shifts left; HF(g) will be removed by reaction with the glass. As HF(g) is removed, the reaction will shift left to produce more HF(g).

    c. Shifts right; As $H_2O(g)$ is removed, the reaction will shift right to produce more $H_2O(g)$.

45. $H^+ + OH^- \rightarrow H_2O$; Sodium hydroxide (NaOH) will react with the H^+ on the product side of the reaction. This effectively removes H^+ from the equilibrium, which will shift the reaction to the right to produce more H^+ and CrO_4^{2-}. Since more CrO_4^{2-} is produced, the solution turns yellow.

47. a. right b. right c. no effect; He(g) is neither a reactant nor a product.

    d. left; Since the reaction is exothermic, heat is a product:

    $$CO(g) + H_2O(g) \rightarrow H_2(g) + CO_2(g) + \text{heat}$$

    Increasing T will add heat. The equilibrium shifts to the left to use up the added heat.

    e. No effect; Since these are equal moles of gaseous reactants as gaseous products (2 mol vs. 2 mol), then a change in volume will have no effect on the equilibrium.

49. a. left b. right c. left

    d. no effect (reactant and product concentrations are unchanged)

    e. no effect; Since there are equal numbers of product and reactant gas molecules, then a change in volume has no effect on the equilibrium position.

f. right; A decrease in temperature will shift the equilibrium to the right since heat is a product in this reaction (as is true in all exothermic reactions).

Additional Exercises

51. a. $N_2(g) + O_2(g) \rightleftharpoons 2\ NO(g)\quad K_p = 1 \times 10^{-31} = \dfrac{P_{NO}^2}{P_{N_2} \times P_{O_2}} = \dfrac{P_{NO}^2}{(0.8)(0.2)},\ P_{NO} = 1 \times 10^{-16}$ atm

In 1.0 cm^3 of air: $n_{NO} = \dfrac{PV}{RT} = \dfrac{(1 \times 10^{-16}\text{ atm})(1.0 \times 10^{-3}\text{ L})}{\left(\dfrac{0.08206\text{ L atm}}{\text{mol K}}\right)(298\text{ K})} = 4 \times 10^{-21}$ mol NO

$$\frac{4 \times 10^{-21}\text{ mol NO}}{\text{cm}^3} \times \frac{6.02 \times 10^{23}\text{ molecules}}{\text{mol NO}} = \frac{2 \times 10^3\text{ molecules NO}}{\text{cm}^3}$$

b. There is more NO in the atmosphere than we would expect from the value of K. The answer must lie in the rates of the reaction. At 25°C the rates of both reactions:

$$N_2 + O_2 \rightarrow 2\ NO \text{ and } 2\ NO \rightarrow N_2 + O_2$$

are so slow that they are essentially zero. Very strong bonds must be broken; the activation energy is very high. Therefore, the reaction essentially doesn't occur at low temperatures. Nitric oxide, however, can be produced in high energy or high temperature environments since the production of NO is endothermic. In nature, some NO is produced by lightning and the primary manmade source is from automobiles. At these high temperatures, K will increase and the rates of the reaction will also increase, resulting in a higher production of NO. Once the NO gets into a more normal temperature environment, it doesn't go back to N_2 and O_2 because of the slow rate.

c. $K_p = P_{NO}^2/(P_{N_2} \times P_{O_2})$

To convert from partial pressures to concentrations in molecules/cm^3, we will have to do the same conversion to all concentrations. All of these conversions will cancel since there are equal product and reactant moles of gas ($\Delta n = 0$). Therefore, $K^* = K_p = 1 \times 10^{-31}$. The equilibrium constant for this reaction is unitless.

53.

$$\begin{array}{ll} O(g) + NO(g) \rightleftharpoons NO_2(g) & K = 1/6.8 \times 10^{-49} = 1.5 \times 10^{48} \\ NO_2(g) + O_2(g) \rightleftharpoons NO(g) + O_3(g) & K = 1/5.8 \times 10^{-34} = 1.7 \times 10^{33} \\ \hline O_2(g) + O(g) \rightleftharpoons O_3(g) & K = (1.5 \times 10^{48})(1.7 \times 10^{33}) = 2.6 \times 10^{81}\text{ L/mol} \end{array}$$

55.

	$3\ H_2(g)$	+	$N_2(g)$	$\rightleftharpoons$	$2\ NH_3(g)$
Initial	$[H_2]_o$		$[N_2]_o$		0
	x mol/L of N_2 reacts to reach equilibrium				
Change	$-3x$		$-x$	$\rightarrow$	$+2x$
Equil	$[H_2]_o - 3x$		$[N_2]_o - x$		$2x$

From the problem:

$[NH_3]_e = 4.0\ M = 2x,\ x = 2.0\ M;\ [H_2]_e = 5.0\ M = [H_2]_o - 3x;\ [N_2]_e = 8.0\ M = [N_2]_o - x$

$5.0\ M = [H_2]_o - 3(2.0\ M),\ [H_2]_o = 11.0\ M;\ 8.0\ M = [N_2]_o - 2.0\ M,\ [N_2]_o = 10.0\ M$

57. a. $PCl_5(g) \rightleftharpoons PCl_3(g) + Cl_2(g) \qquad K_p = P_{PCl_3} \times P_{Cl_2}/P_{PCl_5}$

	$PCl_5(g)$		$PCl_3(g)$	$Cl_2(g)$	
Initial	P_0		0	0	P_0 = initial PCl_5 pressure
Change	$-x$	→	$+x$	$+x$	
Equil.	$P_0 - x$		x	x	

$P_{total} = P_0 - x + x + x = P_0 + x = 358.7$ torr

$$P_0 = \frac{n_{PCl_5}RT}{V} = \frac{\frac{2.4156\ g}{208.22\ g\ mol^{-1}} \times \frac{0.08206\ L\ atm}{mol\ K} \times 523.2\ K}{2.000\ L} = 0.2490\ atm\ or\ 189.2\ torr$$

$x = P_{total} - P_0 = 358.7 - 189.2 = 169.5$ torr

$P_{PCl_3} = P_{Cl_2} = 169.5\ torr = 0.2230\ atm$

$P_{PCl_5} = 189.2 - 169.5 = 19.7\ torr = 0.0259\ atm$

$$K_p = \frac{(0.2230)^2}{0.0259} = 1.92\ atm$$

b. $$P_{Cl_2} = \frac{n_{Cl_2}RT}{V} = \frac{0.250 \times 0.08206 \times 523.2}{2.000} = 5.37\ atm\ Cl_2\ added$$

	$PCl_5(g)$	$\rightleftharpoons$	$PCl_3(g)$	+	$Cl_2(g)$	
Initial	0.0259 atm		0.2230 atm		0.2230 atm	(from a)
	Adding 0.250 mol Cl_2 increases P_{Cl_2} by 5.37 atm.					
Initial'	0.0259		0.2230		5.59	
Change	+0.2230	←	-0.2230		-0.2230	React completely
After	0.2489		0		5.37	New initial
Change	$-x$	→	$+x$		$+x$	
Equil.	$0.2489 - x$		x		$5.37 + x$	

$$\frac{(5.37 + x)(x)}{(0.2489 - x)} = 1.92,\ x^2 + 7.29x - 0.478 = 0$$

Solving using the quadratic formula: $x = 0.0650$ atm

$P_{PCl_3} = 0.0650\ atm;\ P_{PCl_5} = 0.2489 - 0.0650 = 0.1839\ atm;\ P_{Cl_2} = 5.37 + 0.0650 = 5.44\ atm$

59. $N_2O_4(g) \rightleftharpoons 2\ NO_2(g) \qquad K_p = \frac{P^2_{NO_2}}{P_{N_2O_4}} = \frac{(1.20)^2}{0.34} = 4.2$

Doubling the volume decreases each partial pressure by a factor of 2 (P = nRT/V).

P_{NO_2} = 0.600 atm and $P_{N_2O_4}$ = 0.17 atm are the new partial pressures.

$Q = \frac{(0.600)^2}{0.17} = 2.1$, $Q < K$ Equilibrium will shift to the right.

$$N_2O_4(g) \rightleftharpoons 2\ NO_2(g)$$

	$N_2O_4(g)$	$2\ NO_2(g)$
Initial	0.17 atm	0.600 atm
Equil.	0.17 - x	0.600 + 2x

$K_p = 4.2 = \frac{(0.600 + 2x)^2}{(0.17 - x)}$, $4x^2 + 6.6x - 0.354 = 0$ (carrying extra sig. figs.)

Solving using the quadratic formula: $x = 0.052$ atm

$P_{NO_2} = 0.600 + 2(0.052) = 0.704$ atm; $P_{N_2O_4} = 0.17 - 0.052 = 0.12$ atm

61. a. $$2\ NaHCO_3(s) \rightleftharpoons Na_2CO_3(s) + CO_2(g) + H_2O(g) \qquad K_p = 0.25$$

	$2\ NaHCO_3(s)$		$Na_2CO_3(s)$	$CO_2(g)$	$H_2O(g)$
Initial	-		-	0	0
	$NaHCO_3(s)$ decomposes to form x atm each of $CO_2(g)$ and $H_2O(g)$ at equilibrium.				
Change	-	→	-	+x	+x
Equil.	-		-	x	x

$K_p = 0.25 = P_{CO_2} \times P_{H_2O}$, $0.25 = x^2$, $x = P_{CO_2} = P_{H_2O} = 0.50$ atm

b. $n_{CO_2} = \frac{P_{CO_2} \times V}{RT} = \frac{(0.50\ \text{atm})\ (1.00\ \text{L})}{(0.08206\ \text{L atm mol}^{-1}\ \text{K}^{-1})\ (398\ \text{K})} = 1.5 \times 10^{-2}\ \text{mol}\ CO_2$

Mass of Na_2CO_3 produced:

$$1.5 \times 10^{-2}\ \text{mol}\ CO_2 \times \frac{1\ \text{mol}\ Na_2CO_3}{\text{mol}\ CO_2} \times \frac{106.0\ \text{g}\ Na_2CO_3}{\text{mol}\ Na_2CO_3} = 1.6\ \text{g}\ Na_2CO_3$$

Mass of $NaHCO_3$ reacted:

$$1.5 \times 10^{-2}\ \text{mol}\ CO_2 \times \frac{2\ \text{mol}\ NaHCO_3}{1\ \text{mol}\ CO_2} \times \frac{84.01\ \text{g}\ NaHCO_3}{\text{mol}} = 2.5\ \text{g}\ NaHCO_3$$

Mass of $NaHCO_3$ remaining = 10.0 - 2.5 = 7.5 g

c. $$10.0\ \text{g}\ NaHCO_3 \times \frac{1\ \text{mol}\ NaHCO_3}{84.01\ \text{g}\ NaHCO_3} \times \frac{1\ \text{mol}\ CO_2}{2\ \text{mol}\ NaHCO_3} = 5.95 \times 10^{-2}\ \text{mol}\ CO_2$$

When all of the $NaHCO_3$ has just been consumed, we will have 5.95×10^{-2} mol CO_2 gas at a pressure of 0.50 atm (from a).

$$V = \frac{nRT}{P} = \frac{(5.95 \times 10^{-2}\ \text{mol})\ (0.08206\ \text{L atm mol}^{-1}\ \text{K}^{-1})\ (398\ \text{K})}{0.50\ \text{atm}} = 3.9\ \text{L}$$

Challenge Problems

63. a. $2\,NO(g) + Br_2(g) \rightleftharpoons 2\,NOBr(g)$

Initial	98.4 torr	41.3 torr	0
	2*x* torr of NO reacts to reach equilibrium		
Change	$-2x$	$-x$ →	$+2x$
Equil.	$98.4 - 2x$	$41.3 - x$	$2x$

$P_{total} = P_{NO} + P_{Br_2} + P_{NOBr} = (98.4 - 2x) + (41.3 - x) + 2x = 139.7 - x$

$P_{total} = 110.5 = 139.7 - x$, $x = 29.2$ torr; $P_{NO} = 98.4 - 2(29.2) = 40.0$ torr $= 0.0526$ atm

$P_{Br_2} = 41.3 - 29.2 = 12.1$ torr $= 0.0159$ atm; $P_{NOBr} = 2(29.2) = 58.4$ torr $= 0.0768$ atm

$$K_p = \frac{P_{NOBr}^2}{P_{NO}^2 \times P_{Br_2}} = \frac{(0.0768\text{ atm})^2}{(0.0526\text{ atm})^2\,(0.0159\text{ atm})} = 134\text{ atm}^{-1}$$

b. $2\,NO(g) + Br_2(g) \rightleftharpoons 2\,NOBr(g)$

Initial	0.30 atm	0.30 atm	0
	2*x* atm of NO reacts to reach equilibrium		
Change	$-2x$	$-x$ →	$+2x$
Equil.	$0.30 - 2x$	$0.30 - x$	$2x$

This would yield a cubic equation which can be difficult to solve unless you have a graphing calculator. Since K_p is pretty large, so let us approach equilibrium in two steps; assume the reaction goes to completion then solve the back equilibrium problem.

	2 NO +	Br_2 ⇌	2 NOBr	
Before	0.30 atm	0.30 atm	0	
	Let 0.30 atm NO react completely.			
Change	−0.30	−0.15 →	+0.30	React completely
After	0	0.15	0.30	New initial
	2*y* atm of NOBr reacts to reach equilibrium			
Change	$+2y$	$+y$ ←	$-2y$	
Equil.	$2y$	$0.15 + y$	$0.30 - 2y$	

$$K_p = \frac{(0.30 - 2y)^2}{(2y)^2\,(0.15 + y)} = 134, \quad \frac{(0.30 - 2y)^2}{(0.15 + y)} = 134 \times 4\,y^2 = 536\,y^2$$

If $y \ll 0.15$: $\dfrac{(0.30)^2}{0.15} \approx 536\,y^2$ and $y = 0.034$; Assumptions are poor (y is 23% of 0.15).

Use 0.034 as an approximation for y and solve by successive approximations (Appendix 1.4):

$$\frac{(0.30 - 0.068)^2}{0.15 + 0.034} = 536\,y^2,\ y = 0.023; \quad \frac{(0.30 - 0.046)^2}{0.15 + 0.023} = 536\,y^2,\ y = 0.026$$

$$\frac{(0.30 - 0.052)^2}{0.15 + 0.026} = 536\, y^2,\ y = 0.026 \text{ atm}$$ **(We have converged on the correct answer.)**

So: $P_{NO} = 2y = 0.052$ **atm;** $P_{Br_2} = 0.15 + y = 0.18$ **atm;** $P_{NOBr} = 0.30 - 2y = 0.25$ **atm**

65. $P_4(g) \rightleftharpoons 2\ P_2(g)$ $K_p = 0.100 = \frac{P_{P_2}^2}{P_{P_4}}$; $P_{P_4} + P_{P_2} = P_{total} = 1.00$ **atm,** $P_{P_4} = 1.00 - P_{P_2}$

Let $y = P_{P_2}$ **at equilibrium, then** $K_p = \frac{y^2}{1.00 - y} = 0.100$

Solving: $y = 0.270 \text{ atm} = P_{P_2}$; $P_{P_4} = 1.00 - 0.270 = 0.73$ **atm**

To solve for the fraction dissociated, we need the initial pressure of P_4.

	$P_4(g)$	$\rightleftharpoons$	$2\ P_2(g)$	
Initial	P_0		0	P_0 = **initial pressure of** P_4
	x **atm of P_4 reacts to reach equilibrium**			
Change	$-x$	$\rightarrow$	$+2x$	
Equil.	$P_0 - x$		$2x$	

$P_{total} = P_0 - x + 2x = 1.00 \text{ atm} = P_0 + x$

Solving: $0.270 \text{ atm} = P_{P_2} = 2x,\ x = 0.135$ **atm;** $P_0 = 1.00 - 0.135 = 0.87$ **atm**

Fraction dissociation $= \frac{x}{P_0} = \frac{0.135}{0.87} = 0.16$ **or 16% of P_4 is dissociated to reach equilibrium.**

67. $d = \text{density} = \frac{P \times (\text{molar mass})}{RT} = \frac{P_{O_2}(\text{molar mass}_{O_2}) + P_{O_3}(\text{molar mass}_{O_3})}{RT}$

$$0.168 \text{ g/L} = \frac{P_{O_2}(32.00 \text{ g/mol}) + P_{O_3}(48.00 \text{ g/mol})}{\frac{0.08206 \text{ L atm}}{\text{mol K}} \times 448 \text{ K}},\quad 32.00\ P_{O_2} + 48.00\ P_{O_3} = 6.18 \quad (\text{P in atm})$$

$$P_{total} = P_{O_2} + P_{O_3} = 128 \text{ torr} \times \frac{1 \text{ atm}}{760 \text{ torr}} = 0.168 \text{ atm}$$

We have two equations in two unknowns. Solving using simultaneous equations:

$$\begin{aligned} 32.00\ P_{O_2} + 48.00\ P_{O_3} &= 6.18 \\ -32.00\ P_{O_2} - 32.00\ P_{O_3} &= -5.38 \\ \hline 16.00\ P_{O_3} &= 0.80 \end{aligned}$$

$P_{O_3} = \frac{0.80}{16.00} = 0.050$ **atm and** $P_{O_2} = 0.118$ **atm**

$$K_p = \frac{P_{O_3}^2}{P_{O_2}^3} = \frac{(0.050)^2}{(0.118)^3} = 1.5 \text{ atm}^{-1}$$

69. $2\ NOBr(g) \rightleftharpoons 2\ NO(g) + Br_2(g)$

Initial	P_0	0	0	P_0 = initial pressure of NOBr
Equil.	$P_0 - 2x$	$2x$	x	Note: $P_{NO} = 2P_{Br_2}$

$P_{total} = 0.0515\text{ atm} = (P_0 - 2x) + (2x) + (x) = P_0 + x$; $0.0515\text{ atm} = P_{NOBr} + 3\ P_{Br_2}$

$$d = \frac{P \times (\text{molar mass})}{RT} = 0.1861\text{ g/L} = \frac{P_{NOBr}(109.9) + 2P_{Br_2}(30.01) + P_{Br_2}(159.8)}{0.08206 \times 298}$$

$4.55 = 109.9\ P_{NOBr} + 219.8\ P_{Br_2}$

Solving using simultaneous equations:

$$\begin{aligned} 0.0515 &= P_{NOBr} + 3\ P_{Br_2} \\ -0.0414 &= -P_{NOBr} - 2.000\ P_{Br_2} \\ \hline 0.0101 &= P_{Br_2} \end{aligned}$$

$P_{Br_2} = 1.01 \times 10^{-2}$ atm; $P_{NO} = 2\ P_{Br_2} = 2.02 \times 10^{-2}$ atm

$P_{NOBr} = 0.0515 - 3(1.01 \times 10^{-2}) = 2.12 \times 10^{-2}$ atm

$$K_p = \frac{P_{Br_2} \times P_{NO}^2}{P_{NOBr}^2} = \frac{(1.01 \times 10^{-2})(2.02 \times 10^{-2})^2}{(2.12 \times 10^{-2})^2} = 9.17 \times 10^{-3}\text{ atm}$$

71. a. $$P_{PCl_5} = \frac{n_{PCl_5}RT}{V} = \frac{0.100\text{ mol} \times \frac{0.08206\text{ L atm}}{\text{mol K}} \times 480.\text{ K}}{12.0\text{ L}} = 0.328\text{ atm}$$

$PCl_5(g) \rightleftharpoons PCl_3(g) + Cl_2(g)$ $K_p = 0.267$ atm

Initial	0.328 atm		0	0
Change	$-x$	→	$+x$	$+x$
Equil.	$0.328 - x$		x	x

$K_p = \dfrac{x^2}{0.328 - x} = 0.267$, $x^2 + 0.267\ x - 0.08758 = 0$ (carrying extra sig. figs.)

Solving using the quadratic formula: $x = 0.191$ atm

$P_{PCl_3} = P_{Cl_2} = 0.191$ atm; $P_{PCl_5} = 0.328 - 0.191 = 0.137$ atm

b. $PCl_5(g) \rightleftharpoons PCl_3(g) + Cl_2(g)$

Initial	P_0		0	0	P_0 = initial pressure of PCl_5
Change	$-x$	→	$+x$	$+x$	
Equil.	$P_0 - x$		x	x	

$P_{total} = 2.00 \text{ atm} = (P_0 - x) + x + x = P_0 + x, \; P_0 = 2.00 - x$

$K_p = \dfrac{x^2}{P_0 - x} = 0.267; \quad \dfrac{x^2}{2.00 - 2x} = 0.267, \; x^2 = 0.534 - 0.534\,x$

$x^2 + 0.534\,x - 0.534 = 0;$ Solving using the quadratic formula:

$$x = \frac{-0.534 \pm \sqrt{(0.534)^2 + 4(0.534)}}{2} = 0.511 \text{ atm}$$

$P_0 = 2.00 - x = 2.00 - 0.511 = 1.49$ atm; The initial pressure of PCl_5 was 1.49 atm.

$$n_{PCl_5} = \frac{P_{PCl_5} \times V}{RT} = \frac{(1.49 \text{ atm})(5.00 \text{ L})}{(0.08206 \text{ L atm mol}^{-1} \text{ K}^{-1})(480.\text{ K})} = 0.189 \text{ mol } PCl_5$$

0.189 mol PCl_5 × 208.22 g PCl_5/mol = 39.4 g PCl_5 was initially introduced.

73. $N_2(g) + O_2(g) \rightleftharpoons 2\,NO(g)$

Equil. 3.7 p p x

Let:
equilibrium $P_{O_2} = p$
equilibrium $P_{N_2} = 78/21\, P_{O_2} = 3.7\,p$
equilibrium $P_{NO} = x$
equilibrium $P_{NO_2} = y$

$$K_p = 1.5 \times 10^{-4} = \frac{P_{NO}^2}{P_{O_2} \times P_{N_2}}$$

$N_2 + 2\,O_2 \rightleftharpoons 2\,NO_2$

Equil. 3.7 p p y

$$K_p = 1.0 \times 10^{-5} = \frac{P_{NO_2}^2}{P_{O_2}^2 \times P_{N_2}}$$

We want $P_{NO_2} = P_{NO}$ at equilibrium, so $x = y$.

Taking the ratio of the two K_p expressions:

$$\frac{\dfrac{P_{NO}^2}{P_{O_2} \times P_{N_2}}}{\dfrac{P_{NO_2}^2}{P_{O_2}^2 \times P_{N_2}}} = \frac{1.5 \times 10^{-4}}{1.0 \times 10^{-5}}; \; \text{Since } P_{NO} = P_{NO_2},\; P_{O_2} = \frac{1.5 \times 10^{-4}}{1.0 \times 10^{-5}} = 15 \text{ atm}$$

Air is 21 mol % O_2, so:

$$P_{O_2} = 0.21\, P_{total},\; P_{total} = \frac{15 \text{ atm}}{0.21} = 71 \text{ atm}$$

To solve for the equilibrium concentrations of all gases (not required to answer the question), solve one of the K_p expressions where $p = P_{O_2} = 15$ atm.

$$1.5 \times 10^{-4} = \frac{x^2}{15[3.7(15)]}, \ x = P_{NO} = P_{NO_2} = 0.35 \text{ atm}$$

Equilibrium pressures:

$$P_{O_2} = 15 \text{ atm}; \ P_{N_2} = 3.7(15) = 55.5 = 56 \text{ atm}; \ P_{NO} = P_{NO_2} = 0.35 \text{ atm}$$

75. $SO_3(g) \rightleftharpoons SO_2(g) + 1/2\ O_2(g)$

	$SO_3(g)$		$SO_2(g)$	$1/2\ O_2(g)$	
Initial	P_0		0	0	P_0 = initial pressure of SO_3
Change	$-x$	$\rightarrow$	$+x$	$+x/2$	
Equil.	$P_0 - x$		x	$x/2$	

Average molar mass of the mixture is:

$$\text{average molar mass} = \frac{dRT}{P} = \frac{(1.60 \text{ g/L})\ (0.08206 \text{ L atm mol}^{-1} \text{ K}^{-1})\ (873 \text{ K})}{1.80 \text{ atm}} = 63.7 \text{ g/mol}$$

The average molar mass is determined by:

$$\text{average molar mass} = \frac{n_{SO_3}(80.07 \text{ g/mol}) + n_{SO_2}(64.07 \text{ g/mol}) + n_{O_2}(32.00 \text{ g/mol})}{n_{total}}$$

Since χ_A = mol fraction of component A = $n_A/n_{total} = P_A/P_{total}$, then:

$$63.7 \text{ g/mol} = \frac{P_{SO_3}(80.07) + P_{SO_2}(64.07) + P_{O_2}(32.00)}{P_{total}}$$

$P_{total} = P_0 - x + x + x/2 = P_0 + x/2 = 1.80$ atm, $P_0 = 1.80 - x/2$

$$63.7 = \frac{(P_0 - x)\ (80.07) + x(64.07) + \frac{x}{2}(32.00)}{1.80}$$

$$63.7 = \frac{(1.80 - 3/2x)\ (80.07) + x(64.07) + \frac{x}{2}(32.00)}{1.80}$$

$115 = 144 - 120.1\ x + 64.07\ x + 16.00\ x$, $40.0\ x = 29$, $x = 0.73$ atm

$P_{SO_3} = P_0 - x = 1.80 - 3/2\ x = 0.71$ atm; $P_{SO_2} = 0.73$ atm; $P_{O_2} = x/2 = 0.37$ atm

$$K_p = \frac{P_{SO_2} \times P_{O_2}^{1/2}}{P_{SO_3}} = \frac{(0.73)\ (0.37)^{1/2}}{(0.71)} = 0.63 \text{ atm}^{1/2}$$

77. $N_2(g) + 3\ H_2(g) \rightleftharpoons 2\ NH_3(g)$ $\qquad K_p = \frac{P_{NH_3}^2}{P_{N_2} \times P_{H_2}^3} = 6.5 \times 10^{-3}$

1.0 atm	$N_2(g)$	+	$3\ H_2(g)$	$\rightleftharpoons$	$2\ NH_3(g)$
Initial	0.25 atm		0.75 atm		0
Equil.	$0.25 - x$		$0.75 - 3x$		$2x$

$$\frac{(2x)^2}{(0.75 - 3x)^3 (0.25 - x)} = 6.5 \times 10^{-3}$$; Using successive approximations:

$x = 1.2 \times 10^{-2}$ atm; $P_{NH_3} = 2x = 0.024$ atm

10 atm	$N_2(g)$	+	$3\ H_2(g)$	$\rightleftharpoons$	$2\ NH_3(g)$
Initial	2.5 atm		7.5 atm		0
Equil.	$2.5 - x$		$7.5 - 3x$		$2x$

$$\frac{(2x)^2}{(7.5 - 3x)^3 (2.5 - x)} = 6.5 \times 10^{-3}$$; Using successive approximations:

$x = 0.69$ atm; $P_{NH_3} = 1.4$ atm

100 atm Using the same setup as above: $\dfrac{4x^2}{(75 - 3x)^3 (25 - x)} = 6.5 \times 10^{-3}$

Solving by successive approximations: $x = 16$ atm; $P_{NH_3} = 32$ atm

1000 atm

	$N_2(g)$	+	$3\ H_2(g)$	$\rightleftharpoons$	$2\ NH_3(g)$
Initial	250 atm		750 atm		0
	Let 250 atm N_2 react completely.				
New Initial	0		0		5.0×10^2
Equil.	x		$3x$		$5.0 \times 10^2 - 2x$

$$\frac{(5.0 \times 10^2 - 2x)^2}{(3x)^3 x} = 6.5 \times 10^{-3}$$; Assume x is small, then:

$$\frac{(5.0 \times 10^2)^2}{(3x)^3 x} \approx 6.5 \times 10^{-3},\ x = 35$$

Assumption is poor (14% error).

Solving by successive approximations:

$x = 32$ atm

$P_{NH_3} = 5.0 \times 10^2 - 2x = 440$ atm

The results are plotted as log P_{NH_3} vs. log P_{total}. Notice that as P_{total} increases, a larger fraction of N_2 and H_2 is converted to NH_3, i.e., as P_{total} increases (V decreases), the reaction shifts further to the right as predicted by LeChatelier's Principle.

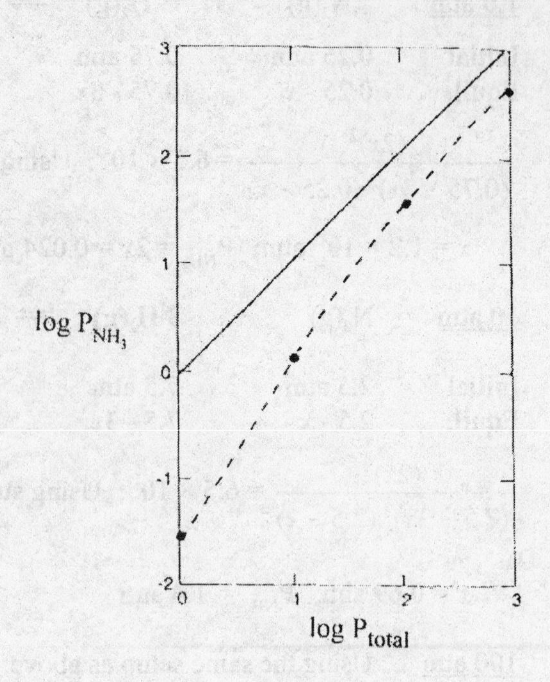

Marathon Problems

79.

	N_2O_4	$\rightleftharpoons$	$2\ NO_2$
Initial	x		0
Change	-0.16x		+0.32x
Equil.	0.84x		0.32x

0.84x + 0.32x = 1.5 atm, x = 1.3 atm

a. $K_p = \dfrac{(0.42)^2}{1.1} = 0.16$ atm

b. $N_2O_4 \rightleftharpoons 2\ NO_2$

Equil. x y

$x + y = 1.0$ atm; $\dfrac{y^2}{x} = 0.16$

Solving, x = 0.67 atm (= $P_{N_2O_4}$) and y = 0.33 atm (= P_{NO_2})

c.

	N_2O_4	$\rightleftharpoons$	$2\ NO_2$
Initial	$P_{N_2O_4}$		0
Change	-x		+2x
Equil.	0.67		0.33

x = 0.165 (using extra sig figs)

$P_{N_2O_4}$ - x = 0.67; Solving: $P_{N_2O_4}$ = 0.84

$\dfrac{0.165}{0.84} \times 100 = 20.\%$ dissociated

CHAPTER SEVEN

ACIDS AND BASES

Nature of Acids and Bases

17. The dissociation reaction (the K_a reaction) of an acid in water commonly omits water as a reactant. We will follow this practice. All dissociation reactions produce H^+ and the conjugate base of the acid that is dissociated.

a. $HC_2H_3O_2(aq) \rightleftharpoons H^+(aq) + C_2H_3O_2^-(aq)$ $\qquad K_a = \dfrac{[H^+][C_2H_3O_2^-]}{[HC_2H_3O_2]}$

b. $Co(H_2O)_6^{3+}(aq) \rightleftharpoons H^+(aq) + Co(H_2O)_5(OH)^{2+}(aq)$ $\qquad K_a = \dfrac{[H^+][Co(H_2O)_5(OH)^{2+}]}{[Co(H_2O)_6^{3+}]}$

c. $CH_3NH_3^+(aq) \rightleftharpoons H^+(aq) + CH_3NH_2(aq)$ $\qquad K_a = \dfrac{[H^+][CH_3NH_2]}{[CH_3NH_3^+]}$

19. The beaker on the left represents a strong acid in solution; the acid, HA, is 100% dissociated into the H^+ and A^- ions. The beaker on the right represents a weak acid in solution; only a little bit of the acid, HB, dissociates into ions, so the acid exists mostly as undissociated HB molecules in water.

a. HNO_2: weak acid beaker
b. HNO_3: strong acid beaker
c. HCl: strong acid beaker
d. HF: weak acid beaker
e. $HC_2H_3O_2$: weak acid beaker

21. The K_a value is directly related to acid strength. As K_a increases, acid strength increases. For water, use K_w when comparing the acid strength of water to other species. The K_a values are:

$HClO_4$: strong acid ($K_a >> 1$); $HClO_2$: $K_a = 1.2 \times 10^{-2}$

HOC_6H_5: $K_a = 1.6 \times 10^{-10}$; H_2O: $K_a = K_w = 1.0 \times 10^{-14}$

From the K_a values, the ordering is: $HClO_4 > HClO_2 > HOC_6H_5 > H_2O$.

23. a. H_2SO_4 is a strong acid and water is a very weak acid with $K_a = K_w = 1.0 \times 10^{-14}$. H_2SO_4 is a much stronger acid than H_2O.

b. H_2O, $K_a = K_w = 1.0 \times 10^{-14}$; HOCl, $K_a = 3.5 \times 10^{-8}$; HOCl is a stronger acid than H_2O since K_a for HOCl > K_a for H_2O.

c. NH_4^+, $K_a = 5.6 \times 10^{-10}$; $HC_2H_2ClO_2$, $K_a = 1.35 \times 10^{-3}$; $HC_2H_2ClO_2$ is a stronger acid than NH_4^+ since K_a for $HC_2H_2ClO_2 > K_a$ for NH_4^+.

25. a. H_2O and $CH_3CO_2^-$

b. An acid-base reaction can be thought of as a competition between two opposing bases. Since this equilibrium lies far to the left ($K_a < 1$), then $CH_3CO_2^-$ is a stronger base than H_2O.

c. The acetate ion is a better base than water and produces basic solutions in water. When we put acetate ion into solution as the only major basic species, the reaction is:

$$CH_3CO_2^- + H_2O \rightleftharpoons CH_3CO_2H + OH^-$$

Now the competition is between $CH_3CO_2^-$ and OH^- for the proton. Hydroxide ion is the strongest base possible in water. The above equilibrium lies far to the left resulting in a K_b value less than one. Those species we specifically call weak bases ($10^{-14} < K_b < 1$) lie between H_2O and OH^- in base strength. Weak bases are stronger bases than water but are weaker bases than OH^-.

27. In deciding whether a substance is an acid or a base, strong or weak, you should keep in mind a couple ideas:

1. There are only a few common strong acids and strong bases all of which should be memorized. Common strong acids = HCl, HBr, HI, HNO_3, $HClO_4$ and H_2SO_4. Common strong bases = LiOH, NaOH, KOH, RbOH, CsOH, $Ca(OH)_2$, $Sr(OH)_2$ and $Ba(OH)_2$.

2. All other acids and bases are weak and will have K_a and K_b values less than one but greater than K_w (10^{-14}). Reference Table 7.2 for K_a values for some weak acids and Table 7.3 for K_b values for some weak bases. There are too many weak acids and weak bases to memorize them all. Therefore, use the tables of K_a and K_b values to help you identify weak acids and weak bases. Appendix 5 contains more complete tables of K_a and K_b values.

a. weak acid ($K_a = 4.0 \times 10^{-4}$)
b. strong acid
c. weak base ($K_a = 4.38 \times 10^{-4}$)
d. strong base
e. weak base ($K_b = 1.8 \times 10^{-5}$)
f. weak acid ($K_a = 7.2 \times 10^{-4}$)
g. weak acid ($K_a = 1.8 \times 10^{-4}$)
h. strong base
i. strong acid

Autoionization of Water and pH Scale

29. At 25 °C, the relationship: $[H^+][OH^-] = K_w = 1.0 \times 10^{-14}$ always holds for aqueous solutions. When $[H^+]$ is greater than 1.0×10^{-7} *M*, then the solution is acidic; when $[H^+]$ is less than 1.0×10^{-7} *M*, then the solution is basic; when $[H^+] = 1.0 \times 10^{-7}$ *M*, then the solution is neutral. In terms of $[OH^-]$, an acidic solution has $[OH^-] < 1.0 \times 10^{-7}$ *M*, a basic solution has $[OH^-] > 1.0 \times 10^{-7}$ *M* and a neutral solution has $[OH^-] = 1.0 \times 10^{-7}$ *M*.

a. $[OH^-] = \frac{K_w}{[H^+]} = \frac{1.0 \times 10^{-14}}{1.0 \times 10^{-7}} = 1.0 \times 10^{-7}\ M$; The solution is neutral.

$pH = -\log [H^+]$; $pOH = -\log [OH^-]$; At 25°C, $pH + pOH = 14.00$

$pH = -\log [H^+] = -\log (1.0 \times 10^{-7}) = 7.00$; $pOH = 14.00 - pH = 14.00 - 7.00 = 7.00$

b. $[OH^-] = \frac{1.0 \times 10^{-14}}{6.7 \times 10^{-4}} = 1.5 \times 10^{-11}\ M$; The solution is acidic.

$pH = -\log (6.7 \times 10^{-4}) = 3.17$; $pOH = 14.00 - 3.17 = 10.83$

c. $[OH^-] = \frac{1.0 \times 10^{-14}}{1.9 \times 10^{-11}} = 5.3 \times 10^{-4}\ M$; The solution is basic.

$pH = -\log (1.9 \times 10^{-11}) = 10.72$; $pOH = 14.00 - 10.72 = 3.28$

d. $[OH^-] = \frac{1.0 \times 10^{-14}}{2.3} = 4.3 \times 10^{-15}\ M$; The solution is acidic.

$pH = -\log (2.3) = -0.36$; $pOH = 14.00 - (-0.36) = 14.36$

Solutions of Acids

31. Strong acids are assumed to completely dissociate in water, e.g., $HCl(aq) + H_2O(l) \rightarrow H_3O^+(aq) + Cl^-(aq)$ or $HCl(aq) \rightarrow H^+(aq) + Cl^-(aq)$.

a. A 0.10 M HCl solution gives 0.10 M H^+ and 0.10 M Cl^- since HCl completely dissociates. The amount of H^+ from H_2O will be insignificant.

$pH = -\log [H^+] = -\log (0.10) = 1.00$

b. 5.0 M H^+ is produced when 5.0 M $HClO_4$ completely dissociates. The amount of H^+ from H_2O will be insignificant. $pH = -\log (5.0) = -0.70$ (Negative pH values just indicate very concentrated acid solutions.)

c. $1.0 \times 10^{-11}\ M$ H^+ is produced when $1.0 \times 10^{-11}\ M$ HI completely dissociates. If you take the negative log of 1.0×10^{-11} this gives pH = 11.00. This is impossible! We dissolved an acid in water and got a basic pH. What we must consider in this problem is that water by itself donates $1.0 \times 10^{-7}\ M$ H^+. We can normally ignore the small amount of H^+ from H_2O except when we have a very dilute solution of an acid (as in the case here). Therefore, the pH is that of neutral water (pH = 7.00) since the amount of HI present is insignificant.

33. a. Major species = $H^+(aq)$, $Cl^-(aq)$ and $H_2O(l)$ (HCl is a strong acid.) $[H^+] = 0.250\ M$

$pH = -\log [H^+] = -\log(0.250) = 0.602$

b. $H^+(aq)$, $Br^-(aq)$ and $H_2O(l)$ (HBr is a strong acid.) pH = 0.602

c. $H^+(aq)$, $ClO_4^-(aq)$ and $H_2O(l)$ ($HClO_4$ is a strong acid.) pH = 0.602

d. $H^+(aq)$, $NO_3^-(aq)$ and $H_2O(l)$ (HNO_3 is a strong acid.) pH = 0.602

e. HNO_2 ($K_a = 4.0 \times 10^{-4}$) and H_2O ($K_a = K_w = 1.0 \times 10^{-14}$) are the major species. HNO_2 is much stronger acid than H_2O so it is the major source of H^+. However, HNO_2 is a weak acid ($K_a < 1$) so it only partially dissociates in water. We must solve an equilibrium problem to determine $[H^+]$. In the Solutions Guide, we will summarize the initial, change and equilibrium concentrations into one table called the ICE table. Solving the weak acid problem:

$$HNO_2(aq) \rightleftharpoons H^+(aq) + NO_2^-(aq)$$

	$HNO_2(aq)$		$H^+(aq)$	$NO_2^-(aq)$
Initial	0.250 M		~0	0
	x mol/L HNO_2 dissociates to reach equilibrium			
Change	$-x$	→	$+x$	$+x$
Equil.	$0.250 - x$		x	x

$$K_a = \frac{[H^+][NO_2^-]}{[HNO_2]} = 4.0 \times 10^{-4} = \frac{x^2}{0.250 - x}$$; If we assume $x \ll 0.250$, then:

$$4.0 \times 10^{-4} \approx \frac{x^2}{0.250}, \ x = \sqrt{4.0 \times 10^{-4}(0.250)} = 0.010\ M$$

We must check the assumption: $\frac{x}{0.250} \times 100 = \frac{0.010}{0.250} \times 100 = 4.0\%$

All the assumptions are good. The H^+ contribution from water ($10^{-7}\ M$) is negligible and x is small compared to 0.250 (percent error = 4.0%). If the percent error is less than 5% for an assumption, we will consider it a valid assumption (called the 5% rule). Finishing the problem: $x = 0.010\ M = [H^+]$; pH = -log $[H^+]$ = -log(0.010) = 2.00

f. CH_3CO_2H ($K_a = 1.8 \times 10^{-5}$) and H_2O ($K_a = K_w = 1.0 \times 10^{-14}$) are the major species. CH_3CO_2H is the major source of H^+. Solving the weak acid problem:

$$CH_3CO_2H \rightleftharpoons H^+ + CH_3CO_2^-$$

	CH_3CO_2H		H^+	$CH_3CO_2^-$
Initial	0.250 M		~0	0
	x mol/L CH_3CO_2H dissociates to reach equilibrium			
Change	$-x$	→	$+x$	$+x$
Equil.	$0.250 - x$		x	x

$$K_a = \frac{[H^+][CH_3CO_2^-]}{[CH_3CO_2H]} = 1.8 \times 10^{-5} = \frac{x^2}{0.250 - x} \approx \frac{x^2}{0.250}$$ (assuming $x \ll 0.250$)

$x = 2.1 \times 10^{-3}\ M$; Checking assumption: $\frac{2.1 \times 10^{-3}}{0.250} \times 100 = 0.84\%$. Assumptions good.

$[H^+] = x = 2.1 \times 10^{-3}\ M$; pH = -log (2.1×10^{-3}) = 2.68

g. HCN ($K_a = 6.2 \times 10^{-10}$) and H_2O are the major species. HCN is the major source of H^+.

	HCN	$\rightleftharpoons$	H^+	+	CN^-
Initial	0.250 *M*		~0		0
	x mol/L HCN dissociates to reach equilibrium				
Change	$-x$	$\rightarrow$	$+x$		$+x$
Equil.	$0.250 - x$		x		x

$$K_a = 6.2 \times 10^{-10} = \frac{[H^+][CN^-]}{[HCN]} = \frac{x^2}{0.250 - x} \approx \frac{x^2}{0.250} \quad (\text{assuming } x \ll 0.250)$$

$x = [H^+] = 1.2 \times 10^{-5}$ *M*; Checking assumption: *x* is 4.8×10^{-3}% of 0.250

Assumptions good. pH = $-\log(1.2 \times 10^{-5}) = 4.92$

35. a. Major species: $HC_2H_3O_2$ ($K_a = 1.8 \times 10^{-5}$) and water; Major source of $H^+ = HC_2H_3O_2$. Since K_a for $HC_2H_3O_2$ is less than one, then $HC_2H_3O_2$ is a weak acid and we must solve an equilibrium problem to determine $[H^+]$. The set-up is:

	$HC_2H_3O_2(aq)$	$\rightleftharpoons$	$H^+(aq)$	+	$C_2H_3O_2^-(aq)$
Initial	0.20 *M*		~0		0
	x mol/L $HC_2H_3O_2$ dissociates to reach equilibrium				
Change	$-x$	$\rightarrow$	$+x$		$+x$
Equil.	$0.20 - x$		x		x

$$K_a = 1.8 \times 10^{-5} = \frac{[H^+][C_2H_3O_2^-]}{[HC_2H_3O_2]} = \frac{x^2}{0.20 - x} \approx \frac{x^2}{0.20} \quad (\text{assuming } x \ll 0.20)$$

$x = [H^+] = 1.9 \times 10^{-3}$ *M*

We have made two assumptions which we must check.

1. $0.20 - x \approx 0.20$

   $(x/0.20) \times 100 = (1.9 \times 10^{-3}/0.20) \times 100 = 0.95\%$. Good assumption (1% error). If the percent error in the assumption is < 5%, then the assumption is valid.

2. Acetic acid is the major source of H^+, i.e., we can ignore 10^{-7} *M* H^+ already present in neutral H_2O.

   $[H^+]$ from $HC_2H_3O_2 = 1.9 \times 10^{-3} \gg 10^{-7}$; This assumption is valid.

In future problems we will always begin the problem solving process by making these assumptions and we will always check them. However, we may not explicitly state that the assumptions are valid. We will always state when the assumptions are not valid and we have to use other techniques to solve the problem. Remember, anytime we make an

assumption, we must check its validity before the solution to the problem is complete. Answering the question:

$$[H^+] = [C_2H_3O_2^-] = 1.9 \times 10^{-3}\ M;\ \ [OH^-] = 5.3 \times 10^{-12}\ M$$

$$[HC_2H_3O_2] = 0.20 - x = 0.198 \approx 0.20\ M;\ \ pH = -\log[H^+] = -\log(1.9 \times 10^{-3}) = 2.72$$

b. HNO_2 ($K_a = 4.0 \times 10^{-4}$) is the dominant producer of H^+. Solving the weak acid problem:

	HNO_2	$\rightleftharpoons$	H^+	+	NO_2^-	$K_a = 4.0 \times 10^{-4}$
Initial	1.5 *M*		~0		0	
	x mol/L HNO_2 dissociates to reach equilibrium					
Change	$-x$	$\rightarrow$	$+x$		$+x$	
Equil.	$1.5 - x$		x		x	

$$K_a = 4.0 \times 10^{-4} = \frac{[H^+][NO_2^-]}{[HNO_2]} = \frac{x^2}{1.5 - x} \approx \frac{x^2}{1.5}\ \ \text{(assuming } x << 1.5)$$

$x = [H^+] = 2.4 \times 10^{-2}\ M$; Assumptions good: $10^{-7} << 2.4 \times 10^{-2} << 1.5$

$$[H^+] = [NO_2^-] = 2.4 \times 10^{-2}\ M;\ \ [OH^-] = 4.2 \times 10^{-13}\ M$$

$$[HNO_2] = 1.5 - x = 1.48 \approx 1.5\ M;\ \ pH = -\log(2.4 \times 10^{-2}) = 1.62$$

c. This is a weak acid in water. Solving the weak acid problem:

	HF	$\rightleftharpoons$	H^+	+	F^-	$K_a = 7.2 \times 10^{-4}$
Initial	0.020 *M*		~0		0	
	x mol/L HF dissociates to reach equilibrium					
Change	$-x$	$\rightarrow$	$+x$		$+x$	
Equil.	$0.020 - x$		x		x	

$$K_a = 7.2 \times 10^{-4} = \frac{[H^+][F^-]}{[HF]} = \frac{x^2}{0.020 - x} \approx \frac{x^2}{0.020}\ \ \text{(assuming } x << 0.020)$$

$x = [H^+] = 3.8 \times 10^{-3}\ M$; Check assumptions: $\frac{x}{0.020} \times 100 = \frac{3.8 \times 10^{-3}}{0.020} \times 100 = 19\%$

The assumption $x << 0.020$ is not good (x is more than 5% of 0.020). We must solve $x^2/(0.020 - x) = 7.2 \times 10^{-4}$ exactly by using either the quadratic formula or by the method of successive approximations (see Appendix 1.4 of text). Using successive approximations, we let 0.016 *M* be a new approximation for [HF]. That is, in the denominator try $x = 0.0038$ (the value of x we calculated making the normal assumption), so $0.020 - 0.0038 = 0.016$, then solve for a new value of x in the numerator.

$$\frac{x^2}{0.020 - x} \approx \frac{x^2}{0.016} = 7.2 \times 10^{-4},\ \ x = 3.4 \times 10^{-3}$$

We use this new value of x to further refine our estimate of [HF], i.e., $0.020 - x = 0.020 - 0.0034 = 0.0166$ (carrying an extra significant figure).

$$\frac{x^2}{0.020 - x} \approx \frac{x^2}{0.0166} = 7.2 \times 10^{-4},\ x = 3.5 \times 10^{-3}$$

We repeat, until we get a self-consistent answer. This would be the same answer we would get solving exactly using the quadratic equation. In this case it is: $x = 3.5 \times 10^{-3}$

So: $[H^+] = [F^-] = x = 3.5 \times 10^{-3}\ M$; $[OH^-] = K_w/[H^+] = 2.9 \times 10^{-12}\ M$

$$[HF] = 0.020 - x = 0.020 - 0.0035 = 0.017\ M;\ pH = 2.46$$

Note: When the 5% assumption fails, use whichever method you are most comfortable with to solve exactly. The method of successive approximations is probably fastest when the percent error is less than ~25% (unless you have a graphing calculator).

37. Major species: HIO_3, H_2O; Major source of H^+: HIO_3 (a weak acid, $K_a = 0.17$)

	HIO_3	$\rightleftharpoons$	H^+	+	IO_3^-
Initial	0.010 M		~0		0
	x mol/L HIO_3 dissociates to reach equilibrium				
Change	$-x$	→	$+x$		$+x$
Equil.	$0.010 - x$		x		x

$$K_a = 0.17 = \frac{[H^+][IO_3^-]}{[HIO_3]} = \frac{x^2}{0.010 - x} \approx \frac{x^2}{0.010},\ x = 0.041;\quad \text{Check assumption.}$$

Assumption is horrible. (x is more than 400% of 0.010). When the assumption is this poor, it is generally quickest to solve exactly using the quadratic formula (see Appendix 1.4 in text). Using the quadratic formula and carrying extra significant figures:

$$0.17 = \frac{x^2}{0.010 - x},\ x^2 = 0.17(0.010 - x),\ x^2 + 0.17x - 1.7 \times 10^{-3} = 0$$

$$x = \frac{-0.17 \pm [(0.17)^2 - 4(1)(-1.7 \times 10^{-3})]^{1/2}}{2(1)} = \frac{-0.17 \pm 0.189}{2},\ x = 9.5 \times 10^{-3}\ M$$

(x must be positive)

$x = 9.5 \times 10^{-3}\ M = [H^+]$; $pH = -\log(9.5 \times 10^{-3}) = 2.02$

39. This is a weak acid in water. We must solve a weak acid problem. Let HBz = $C_6H_5CO_2H$.

$$0.56\text{ g HBz} \times \frac{1\text{ mol HBz}}{122.1\text{ g}} = 4.6 \times 10^{-3}\text{ mol};\ [HBz]_o = 4.6 \times 10^{-3}\ M$$

	HBz	$\rightleftharpoons$	H^+	+	Bz^-
Initial	$4.6 \times 10^{-3}\ M$		~0		0
	x mol/L HBz dissociates to reach equilibrium				
Change	$-x$	→	$+x$		$+x$
Equil.	$4.6 \times 10^{-3} - x$		x		x

$$K_a = 6.4 \times 10^{-5} = \frac{[H^+][Bz^-]}{[HBz]} = \frac{x^2}{4.6 \times 10^{-3} - x} \approx \frac{x^2}{4.6 \times 10^{-3}}$$

$x = [H^+] = 5.4 \times 10^{-4}$; **Check assumptions:** $\frac{x}{4.6 \times 10^{-3}} \times 100 = \frac{5.4 \times 10^{-4}}{4.6 \times 10^{-3}} \times 100 = 12\%$

Assumption is not good (x is 12% of 4.6×10^{-3}). When assumption(s) fail, we must solve exactly using the quadratic formula or the method of successive approximations (see Appendix 1.4 of text). Using successive approximations:

$$\frac{x^2}{(4.6 \times 10^{-4} - 5.4 \times 10^{-4})} = 6.4 \times 10^{-5},\ x = 5.1 \times 10^{-4}$$

$$\frac{x^2}{(4.6 \times 10^{-3} - 5.1 \times 10^{-4})} = 6.4 \times 10^{-5},\ x = 5.1 \times 10^{-4}\ M \text{ (consistent answer)}$$

So: $x = [H^+] = [Bz^-] = [C_6H_5CO_2^-] = 5.1 \times 10^{-4}\ M$

$[HBz] = [C_6H_5CO_2H] = 4.6 \times 10^{-3} - x = 4.1 \times 10^{-3}\ M$

$pH = -\log(5.1 \times 10^{-4}) = 3.29$; $pOH = 14.00 - pH = 10.71$; $[OH^-] = 10^{-10.71} = 1.9 \times 10^{-11}\ M$

41. $20.0 \text{ mL glacial acetic acid} \times \frac{1.05 \text{ g}}{\text{mL}} \times \frac{1 \text{ mol}}{60.05 \text{ g}} = 0.350 \text{ mol } HC_2H_3O_2$

Initial concentration of $HC_2H_3O_2 = \frac{0.350 \text{ mol}}{0.2500 \text{ L}} = 1.40\ M$

	$HC_2H_3O_2$	$\rightleftharpoons$	H^+	+	$C_2H_3O_2^-$	$K_a = 1.8 \times 10^{-5}$
Initial	$1.40\ M$		~0		0	
	x mol/L $HC_2H_3O_2$ dissociates to reach equilibrium					
Change	$-x$	→	$+x$		$+x$	
Equil.	$1.40 - x$		x		x	

$$K_a = 1.8 \times 10^{-5} = \frac{[H^+][C_2H_3O_2^-]}{[HC_2H_3O_2]} = \frac{x^2}{1.40 - x} \approx \frac{x^2}{1.40}$$

$x = [H^+] = 5.0 \times 10^{-3}\ M$; $pH = 2.30$ Assumptions good (x is 0.36% of 1.40).

43. a. HCl is a strong acid. It will produce $0.10\ M\ H^+$. HOCl is a weak acid. Let's consider the equilibrium:

	HOCl	$\rightleftharpoons$	H^+	+	OCl^-	$K_a = 3.5 \times 10^{-8}$
Initial	$0.10\ M$		$0.10\ M$		0	
	x mol/L HOCl dissociates to reach equilibrium					
Change	$-x$	→	$+x$		$+x$	
Equil.	$0.10 - x$		$0.10 + x$		x	

$$K_a = 3.5 \times 10^{-8} = \frac{[H^+][OCl^-]}{[HOCl]} = \frac{(0.10 + x)(x)}{0.10 - x} \approx x,\ x = 3.5 \times 10^{-8}\ M$$

Assumptions are great (x is 3.5×10^{-5}% of 0.10). We are really assuming that HCl is the only important source of H^+, which it is. The $[H^+]$ contribution from HOCl, x, is negligible. Therefore, $[H^+] = 0.10\ M$; pH = 1.00.

b. HNO_3 is a strong acid, giving an initial concentration of H^+ equal to 0.050 *M*. Consider the equilibrium:

	$HC_2H_3O_2$	$\rightleftharpoons$	H^+	+ $C_2H_3O_2^-$	$K_a = 1.8 \times 10^{-5}$
Initial	0.50 *M*		0.050 *M*	0	
	x mol/L $HC_2H_3O_2$ dissociates to reach equilibrium				
Change	$-x$	$\rightarrow$	$+x$	$+x$	
Equil.	$0.50 - x$		$0.050 + x$	x	

$$K_a = 1.8 \times 10^{-5} = \frac{[H^+][C_2H_3O_2^-]}{[HC_2H_3O_2]} = \frac{(0.050 + x)x}{(0.50 - x)} \approx \frac{0.050\,x}{0.50}$$

$x = 1.8 \times 10^{-4}$; Assumptions are good (well within the 5% rule).

$[H^+] = 0.050 + x = 0.050\ M$ and pH = 1.30

45.

	HF	$\rightleftharpoons$	H^+	+ F^-
Initial	0.100 *M*		~0	0
	x mol/L HF dissociates to reach equilibrium			
Change	$-x$	$\rightarrow$	$+x$	$+x$
Equil.	$0.100 - x$		x	x

$$K_a = \frac{[H^+][F^-]}{[HF]} = \frac{x^2}{0.100 - x};\ x = [H^+] = [F^-] = 0.081 \times (0.100\ M) = 8.1 \times 10^{-3}\ M$$

$$[HF] = 0.100 - 8.1 \times 10^{-3} = 0.092\ M;\ K_a = \frac{(8.1 \times 10^{-3})^2}{0.092} = 7.1 \times 10^{-4}$$

47. In all parts of this problem, acetic acid ($HC_2H_3O_2$) is the best weak acid present. We must solve a weak acid problem.

a.

	$HC_2H_3O_2$	$\rightleftharpoons$	H^+	+ $C_2H_3O_2^-$
Initial	0.50 *M*		~0	0
	x mol/L $HC_2H_3O_2$ dissociates to reach equilibrium			
Change	$-x$	$\rightarrow$	$+x$	$+x$
Equil.	$0.50 - x$		x	x

$$K_a = 1.8 \times 10^{-5} = \frac{[H^+][C_2H_3O_2^-]}{[HC_2H_3O_2]} = \frac{x^2}{0.50 - x} \approx \frac{x^2}{0.50}$$

$x = [H^+] = [C_2H_3O_2^-] = 3.0 \times 10^{-3}\ M$ Assumptions good.

$$\text{Percent dissociation} = \frac{[H^+]}{[HC_2H_3O_2]_0} \times 100 = \frac{3.0 \times 10^{-3}}{0.50} \times 100 = 0.60\%$$

b. The set-up for solutions b and c are similar to solution a except the final equation is slightly different, reflecting the new concentration of $HC_2H_3O_2$.

$$K_a = 1.8 \times 10^{-5} = \frac{x^2}{0.050 - x} \approx \frac{x^2}{0.050}$$

$x = [H^+] = [C_2H_3O_2^-] = 9.5 \times 10^{-4}\ M$ Assumptions good.

$$\%\ \text{dissociation} = \frac{9.5 \times 10^{-4}}{0.050} \times 100 = 1.9\%$$

c. $$K_a = 1.8 \times 10^{-5} = \frac{x^2}{0.0050 - x} \approx \frac{x^2}{0.0050}$$

$x = [H^+] = [C_2H_3O_2^-] = 3.0 \times 10^{-4}\ M$; Check assumptions.

Assumption that x is negligible is borderline (6.0% error). We should solve exactly. Using the method of successive approximations (see Appendix 1.4 of text):

$$1.8 \times 10^{-5} = \frac{x^2}{0.0050 - 3.0 \times 10^{-4}} = \frac{x^2}{0.0047},\ x = 2.9 \times 10^{-4}$$

Next trial also gives $x = 2.9 \times 10^{-4}$.

$$\%\ \text{dissociation} = \frac{2.9 \times 10^{-4}}{5.0 \times 10^{-3}} \times 100 = 5.8\%$$

d. As we dilute a solution, all concentrations are decreased. Dilution will shift the equilibrium to the side with the greater number of particles. For example, suppose we double the volume of an equilibrium mixture of a weak acid by adding water, then:

$$Q = \frac{\left(\frac{[H^+]_{eq}}{2}\right)\left(\frac{[X^-]_{eq}}{2}\right)}{\left(\frac{[HX]_{eq}}{2}\right)} = \frac{1}{2} K_a$$

$Q < K_a$, so the equilibrium shifts to the right or towards a greater percent dissociation.

e. $[H^+]$ depends on the initial concentration of weak acid and on how much weak acid dissociates. For solutions a-c the initial concentration of acid decreases more rapidly than the percent dissociation increases. Thus, $[H^+]$ decreases.

49. pH = 2.77, $[H^+] = 10^{-2.77} = 1.7 \times 10^{-3}\ M$

	HOCN	$\rightleftharpoons$	H^+	+	OCN^-
Initial	0.0100		~0		0
Equil.	$0.0100 - x$		x		x

$x = [H^+] = [OCN^-] = 1.7 \times 10^{-3}\ M$; $[HOCN] = 0.0100 - x = 0.0100 - 0.0017 = 0.0083\ M$

$$K_a = \frac{[H^+][OCN^-]}{[HOCN]} = \frac{(1.7 \times 10^{-3})^2}{8.3 \times 10^{-3}} = 3.5 \times 10^{-4}$$

51. Major species: HCOOH and H_2O; Major source of H^+: HCOOH

	HCOOH	$\rightleftharpoons$	H^+	+	$HCOO^-$	
Initial	C		~0		0	where $C = [HCOOH]_o$
	x mol/L HCOOH dissociates to reach equilibrium					
Change	$-x$	$\rightarrow$	$+x$		$+x$	
Equil.	$C - x$		x		x	

$$K_a = 1.8 \times 10^{-4} = \frac{[H^+][HCOO^-]}{[HCOOH]} = \frac{x^2}{C - x} \text{ where } x = [H^+]$$

$$1.8 \times 10^{-4} = \frac{[H^+]^2}{C - [H^+]}; \text{ Since pH} = 2.70, \text{ then: } [H^+] = 10^{-2.70} = 2.0 \times 10^{-3}\ M$$

$$1.8 \times 10^{-4} = \frac{(2.0 \times 10^{-3})^2}{C - (2.0 \times 10^{-3})}, \ C - (2.0 \times 10^{-3}) = \frac{4.0 \times 10^{-6}}{1.8 \times 10^{-4}}, \ C = 2.4 \times 10^{-2}\ M$$

A 0.024 *M* formic acid solution will have pH = 2.70.

53. The reactions are:

$H_3AsO_4 \rightleftharpoons H^+ + H_2AsO_4^-$ $\quad K_{a_1} = 5 \times 10^{-3}$

$H_2AsO_4^- \rightleftharpoons H^+ + HAsO_4^{2-}$ $\quad K_{a_2} = 8 \times 10^{-8}$

$HAsO_4^{2-} \rightleftharpoons H^+ + AsO_4^{3-}$ $\quad K_{a_3} = 6 \times 10^{-10}$

We will deal with the reactions in order of importance, beginning with the largest K_a, K_{a_1}.

$$H_3AsO_4 \rightleftharpoons H^+ + H_2AsO_4^- \quad K_{a_1} = 5 \times 10^{-3} = \frac{[H^+][H_2AsO_4^-]}{[H_3AsO_4]}$$

	H_3AsO_4	H^+	$H_2AsO_4^-$
Initial	0.20 *M*	~0	0
Equil.	$0.20 - x$	x	x

$$5 \times 10^{-3} = \frac{x^2}{0.20 - x} \approx \frac{x^2}{0.20}, \ x = 3 \times 10^{-2}\ M; \text{ Assumption fails the 5\% rule.}$$

Solving by the method of successive approximations:

$5 \times 10^{-3} = x^2 / (0.20 - 0.03)$, $x = 3 \times 10^{-2}$ (consistent answer)

$[H^+] = [H_2AsO_4^-] = 3 \times 10^{-2}\ M$; $[H_3AsO_4] = 0.20 - 0.03 = 0.17\ M$

Since $K_{a_2} = \dfrac{[H^+][HAsO_4^{2-}]}{[H_2AsO_4^-]} = 8 \times 10^{-8}$ is much smaller than the K_{a_1} value, then very little of $H_2AsO_4^-$ (and $HAsO_4^{2-}$) dissociates as compared to H_3AsO_4. Therefore, $[H^+]$ and $[H_2AsO_4^-]$ will not change significantly by the K_{a_2} reaction. Using the previously calculated concentrations of H^+ and $H_2AsO_4^-$ to calculate the concentration of $HAsO_4^{2-}$:

$$8 \times 10^{-8} = \frac{(3 \times 10^{-2})[HAsO_4^{2-}]}{3 \times 10^{-2}}, \quad [HAsO_4^{2-}] = 8 \times 10^{-8}\ M$$

Assumption that the K_{a_2} reaction does not change $[H^+]$ and $[H_2AsO_4^-]$ is good. We repeat the process using K_{a_3} to get $[AsO_4^{3-}]$.

$$K_{a_3} = 6 \times 10^{-10} = \frac{[H^+][AsO_4^{3-}]}{[HAsO_4^{2-}]} = \frac{(3 \times 10^{-2})[AsO_4^{3-}]}{(8 \times 10^{-8})}$$

$[AsO_4^{3-}] = 1.6 \times 10^{-15} \approx 2 \times 10^{-15}\ M$ Assumption good.

So in $0.20\ M$ analytical concentration of H_3AsO_4:

$[H_3AsO_4] = 0.17\ M$; $[H^+] = [H_2AsO_4^-] = 3 \times 10^{-2}\ M$;

$[HAsO_4^{2-}] = 8 \times 10^{-8}\ M$; $[AsO_4^{3-}] = 2 \times 10^{-15}\ M$; $[OH^-] = K_w/[H^+] = 3 \times 10^{-13}\ M$

55. The dominant H^+ producer is the strong acid H_2SO_4. A $2.0\ M\ H_2SO_4$ solution produces $2.0\ M\ HSO_4^-$ and $2.0\ M\ H^+$. However, HSO_4^- is a weak acid which could also add H^+ to the solution.

	HSO_4^-	$\rightleftharpoons$	H^+	+	SO_4^{2-}
Initial	$2.0\ M$		$2.0\ M$		0
	x mol/L HSO_4^- dissociates to reach equilibrium				
Change	$-x$	$\rightarrow$	$+x$		$+x$
Equil.	$2.0 - x$		$2.0 + x$		x

$$K_{a_2} = 1.2 \times 10^{-2} = \frac{[H^+][SO_4^{2-}]}{[HSO_4^-]} = \frac{(2.0 + x)(x)}{2.0 - x} \approx \frac{2.0\,(x)}{2.0}, \quad x = 1.2 \times 10^{-2}\ M$$

Since x is 0.60% of 2.0, then the assumption is valid by the 5% rule. The amount of additional H^+ from HSO_4^- is 1.2×10^{-2}. The total amount of H^+ present is:

$[H^+] = 2.0 + 1.2 \times 10^{-2} = 2.0\ M$; $pH = -\log(2.0) = -0.30$

Note: In this problem, H^+ from HSO_4^- could have been ignored. However, this is not usually the case, especially in more dilute solutions of H_2SO_4.

Solutions of Bases

57. NO_3^-: $K_b << K_w$ since HNO_3 is a strong acid. All conjugate bases of strong acids have no base strength in water. H_2O: $K_b = K_w = 1.0 \times 10^{-14}$; NH_3: $K_b = 1.8 \times 10^{-5}$; C_5H_5N: $K_b = 1.7 \times 10^{-9}$

    Base strength = $NH_3 > C_5H_5N > H_2O > NO_3^-$ (As K_b increases, base strength increases.)

59. a. $C_2H_5NH_2$ b. $C_2H_5NH_2$ c. OH^- d. $C_2H_5NH_2$

    The base with the largest K_b value is the strongest base (K_b for $C_2H_5NH_2 = 5.6 \times 10^{-4}$ and K_b for $C_6H_5NH_2 = 3.8 \times 10^{-10}$). OH^- is the strongest base possible in water.

61. $NaOH(aq) \rightarrow Na^+(aq) + OH^-(aq)$; NaOH is a strong base which completely dissociates into Na^+ and OH^-. The initial concentration of NaOH will equal the concentration of OH^- donated by NaOH.

    a. $[OH^-] = 0.10\ M$; pOH = -log[OH^-] = -log(0.10) = 1.00

    pH = 14.00 - pOH = 14.00 - 1.00 = 13.00

    Note that H_2O is also present but the amount of OH^- produced by H_2O will be insignificant as compared to 0.10 *M* OH^- produced from the NaOH.

    b. The $[OH^-]$ concentration donated by the NaOH is $1.0 \times 10^{-10}\ M$. Water by itself donates $1.0 \times 10^{-7}\ M$. In this problem, water is the major OH^- contributor and $[OH^-] = 1.0 \times 10^{-7}\ M$.

    pOH = -log (1.0×10^{-7}) = 7.00; pH = 14.00 - 7.00 = 7.00

    c. $[OH^-] = 2.0\ M$; pOH = -log (2.0) = -0.30; pH = 14.00 - (-0.30) = 14.30

63. pH = 10.50; pOH = 14.00 - 10.50 = 3.50; $[OH^-] = 10^{-3.50} = 3.2 \times 10^{-4}\ M$

    $Ba(OH)_2(aq) \rightarrow Ba^{2+}(aq) + 2\ OH^-(aq)$; $Ba(OH)_2$ donates two mol OH^- per mol $Ba(OH)_2$.

    $$[Ba(OH)_2] = 3.2 \times 10^{-4}\ M\ OH^- \times \left(\frac{1\ M\ Ba(OH)_2}{2\ M\ OH^-}\right) = 1.6 \times 10^{-4}\ M\ Ba(OH)_2$$

    A $1.6 \times 10^{-4}\ M\ Ba(OH)_2$ solution will produce a pH = 10.50 solution.

65. a. These are all solutions of weak bases in water. We must solve the weak base equilibrium problem.

	$C_2H_5NH_2(aq)$ + $H_2O(l)$	$\rightleftharpoons$	$C_2H_5NH_3^+(aq)$	+ $OH^-(aq)$	$K_b = 5.6 \times 10^{-4}$
Initial	0.20 *M*		0	~0	
	x mol/L $C_2H_5NH_2$ reacts with H_2O to reach equilibrium				
Change	$-x$	$\rightarrow$	$+x$	$+x$	
Equil.	$0.20 - x$		x	x	

$$K_b = \frac{[C_2H_5NH_3^+][OH^-]}{[C_2H_5NH_2]} = \frac{x^2}{0.20 - x} \approx \frac{x^2}{0.20} \quad \text{(assuming } x << 0.20\text{)}$$

$x = 1.1 \times 10^{-2}$; **Checking assumption:** $\dfrac{1.1 \times 10^{-2}}{0.20} \times 100 = 5.5\%$

Assumption fails the 5% rule. We must solve exactly using either the quadratic equation or the method of successive approximations (see Appendix 1.4 of the text). Using successive approximations and carrying extra significant figures:

$$\frac{x^2}{0.20 - 0.011} = \frac{x^2}{0.189} = 5.6 \times 10^{-4},\ x = 1.0 \times 10^{-2}\ M \text{ (consistent answer)}$$

$$x = [OH^-] = 1.0 \times 10^{-2}\ M;\ [H^+] = \frac{K_w}{[OH^-]} = \frac{1.0 \times 10^{-14}}{1.0 \times 10^{-2}} = 1.0 \times 10^{-12}\ M;\ pH = 12.00$$

b. $(C_2H_5)_2NH + H_2O \rightleftharpoons (C_2H_5)_2NH_2^+ + OH^-$ $K_b = 1.3 \times 10^{-3}$

	$(C_2H_5)_2NH$		$(C_2H_5)_2NH_2^+$	OH^-
Initial	$0.20\ M$		0	~0
	x mol/L $(C_2H_5)_2NH$ reacts with H_2O to reach equilibrium			
Change	$-x$	$\rightarrow$	$+x$	$+x$
Equil.	$0.20 - x$		x	x

$$K_b = 1.3 \times 10^{-3} = \frac{[(C_2H_5)_2NH_2^+]\,[OH^-]}{[(C_2H_5)_2NH)]} = \frac{x^2}{0.20 - x} \approx \frac{x^2}{0.20} \text{ (assuming } x \ll 0.20)$$

$x = 1.6 \times 10^{-2}$; **Assumption is bad (x is 8.0% of 0.20).**

Using successive approximations:

$$\frac{x^2}{0.20 - 0.016} = \frac{x^2}{0.184} = 1.3 \times 10^{-3},\ x = 1.55 \times 10^{-2} \text{ (carry extra significant figure)}$$

$$\frac{x^2}{0.185} = 1.3 \times 10^{-3},\ x = 1.55 \times 10^{-2} \text{ (consistent answer)}$$

$[OH^-] = x = 1.55 \times 10^{-2}\ M$; $[H^+] = 6.45 \times 10^{-13}\ M$; **To correct significant figures:**

$$[OH^-] = 1.6 \times 10^{-2}\ M;\ [H^+] = 6.5 \times 10^{-13}\ M;\ pH = -\log [H^+] = 12.19$$

c. $(C_2H_5)_3N + H_2O \rightleftharpoons (C_2H_5)_3NH^+ + OH^-$ $K_b = 4.0 \times 10^{-4}$

	$(C_2H_5)_3N$		$(C_2H_5)_3NH^+$	OH^-
Initial	$0.20\ M$		0	~0
	x mol/L of $(C_2H_5)_3N$ reacts with H_2O to reach equilibrium			
Change	$-x$	$\rightarrow$	$+x$	$+x$
Equil.	$0.20 - x$		x	x

$$K_b = 4.0 \times 10^{-4} = \frac{[(C_2H_5)_3NH^+][OH^-]}{[(C_2H_5)_3N]} = \frac{x^2}{0.20 - x} \approx \frac{x^2}{0.20},\ x = [OH^-] = 8.9 \times 10^{-3}\ M$$

Assumptions good (x is 4.5% of 0.20). $[OH^-] = 8.9 \times 10^{-3}\ M$

$$[H^+] = \frac{K_w}{[OH^-]} = \frac{1.0 \times 10^{-14}}{8.9 \times 10^{-3}} = 1.1 \times 10^{-12}\ M;\ pH = 11.96$$

d. $C_6H_5NH_2 + H_2O \rightleftharpoons C_6H_5NH_3^+ + OH^-$ $K_b = 3.8 \times 10^{-10}$

	$C_6H_5NH_2$		$C_6H_5NH_3^+$	OH^-
Initial	0.20 *M*		0	~0
	x mol/L of $C_6H_5NH_2$ reacts with H_2O to reach equilibrium			
Change	$-x$	→	$+x$	$+x$
Equil.	$0.20 - x$		x	x

$K_b = 3.8 \times 10^{-10} = \frac{x^2}{0.20 - x} \approx \frac{x^2}{0.20}$, $x = [OH^-] = 8.7 \times 10^{-6}$ *M*; Assumptions good.

$[H^+] = K_w/[OH^-] = 1.1 \times 10^{-9}$ *M*; pH = 8.96

e. $C_5H_5N + H_2O \rightleftharpoons C_5H_5NH^+ + OH^-$ $K_b = 1.7 \times 10^{-9}$

	C_5H_5N	$C_5H_5NH^+$	OH^-
Initial	0.20 *M*	0	~0
Equil.	$0.20 - x$	x	x

$K_b = 1.7 \times 10^{-9} = \frac{x^2}{0.20 - x} \approx \frac{x^2}{0.20}$, $x = 1.8 \times 10^{-5}$ *M*; Assumptions good.

$[OH^-] = 1.8 \times 10^{-5}$ *M*; $[H^+] = 5.6 \times 10^{-10}$ *M*; pH = 9.25

f. $HONH_2 + H_2O \rightleftharpoons HONH_3^+ + OH^-$ $K_b = 1.1 \times 10^{-8}$

	$HONH_2$	$HONH_3^+$	OH^-
Initial	0.20 *M*	0	~0
Equil.	$0.20 - x$	x	x

$K_b = 1.1 \times 10^{-8} = \frac{x^2}{0.20 - x} \approx \frac{x^2}{0.20}$, $x = [OH^-] = 4.7 \times 10^{-5}$ *M*; Assumptions good.

$[H^+] = 2.1 \times 10^{-10}$ *M*; pH = 9.68

67. $\frac{5.0 \times 10^{-3}\text{ g}}{0.0100\text{ L}} \times \frac{1\text{ mol}}{299.4\text{ g}} = 1.7 \times 10^{-3}$ *M* = [codeine]$_o$; Let cod = codeine, $C_{18}H_{21}NO_3$

Solving the weak base equilibrium problem:

$cod + H_2O \rightleftharpoons codH^+ + OH^-$ $K_b = 10^{-6.05} = 8.9 \times 10^{-7}$

	cod		$codH^+$	OH^-
Initial	1.7×10^{-3} *M*		0	~0
	x mol/L codeine reacts with H_2O to reach equilibrium			
Change	$-x$	→	$+x$	$+x$
Equil.	$1.7 \times 10^{-3} - x$		x	x

$K_b = 8.9 \times 10^{-7} = \frac{x^2}{1.7 \times 10^{-3} - x} \approx \frac{x^2}{1.7 \times 10^{-3}}$, $x = 3.9 \times 10^{-5}$ Assumptions good.

$[OH^-] = 3.9 \times 10^{-5}$ *M*; $[H^+] = K_w/[OH^-] = 2.6 \times 10^{-10}$ *M*; pH = -log $[H^+]$ = 9.59

69. To solve for percent ionization, just solve the weak base equilibrium problem.

a. $NH_3 + H_2O \rightleftharpoons NH_4^+ + OH^-$ $K_b = 1.8 \times 10^{-5}$

Initial	0.10 *M*		0	~0
Equil.	$0.10 - x$		x	x

$$K_b = 1.8 \times 10^{-5} = \frac{x^2}{0.10 - x} \approx \frac{x^2}{0.10},\ x = [OH^-] = 1.3 \times 10^{-3}\ M;\ \text{Assumptions good.}$$

$$\text{Percent ionization} = \frac{[OH^-]}{[NH_3]_o} \times 100 = \frac{1.3 \times 10^{-3}\ M}{0.10\ M} \times 100 = 1.3\%$$

b. $NH_3 + H_2O \rightleftharpoons NH_4^+ + OH^-$

Initial	0.010 *M*		0	~0
Equil.	$0.010 - x$		x	x

$$1.8 \times 10^{-5} = \frac{x^2}{0.010 - x} \approx \frac{x^2}{0.010},\ x = [OH^-] = 4.2 \times 10^{-4}\ M;\ \text{Assumptions good.}$$

$$\text{Percent ionization} = \frac{4.2 \times 10^{-4}}{0.010} \times 100 = 4.2\%$$

Note: For the same base, the percent ionization increases as the initial concentration of base decreases.

c. $CH_3NH_2 + H_2O \rightleftharpoons CH_3NH_3^+ + OH^-$ $K_b = 4.38 \times 10^{-4}$

Initial	0.10 *M*		0	~0
Equil.	$0.10 - x$		x	x

$$4.38 \times 10^{-4} = \frac{x^2}{0.10 - x} \approx \frac{x^2}{0.10},\ x = 6.6 \times 10^{-3};$$ Assumption fails the 5% rule (x is 6.6% of 0.10).

Using successive approximations and carrying extra significant figures:

$$\frac{x^2}{0.10 - 0.0066} = \frac{x^2}{0.093} = 4.38 \times 10^{-4},\ x = 6.4 \times 10^{-3}\ \text{(consistent answer)}$$

$$\text{Percent ionization} = \frac{6.4 \times 10^{-3}}{0.10} \times 100 = 6.4\%$$

71. Using the K_b reaction to solve where PT = p-toluidine, $CH_3C_6H_4NH_2$:

$PT + H_2O \rightleftharpoons PTH^+ + OH^-$

Initial	0.016 *M*		0	~0
	x mol/L of PT reacts with H_2O to reach equilibrium			
Change	$-x$	$\rightarrow$	$+x$	$+x$
Equil.	$0.016 - x$		x	x

$$K_b = \frac{[PTH^+][OH^-]}{[PT]} = \frac{x^2}{0.016 - x}$$

Since pH = 8.60: pOH = 14.00 - 8.60 = 5.40 and $[OH^-] = x = 10^{-5.40} = 4.0 \times 10^{-6}$ *M*

$$K_b = \frac{(4.0 \times 10^{-6})^2}{0.016 - 4.0 \times 10^{-6}} = 1.0 \times 10^{-9}$$

Acid-Base Properties of Salts

73. One difficult aspect of acid-base chemistry is recognizing what types of species are present in solution, i.e., whether a species is a strong acid, strong base, weak acid, weak base or a neutral species. Below are some ideas and generalizations to keep in mind that will help in recognizing types of species present.

a. Memorize the following strong acids: HCl, HBr, HI, HNO_3, $HClO_4$ and H_2SO_4
b. Memorize the following strong bases: LiOH, NaOH, KOH, RbOH, $Ca(OH)_2$, $Sr(OH)_2$ and $Ba(OH)_2$
c. Weak acids have a K_a value less than 1 but greater than K_w. Some weak acids are in Table 7.2 of the text. Weak bases have a K_b value less than 1 but greater than K_w. Some weak bases are in Table 7.3 of the text.
d. Conjugate bases of weak acids are weak bases, i.e., all have a K_b value less than 1 but greater than K_w. Some examples of these are the conjugate bases of the weak acids in Table 7.2 of the text.
e. Conjugate acids of weak bases are weak acids, i.e., all have a K_a value less than 1 but greater than K_w. Some examples of these are the conjugate acids of the weak bases in Table 7.3 of the text.
f. Alkali metal ions (Li^+, Na^+, K^+, Rb^+, Cs^+) and some alkaline earth metal ions (Ca^{2+}, Sr^{2+}, Ba^{2+}) have no acidic or basic properties in water.
g. Conjugate bases of strong acids (Cl^-, Br^-, I^-, NO_3^-, ClO_4^-, HSO_4^-) have no basic properties in water ($K_b << K_w$) and only HSO_4^- has any acidic properties in water.

Lets apply these ideas to this problem to see what type of species are present. The letters in parenthesis is/are the generalization(s) above which identifies that species.

KOH: strong base (b)
KCl: neutral; K^+ and Cl^- have no acidic/basic properties (f and g).
KCN: CN^- is a weak base, $K_b = 1.0 \times 10^{-14}/6.2 \times 10^{-10} = 1.6 \times 10^{-5}$ (c and d). Ignore K^+(f).
NH_4Cl: NH_4^+ is a weak acid, $K_a = 5.6 \times 10^{-10}$ (c and e). Ignore Cl^-(g).
HCl: strong acid (a)

The most acidic solution will be the strong acid followed by the weak acid. The most basic solution will be the strong base followed by the weak base. The KCl solution will be between the acidic and basic solutions at pH = 7.00.

Most acidic → most basic; HCl > NH_4Cl > KCl > KCN > KOH

75. Reference Table 7.6 of the text and the solution to Exercise 7.73 for some generalizations on acid-base properties of salts. The letters in parenthesis is/are the generalization(s) listed in Exercise 7.73 which identifies that species.

$CaBr_2$: neutral; Ca^{2+} and Br^- have no acidic/basic properties (f and g).
KNO_2: NO_2^- is a weak base, $K_b = 1.0 \times 10^{-14}/4.0 \times 10^{-4} = 2.5 \times 10^{-11}$ (c and d). Ignore K^+(f).

$HClO_4$: strong acid (a)

HNO_2: weak acid, $K_a = 4.0 \times 10^{-4}$ (c)

NH_4ClO_4: NH_4^+ is a weak acid, $K_a = 5.6 \times 10^{-10}$ (c and e). Ignore ClO_4^- (g).

NH_4NO_2: NH_4^+ is a weak acid, $K_a = 5.6 \times 10^{-10}$ (c and e). NO_2^- is a weak base, $K_b = 2.5 \times 10^{-11}$ (c and d). Since the K_a value for NH_4^+ is a slightly larger than K_b for NO_2^-, then the solution will be slightly acidic with a pH a little lower than 7.0.

Using the information above (identity and K_a or K_b values), the ordering is:

most acidic → most basic: $HClO_4 > HNO_2 > NH_4ClO_4 > NH_4NO_2 > CaBr_2 > KNO_2$

77. From the K_a values, acetic acid is a stronger acid than hypochlorous acid. Conversely, the conjugate base of acetic acid, $C_2H_3O_2^-$, will be a weaker base than the conjugate base of hypochlorous acid, OCl^-. Thus, the hypochlorite ion, OCl^-, is a stronger base than the acetate ion, $C_2H_3O_2^-$. In general, the weaker the acid, the stronger the conjugate base. This statement comes from the relationship $K_w = K_a \times K_b$ which holds for all conjugate acid-base pairs.

79. a. $CH_3NH_3Cl \rightarrow CH_3NH_3^+ + Cl^-$; $CH_3NH_3^+$ is a weak acid since it is the conjugate acid of the weak base CH_3NH_2 ($K_b = 4.38 \times 10^{-4}$). Cl^- is the conjugate base of a strong acid. Cl^- has no basic (or acidic) properties. Solving the weak acid problem:

$$CH_3NH_3^+ \rightleftharpoons CH_3NH_2 + H^+ \quad K_a = \frac{[CH_3NH_2][H^+]}{[CH_3NH_3^+]} = \frac{K_w}{K_b} = \frac{1.00 \times 10^{-14}}{4.38 \times 10^{-4}} = 2.28 \times 10^{-11}$$

	$CH_3NH_3^+$	$\rightleftharpoons$	CH_3NH_2	+	H^+
Initial	0.10 *M*		0		~0
	x mol/L $CH_3NH_3^+$ dissociates to reach equilibrium				
Change	$-x$	→	$+x$		$+x$
Equil.	$0.10 - x$		x		x

$$K_a = 2.28 \times 10^{-11} = \frac{x^2}{0.10 - x} \approx \frac{x^2}{0.10} \quad \text{(assuming } x \ll 0.10\text{)}$$

$x = [H^+] = 1.5 \times 10^{-6}$ *M*; pH = 5.82 Assumptions good.

b. $NaCN \rightarrow Na^+ + CN^-$ CN^- is a weak base since it is the conjugate base of the weak acid HCN ($K_a = 6.2 \times 10^{-10}$). Na^+ has no acidic (or basic) properties. Solving the weak base problem:

$$CN^- + H_2O \rightleftharpoons HCN + OH^- \quad K_b = \frac{K_w}{K_a} = \frac{1.0 \times 10^{-14}}{6.2 \times 10^{-10}} = 1.6 \times 10^{-5}$$

	CN^- + H_2O	$\rightleftharpoons$	HCN	+	OH^-
Initial	0.050 *M*		0		~0
	x mol/L CN^- reacts with H_2O to reach equilibrium				
Change	$-x$	→	$+x$		$+x$
Equil.	$0.050 - x$		x		x

$$K_b = 1.6 \times 10^{-5} = \frac{[HCN][OH^-]}{[CN^-]} = \frac{x^2}{0.050 - x} \approx \frac{x^2}{0.050}$$

$x = [OH^-] = 8.9 \times 10^{-4}$ *M*; pOH = 3.05; pH = 10.95 Assumptions good.

c. $Na_2CO_3 \rightarrow 2\ Na^+ + CO_3^{2-}$ CO_3^{2-} is a weak base. Ignore Na^+.

$$CO_3^{2-} + H_2O \rightleftharpoons HCO_3^- + OH^- \quad K_b = \frac{K_w}{K_{a_2}} = \frac{1.0 \times 10^{-14}}{4.8 \times 10^{-11}} = 2.1 \times 10^{-4}$$

Initial	0.20 *M*		0	~0
Equil.	0.20 - *x*		*x*	*x*

$$K_b = 2.1 \times 10^{-4} = \frac{[HCO_3^-][OH^-]}{[CO_3^{2-}]} = \frac{x^2}{0.20 - x} \approx \frac{x^2}{0.20}$$

$x = 6.5 \times 10^{-3}\ M = [OH^-]$; pOH = 2.19; pH = 11.81 Assumptions good.

d. $KNO_2 \rightarrow K^+ + NO_2^-$ NO_2^- is a weak base. Ignore K^+.

$$NO_2^- + H_2O \rightleftharpoons HNO_2 + OH^- \quad K_b = \frac{K_w}{K_a} = \frac{1.0 \times 10^{-14}}{4.0 \times 10^{-4}} = 2.5 \times 10^{-11}$$

Initial	0.12 *M*		0	~0
Equil.	0.12 - *x*		*x*	*x*

$$K_b = 2.5 \times 10^{-11} = \frac{[OH^-][HNO_2]}{[NO_2^-]} = \frac{x^2}{0.12 - x} \approx \frac{x^2}{0.12}$$

$x = [OH^-] = 1.7 \times 10^{-6}\ M$; pOH = 5.77; pH = 8.23 Assumptions good.

e. $NH_4Br \rightarrow NH_4^+ + Br^-$: NH_4^+ is a weak acid. Br^- is the conjugate base of a strong acid. Br^- has no basic (or acidic) properties.

$$NH_4^+ \rightleftharpoons NH_3 + H^+ \quad K_a = \frac{K_w}{K_b} = \frac{1.0 \times 10^{-14}}{1.8 \times 10^{-5}} = 5.6 \times 10^{-10}$$

Initial	0.40 *M*	0	~0
Equil.	0.40 - *x*	*x*	*x*

$$K_a = 5.6 \times 10^{-10} = \frac{[NH_3][H^+]}{[NH_4^+]} = \frac{x^2}{0.40 - x} \approx \frac{x^2}{0.40}$$

$x = [H^+] = 1.5 \times 10^{-5}\ M$; pH = 4.82; Assumptions good.

81. $NaN_3 \rightarrow Na^+ + N_3^-$; Azide, N_3^-, is a weak base since it is the conjugate base of a weak acid. All conjugate bases of weak acids are weak bases ($K_w < K_b < 1$). Ignore Na^+.

$$N_3^- + H_2O \rightleftharpoons HN_3 + OH^- \quad K_b = \frac{K_w}{K_a} = \frac{1.0 \times 10^{-14}}{1.9 \times 10^{-5}} = 5.3 \times 10^{-10}$$

Initial	0.010 *M*		0	~0
	x mol/L of N_3^- reacts with H_2O to reach equilibrium			
Change	-*x*	→	+*x*	+*x*
Equil.	0.010 - *x*		*x*	*x*

$$K_b = \frac{[HN_3][OH^-]}{[N_3^-]} = 5.3 \times 10^{-10} = \frac{x^2}{0.010 - x} \approx \frac{x^2}{0.010} \quad \text{(assuming } x << 0.010\text{)}$$

$$x = [OH^-] = 2.3 \times 10^{-6}\ M;\ [H^+] = \frac{1.0 \times 10^{-14}}{2.3 \times 10^{-6}} = 4.3 \times 10^{-9}\ M \quad \text{Assumptions good.}$$

$[HN_3] = [OH^-] = 2.3 \times 10^{-6}\ M$; $[Na^+] = 0.010\ M$; $[N_3^-] = 0.010 - 2.3 \times 10^{-6} = 0.010\ M$

83. All these salts contain Na^+ which has no acidic/basic properties and a conjugate base of a weak acid (except for NaCl where Cl^- is a neutral species.). All conjugate bases of weak acids are weak bases since K_b for these species are between K_w and 1. To identify the species, we will use the data given to determine the K_b value for the weak conjugate base. From the K_b value and data in Table 7.2 of the text, we can identify the conjugate base present by calculating the K_a value for the weak acid. We will use A^- as an abbreviation for the weak conjugate base.

$$A^- + H_2O \rightleftharpoons HA + OH^-$$

	A^-		HA	OH^-
Initial	0.100 mol/1.00 L		0	~0
	x mol/L A^- reacts with H_2O to reach equilibrium			
Change	$-x$	→	$+x$	$+x$
Equil.	$0.100 - x$		x	x

$$K_b = \frac{[HA][OH^-]}{[A^-]} = \frac{x^2}{0.100 - x}; \text{ From the problem, pH} = 8.07:$$

$pOH = 14.00 - 8.07 = 5.93$; $[OH^-] = x = 10^{-5.93} = 1.2 \times 10^{-6}\ M$

$$K_b = \frac{(1.2 \times 10^{-6})^2}{0.100 - 1.2 \times 10^{-6}} = 1.4 \times 10^{-11} = K_b \text{ value for the conjugate base of a weak acid.}$$

The K_a value for the weak acid equals K_w/K_b: $K_a = \dfrac{1.0 \times 10^{-14}}{1.4 \times 10^{-11}} = 7.1 \times 10^{-4}$

From Table 7.2 of the text, this K_a value is closest to HF. Therefore, the unknown salt is NaF.

85. Major species: $Co(H_2O)_6^{3+}$ ($K_a = 1.0 \times 10^{-5}$), Cl^- (neutral) and H_2O ($K_w = 1.0 \times 10^{-14}$); $Co(H_2O)_6^{3+}$ will determine the pH since it is a stronger acid than water. Solving the weak acid problem in the usual manner:

$$Co(H_2O)_6^{3+} \rightleftharpoons Co(H_2O)_5(OH)^{2+} + H^+ \quad K_a = 1.0 \times 10^{-5}$$

	$Co(H_2O)_6^{3+}$	$Co(H_2O)_5(OH)^{2+}$	H^+
Initial	0.10 M	0	~0
Equil.	$0.10 - x$	x	x

$$K_a = 1.0 \times 10^{-5} = \frac{x^2}{0.10 - x} \approx \frac{x^2}{0.10}, \quad x = [H^+] = 1.0 \times 10^{-3}\ M$$

$pH = -\log(1.0 \times 10^{-3}) = 3.00$; Assumptions good.

Solutions of Dilute Acids and Bases

87. $HBrO \rightleftharpoons H^+ + BrO^-$ $K_a = 2 \times 10^{-9}$

	HBrO		H^+	BrO^-
Initial	$1.0 \times 10^{-6}\ M$		~0	0
	x mol/L HBrO dissociates to reach equilibrium			
Change	$-x$	→	$+x$	$+x$
Equil.	$1.0 \times 10^{-6} - x$		x	x

$$K_a = 2 \times 10^{-9} = \frac{x^2}{1.0 \times 10^{-6} - x} \approx \frac{x^2}{1.0 \times 10^{-6}};\quad x = [H^+] = 4 \times 10^{-8}\ M;\ pH = 7.4$$

Let's check the assumptions. This answer is impossible! We can't add a small amount of an acid to a neutral solution and get a basic solution. The highest pH possible for an acid in water is 7.0. In the correct solution, we would have to take into account the autoionization of water.

89. $HCN \rightleftharpoons H^+ + CN^-$ $K_a = 6.2 \times 10^{-10}$

	HCN	H^+	CN^-
Initial	$5.0 \times 10^{-4}\ M$	~0	0
Equil.	$5.0 \times 10^{-4} - x$	x	x

$$K_a = \frac{x^2}{5.0 \times 10^{-4} - x} \approx \frac{x^2}{5.0 \times 10^{-4}} = 6.2 \times 10^{-10},\ x = 5.6 \times 10^{-7}\ M \quad \text{Check assumptions.}$$

The assumption that the H^+ contribution from water is negligible is poor. Whenever the calculated pH is greater than 6.0 for a weak acid, water contribution to $[H^+]$ must be considered. From Section 7.9 in text:

$$\text{if } \frac{[H^+]^2 - K_w}{[H^+]} << [HCN]_o = 5.0 \times 10^{-4} \text{ then we can use: } [H^+] = (K_a[HCN]_o + K_w)^{1/2}.$$

Using this formula: $[H^+] = [(6.2 \times 10^{-10})(5.0 \times 10^{-4}) + (1.0 \times 10^{-14})]^{1/2}$, $[H^+] = 5.7 \times 10^{-7}\ M$

$$\text{Checking assumptions: } \frac{[H^+]^2 - K_w}{[H^+]} = 5.5 \times 10^{-7} << 5.0 \times 10^{-4}$$

Assumptions good. $pH = -\log(5.7 \times 10^{-7}) = 6.24$

91. We can't neglect the $[H^+]$ contribution from H_2O since this is a very dilute solution of the strong acid. Following the strategy developed in Section 7.10 of the text, we first determine the charge balance equation and then manipulate this equation to get into one unknown.

[positive charge] = [negative charge]

$$[H^+] = [Cl^-] + [OH^-],\ [H^+] = 7.0 \times 10^{-7} + \frac{K_w}{[H^+]} \quad \text{since } [Cl^-] = 7.0 \times 10^{-7} \text{ and } [OH^-] = \frac{K_w}{[H^+]}$$

$$\frac{[H^+]^2 - K_w}{[H^+]} = 7.0 \times 10^{-7},\ [H^+]^2 - 7.0 \times 10^{-7}[H^+] - 1.0 \times 10^{-14} = 0$$

Using the quadratic formula to solve:

$$[H^+] = \frac{-(-7.0 \times 10^{-7}) \pm [(-7.0 \times 10^{-7})^2 - 4(1)(-1.0 \times 10^{-14})]^{1/2}}{2(1)}$$

$[H^+] = 7.1 \times 10^{-7}$ *M*; pH = -log $(7.1 \times 10^{-7}) = 6.15$

Additional Exercises

93. a. $NH_3 + H_3O^+ \rightleftharpoons NH_4^+ + H_2O$

$$K_{eq} = \frac{[NH_4^+]}{[NH_3][H^+]} = \frac{1}{K_a \text{ for } NH_4^+} = \frac{K_b}{K_w} = \frac{1.8 \times 10^{-5}}{1.0 \times 10^{-14}} = 1.8 \times 10^9$$

b. $NO_2^- + H_3O^+ \rightleftharpoons H_2O + HNO_2 \quad K_{eq} = \frac{[HNO_2]}{[NO_2^-][H^+]} = \frac{1}{K_a} = \frac{1}{4.0 \times 10^{-4}} = 2.5 \times 10^3$

c. $NH_4^+ + CH_3CO_2^- \rightleftharpoons NH_3 + CH_3CO_2H \quad K_{eq} = \frac{[NH_3][CH_3CO_2H]}{[NH_4^+][CH_3CO_2^-]} \times \frac{[H^+]}{[H^+]}$

$$K_{eq} = \frac{K_a \text{ for } NH_4^+}{K_a \text{ for HOAc}} = \frac{K_w}{(K_b \text{ for } NH_3)(K_a \text{ for HOAc})} = \frac{1.0 \times 10^{-14}}{(1.8 \times 10^{-5})(1.8 \times 10^{-5})}$$

$K_{eq} = 3.1 \times 10^{-5}$

d. $H_3O^+ + OH^- \rightleftharpoons 2\,H_2O \quad K_{eq} = \frac{1}{K_w} = 1.0 \times 10^{14}$

e. $NH_4^+ + OH^- \rightleftharpoons NH_3 + H_2O \quad K_{eq} = \frac{1}{K_b \text{ for } NH_3} = 5.6 \times 10^4$

f. $HNO_2 + OH^- \rightleftharpoons H_2O + NO_2^-$

$$K_{eq} = \frac{[NO_2^-]}{[HNO_2][OH^-]} \times \frac{[H^+]}{[H^+]} = \frac{K_a \text{ for } HNO_2}{K_w} = \frac{4.0 \times 10^{-4}}{1.0 \times 10^{-14}} = 4.0 \times 10^{10}$$

95. The light bulb is bright because a strong electrolyte is present, i.e., a solute is present that dissolves to produce a lot of ions in solution. The pH meter value of 4.6 indicates that a weak acid is present. (If a strong acid were present, the pH would be close to zero.) Of the possible substances, only HCl (strong acid), NaOH (strong base) and NH_4Cl are strong electrolytes. Of these three substances, only NH_4Cl contains a weak acid (the HCl solution would have a pH close to zero and the NaOH solution would have a pH close to 14.0). NH_4Cl dissociates into NH_4^+ and Cl^- ions when dissolved in water. Cl^- is the conjugate base of a strong acid, so it has no basic (or acidic properties) in water. NH_4^+, however, is the conjugate acid of the weak base NH_3, so NH_4^+ is a weak acid and would produce a solution with a pH = 4.6 when the concentration is ~1.0 *M*.

97. For 0.0010% dissociation: $[NH_4^+] = 1.0 \times 10^{-5} (0.050) = 5.0 \times 10^{-7}\ M$

$NH_3 + H_2O \rightleftharpoons NH_4^+ + OH^-$ $\quad K_b = \dfrac{(5.0 \times 10^{-7})[OH^-]}{0.050 - 5.0 \times 10^{-7}} = 1.8 \times 10^{-5}$

Solving: $[OH^-] = 1.8\ M$; Assuming no volume change:

$$1.0\ L \times \frac{1.8\ mol\ NaOH}{L} \times \frac{40.0\ g\ NaOH}{mol\ NaOH} = 72\ g\ of\ NaOH$$

99. a. $Fe(H_2O)_6^{3+} + H_2O \rightleftharpoons Fe(H_2O)_5(OH)^{2+} + H_3O^+$

	$Fe(H_2O)_6^{3+}$	$Fe(H_2O)_5(OH)^{2+}$	H_3O^+
Initial	0.10 M	0	~0
Equil.	0.10 - x	x	x

$$K_a = \frac{[H_3O^+][Fe(H_2O)_5(OH)^{2+}]}{[Fe(H_2O)_6^{3+}]} = 6.0 \times 10^{-3} = \frac{x^2}{0.10 - x} \approx \frac{x^2}{0.10}$$

$x = 2.4 \times 10^{-2}\ M$; Assumption is poor (24% error).

Using successive approximations:

$$\frac{x^2}{0.10 - 0.024} = 6.0 \times 10^{-3},\ x = 0.021$$

$$\frac{x^2}{0.10 - 0.021} = 6.0 \times 10^{-3},\ x = 0.022; \quad \frac{x^2}{0.10 - 0.022} = 6.0 \times 10^{-3},\ x = 0.022$$

$x = [H^+] = 0.022\ M$; pH = 1.66

b. $\dfrac{[Fe(H_2O)_5(OH)^{2+}]}{[Fe(H_2O)_6^{3+}]} = \dfrac{0.0010}{0.9990}$; $K_a = 6.0 \times 10^{-3} = \dfrac{[H^+](0.0010)}{(0.9990)}$

Solving: $[H^+] = 6.0\ M$; pH = -log (6.0) = -0.78

c. Because of the lower charge, $Fe^{2+}(aq)$ will not be as strong an acid as $Fe^{3+}(aq)$. A solution of iron(II) nitrate will be less acidic (have a higher pH) than a solution with the same concentration of iron(III) nitrate.

101. 0.50 M HA, $K_a = 1.0 \times 10^{-3}$; 0.20 M HB, $K_a = 1.0 \times 10^{-10}$; 0.10 M HC, $K_a = 1.0 \times 10^{-12}$

Major source of H^+ is HA since its K_a value is significantly larger than other K_a values.

$HA \rightleftharpoons H^+ + A^-$

	HA	H^+	A^-
Initial	0.50 M	~0	0
Equil.	0.50 - x	x	x

$$K_a = \frac{x^2}{0.50 - x} = 1.0 \times 10^{-3} \approx \frac{x^2}{0.50},\ x \approx 0.022\ M = [H^+],\ \frac{0.022}{0.50} \times 100 = 4.4\ \%\ error$$

Assumptions good. Let's check out the assumption that only HA is an important source of H^+.

For HB: $1.0 \times 10^{-10} = \dfrac{(0.022)[B^-]}{(0.20)}$, $[B^-] = 9.1 \times 10^{-10}\ M$

At most, HB will produce an additional $9.1 \times 10^{-10}\ M\ H^+$. Even less will be produced by HC. Thus, our original assumption was good. $[H^+] = 0.022\ M$.

103. **Since NH_3 is so concentrated, we need to calculate the OH^- contribution from the weak base NH_3.**

	NH_3 + H_2O	$\rightleftharpoons$	NH_4^+	+	OH^-	$K_b = 1.8 \times 10^{-5}$
Initial	15.0 M		0		0.0100 M	(Assume no volume change.)
Equil.	15.0 - x		x		0.0100 + x	

$K_b = 1.8 \times 10^{-5} = \dfrac{x(0.0100 + x)}{15.0 - x} \approx \dfrac{x(0.0100)}{15.0}$, $x = 0.027$; **Assumption is horrible (x is 270% of 0.0100).**

Using the quadratic formula:

$$1.8 \times 10^{-5}(15.0 - x) = 0.0100\,x + x^2,\quad x^2 + 0.0100\,x - 2.7 \times 10^{-4} = 0$$

$$x = 1.2 \times 10^{-2}\ M,\quad [OH^-] = 1.2 \times 10^{-2} + 0.0100 = 0.022\ M$$

105. **a.** **The initial concentrations are halved since equal volumes of the two solutions are mixed.**

	$HC_2H_3O_2$	$\rightleftharpoons$	H^+	+	$C_2H_3O_2^-$
Initial	0.100 M		$5.00 \times 10^{-4}\ M$		0
Equil.	0.100 - x		$5.00 \times 10^{-4} + x$		x

$$K_a = 1.8 \times 10^{-5} = \frac{x(5.00 \times 10^{-4} + x)}{(0.100 - x)} \approx \frac{x(5.00 \times 10^{-4})}{0.100}$$

$x = 3.6 \times 10^{-3}$; **Assumption is horrible. Using the quadratic formula:**

$$x^2 + 5.18 \times 10^{-4}\,x - 1.8 \times 10^{-6} = 0$$

$$x = 1.1 \times 10^{-3}\ M;\ [H^+] = 5.00 \times 10^{-4} + x = 1.6 \times 10^{-3}\ M;\ pH = 2.80$$

b. $x = [C_2H_3O_2^-] = 1.1 \times 10^{-3}\ M$

107. **For $H_2C_6H_6O_6$. $K_{a_1} = 7.9 \times 10^{-5}$ and $K_{a_2} = 1.6 \times 10^{-12}$. Since $K_{a_1} >> K_{a_2}$, then the amount of H^+ produced by the K_{a_2} reaction will be negligible.**

$$[H_2C_6H_6O_6]_o = \frac{0.500\ g \times \dfrac{1\ mol\ H_2C_6H_6O_6}{176.1\ g}}{0.2000\ L} = 0.0142\ M$$

$$H_2C_6H_6O_6(aq) \rightleftharpoons HC_6H_6O_6^-(aq) + H^+(aq) \quad K_{a_1} = 7.9 \times 10^{-5}$$

	$H_2C_6H_6O_6$	$HC_6H_6O_6^-$	H^+
Initial	0.0142 *M*	0	~0
Equil.	0.0142 - *x*	*x*	*x*

$$K_{a_1} = 7.9 \times 10^{-5} = \frac{x^2}{0.0142 - x} \approx \frac{x^2}{0.0142}, \; x = 1.1 \times 10^{-3};$$ **Assumption fails the 5% rule.**

Solving by the method of successive approximations:

$$7.9 \times 10^{-5} = \frac{x^2}{0.0142 - 1.1 \times 10^{-3}}, \; x = 1.0 \times 10^{-3}\,M \text{ (consistent answer)}$$

Since H^+ produced by the K_{a_2} reaction will be negligible, $[H^+] = 1.0 \times 10^{-3}$ and pH = 3.00.

Challenge Problems

109. a. $HCO_3^- + HCO_3^- \rightleftharpoons H_2CO_3 + CO_3^{2-}$

$$K_{eq} = \frac{[H_2CO_3][CO_3^{2-}]}{[HCO_3^-][HCO_3^-]} \times \frac{[H^+]}{[H^+]} = \frac{K_{a_2}}{K_{a_1}} = \frac{4.8 \times 10^{-11}}{4.3 \times 10^{-7}} = 1.1 \times 10^{-4}$$

b. $[H_2CO_3] = [CO_3^{2-}]$ **since the reaction in part a is the principle equilibrium reaction.**

c. $H_2CO_3 \rightleftharpoons 2\,H^+ + CO_3^{2-} \quad K_{eq} = \frac{[H^+]^2[CO_3^{2-}]}{[H_2CO_3]} = K_{a_1} \times K_{a_2}$

Since $[H_2CO_3] = [CO_3^{2-}]$ from part b, then $[H^+]^2 = K_{a_1} \times K_{a_2}$.

$[H^+] = (K_{a_1} \times K_{a_2})^{1/2}$ **or taking the -log of both sides:** $pH = \frac{pK_{a_1} + pK_{a_2}}{2}$

d. $[H^+] = [(4.3 \times 10^{-7}) \times (4.8 \times 10^{-11})]^{1/2}, \; [H^+] = 4.5 \times 10^{-9}\,M;$ **pH = 8.35**

111. **Major species: BH^+, X^-, H_2O; Since BH^+ is the best acid and X^- is the best base in solution, then the principal equilibrium is:**

$$BH^+ + X^- \rightleftharpoons B + HX$$

	BH^+	X^-	B	HX
Initial	0.100 *M*	0.100 *M*	0	0
Equil.	0.100 - *x*	0.100 - *x*	*x*	*x*

$$K = \frac{K_{a,\,BH^+}}{K_{a,\,HX}} = \frac{[B][HX]}{[BH^+][X^-]}$$ **where [B] = [HX] and $[BH^+] = [X^-]$ (See set-up above.)**

To solve for the K_a of HX, let's use the equilibrium expression to derive a general expression that relates pH to the pK_a for BH^+ and to the pK_a for HX.

$$\frac{K_{a,\,BH^+}}{K_{a,\,HX}} = \frac{[HX]^2}{[X^-]^2}; \; K_{a,\,HX} = \frac{[H^+][X^-]}{[HX]}, \; \frac{[HX]}{[X^-]} = \frac{[H^+]}{K_{a,\,HX}}$$

$$\frac{K_{a,\,BH^+}}{K_{a,\,HX}} = \frac{[HX]^2}{[X^-]^2} = \left(\frac{[H^+]}{K_{a,\,HX}}\right)^2, \; [H^+]^2 = K_{a,\,BH^+} \times K_{a,\,HX}$$

Taking the -log of both sides: $pH = \dfrac{pK_{a, BH^+} + pK_{a, HX}}{2}$

This is a general equation that applies to all BHX type salts. Solving the problem:

K_b for B = 1.0×10^{-3}; K_a for $BH^+ = \dfrac{K_w}{K_b} = 1.0 \times 10^{-11}$

$pH = 8.00 = \dfrac{11.00 + pK_{a, HX}}{2}$, $pK_{a, HX} = 5.00$ and K_a for HX = $10^{-5.00} = 1.0 \times 10^{-5}$

113.

	HA	⇌	H^+	+	A^-	$K_a = 1.00 \times 10^{-6}$
Initial	C		~0		0	C = $[HA]_o$ for pH = 4.000 solution
Equil.	C - 1.00×10^{-4}		1.00×10^{-4}		1.00×10^{-4}	$x = [H^+] = 1.00 \times 10^{-4}$ *M*

$K_a = \dfrac{(1.00 \times 10^{-4})^2}{C - 1.00 \times 10^{-4}} = 1.00 \times 10^{-6}$; Solving: C = 0.0101 *M*

The solution initially contains 50.0×10^{-3} L × 0.0101 mol/L = 5.05×10^{-4} mol HA. We then dilute to a total volume V in liters. The resulting pH = 5.000, so $[H^+] = 1.00 \times 10^{-5}$. In the typical weak acid problem, $x = [H^+]$, so:

	HA	⇌	H^+	+	A^-
Initial	5.05×10^{-4} mol/V		~0		0
Equil.	5.05×10^{-4}/V - 1.00×10^{-5}		1.00×10^{-5}		1.00×10^{-5}

$K_a = \dfrac{(1.00 \times 10^{-5})^2}{5.05 \times 10^{-4}/V - 1.00 \times 10^{-5}} = 1.00 \times 10^{-6}$, $1.00 \times 10^{-4} = 5.05 \times 10^{-4}/V - 1.00 \times 10^{-5}$

V = 4.59 L; 50.0 mL are present initially, so we need to add 4540 mL of water.

115. $\dfrac{0.135 \text{ mol } CO_2}{2.50 \text{ L}} = 5.40 \times 10^{-2}$ mol CO_2/L = 5.40×10^{-2} *M* H_2CO_3; 0.105 *M* CO_3^{2-}

The best acid (H_2CO_3) reacts with the best base present (CO_3^{2-}) for the principle equilibrium.

$H_2CO_3 + CO_3^{2-} \rightarrow 2\ HCO_3^-$ $\quad K = \dfrac{1}{1.3 \times 10^{-4}} = 7.7 \times 10^3$ (See Exercise 7.109)

Since K >> 1, assume all CO_2 (H_2CO_3) is converted into HCO_3^-, i.e., 5.40×10^{-2} mol/L CO_3^{2-} is converted into HCO_3^-.

$[HCO_3^-] = 2(5.40 \times 10^{-2}) = 0.108$ *M*; $[CO_3^{2-}] = 0.105 - 0.0540 = 0.051$ *M*

Note: If we solve for the $[H_2CO_3]$ using these concentrations, we get $[H_2CO_3] = 3.0 \times 10^{-5}$ *M*; our assumption that the reaction goes to completion is good (3.0×10^{-5} is 0.06% of 0.051). Whenever K >> 1, always assumes the reaction goes to completion.

To solve for the $[H^+]$ in equilibrium with HCO_3^- and CO_3^{2-}, use the K_a expression for HCO_3^-.

$HCO_3^- \rightleftharpoons H^+ + CO_3^{2-} \quad K_{a_2} = 4.8 \times 10^{-11}$

$$4.8 \times 10^{-11} = \frac{[H^+][CO_3^{2-}]}{[HCO_3^-]} \approx [H^+]\left(\frac{0.051}{0.108}\right)$$

$[H^+] = 1.0 \times 10^{-10}$; pH = 10.00 Assumptions good.

117. Major species: H_2O, Na^+, NO_2^-; NO_2^- is a weak base. $NO_2^- + H_2O \rightleftharpoons HNO_2 + OH^-$

Since this is a very dilute solution of a weak base, the OH^- contribution from H_2O must be considered. The weak base equations for dilute solutions are analogous to the weak acid equations derived in Section 7.9 of the text. They are:

For $A^- + H_2O \rightleftharpoons HA + OH^-$

$$\text{I.}\quad K_b = \frac{[OH^-]^2 - K_w}{[A^-]_o - \dfrac{[OH^-]^2 - K_w}{[OH^-]}}$$

$$\text{II. When } [A^-]_o >> \frac{[OH^-]^2 - K_w}{[OH^-]},\text{ then } K_b = \frac{[OH^-]^2 - K_w}{[A^-]_o} \text{ and } [OH^-] = (K_b[A^-]_o + K_w)^{1/2}$$

$$\text{Try } [OH^-] = \left(\frac{1.0 \times 10^{-14}}{4.0 \times 10^{-4}} \times (6.0 \times 10^{-4}) + 1.0 \times 10^{-14}\right)^{1/2} = 1.6 \times 10^{-7}\ M$$

$$\text{Checking assumption:}\quad 6.0 \times 10^{-4} >> \frac{(1.6 \times 10^{-7})^2 - 1.0 \times 10^{-14}}{1.6 \times 10^{-7}} = 9.8 \times 10^{-8}$$

Assumption good. $[OH^-] = 1.6 \times 10^{-7}$ *M*; pOH = 6.80; pH = 7.20

119. Major species: NH_3, H^+, Cl^-; The H^+ from the strong acid will react with the best base present, NH_3. Since strong acids are great at donating protons, then the reaction between H^+ and NH_3 essentially goes to completion, i.e., until one or both of the reactants runs out. The reaction is:

$NH_3 + H^+ \rightarrow NH_4^+$

Since equal volumes of 1.0×10^{-4} *M* NH_3 and 1.0×10^{-4} *M* H^+ are mixed, then both reactants are in stoichiometric amounts and both reactants will run out at the same time. After reaction only NH_4^+ and Cl^- remains. Cl^- has no basic properties since it is the conjugate base of a strong acid. Therefore, the only species with acid-base properties is NH_4^+, a weak acid. The initial concentration of NH_4^+ will be exactly one-half of 1.0×10^{-4} *M* since equal volumes of NH_3 and HCl were mixed. Now we must solve the weak acid problem involving 5.0×10^{-5} *M* NH_4^+.

	NH_4^+	$\rightleftharpoons$	H^+	+	NH_3	$K_a = \dfrac{K_w}{K_b} = 5.6 \times 10^{-10}$
Initial	5.0×10^{-5} *M*		~0		0	
Equil.	$5.0 \times 10^{-5} - x$		x		x	

$$K_a = \frac{x^2}{5.0 \times 10^{-5} - x} \approx \frac{x^2}{5.0 \times 10^{-5}} = 5.6 \times 10^{-10},\ x = 1.7 \times 10^{-7}\ M \quad \text{Check assumptions.}$$

We cannot neglect $[H^+]$ that comes from H_2O. As discussed in Section 7.9 of the text, assume $5.0 \times 10^{-5} >> ([H^+]^2 - K_w)/[H^+]$. If this is the case, then:

$[H^+] = (K_a[HA]_o + K_w)^{1/2} = 1.9 \times 10^{-7}$ *M*: **Checking assumption:**

$$\frac{[H^+]^2 - K_w}{[H^+]} = 1.4 \times 10^{-7} << 5.0 \times 10^{-5} \quad \text{Assumption good.}$$

So, $[H^+] = 1.9 \times 10^{-7}$ *M*; **pH = 6.72**

Marathon Problem

121. To determine the pH of solution A, the K_a **value for HX must be determined. Use solution B to determine** K_b **for** X^-**, which can then be used to calculate** K_a **for HX** ($K_a = K_w/K_b$).

Solution B:

$$X^- + H_2O \rightleftharpoons HX + OH^- \qquad K_b = \frac{[HX][OH^-]}{[X^-]}$$

	X^-		HX	OH^-
Initial	0.0500 *M*		0	~0
Change	$-x$	→	$+x$	$+x$
Equil.	$0.0500 - x$		x	x

$K_b = \dfrac{x^2}{0.0500 - x}$; **From the problem, pH = 10.02, so pOH = 3.98 and** $[OH^-] = x = 10^{-3.98}$

$$K_b = \frac{(10^{-3.98})^2}{0.0500 - 10^{-3.98}} = 2.2 \times 10^{-7}$$

Solution A:

$K_{a,\,HX} = K_w/K_{b,\,X^-} = 1.0 \times 10^{-14}/2.2 \times 10^{-7} = 4.5 \times 10^{-8}$

$$HX \rightleftharpoons H^+ + X^- \qquad K_a = 4.5 \times 10^{-8} = \frac{[H^+][X^-]}{[HX]}$$

	HX		H^+	X^-
Initial	0.100 *M*		~0	0
Change	$-x$	→	$+x$	$+x$
Equil.	$0.100 - x$		x	x

$K_a = 4.5 \times 10^{-8} = \dfrac{x^2}{0.100 - x} \approx \dfrac{x^2}{0.100}$, $x = [H^+] = 6.7 \times 10^{-5}$ *M*

Assumptions good (*x* is 6.7×10^{-2} **% of 0.100); pH = 4.17**

Solution C:

Major species: HX ($K_a = 4.5 \times 10^{-8}$), Na^+, OH^-; **The** OH^- **from the strong base is exceptional at accepting protons.** OH^- **will react with the best acid present (HX) and we can assume that** OH^- **will react to completion with HX, i.e., until one (or both) of the reactants runs out. Since we have added one volume of substance to another, we have diluted both solutions from their initial concentrations. What hasn't changed is the moles of each reactant. So let's work with moles of each reactant initially.**

$$\text{mol HX} = 0.0500\text{ L} \times \frac{0.100\text{ mol HX}}{\text{L}} = 5.00 \times 10^{-3}\text{ mol HX}$$

$$\text{mol OH}^- = 0.0150\text{ L} \times \frac{0.250\text{ mol NaOH}}{\text{L}} \times \frac{1\text{ mol OH}^-}{\text{mol NaOH}} = 3.75 \times 10^{-3}\text{ mol OH}^-$$

Now lets determine what is remaining in solution after OH^- reacts completely with HX. Note that OH^- is the limiting reagent.

	HX	+	OH^-	$\rightarrow$	X^-	+	H_2O
Initial	5.00×10^{-3} mol		3.75×10^{-3} mol		0		-----
Change	-3.75×10^{-3}		-3.75×10^{-3}	$\rightarrow$	$+3.75 \times 10^{-3}$		$+3.75 \times 10^{-3}$
After completion	1.25×10^{-3} mol		0		3.75×10^{-3} mol		-----

After reaction, the solution contains HX, X^-, Na^+ and H_2O. The Na^+ (like most +1 metal ions) has no effect on the pH of water. However, HX is a weak acid and its conjugate base, X^-, is a weak base. Since both K_a and K_b reactions refer to these species, we could use either reaction to solve for the pH; we will use the K_b reaction. To solve the equilibrium problem using the K_b reaction, we need to convert to concentration units since K_b is in concentration units of mol/L.

$$[\text{HX}] = \frac{1.25 \times 10^{-3}\text{ mol}}{(0.0500 + 0.0150)\text{ L}} = 0.0192\ M;\ [\text{X}^-] = \frac{3.75 \times 10^{-3}\text{ mol}}{0.0650\text{ L}} = 0.0577\ M$$

$[OH^-] = 0$ (We reacted all of it to completion.)

	X^- + H_2O	$\rightleftharpoons$	HX	+	OH^-	$K_b = 2.2 \times 10^{-7}$
Initial	$0.0577\ M$		$0.0192\ M$		0	
	x mol/L of X^- reacts to reach equilibrium					
Change	$-x$	$\rightarrow$	$+x$		$+x$	
Equil.	$0.0577 - x$		$0.0192 + x$		x	

$$K_b = 2.2 \times 10^{-7} = \frac{(0.0192 + x)(x)}{(0.0577 - x)} \approx \frac{(0.0192)\,x}{(0.0577)} \quad \text{(assuming } x \text{ is} \ll 0.0192\text{)}$$

$$x = [\text{OH}^-] = \frac{2.2 \times 10^{-7}\,(0.0577)}{0.0192} = 6.6 \times 10^{-7}\ M$$ Assumptions great (x is 3.4×10^{-3} % of 0.0192).

$[OH^-] = 6.6 \times 10^{-7}\ M$, pOH = 6.18, pH = 14.00 = 6.18 = 7.82 = pH of solution C

The combination is 4-17-7-82.

CHAPTER EIGHT

APPLICATIONS OF AQUEOUS EQUILIBRIA

Buffers

15. A buffered solution must contain both a weak acid and a weak base. Most buffered solutions are prepared using a weak acid plus the conjugate base of the weak acid (which is a weak base). Buffered solutions are useful for controlling the pH of a solution since they resist pH change.

17. The capacity of a buffer is a measure of how much strong acid or strong base the buffer can neutralize. All the buffers listed have the same pH (= pK_a = 4.74) since they all have a 1:1 concentration ratio between the weak acid and the conjugate base. The 1.0 *M* buffer has the greatest capacity; the 0.01 *M* buffer the least capacity. In general, the larger the concentrations of weak acid and conjugate base, the greater the buffer capacity, i.e., the greater the ability to neutralize added strong acid or strong base.

19. a. This is a weak acid problem. Let $HC_3H_5O_2$ = HOPr and $C_3H_5O_2^-$ = OPr^-.

	HOPr(aq)	$\rightleftharpoons$	H^+(aq)	+	OPr^-(aq)	$K_a = 1.3 \times 10^{-5}$
Initial	0.100 *M*		~0		0	
	x mol/L HOPr dissociates to reach equilibrium					
Change	$-x$	$\rightarrow$	$+x$		$+x$	
Equil.	$0.100 - x$		x		x	

$$K_a = 1.3 \times 10^{-5} = \frac{[H^+][OPr^-]}{[HOPr]} = \frac{x^2}{0.100 - x} \approx \frac{x^2}{0.100}$$

$x = [H^+] = 1.1 \times 10^{-3}$ *M*; pH = 2.96   Assumptions good by the 5% rule.

b. This is a weak base problem.

	OPr^-(aq) + H_2O(l)	$\rightleftharpoons$	HOPr(aq)	+	OH^-(aq)	$K_b = \frac{K_w}{K_a} = 7.7 \times 10^{-10}$
Initial	0.100 *M*		0		~0	
	x mol/L OPr^- reacts with H_2O to reach equilibrium					
Change	$-x$	$\rightarrow$	$+x$		$+x$	
Equil.	$0.100 - x$		x		x	

$$K_b = 7.7 \times 10^{-10} = \frac{[HOPr][OH^-]}{[OPr^-]} = \frac{x^2}{0.100 - x} \approx \frac{x^2}{0.100}$$

$x = [OH^-] = 8.8 \times 10^{-6}$ *M*; pOH = 5.06; pH = 8.94   Assumptions good.

c. pure H_2O, $[H^+] = [OH^-] = 1.0 \times 10^{-7}$ *M*; pH = 7.00

d. This solution contains a weak acid and its conjugate base. This is a buffer solution. We will solve for the pH through the weak acid equilibrium reaction.

$$HOPr(aq) \rightleftharpoons H^+(aq) + OPr^-(aq) \qquad K_a = 1.3 \times 10^{-5}$$

	HOPr		H^+	OPr^-
Initial	0.100 *M*		~0	0.100 *M*
	x mol/L HOPr dissociates to reach equilibrium			
Change	-*x*	→	+*x*	+*x*
Equil.	0.100 - *x*		*x*	0.100 + *x*

$$1.3 \times 10^{-5} = \frac{(0.100 + x)(x)}{0.100 - x} \approx \frac{(0.100)(x)}{0.100} = x = [H^+]$$

$[H^+] = 1.3 \times 10^{-5}$ *M*; pH = 4.89 Assumptions good.

Alternately, we can use the Henderson-Hasselbalch equation to calculate the pH of buffer solutions.

$$pH = pK_a + \log\frac{[Base]}{[Acid]} = pK_a + \log\frac{(0.100)}{(0.100)} = pK_a = -\log(1.3 \times 10^{-5}) = 4.89$$

The Henderson-Hasselbalch equation will be valid when an assumption of the type, $0.1 + x \approx 0.1$, that we just made in this problem is valid. From a practical standpoint, this will almost always be true for useful buffer solutions. If the assumption is not valid, the solution will have such a low buffering capacity it will not be of any use to control the pH. Note: The Henderson-Hasselbalch equation can <u>only</u> be used to solve for the pH of buffer solutions.

21. a. OH^- will react completely with the best acid present, HOPr.

$$HOPr + OH^- \rightarrow OPr^- + H_2O$$

	HOPr	OH^-		OPr^-	
Before	0.100 *M*	0.020 *M*		0	
Change	-0.020	-0.020	→	+0.020	Reacts completely
After	0.080	0		0.020	

A buffer solution results after the reaction. Using the Henderson-Hasselbalch equation:

$$pH = pK_a + \log\frac{[Base]}{[Acid]} = 4.89 + \log\frac{(0.020)}{(0.080)} = 4.29$$

b. We have a weak base and a strong base present at the same time. The amount of OH^- added by the weak base will be negligible. To prove it, lets consider the weak base equilibrium:

$$OPr^- + H_2O \rightleftharpoons HOPr + OH^- \quad K_b = 7.7 \times 10^{-10}$$

	OPr^-		HOPr	OH^-
Initial	0.100 *M*		0	0.020 *M*
	x mol/L OPr^- reacts with H_2O to reach equilibrium			
Change	$-x$	→	$+x$	$+x$
Equil.	$0.100 - x$		x	$0.020 + x$

$[OH^-] = 0.020 + x \approx 0.020$ *M*; pOH = 1.70; pH = 12.30 Assumption good.

Note: The OH^- contribution from the weak base OPr^- was negligible ($x = 3.9 \times 10^{-9}$ *M* as compared to 0.020 *M* OH^- from the strong base). The pH can be determined by only considering the amount of strong base present.

c. This is a strong base in water. $[OH^-] = 0.020$ *M*; pOH = 1.70; pH = 12.30

d. OH^- will react completely with HOPr, the best acid present.

$$HOPr + OH^- \rightarrow OPr^- + H_2O$$

	HOPr	OH^-		OPr^-	
Before	0.100 *M*	0.020 *M*		0.100 *M*	
Change	-0.020	-0.020	→	+0.020	Reacts completely
After	0.080	0		0.120	

Using the Henderson-Hasselbalch equation to solve for the pH of the resulting buffer solution:

$$pH = pK_a + \log\frac{[Base]}{[Acid]} = 4.89 + \log\frac{(0.120)}{(0.080)} = 5.07$$

23. Major species: HF, F^- and K^+ (no acidic/basic properties). The appropriate equilibrium reaction to use is the K_a reaction of HF which contains both HF and F^-.

$$HF \rightleftharpoons F^- + H^+$$

	HF		F^-	H^+
Initial	0.60 *M*		1.00 *M*	~0
	x mol/L HF dissociates to reach equilibrium			
Change	$-x$	→	$+x$	$+x$
Equil.	$0.60 - x$		$1.00 + x$	x

$$K_a = 7.2 \times 10^{-4} = \frac{[F^-][H^+]}{[HF]} = \frac{(1.00 + x)(x)}{(0.60 - x)} \approx \frac{1.00(x)}{0.60} \quad \text{(assuming } x \ll 0.60\text{)}$$

$x = [H^+] = 0.60 \times (7.2 \times 10^{-4}) = 4.3 \times 10^{-4}$ *M*; Assumptions good (x is 7.2×10^{-2}% of 0.60).

$pH = -\log(4.3 \times 10^{-4}) = 3.37$

25. a. $HC_2H_3O_2 \rightleftharpoons H^+ + C_2H_3O_2^-$ $K_a = 1.8 \times 10^{-5}$

	$HC_2H_3O_2$		H^+	$C_2H_3O_2^-$
Initial	0.10 *M*		~0	0.25 *M*
	x mol/L $HC_2H_3O_2$ dissociates to reach equilibrium			
Change	$-x$	→	$+x$	$+x$
Equil.	$0.10 - x$		x	$0.25 + x$

$$1.8 \times 10^{-5} = \frac{x(0.25 + x)}{(0.10 - x)} \approx \frac{x(0.25)}{0.10} \quad \text{(assuming } 0.25 + x \approx 0.25 \text{ and } 0.10 - x \approx 0.10\text{)}$$

$x = [H^+] = 7.2 \times 10^{-6}\ M$; pH = 5.14 Assumptions good by the 5% rule.

Alternatively, we can use the Henderson-Hasselbalch equation:

$$\text{pH} = \text{p}K_a + \log\frac{[\text{Base}]}{[\text{Acid}]} \quad \text{where p}K_a = -\log(1.8 \times 10^{-5}) = 4.74$$

$$\text{pH} = 4.74 + \log\frac{(0.25)}{(0.10)} = 4.74 + 0.40 = 5.14$$

The Henderson-Hasselbalch equation will be valid when assumptions of the type $0.10 - x \approx 0.10$ that we just made are valid. From a practical standpoint, this will almost always be true for useful buffer solutions. Note: The Henderson-Hasselbalch equation can only be used to solve for the pH of buffer solutions.

b. $$\text{pH} = 4.74 + \log\frac{(0.10)}{(0.25)} = 4.74 + (-0.40) = 4.34$$

c. $$\text{pH} = 4.74 + \log\frac{(0.25)}{(0.25)} = 4.74 + 0.00 = 4.74 \quad (\text{pH} = \text{p}K_a \text{ since [acid] = [base]})$$

d. $$\text{pH} = \text{p}K_a + \log\frac{[\text{base}]}{[\text{acid}]}; \quad [\text{base}] = [C_2H_5NH_2] = 0.50\ M; \quad [\text{acid}] = [C_2H_5NH_3^+] = 0.25\ M$$

$$K_a = \frac{K_w}{K_b} = \frac{1.0 \times 10^{-14}}{5.6 \times 10^{-4}} = 1.8 \times 10^{-11}$$

$$\text{pH} = -\log(1.8 \times 10^{-11}) + \log\left(\frac{0.50\ M}{0.25\ M}\right) = 10.74 + 0.30 = 11.04$$

e. $$\text{pH} = 10.74 + \log\left(\frac{0.50\ M}{0.50\ M}\right) = 10.74 + 0.00 = 10.74$$

27. $$C_5H_5NH^+ \rightleftharpoons H^+ + C_5H_5N \quad K_a = \frac{K_w}{K_b} = \frac{1.0 \times 10^{-14}}{1.7 \times 10^{-9}} = 5.9 \times 10^{-6}; \quad \text{p}K_a = -\log(5.9 \times 10^{-6}) = 5.23$$

We will use the Henderson-Hasselbalch equation to calculate the concentration ratio necessary for each buffer.

$$pH = pK_a + \log\frac{[base]}{[acid]},\ pH = 5.23 + \log\frac{[C_5H_5N]}{[C_5H_5NH^+]}$$

a. $4.50 = 5.23 + \log\frac{[C_5H_5N]}{[C_5H_5NH^+]}$

$$\log\frac{[C_5H_5N]}{[C_5H_5NH^+]} = -0.73$$

$$\frac{[C_5H_5N]}{[C_5H_5NH^+]} = 10^{-0.73} = 0.19$$

b. $5.00 = 5.23 + \log\frac{[C_5H_5N]}{[C_5H_5NH^+]}$

$$\log\frac{[C_5H_5N]}{[C_5H_5NH^+]} = -0.23$$

$$\frac{[C_5H_5N]}{[C_5H_5NH^+]} = 10^{-0.23} = 0.59$$

c. $5.23 = 5.23 + \log\frac{[C_5H_5N]}{[C_5H_5NH^+]}$

$$\frac{[C_5H_5N]}{[C_5H_5NH^+]} = 10^{0.0} = 1.0$$

d. $5.50 = 5.23 + \log\frac{[C_5H_5N]}{[C_5H_5NH^+]}$

$$\frac{[C_5H_5N]}{[C_5H_5NH^+]} = 10^{0.27} = 1.9$$

29. $pH = pK_a + \log\frac{[C_2H_3O_2^-]}{[HC_2H_3O_2]}$; $pK_a = -\log(1.8 \times 10^{-5}) = 4.74$

Since the buffer components, $C_2H_3O_2^-$ and $HC_2H_3O_2$, are both in the same volume of water, the concentration ratio of $[C_2H_3O_2^-]/[HC_2H_3O_2]$ will equal the mol ratio of mol $C_2H_3O_2^-$/mol $HC_2H_3O_2$.

$$5.00 = 4.74 + \log\frac{\text{mol } C_2H_3O_2^-}{\text{mol } HC_2H_3O_2};\ \text{mol } HC_2H_3O_2 = 0.5000\ L \times \frac{0.200\ mol}{L} = 0.100\ mol$$

$$0.26 = \log\frac{\text{mol } C_2H_3O_2^-}{0.100\ mol},\ \frac{\text{mol } C_2H_3O_2^-}{0.100} = 10^{0.26} = 1.8,\ \text{mol } C_2H_3O_2^- = 0.18\ mol$$

$$\text{mass } NaC_2H_3O_2 = 0.18\ \text{mol } NaC_2H_3O_2 \times \frac{82.03\ g}{mol} = 15\ g\ NaC_2H_3O_2$$

31. a. pK_b for $C_6H_5NH_2 = -\log(3.8 \times 10^{-10}) = 9.42$; pK_a for $C_6H_5NH_3^+ = 14.00 - 9.42 = 4.58$

$$pH = pK_a + \log\frac{[C_6H_5NH_2]}{[C_6H_5NH_3^+]},\ 4.20 = 4.58 + \log\frac{0.50\ M}{[C_6H_5NH_3^+]}$$

$$-0.38 = \log\frac{0.50\ M}{[C_6H_5NH_3^+]},\ [C_6H_5NH_3^+] = [C_6H_5NH_3Cl] = 1.2\ M$$

b. $4.0\ g\ NaOH \times \frac{1\ mol\ NaOH}{40.00\ g} \times \frac{1\ mol\ OH^-}{mol\ NaOH} = 0.10\ mol\ OH^-,\ [OH^-] = \frac{0.10\ mol}{1.0\ L} = 0.10\ M$

	$C_6H_5NH_3^+$	+ OH^-	$\rightarrow$	$C_6H_5NH_2$	+ H_2O
Before	1.2 *M*	0.10 *M*		0.50 *M*	
Change	-0.10	-0.10	$\rightarrow$	+0.10	
After	1.1	0		0.60	

A buffer solution exists. $pH = 4.58 + \log\left(\frac{0.60}{1.1}\right) = 4.32$

33. a. $pH = pK_a + \log\frac{[Base]}{[Acid]}$, $7.40 = -\log(4.3 \times 10^{-7}) + \log\frac{[HCO_3^-]}{[H_2CO_3]} = 6.37 + \log\frac{[HCO_3^-]}{[H_2CO_3]}$

$$\frac{[HCO_3^-]}{[H_2CO_3]} = 10^{1.03} = 11;\quad \frac{[H_2CO_3]}{[HCO_3^-]} = \frac{[CO_2]}{[HCO_3^-]} = \frac{1}{11} = 0.091$$

b. $7.15 = -\log(6.2 \times 10^{-8}) + \log\frac{[HPO_4^{2-}]}{[H_2PO_4^-]}$, $7.15 = 7.21 + \log\frac{[HPO_4^{2-}]}{[H_2PO_4^-]}$

$$\frac{[HPO_4^{2-}]}{[H_2PO_4^-]} = 10^{-0.06} = 0.9,\quad \frac{[H_2PO_4^-]}{[HPO_4^{2-}]} = \frac{1}{0.9} = 1.1 \approx 1$$

c. A best buffer has approximately equal concentrations of weak acid and conjugate base so that $pH \approx pK_a$ for a best buffer. The pK_a value for a $H_3PO_4/H_2PO_4^-$ buffer is $-\log(7.5 \times 10^{-3}) = 2.12$. A pH of 7.1 is too high for a $H_3PO_4/H_2PO_4^-$ buffer to be effective. At this high of pH, there would be so little H_3PO_4 present that we could hardly consider it a buffer; this solution would not be effective in resisting pH changes, especially when strong base is added.

35. When OH^- is added, it converts $HC_2H_3O_2$ into $C_2H_3O_2^-$: $HC_2H_3O_2 + OH^- \rightarrow C_2H_3O_2^- + H_2O$ From this reaction, the moles of $C_2H_3O_2^-$ produced equal the moles of OH^- added. Also, the total concentration of acetic acid plus acetate ion must equal 2.0 *M* (assuming no volume change on addition of NaOH). Summarizing for each solution:

$[C_2H_3O_2^-] + [HC_2H_3O] = 2.0\ M$ and $[C_2H_3O_2^-]$ produced $= [OH^-]$ added

a. $pH = pK_a + \log\frac{[C_2H_3O_2^-]}{[HC_2H_3O_2]}$; For $pH = pK_a$, $\log\frac{[C_2H_3O_2^-]}{[HC_2H_3O_2]} = 0$

Therefore, $\frac{[C_2H_3O_2^-]}{[HC_2H_3O_2]} = 1.0$ and $[C_2H_3O_2^-] = [HC_2H_3O_2]$

Since $[C_2H_3O_2^-] + [HC_2H_3O_2] = 2.0\ M$, then $[C_2H_3O_2^-] = [HC_2H_3O_2] = 1.0\ M = [OH^-]$ added

To produce a 1.0 *M* $C_2H_3O_2^-$ solution, we need to add 1.0 mol of NaOH to 1.0 L of the 2.0 *M* $HC_2H_3O_2$ solution. The resultant solution will have $pH = pK_a = 4.74$.

b. $4.00 = 4.74 + \log \frac{[C_2H_3O_2^-]}{[HC_2H_3O_2]}, \quad \frac{[C_2H_3O_2^-]}{[HC_2H_3O_2]} = 10^{-0.74} = 0.18$

$[C_2H_3O_2^-] = 0.18\,[HC_2H_3O_2]$ or $[HC_2H_3O_2] = 5.6\,[C_2H_3O_2^-]$; Since $[C_2H_3O_2^-] + [HC_2H_3O_2] =$ 2.0 *M*, then:

$$[C_2H_3O_2^-] + 5.6\,[C_2H_3O_2^-] = 2.0\ M,\ [C_2H_3O_2^-] = \frac{2.0}{6.6} = 0.30\ M = [OH^-] \text{ added}$$

We need to add 0.30 mol of NaOH to 1.0 L of 2.0 *M* $HC_2H_3O_2$ solution to produce 0.30 *M* $C_2H_3O_2^-$. The resultant solution will have pH = 4.00.

c. $5.00 = 4.74 + \log \frac{[C_2H_3O_2^-]}{[HC_2H_3O_2]}, \quad \frac{[C_2H_3O_2^-]}{[HC_2H_3O_2]} = 10^{0.26} = 1.8$

$1.8\,[HC_2H_3O_2] = [C_2H_3O_2^-]$ or $[HC_2H_3O_2] = 0.56\,[C_2H_3O_2^-]$; Since $[HC_2H_3O_2] + [C_2H_3O_2^-] = 2.0$ *M*, then:

$$1.56\,[C_2H_3O_2^-] = 2.0\ M,\ [C_2H_3O_2^-] = 1.3\ M = [OH^-] \text{ added}$$

We need to add 1.3 mol of NaOH to 1.0 L of 2.0 *M* $HC_2H_3O_2$ to produce a solution with pH = 5.00.

37. A best buffer has large and equal quantities of weak acid and conjugate base. Since [acid] = [base] for a best buffer, then $pH = pK_a + \log \frac{[base]}{[acid]} = pK_a + 0 = pK_a$.

The best acid choice for a pH = 7.00 buffer would be the weak acid with a pK_a close to 7.0 or $K_a \approx 1 \times 10^{-7}$. HOCl is the best choice in Table 7.2 ($K_a = 3.5 \times 10^{-8}$; $pK_a = 7.46$). To make this buffer, we need to calculate the [base]/[acid] ratio.

$$7.00 = 7.46 + \log \frac{[base]}{[acid]}, \quad \frac{[OCl^-]}{[HOCl]} = 10^{-0.46} = 0.35$$

Any OCl^-/HOCl buffer in a concentration ratio of 0.35:1 will have a pH = 7.00. One possibility is [NaOCl] = 0.35 *M* and [HOCl] = 1.0 *M*.

39. To solve for [KOCl], we need to use the equation derived in Section 8.3 of the text on the Exact Treatment of Buffered Solutions. The equation is:

$$K_a = \frac{[H^+]\left([A^-]_o + \frac{[H^+]^2 - K_w}{[H^+]}\right)}{[HA]_o - \frac{[H^+]^2 - K_w}{[H^+]}}$$

Since pH = 7.20, then $[H^+] = 10^{-7.20} = 6.3 \times 10^{-8}\ M$.

$$K_a = 3.5 \times 10^{-8} = \frac{6.3 \times 10^{-8}\left([OCl^-] + \dfrac{(6.3 \times 10^{-8})^2 - 1.0 \times 10^{-14}}{6.3 \times 10^{-8}}\right)}{1.0 \times 10^{-6} - \dfrac{(6.3 \times 10^{-8})^2 - 1.0 \times 10^{-14}}{6.3 \times 10^{-8}}}$$

$$3.5 \times 10^{-8} = \frac{6.3 \times 10^{-8}([OCl^-] - 9.57 \times 10^{-8})}{1.0 \times 10^{-6} + 9.57 \times 10^{-8}} \quad \text{(Carrying extra sig. figs.)}$$

$3.83 \times 10^{-14} = 6.3 \times 10^{-8}$ ($[OCl^-] - 9.57 \times 10^{-8}$), $[OCl^-] = [KOCl] = 7.0 \times 10^{-7}\ M$

41. Using regular procedures, $pH = pK_a = -\log(1.6 \times 10^{-7}) = 6.80$ since $[A^-]_o = [HA]_o$ in this buffer solution. However, the pH is very close to that of neutral water so maybe we need to consider the H^+ contribution from water. Another problem with this answer is that x (= $[H^+]$) is not small as compared to $[HA]_o$ and $[A^-]_o$, which was assumed when solving using the regular procedures. Since the concentrations of the buffer components are less than $10^{-6}\ M$, then let us use the expression for the exact treatment of buffers to solve.

$$K_a = 1.6 \times 10^{-7} = \frac{[H^+]\left([A^-]_o + \dfrac{[H^+]^2 - K_w}{[H^+]}\right)}{[HA]_o - \dfrac{[H^+]^2 - K_w}{[H^+]}} = \frac{[H^+]\left(5.0 \times 10^{-7} + \dfrac{[H^+]^2 - 1.0 \times 10^{-14}}{[H^+]}\right)}{5.0 \times 10^{-7} - \dfrac{[H^+]^2 - 1.0 \times 10^{-14}}{[H^+]}}$$

Solving exactly requires solving a cubic equation. Instead, we will use the method of successive approximations where our initial guess for $[H^+] = 1.6 \times 10^{-7}\ M$ (the value obtained using the regular procedures).

$$1.6 \times 10^{-7} = \frac{[H^+]\left(5.0 \times 10^{-7} + \dfrac{(1.6 \times 10^{-7})^2 - 1.0 \times 10^{-14}}{1.6 \times 10^{-7}}\right)}{5.0 \times 10^{-7} - \dfrac{(1.6 \times 10^{-7})^2 - 1.0 \times 10^{-14}}{1.6 \times 10^{-7}}}, \quad [H^+] = 1.1 \times 10^{-7}$$

We continue the process using 1.1×10^{-7} as our estimate for $[H^+]$. This gives $[H^+] = 1.5 \times 10^{-7}$. We continue the process until we get a self consistent answer. After three more iterations, we converge on $[H^+] = 1.3 \times 10^{-7}\ M$. Solving for the pH:

$$pH = -\log(1.3 \times 10^{-7}) = 6.89$$

Note that if we were to solve this problem exactly (using the quadratic formula) while ignoring the H^+ contribution from water, the answer comes out to $[H^+] = 1.0 \times 10^{-7}\ M$. We get a significantly different answer when we consider the H^+ contribution from H_2O.

Acid-Base Titrations

43.

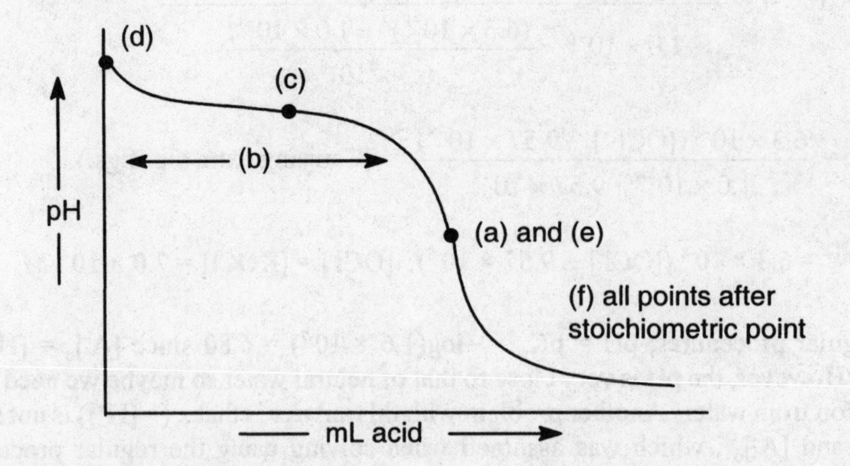

$B + H^+ \rightarrow BH^+$; Added H^+ from the strong acid converts the weak base, B, into its conjugate acid, BH^+. Initially, before any H^+ is added (point d), B is the dominant species present. After H^+ is added, both B and BH^+ are present and a buffered solution results (region b). At the equivalence point (points a and e), exactly enough H^+ has been added to convert all of the weak base present initially into its conjugate acid, BH^+. Past the equivalence point (region f), excess H^+ is present. For the answer to b, we included almost the entire buffer region. The maximum buffer region is around the halfway point to equivalence (point c) where [B] = [BH^+]. Here, pH = pK_a which is a characteristic of a best buffer.

45. **This is a strong acid ($HClO_4$) titrated by a strong base (KOH). Added OH^- from the strong base will react completely with the H^+ present from the strong acid to produce H_2O.**

a. **Only strong acid present. [H^+] = 0.200 *M*; pH = 0.699**

b. **mmol OH^- added = 10.0 mL × $\frac{0.100 \text{ mmol OH}^-}{\text{mL}}$ = 1.00 mmol OH^-**

mmol H^+ present = 40.0 mL × $\frac{0.200 \text{ mmol H}^+}{\text{mL}}$ = 8.0 mmol H^+

Note: The units mmoles are usually easier numbers to work with. The units for molarity are moles/L but are also equal to mmoles/mL.

	H^+	+	OH^-	→	H_2O
Before	**8.00 mmol**		**1.00 mmol**		
Change	**-1.00 mmol**		**-1.00 mmol**		**Reacts completely**
After	**7.00 mmol**		**0**		

The excess H^+ determines the pH. $[H^+]_{excess} = \frac{7.00 \text{ mmol H}^+}{40.0 \text{ mL} + 10.0 \text{ mL}}$ = 0.140 *M*; pH = 0.854

c. mmol OH^- added = 40.0 mL × 0.100 *M* = 4.00 mmol OH^-

	H^+	+	OH^-	→	H_2O
Before	8.00 mmol		4.00 mmol		
After	4.00 mmol		0		

$$[H^+]_{excess} = \frac{4.00 \text{ mmol}}{(40.0 + 40.0) \text{ mL}} = 0.0500\ M; \ \text{pH} = 1.301$$

d. mmol OH^- added = 80.0 mL × 0.100 *M* = 8.00 mmol OH^-; This is the equivalence point since we have added just enough OH^- to react with all the acid present. For a strong acid-strong base titration, pH = 7.00 at the equivalence point since only neutral species are present (K^+, ClO_4^-, H_2O).

e. mmol OH^- added = 100.0 mL × 0.100 *M* = 10.0 mmol OH^-

	H^+	+	OH^-	→	H_2O
Before	8.00 mmol		10.0 mmol		
After	0		2.0 mmol		

Past the equivalence point, the pH is determined by the excess OH^- present.

$$[OH^-]_{excess} = \frac{2.0 \text{ mmol}}{(40.0 + 100.0) \text{ mL}} = 0.014\ M; \ \text{pOH} = 1.85; \ \text{pH} = 12.15$$

47. This is a weak acid ($HC_2H_3O_2$) titrated by a strong base (KOH).

a. Only weak acid is present. Solving the weak acid problem:

	$HC_2H_3O_2$	⇌	H^+	+	$C_2H_3O_2^-$
Initial	0.200 *M*		~0		0
	x mol/L $HC_2H_3O_2$ dissociates to reach equilibrium				
Change	-*x*	→	+*x*		+*x*
Equil.	0.200 - *x*		*x*		*x*

$$K_a = 1.8 \times 10^{-5} = \frac{x^2}{0.200 - x} \approx \frac{x^2}{0.200}, \ x = [H^+] = 1.9 \times 10^{-3}\ M$$

pH = 2.72; Assumptions good.

b. The added OH^- will react completely with the best acid present, $HC_2H_3O_2$.

$$\text{mmol } HC_2H_3O_2 \text{ present} = 100.0 \text{ mL} \times \frac{0.200 \text{ mmol } HC_2H_3O_2}{\text{mL}} = 20.0 \text{ mmol } HC_2H_3O_2$$

$$\text{mmol } OH^- \text{ added} = 50.0 \text{ mL} \times \frac{0.100 \text{ mmol } OH^-}{\text{mL}} = 5.00 \text{ mmol } OH^-$$

	$HC_2H_3O_2$	+ OH^-	$\rightarrow$	$C_2H_3O_2^-$ + H_2O	
Before	20.0 mmol	5.00 mmol		0	
Change	-5.00 mmol	-5.00 mmol	$\rightarrow$	+5.00 mmol	Reacts completely
After	15.0 mmol	0		5.00 mmol	

After reaction of all the strong base, we have a buffer solution containing a weak acid ($HC_2H_3O_2$) and its conjugate base ($C_2H_3O_2^-$). We will use the Henderson-Hasselbalch equation to solve for the pH.

$$pH = pK_a + \log \frac{[C_2H_3O_2^-]}{[HC_2H_3O_2]} = -\log(1.8 \times 10^{-5}) + \log\left(\frac{5.00\ \text{mmol}/V_T}{15.0\ \text{mmol}/V_T}\right) \quad \text{where } V_T = \text{total volume}$$

$$pH = 4.74 + \log\left(\frac{5.00}{15.0}\right) = 4.74 + (-0.477) = 4.26$$

Note that the total volume cancels in the Henderson-Hasselbalch equation. For the [base]/[acid] term, the mole ratio equals the concentration ratio since the components of the buffer are always in the same volume of solution.

c. mmol OH^- added = 100.0 mL × 0.100 mmol OH^-/mL = 10.0 mmol OH^-; The same amount (20.0 mmol) of $HC_2H_3O_2$ is present as before (it never changes). As before, let the OH^- react to completion, then see what is remaining in solution after this reaction.

	$HC_2H_3O_2$	+ OH^-	$\rightarrow$	$C_2H_3O_2^-$ + H_2O
Before	20.0 mmol	10.0 mmol		0
After	10.0 mmol	0		10.0 mmol

A buffer solution results after reaction. Since $[C_2H_3O_2^-] = [HC_2H_3O_2] = 10.0$ mmol/total volume, then pH = pK_a. This is always true at the halfway point to equivalence for a weak acid/strong base titration, pH = pK_a.

$$pH = -\log(1.8 \times 10^{-5}) = 4.74$$

d. mmol OH^- added = 150.0 mL × 0.100 *M* = 15.0 mmol OH^-. Added OH^- reacts completely with the weak acid.

	$HC_2H_3O_2$	+ OH^-	$\rightarrow$	$C_2H_3O_2^-$ + H_2O
Before	20.0 mmol	15.0 mmol		0
After	5.0 mmol	0		15.0 mmol

We have a buffer solution after all the OH^- reacts to completion. Using the Henderson-Hasselbalch equation:

$$pH = 4.74 + \log \frac{[C_2H_3O_2^-]}{[HC_2H_3O_2]} = 4.74 + \log\left(\frac{15.0\ \text{mmol}}{5.0\ \text{mmol}}\right) \quad \text{(Total volume cancels, so we can use mol ratios.)}$$

$$pH = 4.74 + 0.48 = 5.22$$

e. mmol OH^- added = 200.00 mL × 0.100 *M* = 20.0 mmol OH^-; As before, let the added OH^- react to completion with the weak acid, then see what is in solution after this reaction.

	$HC_2H_3O_2$	+	OH^-	→	$C_2H_3O_2^-$ + H_2O
Before	20.0 mmol		20.0 mmol		0
After	0		0		20.0 mmol

This is the equivalence point. Enough OH^- has been added to exactly neutralize all the weak acid present initially. All that remains that affects the pH at the equivalence point is the conjugate base of the weak acid, $C_2H_3O_2^-$. This is a weak base equilibrium problem.

$$C_2H_3O_2^- + H_2O \rightleftharpoons HC_2H_3O_2 + OH^- \quad K_b = \frac{K_w}{K_a} = \frac{1.0 \times 10^{-14}}{1.8 \times 10^{-5}} = 5.6 \times 10^{-10}$$

	$C_2H_3O_2^-$		$HC_2H_3O_2$	OH^-
Initial	20.0 mmol/300.0 mL		0	0
	x mol/L $C_2H_3O_2^-$ reacts with H_2O to reach equilibrium			
Change	$-x$	→	$+x$	$+x$
Equil.	$0.0667 - x$		x	x

$$K_b = 5.6 \times 10^{-10} = \frac{x^2}{0.0667 - x} \approx \frac{x^2}{0.0667}, \quad x = [OH^-] = 6.1 \times 10^{-6}\ M$$

pOH = 5.21; pH = 8.79; Assumptions good.

f. mmol OH^- added = 250.0 mL × 0.100 *M* = 25.0 mmol OH^-

	$HC_2H_3O_2$	+	OH^-	→	$C_2H_3O_2^-$ + H_2O
Before	20.0 mmol		25.0 mmol		0
After	0		5.0 mmol		20.0 mmol

After the titration reaction, we have a solution containing excess OH^- and a weak base, $C_2H_3O_2^-$. When a strong base and a weak base are both present, assume the amount of OH^- added from the weak base will be minimal, i.e., the pH past the equivalence point is determined by the amount of excess base.

$$[OH^-]_{excess} = \frac{5.0\ \text{mmol}}{100.0\ \text{mL} + 250.0\ \text{mL}} = 0.014\ M;\ \text{pOH} = 1.85;\ \text{pH} = 12.15$$

49. We will do sample calculations for the various parts of the titration. All results are summarized in Table 8.1 at the end of Exercise 8.51.

At the beginning of the titration, only the weak acid $HC_3H_5O_3$ is present.

$$HLac \rightleftharpoons H^+ + Lac^- \quad K_a = 10^{-3.86} = 1.4 \times 10^{-4} \quad HLac = HC_3H_5O_3,\ Lac^- = C_3H_5O_3^-$$

	HLac		H^+	Lac^-
Initial	0.100 *M*		~0	0
	x mol/L HLac dissociates to reach equilibrium			
Change	$-x$	→	$+x$	$+x$
Equil.	$0.100 - x$		x	x

$1.4 \times 10^{-4} = \frac{x^2}{0.100 - x} \approx \frac{x^2}{0.100}$, $x = [H^+] = 3.7 \times 10^{-3}$ *M*; pH = 2.43 **Assumptions good.**

Up to the stoichiometric point, we calculate the pH using the Henderson-Hasselbalch equation. This is the buffer region. For example, at 4.0 mL of NaOH added:

initial mmol HLac present $= 25.0 \text{ mL} \times \frac{0.100 \text{ mmol}}{\text{mL}} = 2.50$ **mmol HLac**

mmol OH^- **added** $= 4.0 \text{ mL} \times \frac{0.100 \text{ mmol}}{\text{mL}} = 0.40$ **mmol** OH^-

Note: The units mmol are usually easier numbers to work with. The units for molarity are moles/L but are also equal to mmoles/mL.

The 0.40 mmol added OH^- **converts 0.40 mmoles HLac to 0.40 mmoles** Lac^- **according to the equation:**

$HLac + OH^- \rightarrow Lac^- + H_2O$ **Reacts completely**

mmol HLac remaining = 2.50 - 0.40 = 2.10 mmol; mmol Lac^- **produced = 0.40 mmol**

We have a buffer solution. Using the Henderson-Hasselbalch equation where $pK_a = 3.86$:

$$pH = pK_a + \log \frac{[Lac^-]}{[HLac]} = 3.86 + \log \frac{(0.40)}{(2.10)}$$

(Total volume cancels, so we can use use the ratio of moles or mmoles.)

$$pH = 3.86 - 0.72 = 3.14$$

Other points in the buffer region are calculated in a similar fashion. Perform a stoichiometry problem first, followed by a buffer problem. The buffer region includes all points up to 24.9 mL OH^- **added.**

At the stoichiometric point (25.0 mL OH^- **added), we have added enough** OH^- **to convert all of the HLac (2.50 mmol) into its conjugate base,** Lac^-**. All that is present is a weak base. To determine the pH, we perform a weak base calculation.**

$$[Lac^-]_o = \frac{2.50 \text{ mmol}}{25.0 \text{ mL} + 25.0 \text{ mL}} = 0.0500\ M$$

$$Lac^- + H_2O \rightleftharpoons HLac + OH^- \quad K_b = \frac{1.0 \times 10^{-14}}{1.4 \times 10^{-4}} = 7.1 \times 10^{-11}$$

	Lac^-		$HLac$	OH^-
Initial	0.0500 *M*		0	0
	x mol/L Lac^- reacts with H_2O to reach equilibrium			
Change	$-x$	→	$+x$	$+x$
Equil.	$0.0500 - x$		x	x

$$K_b = \frac{x^2}{0.0500 - x} \approx \frac{x^2}{0.0500} = 7.1 \times 10^{-11}$$

$x = [OH^-] = 1.9 \times 10^{-6}\ M$; pOH = 5.72; pH = 8.28 Assumptions good.

Past the stoichiometric point, we have added more than 2.50 mmol of NaOH. The pH will be determined by the excess OH^- ion present. An example of this calculation follows.

At 25.1 mL: OH^- added $= 25.1\ mL \times \frac{0.100\ mmol}{mL} = 2.51$ mmol OH^-; 2.50 mmol OH^- neutralizes all the weak acid present. The remainder is excess OH^-.

Excess $OH^- = 2.51 - 2.50 = 0.01$ mmol OH^-

$$[OH^-]_{excess} = \frac{0.01\ mmol}{(25.0 + 25.1)\ mL} = 2 \times 10^{-4}\ M;\ pOH = 3.7;\ pH = 10.3$$

All results are listed in Table 8.1 at the end of the solution to Exercise 8.51.

51. At beginning of the titration, only the weak base NH_3 is present. As always, solve for the pH using the K_b reaction for NH_3.

	NH_3 + H_2O	$\rightleftharpoons$	NH_4^+	+	OH^-	$K_b = 1.8 \times 10^{-5}$
Initial	0.100 *M*		0		~0	
Equil.	0.100 - x		x		x	

$$K_b = \frac{x^2}{0.100 - x} \approx \frac{x^2}{0.100} = 1.8 \times 10^{-5}$$

$x = [OH^-] = 1.3 \times 10^{-3}\ M$; pOH = 2.89; pH = 11.11 Assumptions good.

In the buffer region (4.0 - 24.9 mL), we can use the Henderson-Hasselbalch equation:

$$K_a = \frac{1.0 \times 10^{-14}}{1.8 \times 10^{-5}} = 5.6 \times 10^{-10};\ pK_a = 9.25;\ pH = 9.25 + \log\frac{[NH_3]}{[NH_4^+]}$$

We must determine the amounts of NH_3 and NH_4^+ present after the added H^+ reacts completely with the NH_3. For example, after 8.0 mL HCl added:

$$\text{initial mmol } NH_3 \text{ present} = 25.0\ mL \times \frac{0.100\ mmol}{mL} = 2.50\ mmol\ NH_3$$

$$\text{mmol } H^+ \text{ added} = 8.0\ mL \times \frac{0.100\ mmol}{mL} = 0.80\ mmol\ H^+$$

Added H^+ reacts with NH_3 to completion: $NH_3 + H^+ \rightarrow NH_4^+$

mmol NH_3 remaining = 2.50 - 0.80 = 1.70 mmol; mmol NH_4^+ produced = 0.80 mmol

$$pH = 9.25 + \log\frac{1.70}{0.80} = 9.58$$ (Mole ratios can be used since the total volume cancels.)

Other points in the buffer region are calculated in similar fashion. Results are summarized in Table 8.1 at the end of Exercise 8.51.

At the stoichiometric point (25.0 mL H^+ added), just enough HCl has been added to convert all of

the weak base (NH_3) into its conjugate acid (NH_4^+). Perform a weak acid calculation. $[NH_4^+]_o$ = 2.50 mmol/50.0 mL = 0.0500 *M*

	NH_4^+	$\rightleftharpoons$	H^+	+	NH_3	$K_a = 5.6 \times 10^{-10}$
Initial	**0.0500 *M***		**0**		**0**	
Equil.	**0.0500 - *x***		***x***		***x***	

$$5.6 \times 10^{-10} = \frac{x^2}{0.0500 - x} = \frac{x^2}{0.0500}, \quad x = [H^+] = 5.3 \times 10^{-6}\ M;\ pH = 5.28$$ **Assumptions good.**

Beyond the stoichiometric point, the pH is determined by the excess H^+. For example, at 28.0 mL of H^+ added:

$$H^+ \text{ added} = 28.0\ mL \times \frac{0.100\ mmol}{mL} = 2.80\ mmol\ H^+$$

$$\text{Excess } H^+ = 2.80\ mmol - 2.50\ mmol = 0.30\ mmol \text{ excess } H^+$$

$$[H^+]_{excess} = \frac{0.30\ mmol}{(25.0 + 28.0)\ mL} = 5.7 \times 10^{-3}\ M;\ pH = 2.24$$

All results are summarized in Table 8.1.

Table 8.1: Summary of pH Results for Exercises 8.49 and 8.51 (Graph follows)

titrant mL	Exercise 8.49	Exercise 8.51
0.0	2.43	11.11
4.0	3.14	9.97
8.0	3.53	9.58
12.5	3.86	9.25
20.0	4.46	8.65
24.0	5.24	7.87
24.5	5.6	7.6
24.9	6.3	6.9
25.0	8.28	5.28
25.1	10.3	3.7
26.0	11.29	2.71
28.0	11.75	2.24
30.0	11.96	2.04

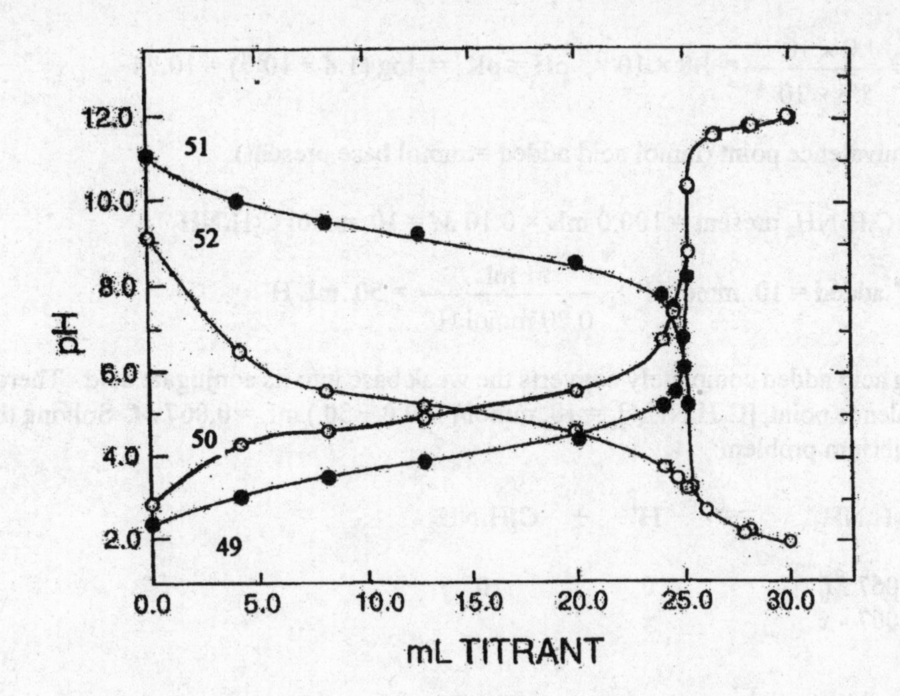

53. a. **This is a weak acid/strong base titration. At the halfway point to equivalence, [weak acid] = [conjugate base], so pH = pK_a (always for a weak acid/strong base titration).**

pH = -log (6.4 × 10^{-5}) = 4.19

mmol $HC_7H_5O_2$ present = 100. mL × 0.10 *M* = 10. mmol $HC_7H_5O_2$. For the equivalence point, 10. mmol of OH^- must be added. The volume of OH^- added to reach the equivalence point is:

$$10.\text{ mmol OH}^- \times \frac{1\text{ mL}}{0.10\text{ mmol OH}^-} = 1.0 \times 10^2\text{ mL OH}^-$$

At the equivalence point, 10. mmol of $HC_7H_5O_2$ is neutralized by 10. mmol of OH^- to produce 10. mmol of $C_7H_5O_2^-$. This is a weak base. The total volume of the solution is 100.0 mL + 1.0 × 10^2 mL = 2.0 × 10^2 mL. Solving the weak base equilibrium problem:

$$C_7H_5O_2^- + H_2O \rightleftharpoons HC_7H_5O_2 + OH^- \quad K_b = \frac{K_w}{K_a} = \frac{1.0\times10^{-14}}{6.4\times10^{-5}} = 1.6\times10^{-10}$$

	$C_7H_5O_2^-$	$HC_7H_5O_2$	OH^-
Initial	10. mmol/2.0 × 10^2 mL	0	0
Equil.	0.050 - x	x	x

$$K_b = 1.6\times10^{-10} = \frac{x^2}{0.050 - x} \approx \frac{x^2}{0.050},\ x = [OH^-] = 2.8\times10^{-6}\ M$$

pOH = 5.55; pH = 8.45 Assumptions good.

b. **At the halfway point to equivalence for a weak base/strong acid titration, pH = pK_a since [weak base] = [conjugate acid].**

$$K_a = \frac{K_w}{K_b} = \frac{1.0 \times 10^{-14}}{5.6 \times 10^{-4}} = 1.8 \times 10^{-11};\ pH = pK_a = -\log(1.8 \times 10^{-11}) = 10.74$$

For the equivalence point (mmol acid added = mmol base present):

$$\text{mmol } C_2H_5NH_2 \text{ present} = 100.0 \text{ mL} \times 0.10\ M = 10. \text{ mmol } C_2H_5NH_2$$

$$\text{mL } H^+ \text{ added} = 10. \text{ mmol } H^+ \times \frac{1 \text{ mL}}{0.20 \text{ mmol } H^+} = 50. \text{ mL } H^+$$

The strong acid added completely converts the weak base into its conjugate acid. Therefore, at the equivalence point, $[C_2H_5NH_3^+]_o = 10.$ mmol/(100.0 + 50.) mL = 0.067 M. Solving the weak acid equilibrium problem:

$$C_2H_5NH_3^+ \rightleftharpoons H^+ + C_2H_5NH_2$$

	$C_2H_5NH_3^+$	H^+	$C_2H_5NH_2$
Initial	0.067 M	0	0
Equil.	0.067 - x	x	x

$$K_a = 1.8 \times 10^{-11} = \frac{x^2}{0.067 - x} \approx \frac{x^2}{0.067},\ x = [H^+] = 1.1 \times 10^{-6}\ M$$

pH = 5.96; Assumptions good.

c. **In a strong acid/strong base titration, the halfway point has no special significance other than exactly one-half of the original amount of acid present has been neutralized.**

$$\text{mmol } H^+ \text{ present} = 100.0 \text{ mL} \times 0.50\ M = 50. \text{ mmol } H^+$$

$$\text{mL } OH^- \text{ added} = 25. \text{ mmol } OH^- \times \frac{1 \text{ mL}}{0.25 \text{ mmol}} = 1.0 \times 10^2 \text{ mL } OH^-$$

$$H^+ + OH^- \rightarrow H_2O$$

	H^+	OH^-
Before	50. mmol	25 mmol
After	25 mmol	0

$$[H^+]_{excess} = \frac{25 \text{ mmol}}{(100.0 + 1.0 \times 10^2) \text{ mL}} = 0.13\ M;\ pH = 0.89$$

At the equivalence point of a strong acid/strong base titration, only neutral species are present (Na^+, Cl^-, H_2O) so the pH = 7.00.

55. a. **1.00 L × 0.100 mol/L = 0.100 mol HCl added to reach stoichiometric point.**

The 10.00 g sample must have contained 0.100 mol of NaA. $\frac{10.00 \text{ g}}{0.100 \text{ mol}} = 100.$ g/mol

b. **500.0 mL of HCl added represents the halfway point to equivalence. So, pH = pK_a = 5.00 and $K_a = 1.0 \times 10^{-5}$. At the equivalence point, enough H^+ has been added to convert all the A^- present initially into HA. The concentration of HA at the equivalence point is:**

$$[HA]_0 = \frac{0.100 \text{ mol}}{1.10 \text{ L}} = 0.0909 \, M$$

	HA	⇌	H^+	+	A^-	$K_a = 1.0 \times 10^{-5}$
Initial	0.0909 M		0		0	
Equil.	0.0909 - x		x		x	

$$K_a = 1.0 \times 10^{-5} = \frac{x^2}{0.0909 - x} \approx \frac{x^2}{0.0909}$$

$x = 9.5 \times 10^{-4} \, M = [H^+]$; pH = 3.02 Assumptions good.

57. a. yellow b. green (Both yellow and blue forms are present.)

c. yellow d. blue

59. Equivalence point: when enough titrant has been added to react exactly with the substance in solution being titrated. End point: indicator changes color. We want the indicator to tell us when we have reached the equivalence point. We can detect the end point visually and assume it is the equivalence point for doing stoichiometric calculations. They don't have to be as close as 0.01 pH unit since at the equivalence point the pH is changing very rapidly with added titrant. The range over which an indicator changes color only needs to be close to the pH of the equivalence point.

61. When choosing an indicator, we want the color change of the indicator to occur approximately at the pH of the equivalence point. Since the pH generally changes very rapidly at the equivalence point, we don't have to be exact. This is especially true for strong acid/strong base titrations. Some choices where color change occurs at about the pH of the equivalence point are:

Exercise	pH at eq. pt.	Indicator
8.45	7.00	bromthymol blue or phenol red
8.47	8.79	o-cresolphthalein or phenolphthalein

63.

Exercise	pH at eq. pt.	Indicator
8.49	8.28	o-cresolphthalein or phenolphthalein
8.51	5.28	bromcresol green

65. The color of the indicator will change over the approximate range of pH = $pK_a \pm 1 = 5.3 \pm 1$. Therefore, the useful pH range of methyl red where it changes color would be about: 4.3 (red) - 6.3 (yellow). Note that at pH < 4.3, the HIn form of the indicator dominates and the color of the solution is the color of HIn (red). At pH > 6.3, the In^- form of the indicator dominates and the color of the solution is the color of In^- (yellow). In titrating a weak acid with base, we start off with an acidic solution with pH < 4.3 so the color would change from red to reddish-orange at pH ~ 4.3. In titrating a weak base with acid, the color change would be from yellow to yellowish-orange at pH ~ 6.3. Only a weak base/strong acid titration would have an acidic pH at the equivalence point so only in this type of titration would the color change of methyl red indicate the approximate endpoint.

67. 100.0 mL × 0.0500 M = 5.00 mmol H_3X initially

a. Since $K_{a_1} >> K_{a_2} >> K_{a_3}$, pH initially determined by H_3X.

	H_3X	$\rightleftharpoons$	H^+	+	H_2X^-
Initial	0.0500 M		~0		0
Equil.	0.0500 - x		x		x

$$K_{a_1} = 1.0 \times 10^{-3} = \frac{x^2}{0.0500 - x} \approx \frac{x^2}{0.0500},\ x = 7.1 \times 10^{-3} \quad \text{Assumption poor.}$$

Using the quadratic formula:

$$x^2 + 1.0 \times 10^{-3}\, x - 5.0 \times 10^{-5} = 0, \quad x = 6.6 \times 10^{-3}\ M = [H^+];\ pH = 2.18$$

b. 1.00 mmol OH^- added converts H_3X into H_2X^-. After this reaction goes to completion, 4.00 mmol H_3X and 1.00 mmol H_2X^- are in a total volume of 110.0 mL. Solving the buffer problem:

	H_3X	$\rightleftharpoons$	H^+	+	H_2X^-
Initial	0.0364 M		~0		0.00909 M
Equil.	0.0364 - x		x		0.00909 + x

$$K_{a_1} = 1.0 \times 10^{-3} = \frac{x(0.00909 + x)}{0.0364 - x} \quad \text{Assumption that } x \text{ is small does not work here.}$$

Using the quadratic formula and carrying extra sig. figs: $x^2 + 1.01 \times 10^{-2}\, x - 3.64 \times 10^{-5} = 0$

$$x = 2.8 \times 10^{-3}\ M = [H^+];\ pH = 2.55$$

c. 2.50 mmol OH^- added, results in 2.50 mmol H_3X and 2.50 mmol H_2X^- after OH^- reacts completely with H_3X. This is the first halfway point to equivalence. $pH = pK_{a_1} = 3.00$; Assumptions good (5% error).

d. 5.00 mmol OH^- added, results in 5.00 mmol H_2X^- after OH^- reacts completely with H_3X. This is the 1st stoichiometric point.

$$pH = \frac{pK_{a_1} + pK_{a_2}}{2} = \frac{3.00 + 7.00}{2} = 5.00$$

e. 6.0 mmol OH^- added, results in 4.00 mmol H_2X^- and 1.00 mmol HX^{2-} after OH^- reacts completely with H_3X and then reacts completely with H_2X^-.

Using the $H_2X^- \rightleftharpoons H^+ + HX^{2-}$ reaction:

$$pH = pK_{a2} + \log\frac{[HX^{2-}]}{[H_2X^-]} = 7.00 - \log(1.00/4.00) = 6.40 \qquad \text{Assumptions good.}$$

f. 7.50 mmol KOH added, results in 2.50 mmol H_2X^- and 2.50 mmol HX^{2-} after OH^- reacts completely. This is the second halfway point to equivalence.

$pH = pK_{a2} = 7.00$ Assumptions good.

g. 10.0 mmol OH^- added, results in 5.0 mmol HX^{2-} after OH^- reacts completely. This is the 2nd stoichiometric point.

$$pH = \frac{pK_{a_2} + pK_{a_3}}{2} = \frac{7.00 + 12.00}{2} = 9.50$$

h. 12.5 mmol OH^- added, results in 2.5 mmol HX^{2-} and 2.5 mmol X^{3-} after OH^- reacts completely with H_3X first, then H_2X^- and finally HX^{2-}. This is the third halfway point to equivalence. Usually $pH = pK_{a_3}$ but normal assumptions don't hold. We must solve for the pH exactly.

$[X^{3-}] = [HX^{2-}] = 2.5 \text{ mmol}/225.0 \text{ mL} = 1.1 \times 10^{-2}\ M$

	X^{3-} + H_2O	⇌	HX^{2-} +	OH^-	$K_b = \frac{K_w}{K_{a_3}} = 1.0 \times 10^{-2}$
Initial	0.011 *M*		0.011 *M*	0	
Equil.	0.011 - *x*		0.011 + *x*	*x*	

$K_b = 1.0 \times 10^{-2} = \frac{x(0.011 + x)}{(0.011 - x)}$; Using the quadratic formula:

$x^2 + 2.1 \times 10^{-2}x - 1.1 \times 10^{-4} = 0$, $x = 4.3 \times 10^{-3}\ M = [OH^-]$; pH = 11.63

i. 15.0 mmol OH^- added, results in 5.0 mmol X^{3-} after OH^- reacts completely. This is the 3rd stoichiometric point.

	X^{3-} + H_2O	⇌	HX^{2-} +	OH^-	$K_b = \frac{K_w}{K_{a_3}} = 1.0 \times 10^{-2}$
Initial	$\frac{5.0 \text{ mmol}}{250.0 \text{ mL}} = 0.020\ M$		0	0	
Equil.	0.020 - *x*		*x*	*x*	

$K_b = \frac{x^2}{0.020 - x} = 1.0 \times 10^{-2} \approx \frac{x^2}{0.020}$, $x = 1.4 \times 10^{-2}$ Assumption poor.

Using the quadratic formula: $x^2 + 1.0 \times 10^{-2}x - 2.0 \times 10^{-4} = 0$

$x = [OH^-] = 1.0 \times 10^{-2}\ M$; pH = 12.00

j. 20.0 mmol OH^- added, results in 5.0 mmol X^{3-} and 5.0 mmol OH^- excess after OH^- reacts completely. Since K_b for X^{3-} is fairly large for a weak base, we have to worry about the OH^- contribution from X^{3-}.

$[X^{3-}] = [OH^-] = \frac{5.0 \text{ mmol}}{300.0 \text{ mL}} = 1.7 \times 10^{-2}\ M$

	X^{3-} + H_2O	$\rightleftharpoons$	OH^-	+	HX^{2-}
Initial	$1.7 \times 10^{-2}\ M$		$1.7 \times 10^{-2}\ M$		0
Equil.	$1.7 \times 10^{-2} - x$		$1.7 \times 10^{-2} + x$		x

$$K_b = \frac{[OH^-][HX^{2-}]}{[X^{3-}]} = 1.0 \times 10^{-2} = \frac{(1.7 \times 10^{-2} + x)x}{(1.7 \times 10^{-2} - x)}$$

Using the quadratic formula: $x^2 + 2.7 \times 10^{-2}\ x - 1.7 \times 10^{-4} = 0,\ x = 5.3 \times 10^{-3}\ M$

$[OH^-] = 1.7 \times 10^{-2} + x = 1.7 \times 10^{-2} + 5.3 \times 10^{-3} = 2.2 \times 10^{-2}\ M$; pH = 12.34

69. a. Na^+ is present in all solutions. The added H^+ from HCl reacts completely with CO_3^{2-} to convert it into HCO_3^-. After all CO_3^{2-} is reacted (after point C, the first equivalence point), then H^+ reacts completely with the next best base present, HCO_3^-. Point E represents the second equivalence point. The major species present at the various points after H^+ reacts completely follow.

A. CO_3^{2-}, H_2O B. CO_3^{2-}, HCO_3^-, H_2O, Cl^-

C. HCO_3^-, H_2O, Cl^- D. HCO_3^-, CO_2 (H_2CO_3), H_2O, Cl^-

E. CO_2 (H_2CO_3), H_2O, Cl^- F. H^+ (excess), CO_2 (H_2CO_3), H_2O, Cl^-

b. Point A (initially):

$$CO_3^{2-} + H_2O \rightleftharpoons HCO_3^- + OH^- \quad K_b(CO_3^{2-}) = \frac{K_w}{K_{a_2}} = \frac{1.0 \times 10^{-14}}{4.8 \times 10^{-11}} = 2.1 \times 10^{-4}$$

	CO_3^{2-}	HCO_3^-	OH^-
Initial	$0.100\ M$	0	~0
Equil.	$0.100 - x$	x	x

$$K_b = 2.1 \times 10^{-4} = \frac{[HCO_3^-][OH^-]}{[CO_3^{2-}]} = \frac{x^2}{0.100 - x} \approx \frac{x^2}{0.100}$$

$x = 4.6 \times 10^{-3}\ M = [OH^-]$; pH = 11.66 Assumptions good.

Point B: The first halfway point where $[CO_3^{2-}] = [HCO_3^-]$.

$pH = pK_{a_2} = -\log(4.8 \times 10^{-11}) = 10.32$ Assumptions good.

Point C: First equivalence point (25.00 mL of 0.100 M HCl added). The amphoteric HCO_3^- is the major acid/base species present.

$$pH = \frac{pK_{a_1} + pK_{a_2}}{2};\ pK_{a_1} = -\log(4.3 \times 10^{-7}) = 6.37$$

$$pH = \frac{6.37 + 10.32}{2} = 8.35$$

Point D: The second halfway point where $[HCO_3^-] = [H_2CO_3]$.

$pH = pK_{a_1} = 6.37$ Assumptions good.

Point E: This is the second equivalence point where all of the CO_3^{2-} present initially has been converted into H_2CO_3 by the added strong acid. 50.0 mL HCl added. $[H_2CO_3] = 2.50$ mmol/75.0 mL = 0.0333 M

	H_2CO_3	$\rightleftharpoons$	H^+	+	HCO_3^-	$K_{a_1} = 4.3 \times 10^{-7}$
Initial	0.0333 M		0		0	
Equil.	0.0333 - x		x		x	

$$K_{a_1} = 4.3 \times 10^{-7} = \frac{x^2}{0.0333 - x} \approx \frac{x^2}{0.0333}$$

$x = [H^+] = 1.2 \times 10^{-4}$ M; pH = 3.92 Assumptions good.

Solubility Equilibria

71. In our set-ups, s = solubility in mol/L. Since solids do not appear in the K_{sp} expression, we do not need to worry about their initial or equilibrium amounts.

a.

	$Ag_3PO_4(s)$	$\rightleftharpoons$	$3\ Ag^+(aq)$	+	$PO_4^{3-}(aq)$
Initial			0		0
	s mol/L of $Ag_3PO_4(s)$ dissolves to reach equilibrium				
Change	$-s$	$\rightarrow$	$+3s$		$+s$
Equil.			$3s$		s

$K_{sp} = 1.8 \times 10^{-18} = [Ag^+]^3\ [PO_4^{3-}] = (3s)^3(s) = 27s^4$

$27s^4 = 1.8 \times 10^{-18}$, $s = (6.7 \times 10^{-20})^{1/4} = 1.6 \times 10^{-5}$ mol/L = molar solubility

$$\frac{1.6 \times 10^{-5}\ \text{mol}\ Ag_3PO_4}{\text{L}} \times \frac{418.7\ \text{g}\ Ag_3PO_4}{\text{mol}\ Ag_3PO_4} = 6.7 \times 10^{-3}\ \text{g/L}$$

b.

	$CaCO_3(s)$	$\rightleftharpoons$	$Ca^{2+}(aq)$	+	$CO_3^{2-}(aq)$
Initial	s = solubility (mol/L)		0		0
Equil.			s		s

$K_{sp} = 8.7 \times 10^{-9} = [Ca^{2+}]\ [CO_3^{2-}] = s^2$, $s = 9.3 \times 10^{-5}$ mol/L

$$\frac{9.3 \times 10^{-5}\ \text{mol}}{\text{L}} \times \frac{100.1\ \text{g}}{\text{mol}} = 9.3 \times 10^{-3}\ \text{g/L}$$

c.	$Hg_2Cl_2(s)$	$\rightleftharpoons$	$Hg_2^{2+}(aq)$	+	$2\ Cl^-(aq)$
Initial	s = solubility (mol/L)		0		0
Equil.			s		$2s$

$K_{sp} = 1.1 \times 10^{-18} = [Hg_2^{2+}]\,[Cl^-]^2 = (s)(2s)^2 = 4s^3,\ s = 6.5 \times 10^{-7}$ mol/L

$$\frac{6.5 \times 10^{-7}\ \text{mol}}{\text{L}} \times \frac{472.1\ \text{g}}{\text{mol}} = 3.1 \times 10^{-4}\ \text{g/L}$$

73. In our set-up, s = solubility of the ionic solid in mol/L. This is defined as the maximum amount of a salt which can dissolve. Since solids do not appear in the K_{sp} expression, we do not need to worry about their initial and equilibrium amounts.

a.	$CaC_2O_4(s)$	$\rightleftharpoons$	$Ca^{2+}(aq)$	+	$C_2O_4^{2-}(aq)$
Initial			0		0
	s mol/L of $CaC_2O_4(s)$ dissolves to reach equilibrium				
Change	$-s$	$\rightarrow$	$+s$		$+s$
Equil.			s		s

From the problem, $s = 4.8 \times 10^{-5}$ mol/L.

$K_{sp} = [Ca^{2+}]\,[C_2O_4^{2-}] = (s)(s) = s^2,\ K_{sp} = (4.8 \times 10^{-5})^2 = 2.3 \times 10^{-9}$

b.	$PbBr_2(s)$	$\rightleftharpoons$	$Pb^{2+}(aq)$	+	$2\ Br^-(aq)$
Initial			0		0
	s mol/L of $PbBr_2(s)$ dissolves to reach equilibrium				
Change	$-s$	$\rightarrow$	$+s$		$+2s$
Equil.			s		$2s$

From the problem, $s = [Pb^{2+}] = 2.14 \times 10^{-2}$ *M*. So:

$K_{sp} = [Pb^{2+}]\,[Br^-]^2 = s(2s)^2 = 4s^3,\ K_{sp} = 4(2.14 \times 10^{-2})^3 = 3.92 \times 10^{-5}$

c.	$BiI_3(s)$	$\rightleftharpoons$	$Bi^{3+}(aq)$	+	$3\ I^-(aq)$
Initial			0		0
	s mol/L of $BiI_3(s)$ dissolves to reach equilibrium				
Change	$-s$	$\rightarrow$	$+s$		$+3s$
Equil.			s		$3s$

$K_{sp} = [Bi^{3+}]\,[I^-]^3 = (s)(3s)^3 = 27\,s^4,\ K_{sp} = 27(1.32 \times 10^{-5})^4 = 8.20 \times 10^{-19}$

d.	$FeC_2O_4(s)$	$\rightleftharpoons$	Fe^{2+}	+	$C_2O_4^{2-}$
Initial	s = solubility (mol/L)		0		0
Equil.			s		s

$$s = \frac{65.9 \times 10^{-3}\ g}{L} \times \frac{1\ mol}{143.9\ g} = 4.58 \times 10^{-4}\ mol/L$$

$[Fe^{2+}] = [C_2O_4^{2-}] = 4.58 \times 10^{-4}\ M$; $K_{sp} = [Fe^{2+}]\,[C_2O_4^{2-}] = (4.58 \times 10^{-4})^2 = 2.10 \times 10^{-7}$

e. $Cu(IO_4)_2(s) \rightleftharpoons Cu^{2+} + 2\ IO_4^-$

		Cu^{2+}	$2\ IO_4^-$
Initial	s = solubility (mol/L	0	0
Equil.		s	$2s$

$$s = \frac{0.146\ g\ Cu(IO_4)_2}{0.100\ L} \times \frac{1\ mol}{445.4\ g} = 3.28 \times 10^{-3}\ mol/L$$

$K_{sp} = [Cu^{2+}]\,[IO_4^-]^2 = (3.28 \times 10^{-3})\,(6.56 \times 10^{-3})^2 = 1.41 \times 10^{-7}$

75. a. Since both solids dissolve to produce 3 ions in solution, then we can compare values of K_{sp} to determine relative solubility. Since the K_{sp} for CaF_2 is the smallest, then $CaF_2(s)$ has the smallest molar solubility.

b. We must calculate molar solubilities since each salt yields a different number of ions when it dissolves.

$Ca_3(PO_4)_2(s) \rightleftharpoons 3\ Ca^{2+}(aq) + 2\ PO_4^{3-}(aq)$ $\quad K_{sp} = 1.3 \times 10^{-32}$

		$3\ Ca^{2+}(aq)$	$2\ PO_4^{3-}(aq)$
Initial	s = solubility (mol/L)	0	0
Equil.		$3s$	$2s$

$K_{sp} = [Ca^{2+}]^3\,[PO_4^{3-}]^2 = (3s)^3(2s)^2 = 108s^5$, $\quad s = (1.3 \times 10^{-32}/108)^{1/5} = 1.6 \times 10^{-7}\ mol/L$

$FePO_4(s) \rightleftharpoons Fe^{3+}(aq) + PO_4^{3-}(aq)$ $\quad K_{sp} = 1.0 \times 10^{-22}$

		$Fe^{3+}(aq)$	$PO_4^{3-}(aq)$
Initial	s = solubility (mol/L)	0	0
Equil.		s	s

$K_{sp} = [Fe^{3+}]\,[PO_4^{3-}] = s^2$, $\quad s = \sqrt{1.0 \times 10^{-22}} = 1.0 \times 10^{-11}\ mol/L$

$FePO_4$ has the smallest molar solubility.

77. a. $Fe(OH)_3(s) \rightleftharpoons Fe^{3+} + 3\ OH^-$ $\qquad$ pH = 7.0 so $[OH^-] = 1 \times 10^{-7}\ M$

	$Fe(OH)_3(s)$		Fe^{3+}	$3\ OH^-$
Initial			0	$1 \times 10^{-7}\ M$

s mol/L of $Fe(OH)_3(s)$ dissolves to reach equilibrium = molar solubility

	$Fe(OH)_3(s)$		Fe^{3+}	$3\ OH^-$
Change	$-s$	$\rightarrow$	$+s$	$+3s$
Equil.			s	$3s + 1 \times 10^{-7}$

$K_{sp} = 4 \times 10^{-38} = [Fe^{3+}]\,[OH^-]^3 = (s)(3s + 1 \times 10^{-7})^3 \approx s(1 \times 10^{-7})^3$

$s = 4 \times 10^{-17}$ mol/L $\qquad$ Assumption good ($3s << 1 \times 10^{-7}$).

b. $Fe(OH)_3(s) \rightleftharpoons Fe^{3+} + 3\ OH^-$ pH = 5.0 so $[OH^-] = 1 \times 10^{-9}\ M$

	$Fe(OH)_3(s)$		Fe^{3+}	OH^-	
Initial			0	$1 \times 10^{-9}\ M$	(buffered)
	s mol/L dissolves to reach equilibrium				
Change	$-s$	→	$+s$	-----	(assume no pH change in buffer)
Equil.			s	1×10^{-9}	

$K_{sp} = 4 \times 10^{-38} = [Fe^{3+}]\ [OH^-]^3 = (s)(1 \times 10^{-9})^3$, $s = 4 \times 10^{-11}$ mol/L = molar solubility

c. $Fe(OH)_3(s) \rightleftharpoons Fe^{3+} + 3\ OH^-$ pH = 11.0 so $[OH^-] = 1 \times 10^{-3}\ M$

	$Fe(OH)_3(s)$		Fe^{3+}	OH^-	
Initial			0	0.001 M	(buffered)
	s mol/L dissolves to reach equilibrium				
Change	$-s$	→	$+s$	-----	(assume no pH change)
Equil.			s	0.001	

$K_{sp} = 4 \times 10^{-38} = [Fe^{3+}]\ [OH^-]^3 = (s)(0.001)^3$, $s = 4 \times 10^{-29}$ mol/L = molar solubility

Note: As $[OH^-]$ increases, solubility decreases. This is the common ion effect.

79. If the anion in the salt can act as a base in water, then the solubility of the salt will increase as the solution becomes more acidic. Added H^+ will react with the base forming the conjugate acid. As the basic anion is removed, more of the salt will dissolve to replenish the basic anion. The salts with basic anions are Ag_3PO_4, $CaCO_3$, $CdCO_3$ and $Sr_3(PO_4)_2$. Hg_2Cl_2 and PbI_2 do not have any pH dependence since Cl^- and I^- are terrible bases (the conjugate bases of a strong acids).

$$Ag_3PO_4(s) + H^+(aq) \rightarrow 3\ Ag^+(aq) + HPO_4^{2-}(aq) \xrightarrow{\text{excess } H^+} 3\ Ag^+(aq) + H_3PO_4(aq)$$

$$CaCO_3(s) + H^+ \rightarrow Ca^{2+} + HCO_3^- \xrightarrow{\text{excess } H^+} Ca^{2+} + H_2CO_3\ \ [H_2O(l) + CO_2(g)]$$

$$CdCO_3(s) + H^+ \rightarrow Cd^{2+} + HCO_3^- \xrightarrow{\text{excess } H^+} Cd^{2+} + H_2CO_3\ \ [H_2O(l) + CO_2(g)]$$

$$Sr_3(PO_4)_2(s) + 2\ H^+ \rightarrow 3\ Sr^{2+} + 2\ HPO_4^{2-} \xrightarrow{\text{excess } H^+} 3\ Sr^{2+} + 2\ H_3PO_4$$

81. The formation of $Mg(OH)_2(s)$ is the only possible precipitate. $Mg(OH)_2(s)$ will form if $Q > K_{sp}$.

$$Mg(OH)_2(s) \rightleftharpoons Mg^{2+}(aq) + 2\ OH^-(aq)\quad K_{sp} = [Mg^{2+}][OH^-]^2 = 8.9 \times 10^{-12}$$

$$[Mg^{2+}]_o = \frac{100.0\ \text{mL} \times 4.0 \times 10^{-4}\ \text{mmol}\ Mg^{2+}/\text{mL}}{100.0\ \text{mL} + 100.0\ \text{mL}} = 2.0 \times 10^{-4}\ M$$

$$[OH^-]_o = \frac{100.0\ \text{mL} \times 2.0 \times 10^{-4}\ \text{mmol}\ OH^-/\text{mL}}{200.0\ \text{mL}} = 1.0 \times 10^{-4}\ M$$

$Q = [Mg^{2+}]_o[OH^-]_o^2 = (2.0 \times 10^{-4}\ M)(1.0 \times 10^{-4})^2 = 2.0 \times 10^{-12}$

Since $Q < K_{sp}$, then $Mg(OH)_2(s)$ will not precipitate, so no precipitate forms.

83. The concentrations of ions are large, so Q will be greater than K_{sp} and $BaC_2O_4(s)$ will form. To solve this problem, we will assume that the precipitation reaction goes to completion; then we will solve an equilibrium problem to get the actual ion concentrations.

$$100.\ mL \times \frac{0.200\ mmol\ K_2C_2O_4}{mL} = 20.0\ mmol\ K_2C_2O_4$$

$$150.\ mL \times \frac{0.250\ mmol\ BaBr_2}{mL} = 37.5\ mmol\ BaBr_2$$

	$Ba^{2+}(aq)$ +	$C_2O_4^{2-}(aq)$	$\rightarrow$	$BaC_2O_4(s)$	$K = 1/K_{sp} >> 1$
Before	37.5 mmol	20.0 mmol		0	
Change	-20.0	-20.0	$\rightarrow$	+20.0	Reacts completely (K is large)
After	17.5	0		20.0	

New initial concentrations (after complete precipitation) are: $[Ba^{2+}] = \frac{17.5\ mmol}{250.\ mL} = 7.00 \times 10^{-2}\ M$

$[K^+] = \frac{2(20.0\ mmol)}{250.\ mL} = 0.160\ M$; $[Br^-] = \frac{2(37.5\ mmol)}{250.\ mL} = 0.300\ M$

For K^+ and Br^-, these are also the final concentrations. For Ba^{2+} and $C_2O_4^{2-}$, we need to perform an equilibrium calculation.

	$BaC_2O_4(s)$ $\rightleftharpoons$	$Ba^{2+}(aq)$ +	$C_2O_4^{2-}(aq)$	$K_{sp} = 2.3 \times 10^{-8}$
Initial		0.0700 *M*	0	
	s mol/L of $BaC_2O_4(s)$ dissolves to reach equilibrium			
Equil.		0.0700 + *s*	*s*	

$K_{sp} = 2.3 \times 10^{-8} = [Ba^{2+}][C_2O_4^{2-}] = (0.0700 + s)(s) \approx 0.0700\ s$

$s = [C_2O_4^{2-}] = 3.3 \times 10^{-7}$ mol/L; $[Ba^{2+}] = 0.0700\ M$ Assumption good ($s << 0.0700$).

85. $Ag_3PO_4(s) \rightleftharpoons 3\ Ag^+(aq) + PO_4^{3-}(aq)$; When Q is greater than K_{sp}, then precipitation will occur. We will calculate the $[Ag^+]_o$ necessary for $Q = K_{sp}$. Any $[Ag^+]_o$ greater than this calculated number will cause precipitation of $Ag_3PO_4(s)$. In this problem, $[PO_4^{3-}]_o = [Na_3PO_4]_o = 1.0 \times 10^{-5}\ M$.

$K_{sp} = 1.8 \times 10^{-18}$; $Q = 1.8 \times 10^{-18} = [Ag^+]_o^3[PO_4^{3-}]_o = [Ag^+]_o^3(1.0 \times 10^{-5}\ M)$

$$[Ag^+]_o = \left(\frac{1.8 \times 10^{-18}}{1.0 \times 10^{-5}}\right)^{1/3},\quad [Ag^+]_o = 5.6 \times 10^{-5}\ M$$

When $[Ag^+]_o = [AgNO_3]_o$ is greater than $5.6 \times 10^{-5}\ M$, then $Ag_3PO_4(s)$ will precipitate.

87.

a.

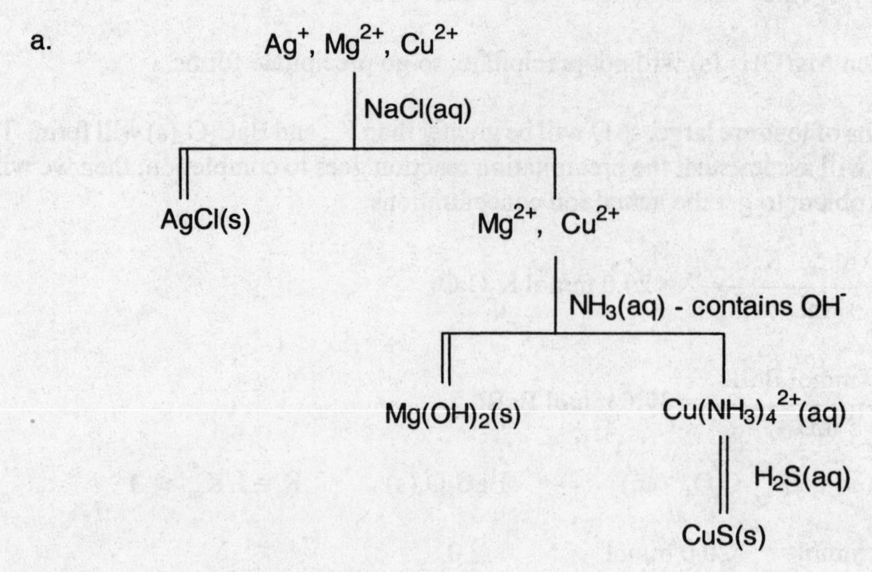
Ag+, Mg2+, Cu2+
NaCl(aq)
AgCl(s)
Mg2+, Cu2+
NH3(aq) - contains OH-
Mg(OH)2(s)
Cu(NH3)4 2+(aq)
H2S(aq)
CuS(s)

b.

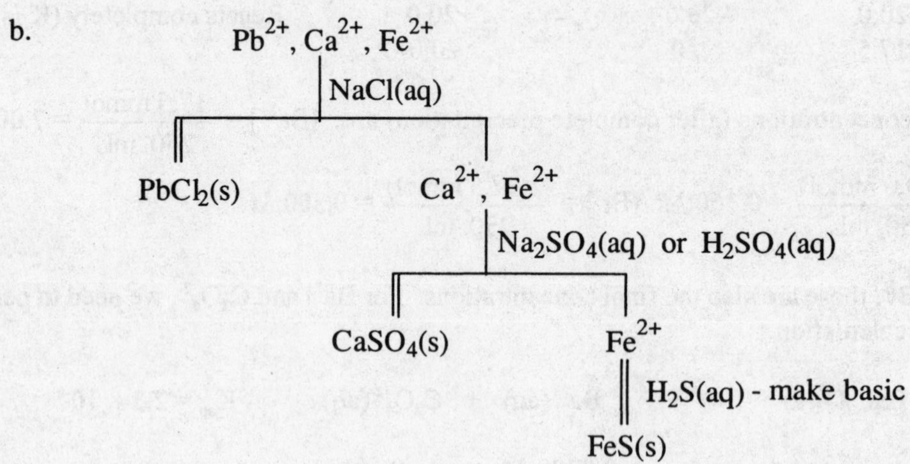
Pb2+, Ca2+, Fe2+
NaCl(aq)
PbCl2(s)
Ca2+, Fe2+
Na2SO4(aq) or H2SO4(aq)
CaSO4(s)
Fe2+
H2S(aq) - make basic
FeS(s)

c.

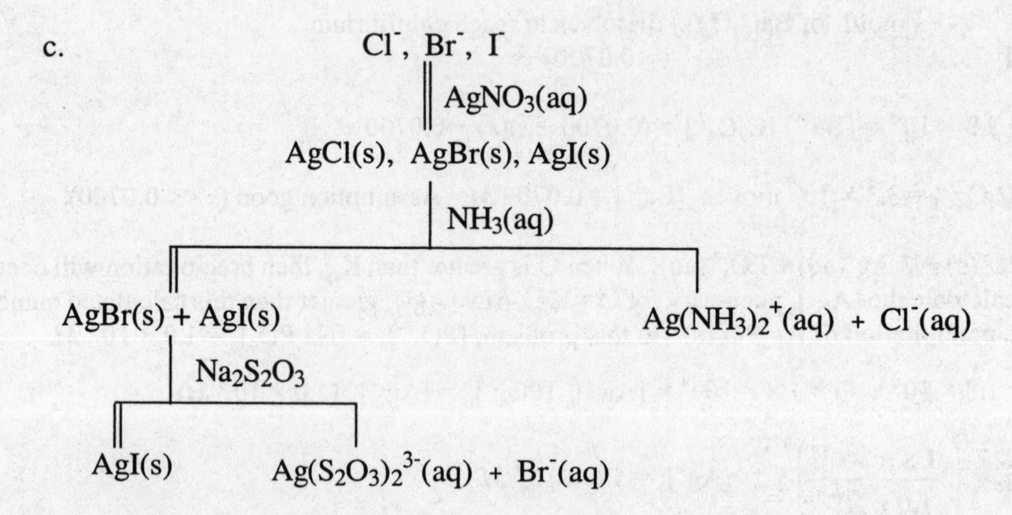
Cl-, Br-, I-
AgNO3(aq)
AgCl(s), AgBr(s), AgI(s)
NH3(aq)
AgBr(s) + AgI(s)
Ag(NH3)2+(aq) + Cl-(aq)
Na2S2O3
AgI(s)
Ag(S2O3)2 3-(aq) + Br-(aq)

d. Pb^{2+}, Bi^{3+}

$Na_2SO_4(aq)$ or $H_2SO_4(aq)$

$PbSO_4(s)$

Bi^{3+}

$H_2S(aq)$- make basic

$Bi_2S_3(s)$

Complex Ion Equilibria

89. $Hg^{2+}(aq) + 2\ I^-(aq) \rightarrow HgI_2(s)$, orange ppt; $HgI_2(s) + 2\ I^-(aq) \rightarrow HgI_4^{2-}(aq)$, soluble complex ion

91. $$\frac{65\text{ g KI}}{0.500\text{ L}} \times \frac{1\text{ mol KI}}{166.0\text{ g KI}} = 0.78\ M\text{ KI}$$

The formation constant for HgI_4^{2-} is an extremely large number. Because of this, we will let the Hg^{2+} and I^- ions present initially react to completion, and then solve an equilibrium problem to determine the Hg^{2+} concentration.

	Hg^{2+}	+ $4\ I^-$	$\rightleftharpoons$	HgI_4^{2-}	$K = 1.0 \times 10^{30}$
Before	0.010 M	0.78 M		0	
Change	-0.010	-0.040	$\rightarrow$	+0.010	Reacts completely (K large)
After	0	0.74		0.010	New initial
	x mol/L HgI_4^{2-} dissociates to reach equilibrium				
Change	$+x$	$+4x$	$\leftarrow$	$-x$	
Equil.	x	$0.74 + 4x$		$0.010 - x$	

$$K = 1.0 \times 10^{30} = \frac{[HgI_4^{2-}]}{[Hg^{2+}][I^-]^4} = \frac{(0.010 - x)}{(x)(0.74 + 4x)^4};\ \text{Making normal assumptions:}$$

$$1.0 \times 10^{30} \approx \frac{(0.010)}{(x)(0.74)^4},\ x = [Hg^{2+}] = 3.3 \times 10^{-32}\ M \quad \text{Assumptions good.}$$

Note: 3.3×10^{-32} mol/L corresponds to one Hg^{2+} ion per 5×10^7 L. It is very reasonable to approach the equilibrium in two steps. The reaction does essentially go to completion.

93. a.

$$Cu(OH)_2 \rightleftharpoons Cu^{2+} + 2\ OH^- \qquad K_{sp} = 1.6 \times 10^{-19}$$

$$Cu^{2+} + 4\ NH_3 \rightleftharpoons Cu(NH_3)_4^{2+} \qquad K_f = 1.0 \times 10^{13}$$

$$Cu(OH)_2(s) + 4\ NH_3(aq) \rightleftharpoons Cu(NH_3)_4^{2+}(aq) + 2\ OH^-(aq) \qquad K = K_{sp}K_f = 1.6 \times 10^{-6}$$

b. $Cu(OH)_2(s) + 4\,NH_3 \rightleftharpoons Cu(NH_3)_4^{2+} + 2\,OH^-$ $K = 1.6 \times 10^{-6}$

	$Cu(OH)_2(s)$	$4\,NH_3$	$Cu(NH_3)_4^{2+}$	$2\,OH^-$
Initial		5.0 M	0	0.0095 M
	s mol/L $Cu(OH)_2$ dissolves to reach equilibrium			
Equil.		$5.0 - 4s$	s	$0.0095 + 2s$

$$K = 1.6 \times 10^{-6} = \frac{[Cu(NH_3)_4^{2+}][OH^-]^2}{[NH_3]^4} = \frac{s(0.0095 + 2s)^2}{(5.0 - 4s)^4}$$

If s is small: $1.6 \times 10^{-6} = \dfrac{s(0.0095)^2}{(5.0)^4}$, $s = 11.$ mol/L

Assumptions are not good. We will solve the problem by successive approximations.

$$s_{calc} = \frac{1.6 \times 10^{-6}\,(5.0 - 4s_{guess})^4}{(0.0095 + 2s_{guess})^2}$$, The results from six trials are:

s_{guess}: 0.10, 0.050, 0.060, 0.055, 0.056

s_{calc}: 1.6×10^{-2}, 0.071, 0.049, 0.058, 0.056

Thus, the solubility of $Cu(OH)_2$ is 0.056 mol/L in 5.0 M NH_3.

95. $AgBr(s) \rightleftharpoons Ag^+ + Br^-$ $K_{sp} = 5.0 \times 10^{-13}$

$Ag^+ + 2\,S_2O_3^{2-} \rightleftharpoons Ag(S_2O_3)_2^{3-}$ $K_f = 2.9 \times 10^{13}$

$AgBr(s) + 2\,S_2O_3^{2-} \rightleftharpoons Ag(S_2O_3)_2^{3-} + Br^-$ $K = K_{sp} \times K_f = 14.5$ (Carry extra sig. figs.)

	$AgBr(s)$	$2\,S_2O_3^{2-}$	$Ag(S_2O_3)_2^{3-}$	Br^-
Initial		0.500 M	0	0
	s mol/L $AgBr(s)$ dissolves to reach equilibrium			
Change	$-s$	$-2s$ →	$+s$	$+s$
Equil.		$0.500 - 2s$	s	s

$K = \dfrac{s^2}{(0.500 - 2s)^2} = 14.5$; Taking the square root of both sides:

$\dfrac{s}{0.500 - 2s} = 3.81$, $s = 1.91 - 7.62\,s$, $s = 0.222$ mol/L

$$1.00\ L \times \frac{0.222\ mol\ AgBr}{L} \times \frac{187.8\ g\ AgBr}{mol\ AgBr} = 41.7\ g\ AgBr = 42\ g\ AgBr$$

97. Test tube 1: added Cl^- reacts with Ag^+ to form a silver chloride precipitate. The net ionic equation is $Ag^+(aq) + Cl^-(aq) \rightarrow AgCl(s)$. Test tube 2: added NH_3 reacts with Ag^+ ions to form a soluble complex ion, $Ag(NH_3)_2^+$. As this complex ion forms, Ag^+ is removed from the solution, which

causes the AgCl(s) to dissolve. When enough NH_3 is added, all of the silver chloride precipitate will dissolve. The equation is $AgCl(s) + 2\ NH_3(aq) \rightarrow Ag(NH_3)_2^+(aq) + Cl^-(aq)$. Test tube 3: added H^+ reacts with the weak base, NH_3, to form NH_4^+. As NH_3 is removed from the $Ag(NH_3)_2^+$ complex ion, Ag^+ ions are released to solution and can then react with Cl^- to reform AgCl(s). The equations are $Ag(NH_3)_2^+(aq) + 2\ H^+(aq) \rightarrow Ag^+(aq) + 2\ NH_4^+(aq)$ and $Ag^+(aq) + Cl^-(aq) \rightarrow AgCl(s)$.

Additional Exercises

99. a. The optimum pH for a buffer is when pH = pK_a. At this pH a buffer will have equal neutralization capacity for both added acid and base. As shown below, since the pK_a for $TRISH^+$ is 8.1, then the optimal buffer pH is about 8.1.

$$K_b = 1.19 \times 10^{-6};\ K_a = K_w/K_b = 8.40 \times 10^{-9};\ pK_a = -\log(8.40 \times 10^{-9}) = 8.076$$

b. $$pH = pK_a + \log\frac{[TRIS]}{[TRISH^+]},\ 7.00 = 8.076 + \log\frac{[TRIS]}{[TRISH^+]}$$

$$\frac{[TRIS]}{[TRISH^+]} = 10^{-1.08} = 0.083 \quad \text{(at pH = 7.00)}$$

$$9.00 = 8.076 + \log\frac{[TRIS]}{[TRISH^+]},\ \frac{[TRIS]}{[TRISH^+]} = 10^{0.92} = 8.3 \quad \text{(at pH = 9.00)}$$

c. $$\frac{50.0\text{ g TRIS}}{2.0\text{ L}} \times \frac{1\text{ mol}}{121.14\text{ g}} = 0.206\ M = 0.21\ M = [TRIS]$$

$$\frac{65.0\text{ g TRISHCl}}{2.0\text{ L}} \times \frac{1\text{ mol}}{157.60\text{ g}} = 0.206\ M = 0.21\ M = [TRISHCl] = [TRISH^+]$$

$$pH = pK_a + \log\frac{[TRIS]}{[TRISH^+]} = 8.076 + \log\frac{(0.21)}{(0.21)} = 8.08$$

The amount of H^+ added from HCl is: $0.50 \times 10^{-3}\text{ L} \times \frac{12\text{ mol}}{\text{L}} = 6.0 \times 10^{-3}\text{ mol H}^+$

The H^+ from HCl will convert TRIS into $TRISH^+$. The reaction is:

	TRIS	+	H^+	→	$TRISH^+$	
Before	0.21 M		$\frac{6.0 \times 10^{-3}}{0.2005} = 0.030\ M$		0.21 M	
Change	-0.030		-0.030	→	+0.030	Reacts completely
After	0.18		0		0.24	

Now use the Henderson-Hasselbalch equation to solve the buffer problem.

$$pH = 8.076 + \log\left(\frac{0.18}{0.24}\right) = 7.95$$

101.

Acid	pK_a
$(CH_3)_2AsO_2H$	6.19
TRISHCl	8.08
benzoic acid	4.19
acetic acid	4.74
HF	3.14
NH_4Cl	9.26

Best buffer is when $pH = pK_a$ which requires equal concentrations of weak acid and conjugate base. Choose combinations that yield a buffer where $pH \approx pK_a$. We will use HCl or NaOH when necessary to convert either weak base into its conjugate acid or weak acid into its conjugate base. Remember that a buffer must have both the weak acid and weak base present at the same time.

a. potassium fluoride + HCl

b. benzoic acid + NaOH

c. sodium acetate + acetic acid

d. $(CH_3)_2AsO_2Na$ + HCl: This is a good choice to produce a conjugate acid/base pair for a pH = 7.0 buffer. Actually the best choice is an equimolar mixture of ammonium chloride and sodium acetate. NH_4^+ is a weak acid ($K_a = 5.6 \times 10^{-10}$). $C_2H_3O_2^-$ is a weak base ($K_b = 5.6 \times 10^{-10}$). A mixture of the two will give a buffer at pH = 7.0 since the weak acid and weak base are the same strengths. $NH_4C_2H_3O_2$ is commercially available and its solutions are used as pH = 7.0 buffers.

e. ammonium chloride + NaOH

103. $$\text{NaOH added} = 50.0\text{ mL} \times \frac{0.500\text{ mmol}}{\text{mL}} = 25.0\text{ mmol NaOH}$$

$$\text{NaOH left unreacted} = 31.92\text{ mL HCl} \times \frac{0.289\text{ mmol}}{\text{mL}} \times \frac{1\text{ mmol NaOH}}{\text{mmol HCl}} = 9.22\text{ mmol NaOH}$$

NaOH reacted with aspirin = 25.0 - 9.22 = 15.8 mmol NaOH

$$15.8\text{ mmol NaOH} \times \frac{1\text{ mmol aspirin}}{2\text{ mmol NaOH}} \times \frac{180.2\text{ mg}}{\text{mmol}} = 1420\text{ mg} = 1.42\text{ g aspirin}$$

$$\text{Purity} = \frac{1.42\text{ g}}{1.427\text{ g}} \times 100 = 99.5\%$$

Here, a strong base is titrated by a strong acid. The equivalence point will be at pH = 7.0. Bromthymol blue would be the best indicator since it changes color at pH ≈ 7 (from base color to acid color). See Figure 8.8 of the text.

105. At the equivalence point, P^{2-} is the major species. It is a weak base in water since it is the conjugate base of a weak acid.

	P^{2-}	+	H_2O	$\rightleftharpoons$	HP^-	+	OH^-	
Initial	$\frac{0.5\text{ g}}{0.1\text{ L}} \times \frac{1\text{ mol}}{204.2\text{ g}} = 0.024\ M$				0		~0	(carry extra sig. fig.)
Equil.	$0.024 - x$				x		x	

$$K_b = \frac{[HP^-][OH^-]}{[P^{2-}]} = \frac{K_w}{K_a} = \frac{1.0 \times 10^{-14}}{10^{-5.51}} = 3.2 \times 10^{-9} = \frac{x^2}{0.024 - x} \approx \frac{x^2}{0.024}$$

$x = [OH^-] = 8.8 \times 10^{-6}\ M$; pOH = 5.1; pH = 8.9 Assumptions good.

Phenolphthalein would be the best indicator for this titration because it changes color at pH ~ 9.

107. a.

	$Pb(OH)_2(s)$	$\rightleftharpoons$	Pb^{2+}	+	$2\ OH^-$
Initial	s = solubility (mol/L)		0		$1.0 \times 10^{-7}\ M$ from water
Equil.			s		$1.0 \times 10^{-7} + 2s$

$K_{sp} = 1.2 \times 10^{-15} = [Pb^{2+}][OH^-]^2 = s(1.0 \times 10^{-7} + 2s)^2 \approx s(2s^2) = 4s^3$

$s = [Pb^{2+}] = 6.7 \times 10^{-6}\ M$; Assumption to ignore OH^- from water is good by the 5% rule.

b.

	$Pb(OH)_2(s)$	$\rightleftharpoons$	Pb^{2+}	+	$2\ OH^-$	
Initial			0		0.10 M	pH = 13.00, $[OH^-] = 0.10\ M$
	s mol/L $Pb(OH)_2(s)$ dissolves to reach equilibrium					
Equil.			s		0.10	(buffered solution)

$1.2 \times 10^{-15} = (s)(0.10)^2$, $s = [Pb^{2+}] = 1.2 \times 10^{-13}\ M$

c. We need to calculate the Pb^{2+} concentration in equilibrium with $EDTA^{4-}$. Since K is large for the formation of $PbEDTA^{2-}$, let the reaction go to completion then solve an equilibrium problem to get the Pb^{2+} concentration.

	Pb^{2+}	+	$EDTA^{4-}$	$\rightleftharpoons$	$PbEDTA^{2-}$	$K = 1.1 \times 10^{18}$
Before	0.010 M		0.050 M		0	
	0.010 mol/L Pb^{2+} reacts completely (large K)					
Change	-0.010		-0.010	→	+0.010	Reacts completely
After	0		0.040		0.010	New initial
	x mol/L $PbEDTA^{2-}$ dissociates to reach equilibrium					
Equil.	x		$0.040 + x$		$0.010 - x$	

$$1.1 \times 10^{18} = \frac{(0.010 - x)}{(x)(0.040 + x)} \approx \frac{(0.010)}{x(0.040)},\ x = [Pb^{2+}] = 2.3 \times 10^{-19}\ M \quad \text{Assumptions good.}$$

Now calculate the solubility quotient for $Pb(OH)_2$ to see if precipitation occurs. The concentration of OH^- is 0.10 M since we have a solution buffered at pH = 13.00.

$$Q = [Pb^{2+}]_o[OH^-]_o^2 = (2.3 \times 10^{-19})(0.10)^2 = 2.3 \times 10^{-21} < K_{sp}\ (1.2 \times 10^{-15})$$

$Pb(OH)_2(s)$ will not form since Q is less than K_{sp}.

109. It will be more soluble in base. Addition of OH^- will react with H^+ to produce H_2O. This drives the equilibrium $H_4SiO_4 \rightleftharpoons H_3SiO_4^- + H^+$ to the right. This, in turn, drives the solubility reaction $SiO_2(s) + 2\ H_2O \rightleftharpoons H_4SiO_4$ to the right, increasing the solubility of the silica.

111. $$[BaBr_2]_o = \frac{0.150(1.0 \times 10^{-4})}{0.250} = 6.0 \times 10^{-5}\ M$$

$$[K_2C_2O_4] = \frac{0.100(6.0 \times 10^{-4})}{0.250} = 2.4 \times 10^{-4}\ M$$

$$Q = [Ba^{2+}][C_2O_4^{2-}] = (6.0 \times 10^{-5})(2.4 \times 10^{-4}) = 1.5 \times 10^{-8}\ M$$

Since $Q < K_{sp}$, then $BaC_2O_4(s)$ will not precipitate. The final concentration of ions will be:

$[Ba^{2+}] = 6.0 \times 10^{-5}\ M$, $[Br^-] = 1.2 \times 10^{-4}\ M$,

$[K^+] = 4.8 \times 10^{-4}\ M$, $[C_2O_4^{2-}] = 2.4 \times 10^{-4}\ M$

113. $HC_2H_3O_2 \rightleftharpoons H^+ + C_2H_3O_2^-$; Let C_o = initial concentration of $HC_2H_3O_2$

From normal weak-acid setup: $K_a = 1.8 \times 10^{-5} = \dfrac{[H^+][C_2H_3O_2^-]}{[HC_2H_3O_2]} = \dfrac{[H^+]^2}{C_o - [H^+]}$

$[H^+] = 10^{-2.68} = 2.1 \times 10^{-3}\ M$; $1.8 \times 10^{-5} = \dfrac{(2.1 \times 10^{-3})^2}{C_o - 2.1 \times 10^{-3}}$, $C_o = 0.25\ M$

25.0 mL × 0.25 mmol/mL = 6.3 mmol $HC_2H_3O_2$

Need 6.3 mmol KOH = V_{KOH} × 0.0975 mmol/mL, V_{KOH} = 65 mL

115. 0.400 mol/L × V_{NH_3} = mol NH_3 = mol NH_4^+ after reaction with HCl at the equivalence point.

At the equivalence point: $[NH_4^+]_o = \dfrac{\text{mol } NH_4^+}{\text{total volume}} = \dfrac{0.400 \times V_{NH_3}}{1.50 \times V_{NH_3}} = 0.267\ M$

	NH_4^+	$\rightleftharpoons$	H^+	+	NH_3
Initial	0.267 *M*		0		0
Equil.	0.267 - *x*		*x*		*x*

$$K_a = \frac{K_w}{K_b} = \frac{1.0 \times 10^{-14}}{1.8 \times 10^{-5}} = 5.6 \times 10^{-10} = \frac{x^2}{0.267 - x} \approx \frac{x^2}{0.267}$$

$x = [H^+] = 1.2 \times 10^{-5}\ M$; pH = 4.92 Assumptions good.

117. $HA + OH^- \rightarrow A^- + H_2O$ where HA = acetylsalicylic acid

$$\text{mmol HA present} = 27.36 \text{ mL } OH^- \times \frac{0.5106 \text{ mmol } OH^-}{\text{mL } OH^-} \times \frac{1 \text{ mmol HA}}{\text{mmol } OH^-} = 13.97 \text{ mmol HA}$$

$$\text{molar mass of HA} = \frac{\text{grams}}{\text{moles}} = \frac{2.51 \text{ g HA}}{13.97 \times 10^{-3} \text{ mol HA}} = 180.\ \text{g/mol}$$

To determine the K_a value, use the pH data. After complete neutralization of acetylsalicylic acid by OH^-, we have 13.97 mmol of A^- produced from the neutralization reaction. A^- will react completely with the added H^+ and reform acetylsalicylic acid, HA.

$$\text{mmol H}^+ \text{ added} = 15.44 \text{ mL} \times \frac{0.4524 \text{ mmol H}^+}{\text{mL}} = 6.985 \text{ mmol H}^+$$

	A^-	+	H^+	→	HA	
Before	13.97 mmol		6.985 mmol		0	
Change	-6.985		-6.985	→	+6.985	Reacts completely
After	6.985 mmol		0		6.985 mmol	

We have back titrated this solution to the halfway point to equivalence where pH = pK_a (assuming HA is a weak acid). This is true because after H^+ reacts completely, equal mmol of HA and A^- are present which only occurs at the halfway point to equivalence. Assuming acetylsalicylic acid is a weak acid, then pH = pK_a = 3.48. $K_a = 10^{-3.48} = 3.3 \times 10^{-4}$

119. $[X^-]_0 = 5.00\ M$ and $[Cu^+]_0 = 1.0 \times 10^{-3}\ M$ since equal volumes of each reagent are mixed.

Since the K values are large, assume the reaction goes completely to CuX_3^{2-}, then solve an equilibrium problem.

	Cu^+	+	$3X^-$	⇌	CuX_3^{2-}	$K = K_1 \times K_2 \times K_3 = 1.0 \times 10^9$
Before	$1.0 \times 10^{-3}\ M$		$5.00\ M$		0	
After	0		$5.00 - 3(10^{-3}) \approx 5.00$		1.0×10^{-3}	
Equil.	x		$5.00 + 3x$		$1.0 \times 10^{-3} - x$	

$$K = \frac{(1.0 \times 10^{-3} - x)}{x(5.00 + 3x)^3} = 1.0 \times 10^9 \approx \frac{1.0 \times 10^{-3}}{x(5.00)^3}, \quad x = [Cu^+] = 8.0 \times 10^{-15}\ M \quad \text{Assumptions good.}$$

$$[CuX_3^{2-}] = 1.0 \times 10^{-3} - 8.0 \times 10^{-15} = 1.0 \times 10^{-3}\ M$$

$$K_3 = \frac{[CuX_3^{2-}]}{[CuX_2^-][X^-]} = 1.0 \times 10^3 = \frac{(1.0 \times 10^{-3})}{[CuX_2^-](5.00)}, \quad [CuX_2^-] = 2.0 \times 10^{-7}\ M$$

Summarizing:

$[CuX_3^{2-}] = 1.0 \times 10^{-3}\ M$ (answer a)
$[CuX_2^-] = 2.0 \times 10^{-7}\ M$ (answer b)
$[Cu^{2+}] = 8.0 \times 10^{-15}\ M$ (answer c)

121. $$50.0 \text{ mL} \times \frac{0.10 \text{ mmol } H_2A}{\text{mL}} = 5.0 \text{ mmol } H_2A \text{ initially}$$

To reach the first equivalence point, 5.0 mmol OH^- must be added. This occurs after addition of 50.0 mL of 0.10 M NaOH. At the first equivalence point for a diprotic acid, pH = $(pK_{a_1} + pK_{a_2})/2$ = 8.00. Addition of 25.0 mL of 0.10 M NaOH will be the first halfway point to equivalence where $[H_2A] = [HA^-]$ and pH = pK_{a_1} = 6.70. Solving for the K_a values:

$pK_{a_1} = 6.70,\ K_{a_1} = 10^{-6.70} = 2.0 \times 10^{-7}$

$\frac{pK_{a_1} + pK_{a_2}}{2} = 8.00,\quad \frac{6.70 + pK_{a_2}}{2} = 8.00,\ pK_{a_2} = 9.30,\ K_{a_2} = 10^{-9.30} = 5.0 \times 10^{-10}$

Challenge Problems

123. mmol $HC_3H_5O_2$ present initially = 45.0 mL × $\frac{0.750 \text{ mmol}}{\text{mL}}$ = 33.8 mmol $HC_3H_5O_2$

mmol $C_3H_5O_2^-$ present initially = 55.0 mL × $\frac{0.700 \text{ mmol}}{\text{mL}}$ = 38.5 mmol $C_3H_5O_2^-$

The initial pH of the buffer is:

$$pH = pK_a + \log\frac{[C_3H_5O_2^-]}{[HC_3H_5O_2]} = -\log(1.3 \times 10^{-5}) + \log\frac{\frac{38.5 \text{ mmol}}{100.0 \text{ mL}}}{\frac{33.8 \text{ mmol}}{100.0 \text{ mL}}} = 4.89 + \log\frac{38.5}{33.8} = 4.95$$

Note: Since the buffer components are in the same volume of solution, we can use the mol (or mmol) ratio in the Henderson-Hasselbalch equation to solve for pH instead of using the concentration ratio of $[C_3H_5O_2^-]/[HC_3H_5O_2]$. The total volume always cancels for buffer solutions.

When NaOH is added, the pH will increase and the added OH^- will convert $HC_3H_5O_2$ into $C_3H_5O_2^-$. The pH after addition OH^- increases by 2.5%, so the resulting pH is:

$$4.95 + 0.025\,(4.95) = 5.07$$

At this pH, a buffer solution still exists and the mmol ratio between $C_3H_5O_2^-$ and $HC_3H_5O_2$ is:

$$pH = pK_a + \log\frac{\text{mmol } C_3H_5O_2^-}{\text{mmol } HC_3H_5O_2},\quad 5.07 = 4.89 + \log\frac{\text{mmol } C_3H_5O_2^-}{\text{mmol } HC_3H_5O_2}$$

$$\frac{\text{mmol } C_3H_5O_2^-}{\text{mmol } HC_3H_5O_2} = 10^{0.18} = 1.5$$

Let x = mmol OH^- added to increase pH to 5.07. Since OH^- will essentially react to completion with $HC_3H_5O_2$ then the set-up to the problem using mmol is:

	$HC_3H_5O_2$	+	OH^-	→	$C_3H_5O_2^-$	
Before	33.8 mmol		x mmol		38.5 mmol	
Change	$-x$		$-x$	→	$+x$	Reacts completely
After	$33.8 - x$		0		$38.5 + x$	

Solving for x:

$$\frac{\text{mmol } C_3H_5O_2^-}{\text{mmol } HC_3H_5O_2} = 1.5 = \frac{38.5 + x}{33.8 - x}, \quad 1.5\,(33.8 - x) = 38.5 + x, \quad x = 4.9 \text{ mmol OH}^- \text{ added}$$

The volume of NaOH necessary to raise the pH by 2.5% is:

$$4.9 \text{ mmol NaOH} \times \frac{1 \text{ mL}}{0.10 \text{ mmol NaOH}} = 49 \text{ mL}$$

49 mL of 0.10 *M* NaOH must be added to increase the pH by 2.5%.

125. a. Best acid will react with the best base present, so the dominate equilibrium is:

$$NH_4^+ + X^- \rightleftharpoons NH_3 + HX \qquad K_{eq} = \frac{[NH_3]\,[HX]}{[NH_4^+]\,[X^-]} = \frac{K_a(NH_4^+)}{K_a(HX)}$$

Since initially $[NH_4^+]_o = [X^-]_o$ and $[NH_3]_o = [HX]_o = 0$, then at equilibrium $[NH_4^+] = [X^-]$ and $[NH_3] = [HX]$.

Therefore, $K_{eq} = \dfrac{K_a(NH_4^+)}{K_a(HX)} = \dfrac{[HX]^2}{[X^-]^2}$

The K_a expression for HX is: $K_a\,(HX) = \dfrac{[H^+]\,[X^-]}{[HX]}, \quad \dfrac{[HX]}{[X^-]} = \dfrac{[H^+]}{K_a(HX)}$

Substituting into the K_{eq} expression: $K_{eq} = \dfrac{K_a(NH_4^+)}{K_a(HX)} = \dfrac{[HX]^2}{[X^-]^2} = \left(\dfrac{[H^+]}{K_a(HX)}\right)^2$

Rearranging: $[H^+]^2 = K_a(NH_4^+) \times K_a\,(HX)$ or taking the -log of both sides:

$$pH = \frac{pK_a(NH_4^+) + pK_a(HX)}{2}$$

b. Ammonium formate = $NH_4(HCO_2)$

$$K_a(NH_4^+) = \frac{1.0 \times 10^{-14}}{1.8 \times 10^{-5}} = 5.6 \times 10^{-10}; \; K_a(HCO_2H) = 1.8 \times 10^{-4}; \; pK_a = 3.74$$

$$pH = \frac{pK_a(NH_4^+) + pK_a(HCO_2H)}{2} = \frac{9.25 + 3.74}{2} = 6.50$$

Ammonium acetate = $NH_4(C_2H_3O_2)$; $K_a\,(HC_2H_3O_2) = 1.8 \times 10^{-5}$; $pK_a = 4.74$

$$pH = \frac{9.25 + 4.74}{2} = 7.00$$

Ammonium bicarbonate = $NH_4(HCO_3)$; $K_a(H_2CO_3) = 4.3 \times 10^{-7}$; $pK_a = 6.37$

$$pH = \frac{9.25 + 6.37}{2} = 7.81$$

c. $NH_4^+(aq) + OH^-(aq) \rightarrow NH_3(aq) + H_2O(l)$; $C_2H_3O_2^-(aq) + H^+(aq) \rightarrow HC_2H_3O_2(aq)$

127. a. $SrF_2(s) \rightleftharpoons Sr^{2+}(aq) + 2\,F^-(aq)$

		$Sr^{2+}(aq)$	$2\,F^-(aq)$
Initial		0	0
	s mol/L SrF_2 dissolves to reach equilibrium		
Equil.		*s*	2*s*

$[Sr^{2+}][F^-]^2 = K_{sp} = 7.9 \times 10^{-10} = 4s^3$, $s = 5.8 \times 10^{-4}$ mol/L

b. Greater, because some of the F^- would react with water:

$$F^- + H_2O \rightleftharpoons HF + OH^- \quad K_b = \frac{K_w}{K_a(HF)} = 1.4 \times 10^{-11}$$

This lowers the concentration of F^-, forcing more SrF_2 to dissolve.

c. $SrF_2(s) \rightleftharpoons Sr^{2+} + 2\,F^- \quad K_{sp} = 7.9 \times 10^{-10} = [Sr^{2+}][F^-]^2$

Let s = solubility = $[Sr^{2+}]$, then $2s$ = total F^- concentration. Since F^- is a weak base, some of the F^- is converted into HF. Therefore:

total F^- concentration = $2s = [F^-] + [HF]$.

$$HF \rightleftharpoons H^+ + F^- \quad K_a = 7.2 \times 10^{-4} = \frac{[H^+][F^-]}{[HF]} = \frac{1.0 \times 10^{-2}[F^-]}{[HF]} \quad \text{(since pH = 2.00 buffer)}$$

$$7.2 \times 10^{-2} = \frac{[F^-]}{[HF]}, \quad [HF] = 14\,[F^-]; \text{ Solving:}$$

$[Sr^{2+}] = s$; $2s = [F^-] + [HF] = [F^-] + 14\,[F^-]$, $2s = 15\,[F^-]$, $[F^-] = 2s/15$

$$7.9 \times 10^{-10} = [Sr^{2+}][F^-]^2 = (s)\left(\frac{2s}{15}\right)^2, \quad s = 3.5 \times 10^{-3} \text{ mol/L}$$

129. a. $Al(OH)_3(s) \rightleftharpoons Al^{3+} + 3\,OH^-$; $Al(OH)_3(s) + OH^- \rightleftharpoons Al(OH)_4^-$

S = solubility = total Al^{3+} concentration = $[Al^{3+}] + [Al(OH)_4^-]$

$$[Al^{3+}] = \frac{K_{sp}}{[OH^-]^3} = K_{sp} \times \frac{[H^+]^3}{K_w^3} \quad \text{since } [OH^-]^3 = (K_w/[H^+])^3$$

$$\frac{[Al(OH)_4^-]}{[OH^-]} = K; \quad [OH^-] = \frac{K_w}{[H^+]}; \quad [Al(OH)_4^-] = K\,[OH^-] = \frac{KK_w}{[H^+]}$$

$$S = [Al^{3+}] + [Al(OH)_4^-] = [H^+]^3 K_{sp}/K_w^3 + KK_w/[H^+]$$

b. $K_{sp} = 2 \times 10^{-32}$; $K_w = 1.0 \times 10^{-14}$; $K = 40.0$

$$S = \frac{[H^+]^3(2 \times 10^{-32})}{(1.0 \times 10^{-14})^3} + \frac{40.0(1.0 \times 10^{-14})}{[H^+]} = [H^+]^3(2 \times 10^{10}) + \frac{4.0 \times 10^{-13}}{[H^+]}$$

pH	solubility (S, mol/L)	log S
4.0	2×10^{-2}	-1.7
5.0	2×10^{-5}	-4.7
6.0	4.2×10^{-7}	-6.38
7.0	4.0×10^{-6}	-5.40
8.0	4.0×10^{-5}	-4.40
9.0	4.0×10^{-4}	-3.40
10.0	4.0×10^{-3}	-2.40
11.0	4.0×10^{-2}	-1.40
12.0	4.0×10^{-1}	-0.40

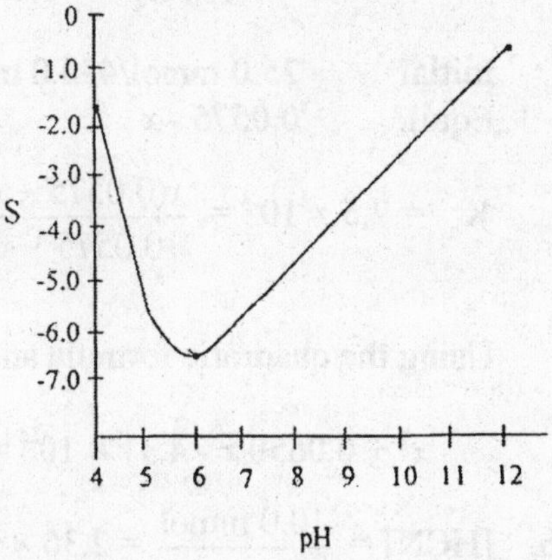

As expected, the solubility of $Al(OH)_3(s)$ is increased by very acidic solutions and by very basic solutions.

131. **For HOCl, $K_a = 3.5 \times 10^{-8}$ and $pK_a = -\log(3.5 \times 10^{-8}) = 7.46$. This will be a buffer solution since the pH is close to the pK_a value.**

$$pH = pK_a + \log\frac{[OCl^-]}{[HOCl]},\quad 8.00 = 7.46 + \log\frac{[OCl^-]}{[HOCl]},\quad \frac{[OCl^-]}{[HOCl]} = 10^{0.54} = 3.5$$

1.00 L × 0.0500 *M* = 0.0500 mol HOCl initially. Added OH^- converts HOCl into OCl^-. The total moles of OCl^- and HOCl must equal 0.0500 mol. Solving where n = moles:

$$n_{OCl^-} + n_{HOCl} = 0.0500 \text{ and } n_{OCl^-} = 3.5\, n_{HOCl}$$

$$4.5\, n_{HOCl} = 0.0500,\quad n_{HOCl} = 0.011 \text{ mol};\quad n_{OCl^-} = 0.039 \text{ mol}$$

Need to add 0.039 mol NaOH to produce 0.039 mol OCl^-.

0.039 mol = V × 0.0100 *M*, V = 3.9 L NaOH

Note: Normal buffer assumptions hold.

133. **a.** **200.0 mL × 0.250 mmol Na_3PO_4/mL = 50.0 mmol Na_3PO_4**

135.0 mL × 1.000 mmol HCl/mL = 135.0 mmol HCl

100.0 mL × 0.100 mmol NaCN/mL = 10.0 mmol NaCN

Let H^+ from the HCl react to completion with the bases in solution. In general, react the strongest base first and so on. Here, 110.0 mmol of HCl reacts to convert all CN^- to HCN and all PO_4^{3-} to $H_2PO_4^-$. At this point 10.0 mmol HCN, 50.0 mmol $H_2PO_4^-$ and 25.0 mmol HCl are in solution. The remaining HCl reacts completely with $H_2PO_4^-$, converting 25.0 mmol to H_3PO_4. Final solution contains: 25.0 mmol H_3PO_4, (50.0 - 25.0 =) 25.0 mmol $H_2PO_4^-$ and 10.0 mmol HCN. HCN ($K_a = 6.2 \times 10^{-10}$) is a much weaker acid than either H_3PO_4 ($K_{a_1} = 7.5 \times 10^{-3}$) or $H_2PO_4^-$ ($K_{a_2} = 6.2 \times 10^{-8}$), so ignore it. Principle equilibrium reaction is:

	H_3PO_4	$\rightleftharpoons$	H^+	+	$H_2PO_4^-$	$K_{a_1} = 7.5 \times 10^{-3}$
Initial	25.0 mmol/435.0 mL		0		25.0/435.0	
Equil.	$0.0575 - x$		x		$0.0575 + x$	

$K_{a_1} = 7.5 \times 10^{-3} = \dfrac{x(0.0575 + x)}{0.0575 - x}$; Normal assumptions don't hold here.

Using the quadratic formula and carrying extra sig. figs.:

$x^2 + 0.0650\,x - 4.31 \times 10^{-4} = 0,\ x = 0.0061\ M = [H^+];\ pH = 2.21$

b. $[HCN] = \dfrac{10.0 \text{ mmol}}{435.0 \text{ mL}} = 2.30 \times 10^{-2}\ M$; HCN dissociation will be minimal.

135. H_3A: $pK_{a1} = 3.00,\ pK_{a_2} = 7.30,\ pK_{a_3} = 11.70$

$pH = 9.50 = \dfrac{pK_{a_2} + pK_{a_3}}{2} = \dfrac{7.30 + 11.70}{2}$ = pH at 2nd equivalence point where HA^{2-} dominates. See Section 8.7 of text on calculating the pH of amphoteric species like HA^{2-} or H_2A^-.

100.0 mL × 0.0500 M = 5.00 mmol H_3A initially. To reach the 2nd stoichiometric point, need 10.0 mmol OH^- = 1.00 mmol/mL × V_{NaOH}. Solving for V_{NaOH}:

$V_{NaOH} = 10.0$ mL (to reach pH = 9.50)

pH = 4.00 is between the first halfway point to equivalence ($pH = pK_{a_1} = 3.00$) and the first stoichiometric point ($pH = \dfrac{pK_{a_1} + pK_{a_2}}{2} = 5.15$).

This is the buffer region controlled by: $H_3A \rightleftharpoons H_2A^- + H^+$

$pH = pK_{a_1} + \log \dfrac{[H_2A^-]}{[H_3A]},\ 4.00 = 3.00 + \log \dfrac{[H_2A^-]}{[H_3A]},\ \dfrac{[H_2A^-]}{[H_3A]} = 10.$

Since both species are in the same volume, the mole ratio also equals 10. Let n = mmol:

$\dfrac{n_{H_2A^-}}{n_{H_3A}} = 10.$ and $n_{H_2A^-} + n_{H_3A} = 5.00$ mmol (mass balance)

$11\ n_{H_3A} = 5.00,\ n_{H_3A} = 0.45$ mmol; $n_{H_2A^-} = 4.55$ mmol

We need to add 4.55 mmol OH^- to get 4.55 mmol H_2A^- from the original H_3A present.

4.55 mmol = 1.00 mmol/mL × V_{NaOH}, V_{NaOH} = 4.55 mL of NaOH (to reach pH = 4.00)

Note: Normal buffer assumptions are good.

137. a. V_1 corresponds to the titration reaction of $CO_3^{2-} + H^+ \rightarrow HCO_3^-$; V_2 corresponds to the titration reaction of $HCO_3^- + H^+ \rightarrow H_2CO_3$.

Here, there are two sources of HCO_3^-: $NaHCO_3$ and the titration of Na_2CO_3. So, $V_2 > V_1$

b. V_1 corresponds to the titration reactions of $OH^- + H^+ \rightarrow H_2O$ and $CO_3^{2-} + H^+ \rightarrow HCO_3^-$. V_2 corresponds to the titration reaction of $HCO_3^- + H^+ \rightarrow H_2CO_3$.

Here, $V_1 > V_2$ due to the presence of OH^- which is titrated in the V_1 region.

c. 0.100 mmol HCl/mL × 18.9 mL = 1.89 mmol H^+

Since the first stoichiometric point only involves the titration of Na_2CO_3 by H^+, then 1.89 mmol of CO_3^{2-} has been converted into HCO_3^-. The sample contains 1.89 mmol Na_2CO_3 × 105.99 mg/mmol = 2.00×10^2 mg = 0.200 g Na_2CO_3.

The second stoichiometric point involves the titration of HCO_3^- by H^+.

$$\frac{0.100 \text{ mmol } H^+}{\text{mL}} \times 36.7 \text{ mL} = 3.67 \text{ mmol } H^+ = 3.67 \text{ mmol } HCO_3^-$$

1.89 mmol $NaHCO_3$ came from the first stoichiometric point of the Na_2CO_3 titration. 3.67 - 1.89 = 1.78 mmol HCO_3^- came from $NaHCO_3$.

1.78 mmol $NaHCO_3$ × 84.01 mg $NaHCO_3$/mmol = 1.50×10^2 mg $NaHCO_3$ = 0.150 g $NaHCO_3$

$$\% \ Na_2CO_3 = \frac{0.200 \text{ g}}{(0.200 + 0.150) \text{ g}} \times 100 = 57.1 \ \% \ Na_2CO_3 \text{ by mass}$$

$$\% \ NaHCO_3 = \frac{0.150 \text{ g}}{0.350 \text{ g}} \times 100 = 42.9 \ \% \ NaHCO_3 \text{ by mass}$$

Marathon Problem

139. a. Since $K_{a1} >> K_{a2}$, then the amount of H^+ contributed by the K_{a2} reaction will be negligible. The $[H^+]$ donated by the K_{a1} reaction is $10^{-2.06} = 8.7 \times 10^{-3}$ M H^+.

	H_2A	$\rightleftharpoons$	H^+	+	HA^-	$K_{a1} = 5.90 \times 10^{-2}$
Initial	$[H_2A]_o$		~0		0	$[H_2A]_o$ = initial concentration
Equil.	$[H_2A]_o - x$		x		x	

$$K_{a1} = 5.90 \times 10^{-2} = \frac{x^2}{[H_2A]_o - x} = \frac{(8.7 \times 10^{-3})^2}{[H_2A]_o - 8.7 \times 10^{-3}}, \ [H_2A]_o = 1.0 \times 10^{-2} \ M$$

$$\text{mol } H_2A \text{ present initially} = 0.250 \text{ L} \times \frac{1.0 \times 10^{-2} \text{ mol } H_2A}{\text{L}} = 2.5 \times 10^{-3} \text{ mol } H_2A$$

$$\text{molar mass } H_2A = \frac{0.225 \text{ g } H_2A}{2.5 \times 10^{-3} \text{ mol } H_2A} = 90. \text{ g/mol}$$

b. $H_2A + 2\ OH^- \rightarrow A^{2-} + H_2O$; At the second equivalence point, the added OH^- has converted all the H_2A into A^{2-}, so A^{2-} is the major species present that determines the pH. The mmol of A^{2-} present at the equivalence point equals the mmol of H_2A present initially (2.5 mmol) and the mmol of OH^- added to reach the second equivalence point is 2(2.5 mmol) = 5.0 mmol OH^- added. The only information we need now in order to calculate the K_{a_2} value is the volume of $Ca(OH)_2$ added in order to reach the second equivalent point. We will use the K_{sp} value for $Ca(OH)_2$ to help solve for the volume of $Ca(OH)_2$ added.

	$Ca(OH)_2(s)$	$\rightleftharpoons$	Ca^{2+}	+	$2\ OH^-$	$K_{sp} = 1.3 \times 10^{-6} = [Ca^{2+}]\,[OH^-]^2$
Initial	s = solubility (mol/L)		0		~0	
Equil.			s		$2s$	

$K_{sp} = 1.3 \times 10^{-6} = (s)\,(2s)^2 = 4s^3$, $s = 6.9 \times 10^{-3}\ M\ Ca(OH)_2$ Assumptions good.

The volume of $Ca(OH)_2$ required to deliver 5.0 mmol OH^- (the amount of OH^- necessary to reach the second equivalence point) is:

$$5.0\ \text{mmol}\ OH^- \times \frac{1\ \text{mmol}\ Ca(OH)_2}{2\ \text{mmol}\ OH^-} \times \frac{1\ \text{mL}}{6.9 \times 10^{-3}\ \text{mmol}\ Ca(OH)_2}$$

$$= 362\ \text{mL} = 360\ \text{mL}\ Ca(OH)_2$$

At the second equivalence point, the total volume of solution is:

250. mL + 360 mL = 610 mL

Now we can solve for K_{a_2} using the pH data at the second equivalence point. Since the only species present which has any effect on pH is the weak base, A^{2-}, then the set-up to the problem requires the K_b reaction for A^{2-}.

	A^{2-}	+	H_2O	$\rightleftharpoons$	HA^-	+	OH^-	$K_b = \frac{K_w}{K_{a_2}} = \frac{1.0 \times 10^{-14}}{K_{a_2}}$
Initial	$\frac{2.5\ \text{mmol}}{610\ \text{mL}}$				0		0	
Equil.	$4.1 \times 10^{-3}\ M - x$				x		x	

$$K_b = \frac{1.0 \times 10^{-14}}{K_{a_2}} = \frac{x^2}{4.1 \times 10^{-3} - x}$$

From the problem, pH = 7.96 so $[OH^-] = 10^{-6.04} = 9.1 \times 10^{-7}\ M = x$

$$K_b = \frac{1.0 \times 10^{-14}}{K_{a_2}} = \frac{(9.1 \times 10^{-7})^2}{4.1 \times 10^{-3} - 9.1 \times 10^{-7}} = 2.0 \times 10^{-10},\ K_{a_2} = 5.0 \times 10^{-5}$$

Note: The amount of OH^- donated by the weak base HA^- will be negligible since the K_b value for A^{2-} is more than a 1000 times the K_b value for HA^-. In addition, since the pH is less than 8.0 at the second equivalence point, then the amount of OH^- added by H_2O may need to be considered. Using the equation derived in Exercise 7.117, we get the same K_{a_2} value as calculated above by ignoring the OH^- contribution from H_2O.

CHAPTER NINE

ENERGY, ENTHALPY, AND THERMOCHEMISTRY

The Nature of Energy

15. Ball A: $PE = mgz = 2.00\ kg \times \frac{9.80\ m}{s^2} \times 10.0\ m = \frac{196\ kg\ m^2}{s^2} = 196\ J$

At Point I: All of this energy is transferred to Ball B. All of B's energy is kinetic energy at this point. $E_{total} = KE = 196\ J$. At point II, the sum of the total energy will equal 196 J.

At Point II: $PE = mgz = 4.00\ kg \times \frac{9.80\ m}{s^2} \times 3.00\ m = 118\ J$

$KE = E_{total} - PE = 196\ J - 118\ J = 78\ J$

17. Step 1: $\Delta E_1 = q + w = 72\ J + 35\ J = 107\ J$; Step 2: $\Delta E_2 = 35\ J - 72\ J = -37\ J$

$\Delta E_{overall} = \Delta E_1 + \Delta E_2 = 107\ J - 37\ J = 70.\ J$

19. $q = \text{molar heat capacity} \times \text{mol} \times \Delta T = \frac{20.8\ J}{°C\ mol} \times 39.1\ mol \times (38.0 - 0.0)\ °C = 30{,}900\ J = 30.9\ kJ$

$w = -P\Delta V = -1.00\ atm \times (998\ L - 876\ L) = -122\ L\ atm \times \frac{101.3\ J}{L\ atm} = -12{,}400\ J = -12.4\ kJ$

$\Delta E = q + w = 30.9\ kJ + (-12.4\ kJ) = 18.5\ kJ$

21. $H_2O(g) \rightarrow H_2O(l)$; $\Delta E = q + w$; $q = -40.66\ kJ$; $w = -P\Delta V$

Volume of one mol $H_2O(l) = 1.000\ mol\ H_2O(l) \times \frac{18.02\ g}{mol} \times \frac{1 cm^3}{0.996\ g} = 18.1\ cm^3 = 18.1\ mL$

$w = -P\Delta V = -\ 1.00\ atm \times (0.0181\ L - 30.6\ L) = 30.6\ L\ atm \times \frac{101.3\ J}{L\ atm} = 3.10 \times 10^3\ J = 3.10\ kJ$

$\Delta E = q + w = -40.66\ kJ + 3.10\ kJ = -37.56\ kJ$

Properties of Enthalpy

23. One should try to cool the reaction mixture or provide some means of removing heat since the reaction is very exothermic (heat is released). The $H_2SO_4(aq)$ will get very hot and possibly boil, unless cooling is provided.

25. a. $1.00 \text{ mol } H_2O \times \dfrac{-572 \text{ kJ}}{2 \text{ mol } H_2O} = -286$ kJ heat released

b. $4.03 \text{ g } H_2 \times \dfrac{1 \text{ mol } H_2}{2.016 \text{ g } H_2} \times \dfrac{-572 \text{ kJ}}{2 \text{ mol } H_2} = -572$ kJ heat released

c. $186 \text{ g } O_2 \times \dfrac{1 \text{ mol } O_2}{32.00 \text{ g } O_2} \times \dfrac{-572 \text{ kJ}}{\text{mol } O_2} = -3320$ kJ heat released

d. $n_{H_2} = \dfrac{PV}{RT} = \dfrac{1.0 \text{ atm} \times 2.0 \times 10^8 \text{ L}}{\dfrac{0.08206 \text{ L atm}}{\text{mol K}} \times 298 \text{ K}} = 8.2 \times 10^6 \text{ mol } H_2$

$8.2 \times 10^6 \text{ mol } H_2 \times \dfrac{-572 \text{ kJ}}{2 \text{ mol } H_2} = -2.3 \times 10^9$ kJ heat released

The Thermodynamics of Ideal Gases

27. Consider the constant volume process first.

$n = 1.00 \times 10^3 \text{ g} \times \dfrac{1 \text{ mol}}{30.07 \text{ g}} = 33.3 \text{ mol } C_2H_6;\ C_v = \dfrac{44.60 \text{ J}}{\text{K mol}} = \dfrac{44.60 \text{ J}}{°\text{C mol}}$

$\Delta E = nC_v\Delta T = (33.3 \text{ mol})(44.60 \text{ J °C}^{-1} \text{ mol}^{-1})(75.0 - 25.0°\text{C}) = 74{,}300 \text{ J} = 74.3 \text{ kJ}$

$\Delta E = q + w$; Since $\Delta V = 0$, $w = 0$; $\Delta E = q_v = 74.3$ kJ

$\Delta H = \Delta E + \Delta PV = \Delta E + nR\Delta T$

$\Delta H = 74.3 \text{ kJ} + (33.3 \text{ mol})(8.3145 \text{ J mol}^{-1} \text{ K}^{-1})(50.0 \text{ K})(1 \text{ kJ}/1000 \text{ J})$

$\Delta H = 74.3 \text{ kJ} + 13.8 \text{ kJ} = 88.1 \text{ kJ}$

Now consider the constant pressure process.

$q_p = \Delta H = nC_p\Delta T = (33.3 \text{ mol})(52.92 \text{ J mol}^{-1} \text{ K}^{-1})(50.0 \text{ K})$

$q_p = 88{,}100 \text{ J} = 88.1 \text{ kJ} = \Delta H$

$w = -P\Delta V = -nR\Delta T = -(33.3 \text{ mol})(8.3145 \text{ J mol}^{-1} \text{ K}^{-1})(50.0 \text{ K}) = -13{,}800 \text{ J} = -13.8 \text{ kJ}$

$\Delta E = q + w = 88.1 \text{ kJ} - 13.8 \text{ kJ} = 74.3 \text{ kJ}$

Summary:	Constant V	Constant P
q	74.3 kJ	88.1 kJ
ΔE	74.3 kJ	74.3 kJ
ΔH	88.1 kJ	88.1 kJ
w	0	-13.8 kJ

29. Pathway I:

Step 1: (5.00 mol, 3.00 atm, 15.0 L) → (5.00 mol, 3.00 atm, 55.0 L)

$$w = -P\Delta V = -(3.00 \text{ atm})(55.0 - 15.0 \text{ L}) = -120. \text{ L atm}$$

$$w = -120. \text{ L atm} \times \frac{101.3 \text{ J}}{\text{L atm}} \times \frac{1 \text{ kJ}}{1000 \text{ J}} = -12.2 \text{ kJ}$$

$$\Delta H = q_p = nC_p\Delta T = nC_p \frac{\Delta(PV)}{nR} = \frac{C_p\Delta(PV)}{R}; \ \Delta(PV) = (P_2V_2 - P_1V_1)$$

For an ideal monatomic gas: $C_p = \frac{5}{2}R$

$$\Delta H = q_p = \frac{5}{2}\Delta(PV) = \frac{5}{2}(165 - 45.0) \text{ L atm} = 300. \text{ L atm}$$

$$\Delta H = q_p = 300. \text{ L atm} \times \frac{101.3 \text{ J}}{\text{L atm}} \times \frac{1 \text{ kJ}}{1000 \text{ J}} = 30.4 \text{ kJ}$$

$$\Delta E = q + w = 30.4 \text{ kJ} - 12.2 \text{ kJ} = 18.2 \text{ kJ}$$

Step 2: (5.00 mol, 3.00 atm, 55.0 L) → (5.00 mol, 6.00 atm, 20.0 L)

$$\Delta E = nC_v\Delta T = n\left(\frac{3}{2}R\right)\left(\frac{\Delta(PV)}{nR}\right) = \frac{3}{2}\Delta PV$$

$$\Delta E = \frac{3}{2}(120. - 165) \text{ L atm} = -67.5 \text{ L atm}$$ (Carry extra significant figure.)

$$\Delta E = -67.5 \text{ L atm} \times \frac{101.3 \text{ J}}{\text{L atm}} \times \frac{1 \text{ kJ}}{1000 \text{ J}} = -6.8 \text{ kJ}$$

$$\Delta H = nC_p\Delta T = n\left(\frac{5}{2}R\right)\left(\frac{\Delta(PV)}{nR}\right) = \frac{5}{2}\Delta(PV)$$

$$\Delta H = \frac{5}{2}(-45 \text{ L atm}) = -113 \text{ L atm}$$ (Carry extra significant figure.)

$$\Delta H = -113 \text{ L atm} \times \frac{101.3 \text{ J}}{\text{L atm}} \times \frac{1 \text{ kJ}}{1000 \text{ J}} = -11.4 = -11 \text{ kJ}$$

$$w = -P_{ext}\Delta V = -(6.00 \text{ atm})(20.0 - 55.0)\text{L} = 210. \text{ L atm}$$

$$w = 210. \text{ L atm} \times \frac{101.3 \text{ J}}{\text{L atm}} \times \frac{1 \text{ kJ}}{1000 \text{ J}} = 21.3 \text{ kJ}$$

$$\Delta E = q + w, \ -6.8 \text{ kJ} = q + 21.3 \text{ kJ}, \ q = -28.1 \text{ kJ}$$

Summary:	Path I	Step 1	Step 2	Total
	q	30.4 kJ	-28.1 kJ	2.3 kJ
	w	-12.2 kJ	21.3 kJ	9.1 kJ
	ΔE	18.2 kJ	-6.8 kJ	11.4 kJ
	ΔH	30.4 kJ	-11 kJ	19 kJ

Pathway II:

Step 3: (5.00 mol, 3.00 atm, 15.0 L) → (5.00 mol, 6.00 atm, 15.0 L)

$$\Delta E = q_v = \frac{3}{2}\Delta(PV) = \frac{3}{2}(90.0 - 45.0)\text{L atm} = 67.5 \text{ L atm}$$

$$\Delta E = q_v = 67.5 \text{ L atm} \times \frac{101.3 \text{ J}}{\text{L atm}} \times \frac{1 \text{ kJ}}{1000 \text{ J}} = 6.84 \text{ kJ}$$

$w = -P\Delta V = 0$ since $\Delta V = 0$

$$\Delta H = \Delta E + \Delta(PV) = 67.5 \text{ L atm} + 45.0 \text{ L atm} = 112.5 \text{ L atm} = 11.40 \text{ kJ}$$

Step 4: (5.00 mol, 6.00 atm, 15.0 L) → (5.00 mol, 6.00 atm, 20.0 L)

$$\Delta H = q_p = nC_p\Delta T = n\left(\frac{5}{2}R\right)\left(\frac{\Delta(PV)}{nR}\right) = \frac{5}{2}\Delta PV$$

$$\Delta H = \frac{5}{2}(120. - 90.0) \text{ L atm} = 75 \text{ L atm}$$

$$\Delta H = q_p = 75 \text{ L atm} \times \frac{101.3 \text{ J}}{\text{L atm}} \times \frac{1 \text{ kJ}}{1000 \text{ J}} = 7.6 \text{ kJ}$$

$$w = -P\Delta V = -(6.00 \text{ atm})(20.0 - 15.0)\text{L} = -30. \text{ L atm}$$

$$w = -30. \text{ L atm} \times \frac{101.3 \text{ J}}{\text{L atm}} \times \frac{1 \text{ kJ}}{1000 \text{ J}} = -3.0 \text{ kJ}$$

$$\Delta E = q + w = 7.6 \text{ kJ} - 3.0 \text{ kJ} = 4.6 \text{ kJ}$$

Summary:	Path II	Step 3	Step 4	Total
	q	6.84 kJ	7.6 kJ	14.4 kJ
	w	0	-3.0 kJ	-3.0 kJ
	ΔE	6.84 kJ	4.6 kJ	11.4 kJ
	ΔH	11.40 kJ	7.6 kJ	19.0 kJ

State functions are independent of the particular pathway taken between two states; path functions are dependent on the particular pathway. In this problem, the overall values of ΔH and ΔE for the two pathways are the same; hence, ΔH and ΔE are state functions. The overall values of q and w for the two pathways are different; hence, q and w are path functions.

Calorimetry and Heat Capacity

31. A coffee-cup calorimeter is at constant (atmospheric) pressure. The heat released or gained at constant pressure is ΔH. A bomb calorimeter is at constant volume. The heat released or gained at constant volume is ΔE.

33. Specific heat capacity is defined as the amount of heat necessary to raise the temperature of one gram of substance by one degree Celsius. Therefore, $H_2O(l)$ with the largest heat capacity value requires the largest amount of heat for this process. The amount of heat for $H_2O(l)$ is:

$$\text{energy} = s \times m \times \Delta T = \frac{4.18\ \text{J}}{\text{g °C}} \times 25.0\ \text{g} \times (37.0\text{°C} - 15.0\text{°C}) = 2.30 \times 10^3\ \text{J}$$

The largest temperature change when a certain amount of energy is added to a certain mass of substance will occur for the substance with the smallest specific heat capacity. This is Hg(l), and the temperature change for this process is:

$$\Delta T = \frac{\text{energy}}{s \times m} = \frac{10.7\ \text{kJ} \times \frac{1000\ \text{J}}{\text{kJ}}}{\frac{0.14\ \text{J}}{\text{g °C}} \times 550.\ \text{g}} = 140\text{°C}$$

35. Heat gained by water = Heat lost by copper; Let s = specific capacity of copper.

$$\frac{4.18\ \text{J}}{\text{g °C}} \times 75.0\ \text{g} \times 2.2\text{°C} = s \times 46.2\ \text{g} \times 73.6\text{°C},\ \ s = 0.20\ \text{J °C}^{-1}\ \text{g}^{-1}$$

37. 50.0×10^{-3} L × 0.100 mol/L = 5.00×10^{-3} mol of both $AgNO_3$ and HCl are reacted. Thus, 5.00×10^{-3} mol of AgCl will be produced since there is a 1:1 mol ratio between reactants.

Heat lost by chemicals = Heat gained by solution

$$\text{Heat gain} = \frac{4.18\ \text{J}}{\text{g °C}} \times 100.0\ \text{g} \times (23.40 - 22.60)\text{°C} = 330\ \text{J}$$

Heat loss = 330 J; This is the heat evolved (exothermic reaction) when 5.00×10^{-3} mol of AgCl is produced. So q = -330 J and ΔH (heat per mol AgCl formed) is negative with a value of:

$$\Delta H = \frac{-330\ \text{J}}{5.00 \times 10^{-3}\ \text{mol}} \times \frac{1\ \text{kJ}}{1000\ \text{J}} = -66\ \text{kJ/mol}$$

Note: Sign errors are common with calorimetry problems. However, the correct sign for ΔH can easily be determined from the ΔT data, i.e., if ΔT of the solution increases, then the reaction is exothermic since heat was released, and if ΔT of the solution decreases, then the reaction is endothermic since the reaction absorbed heat from the water. For calorimetry problems, keep all quantities positive until the end of the calculation, then decide the sign for ΔH. This will help eliminate sign errors.

39. Since ΔH is exothermic, then the temperature of the solution will increase as $CaCl_2(s)$ dissolves. Keeping all quantities positive:

$$\text{Heat loss as } CaCl_2 \text{ dissolves} = 11.0 \text{ g } CaCl_2 \times \frac{1 \text{ mol } CaCl_2}{110.98 \text{ g } CaCl_2} \times \frac{81.5 \text{ kJ}}{\text{mol } CaCl_2} = 8.08 \text{ kJ}$$

$$\text{Heat gained by solution} = 8.08 \times 10^3 \text{ J} = \frac{4.18 \text{ J}}{\text{g °C}} \times (125 + 11.0) \text{ g} \times (T_f - 25.0°C)$$

$$T_f - 25.0°C = \frac{8.08 \times 10^3}{4.18 \times 136} = 14.2°C, \quad T_f = 14.2°C + 25.0°C = 39.2°C$$

41. $\text{Heat gain by calorimeter} = \frac{1.56 \text{ kJ}}{°C} \times 3.2°C = 5.0 \text{ kJ} = \text{heat loss by quinone}$

Heat loss = 5.0 kJ, which is the heat evolved (exothermic reaction) by the combustion of 0.1964 g of quinone.

$$\Delta E_{comb} = \frac{-5.0 \text{ kJ}}{0.1964 \text{ g}} = -25 \text{ kJ/g}; \qquad \Delta E_{comb} = \frac{-25 \text{ kJ}}{\text{g}} \times \frac{108.09 \text{ g}}{\text{mol}} = -2700 \text{ kJ/mol}$$

43. a. $C_{12}H_{22}O_{11}(s) + 12\ O_2(g) \rightarrow 12\ CO_2(g) + 11\ H_2O(l)$

b. A bomb calorimeter is at constant volume, so heat released = $q_v = \Delta E$:

$$\Delta E = \frac{-24.00 \text{ kJ}}{1.46 \text{ g}} \times \frac{342.30 \text{ g}}{\text{mol}} = -5630 \text{ kJ/mol } C_{12}H_{22}O_{11}$$

c. $\Delta H = \Delta E + \Delta(PV) = \Delta E + \Delta(nRT) = \Delta E + \Delta nRT$ where Δn = mol gaseous products - mol gaseous reactants.

For this reaction $\Delta n = 12 - 12 = 0$, so $\Delta H = \Delta E = -5630$ kJ/mol.

Hess's Law

45.

$2\ C + 2\ O_2 \rightarrow 2\ CO_2$	$\Delta H = 2(-394 \text{ kJ})$
$H_2 + 1/2\ O_2 \rightarrow H_2O$	$\Delta H = -286$ kJ
$2\ CO_2 + H_2O \rightarrow C_2H_2 + 5/2\ O_2$	$\Delta H = -(-1300.\text{kJ})$
$2\ C(s) + H_2(g) \rightarrow C_2H_2(g)$	$\Delta H = 226$ kJ

Note: The enthalpy change for a reaction that is reversed is the negative quantity of the enthalpy change for the original reaction. If the coefficients in a balanced reaction are multiplied by an integer, then the value of ΔH is multiplied by the same integer.

47.

$$4\ HNO_3 \rightarrow 2\ N_2O_5 + 2\ H_2O \qquad \Delta H = -2(-76.6\ kJ)$$
$$2\ N_2 + 6\ O_2 + 2\ H_2 \rightarrow 4\ HNO_3 \qquad \Delta H = 4(-174.1\ kJ)$$
$$2\ H_2O \rightarrow 2\ H_2 + O_2 \qquad \Delta H = -2(-285.8\ kJ)$$

$$2\ N_2(g) + 5\ O_2(g) \rightarrow 2\ N_2O_5(g) \qquad \Delta H = 28.4\ kJ$$

49.

$$C_4H_4(g) + 5\ O_2(g) \rightarrow 4\ CO_2(g) + 2\ H_2O(l) \qquad \Delta H_{comb} = -2341\ kJ$$
$$C_4H_8(g) + 6\ O_2(g) \rightarrow 4\ CO_2(g) + 4\ H_2O(l) \qquad \Delta H_{comb} = -2755\ kJ$$
$$H_2(g) + 1/2\ O_2(g) \rightarrow H_2O(l) \qquad \Delta H_{comb} = -286\ kJ$$

By convention, $H_2O(l)$ is produced when enthalpies of combustion are given and, since per mol quantities are given, the combustion reaction refers to 1 mol of that quantity reacting with $O_2(g)$.

Using Hess's Law to solve:

$$C_4H_4(g) + 5\ O_2(g) \rightarrow 4\ CO_2(g) + 2\ H_2O(l) \qquad \Delta H_1 = -2341\ kJ$$
$$4\ CO_2(g) + 4\ H_2O(l) \rightarrow C_4H_8(g) + 6\ O_2(g) \qquad \Delta H_2 = -(-2755\ kJ)$$
$$2\ H_2(g) + O_2(g) \rightarrow 2\ H_2O(l) \qquad \Delta H_3 = 2(-286\ kJ)$$

$$C_4H_4(g) + 2\ H_2(g) \rightarrow C_4H_8(g) \qquad \Delta H = \Delta H_1 + \Delta H_2 + \Delta H_3 = -158\ kJ$$

51.

$$C_6H_4(OH)_2 \rightarrow C_6H_4O_2 + H_2 \qquad \Delta H = 177.4\ kJ$$
$$H_2O_2 \rightarrow H_2 + O_2 \qquad \Delta H = -(-191.2\ kJ)$$
$$2\ H_2 + O_2 \rightarrow 2\ H_2O(g) \qquad \Delta H = 2(-241.8\ kJ)$$
$$2\ H_2O(g) \rightarrow 2\ H_2O(l) \qquad \Delta H = 2(-43.8\ kJ)$$

$$C_6H_4(OH)_2(aq) + H_2O_2(aq) \rightarrow C_6H_4O_2(aq) + 2\ H_2O(l) \qquad \Delta H = -202.6\ kJ$$

53.

$$2\ N_2(g) + 6\ H_2(g) \rightarrow 4\ NH_3(g) \qquad \Delta H = -4(46\ kJ)$$
$$6\ H_2O(g) \rightarrow 6\ H_2(g) + 3\ O_2(g) \qquad \Delta H = -3(-484\ kJ)$$

$$2\ N_2(g) + 6\ H_2O(g) \rightarrow 3\ O_2(g) + 4\ NH_3(g) \qquad \Delta H = 1268\ kJ$$

No, since the reaction is very endothermic (requires a lot of heat), it would not be a practical way of making ammonia due to the high energy costs required.

Standard Enthalpies of Formation

55. **In general: $\Delta H° = \Sigma n_p \Delta H^\circ_{f,\ products} - \Sigma n_r \Delta H^\circ_{f,\ reactants}$ and all elements in their standard state have $\Delta H^\circ_f = 0$ by definition.**

a. The balanced equation is: $2\ NH_3(g) + 3\ O_2(g) + 2\ CH_4(g) \rightarrow 2\ HCN(g) + 6\ H_2O(g)$

$$\Delta H° = [2 \text{ mol HCN} \times \Delta H^\circ_{f,\,HCN} + 6 \text{ mol } H_2O(g) \times \Delta H^\circ_{f,\,H_2O}]$$

$$- [2 \text{ mol } NH_3 \times \Delta H^\circ_{f,\,NH_3} + 2 \text{ mol } CH_4 \times \Delta H^\circ_{f,\,CH_4}]$$

$$\Delta H° = [2(135.1) + 6(-242)] - [2(-46) + 2(-75)] = -940.\text{ kJ}$$

b. $Ca_3(PO_4)_2(s) + 3\ H_2SO_4(l) \rightarrow 3\ CaSO_4(s) + 2\ H_3PO_4(l)$

$$\Delta H° = \left[3 \text{ mol } CaSO_4\left(\frac{-1433 \text{ kJ}}{\text{mol}}\right) + 2 \text{ mol } H_3PO_4(l)\left(\frac{-1267 \text{ kJ}}{\text{mol}}\right)\right]$$

$$- \left[1 \text{ mol } Ca_3(PO_4)_2\left(\frac{-4126 \text{ kJ}}{\text{mol}}\right) + 3 \text{ mol } H_2SO_4(l)\left(\frac{-814 \text{ kJ}}{\text{mol}}\right)\right]$$

$$\Delta H° = -6833 \text{ kJ} - (-6568 \text{ kJ}) = -265 \text{ kJ}$$

c. $NH_3(g) + HCl(g) \rightarrow NH_4Cl(s)$

$$\Delta H° = [1 \text{ mol } NH_4Cl \times \Delta H^\circ_{f,\,NH_4Cl}] - [1 \text{ mol } NH_3 \times \Delta H^\circ_{f,\,NH_3} + 1 \text{ mol HCl} \times \Delta H^\circ_{f,\,HCl}]$$

$$\Delta H° = \left[1 \text{ mol}\left(\frac{-314 \text{ kJ}}{\text{mol}}\right)\right] - \left[1 \text{ mol}\left(\frac{-46 \text{ kJ}}{\text{mol}}\right) + 1 \text{ mol}\left(\frac{-92 \text{ kJ}}{\text{mol}}\right)\right]$$

$$\Delta H° = -314 \text{ kJ} + 138 \text{ kJ} = -176 \text{ kJ}$$

d. The balanced equation is: $C_2H_5OH(l) + 3\ O_2(g) \rightarrow 2\ CO_2(g) + 3\ H_2O(g)$

$$\Delta H° = \left[2 \text{ mol}\left(\frac{-393.5 \text{ kJ}}{\text{mol}}\right) + 3 \text{ mol}\left(\frac{-242 \text{ kJ}}{\text{mol}}\right)\right] - \left[1 \text{ mol}\left(\frac{-278 \text{ kJ}}{\text{mol}}\right)\right]$$

$$= -1513 \text{ kJ} - (-278 \text{ kJ}) = -1235 \text{ kJ}$$

e. $SiCl_4(l) + 2\ H_2O(l) \rightarrow SiO_2(s) + 4\ HCl(aq)$

Since HCl(aq) is $H^+(aq) + Cl^-(aq)$, then $\Delta H^\circ_f = 0 - 167 = -167$ kJ/mol.

$$\Delta H° = \left[4 \text{ mol}\left(\frac{-167 \text{ kJ}}{\text{mol}}\right) + 1 \text{ mol}\left(\frac{-911 \text{ kJ}}{\text{mol}}\right)\right] - \left[1 \text{ mol}\left(\frac{-687 \text{ kJ}}{\text{mol}}\right) + 2 \text{ mol}\left(\frac{-286 \text{ kJ}}{\text{mol}}\right)\right]$$

$$\Delta H° = -1579 \text{ kJ} - (-1259 \text{ kJ}) = -320.\text{ kJ}$$

f. $MgO(s) + H_2O(l) \rightarrow Mg(OH)_2(s)$

$$\Delta H° = \left[1 \text{ mol}\left(\frac{-925 \text{ kJ}}{\text{mol}}\right)\right] - \left[1 \text{ mol}\left(\frac{-602 \text{ kJ}}{\text{mol}}\right) + 1 \text{ mol}\left(\frac{-286 \text{ kJ}}{\text{mol}}\right)\right]$$

$$\Delta H° = -925 \text{ kJ} - (-888 \text{ kJ}) = -37 \text{ kJ}$$

57. $4\ Na(s) + O_2(g) \rightarrow 2\ Na_2O(s)$, $\Delta H° = 2\text{ mol}\left(\frac{-416\text{ kJ}}{\text{mol}}\right) = -832\text{ kJ}$

$2\ Na(s) + 2\ H_2O(l) \rightarrow 2\ NaOH(aq) + H_2(g)$

$$\Delta H° = \left[2\text{ mol}\left(\frac{-470.\text{ kJ}}{\text{mol}}\right)\right] - \left[2\text{ mol}\left(\frac{-286\text{ kJ}}{\text{mol}}\right)\right] = -368\text{ kJ}$$

$2Na(s) + CO_2(g) \rightarrow Na_2O(s) + CO(g)$

$$\Delta H° = \left[1\text{ mol}\left(\frac{-416\text{ kJ}}{\text{mol}}\right) + 1\text{ mol}\left(\frac{-110.5\text{ kJ}}{\text{mol}}\right)\right] - \left[1\text{ mol}\left(\frac{-393.5\text{ kJ}}{\text{mol}}\right)\right] = -133\text{ kJ}$$

In reactions 2 and 3, sodium metal reacts with the "extinguishing agent." Both reactions are exothermic and each reaction produces a flammable gas, H_2 and CO, respectively.

59. $5\ N_2O_4(l) + 4\ N_2H_3CH_3(l) \rightarrow 12\ H_2O(g) + 9\ N_2(g) + 4\ CO_2(g)$

$$\Delta H° = \left[12\text{ mol}\left(\frac{-242\text{ kJ}}{\text{mol}}\right) + 4\text{ mol}\left(\frac{-393.5\text{ kJ}}{\text{mol}}\right)\right]$$

$$- \left[5\text{ mol}\left(\frac{-20.\text{ kJ}}{\text{mol}}\right) + 4\text{ mol}\left(\frac{54\text{ kJ}}{\text{mol}}\right)\right] = -4594\text{ kJ}$$

61. a. $\Delta H° = 3\text{ mol }(227\text{ kJ/mol}) - 1\text{ mol }(49\text{ kJ/mol}) = 632\text{ kJ}$

b. Since 3 $C_2H_2(g)$ is higher in energy than $C_6H_6(l)$, then acetylene will release more energy per gram when burned in air.

63. $2\ ClF_3(g) + 2\ NH_3(g) \rightarrow N_2(g) + 6\ HF(g) + Cl_2(g)$ $\Delta H° = -1196\text{ kJ}$

$\Delta H° = [6\ \Delta H°_{f,\ HF}] - [2\ \Delta H°_{f,\ ClF_3} + 2\ \Delta H°_{f,\ NH_3}]$

$$-1196\text{ kJ} = 6\text{ mol}\left(\frac{-271\text{ kJ}}{\text{mol}}\right) - 2\ \Delta H°_{f,\ ClF_3} - 2\text{ mol}\left(\frac{-46\text{ kJ}}{\text{mol}}\right)$$

$$-1196\text{ kJ} = -1626\text{ kJ} - 2\ \Delta H°_{f,\ ClF_3} + 92\text{ kJ},\ \Delta H°_{f,\ ClF_3} = \frac{(-1626 + 92 + 1196)\text{ kJ}}{2\text{ mol}} = \frac{-169\text{ kJ}}{\text{mol}}$$

Energy Consumption and Sources

65. Mass of $H_2O = 1.00 \text{ gal} \times \frac{3.785 \text{ L}}{\text{gal}} \times \frac{1000 \text{ mL}}{\text{L}} \times \frac{1.00 \text{ g}}{\text{mL}} = 3790 \text{ g } H_2O$

Energy required (theoretical) $= s \times m \times \Delta T = \frac{4.18 \text{ J}}{\text{g °C}} \times 3790 \text{ g} \times 10.0 \text{ °C} = 1.58 \times 10^5 \text{ J}$

For an actual (80.0% efficient) process, more than this quantity of energy is needed since heat is always lost in any transfer of energy. The energy required is:

$$1.58 \times 10^5 \text{ J} \times \frac{100. \text{ J}}{80.0 \text{ J}} = 1.98 \times 10^5 \text{ J}$$

Mass of $C_2H_2 = 1.98 \times 10^5 \text{ J} \times \frac{1 \text{ mol } C_2H_2}{1300. \times 10^3 \text{ J}} \times \frac{26.04 \text{ g } C_2H_2}{\text{mol } C_2H_2} = 3.97 \text{ g } C_2H_2$

67. $CO(g) + 2\,H_2(g) \rightarrow CH_3OH(l)$ $\Delta H° = (-239 \text{ kJ}) - (-110.5 \text{ kJ}) = -129 \text{ kJ}$

Additional Exercises

69. $\Delta E_{overall} = \Delta E_{step\ 1} + \Delta E_{step\ 2}$; This is a cyclic process which means that the overall initial state and final state are the same. Since ΔE is a state function, then $\Delta E_{overall} = 0$ and $\Delta E_{step\ 1} = -\Delta E_{step\ 2}$.

$\Delta E_{step\ 1} = q + w = 45 \text{ J} + (-10. \text{ J}) = 35 \text{ J}$

$\Delta E_{step\ 2} = -\Delta E_{step\ 1} = -35 \text{ J} = q + w$, $-35 \text{ J} = -60 \text{ J} + w$, $w = 25 \text{ J}$

71. $H_2(g) + 1/2\,O_2(g) \rightarrow H_2O(l)$ $\Delta H° = \Delta H°_{f,\,H_2O(l)} = -285.8 \text{ kJ}$

$H_2O(l) \rightarrow H_2(g) + 1/2\,O_2(g)$ $\Delta H° = 285.8 \text{ kJ}$

$\Delta E° = \Delta H° - P\Delta V = \Delta H° - \Delta nRT = 285.8 \text{ kJ} - (1.50 - 0 \text{ mol})(8.3145 \text{ J mol}^{-1} \text{ K}^{-1})(298 \text{ K})\left(\frac{1 \text{ kJ}}{1000 \text{ J}}\right)$

$\Delta E° = 285.8 \text{ kJ} - 3.72 \text{ kJ} = 282.1 \text{ kJ}$

73. The specific heat of water is $4.18 \text{ J °C}^{-1} \text{ g}^{-1}$, which is equal to $4.18 \text{ kJ °C}^{-1} \text{ kg}^{-1}$

We have 1.00 kg of H_2O, so: $1.00 \text{ kg} \times \frac{4.18 \text{ kJ}}{\text{kg °C}} = 4.18 \text{ kJ/°C}$

This is the portion of the heat capacity that can be attributed to H_2O.

Total heat capacity $= C_{cal} + C_{H_2O}$, $C_{cal} = 10.84 - 4.18 = 6.66 \text{ kJ/°C}$

75. $N_2H_4(l) + O_2(g) \rightarrow N_2(g) + 2\,H_2O(g)$ $\Delta H° = 2(-242 \text{ kJ}) - (51 \text{ kJ}) = -535 \text{ kJ}$

$2\,N_2H_4(l) + N_2O_4(l) \rightarrow 3\,N_2(g) + 4\,H_2O(g)$ $\Delta H° = 4(-242 \text{ kJ}) - [2(51 \text{ kJ}) + (-20. \text{ kJ})] = -1050. \text{ kJ}$

For hydrazine plus oxygen the stoichiometric reactant mixture contains 1 mol N_2H_4 for every mol O_2. This is a total mass of 64.05 g or 0.06405 kg.

$$\Delta H = \frac{-535 \text{ kJ}}{0.06405 \text{ kg}} = -8.35 \times 10^3 \text{ kJ/kg for } N_2H_4(l) + O_2(g)$$

For hydrazine plus N_2O_4 the optimum mixture contains 2 mol of N_2H_4 for every mol N_2O_4. This is a total mass of 156.12 g or 0.15612 kg.

$$\Delta H = \frac{-1050. \text{ kJ}}{0.15612 \text{ kg}} = -6.726 \times 10^3 \text{ kJ/kg for } 2\ N_2H_4(l) + N_2O_4(l)$$

From Exercise 60, the most efficient fuel was methyl hydrazine ($N_2H_3CH_3$) plus N_2O_4, with a fuel value of -7129 kJ/kg. Therefore, the hydrazine plus oxygen combination is the most efficient fuel.

77. $w = -P\Delta V$; Δn = mol gaseous products - mol gaseous reactants. Only gases can do PV work (we ignore solids and liquids). When a balanced reaction has more mol of product gases than mol of reactant gases (Δn positive), the reaction will expand in volume (ΔV positive) and the system will do work on the surroundings. For example, in reaction c, $\Delta n = 2 - 0 = 2$ mol, and this reaction would do expansion work against the surroundings. When a balanced reaction has a decrease in the mol of gas from reactants to products (Δn negative), the reaction will contract in volume (ΔV negative) and the surroundings will do compression work on the system, e.g., reaction a where $\Delta n = 0 - 1 = -1$. When there is no change in the mol of gas from reactants to products, $\Delta V = 0$ and $w = 0$, e.g., reaction b where $\Delta n = 2 - 2 = 0$.

When $\Delta V > 0$ ($\Delta n > 0$), then $w < 0$ and system does work on the surroundings (c and e).

When $\Delta V < 0$ ($\Delta n < 0$), then $w > 0$ and the surroundings do work on the system (a and d).

When $\Delta V = 0$ ($\Delta n = 0$), then $w = 0$ (b).

79. a. aluminum oxide = Al_2O_3; $2\ Al(s) + 3/2\ O_2(g) \rightarrow Al_2O_3(s)$

b. $C_2H_5OH(l) + 3\ O_2(g) \rightarrow 2\ CO_2(g) + 3\ H_2O(l)$

c. $Ba(OH)_2(aq) + 2\ HCl(aq) \rightarrow 2\ H_2O(l) + BaCl_2(aq)$

d. $2\ C(\text{graphite, s}) + 3/2\ H_2(g) + 1/2\ Cl_2(g) \rightarrow C_2H_3Cl(g)$

e. $C_6H_6(l) + 15/2\ O_2(g) \rightarrow 6\ CO_2(g) + 3\ H_2O(l)$

Note: ΔH_{comb} values assume one mol of compound combusted.

f. $NH_4Br(s) \rightarrow NH_4^+(aq) + Br^-(aq)$

Challenge Problems

81. If the gas is monoatomic, then $C_v = \frac{3}{2}R = 12.47\ J\ mol^{-1}\ K^{-1}$ and $C_p = \frac{5}{2}R = 20.79\ J\ mol^{-1}\ K^{-1}$.
If the gas is behaving ideally, then $C_p - C_v = R = 8.3145\ J\ mol^{-1}\ K^{-1}$.

At constant volume: $q_v = 2079\ J = nC_v\Delta T$

$$C_v = \frac{2079\ J}{n\Delta T} = \frac{2079\ J}{(1\ mol)\ (400.0-300.0\ K)} = 20.79\ J\ mol^{-1}\ K^{-1}$$

Since $C_v \neq 3/2\ R = 12.47\ J$, then the gas is not a monoatomic gas.

At constant pressure: $q_p = nC_p\Delta T$

$q_p = \Delta E - w = 1305\ J - (-150.\ J) = 1455\ J$ (Gas expansion so system does work or surroundings.)

$$C_p = \frac{q_p}{n\Delta T} = \frac{1455\ J}{(1\ mol)\ (600.0 - 550.0\ K)} = 29.1\ J\ mol^{-1}\ K^{-1}$$

$C_p - C_v = 29.1 - 20.79 = 8.31\ J\ mol^{-1}\ K^{-1} = R$

The gas is behaving ideally since $C_p - C_v = R$.

83. $H_2O(s) \rightarrow H_2O(l)$ $\Delta H = \Delta H_{fus}$; For 1 mol of supercooled water at -15.0°C (or 258.2 K), $\Delta H_{fus,\ 258.2\ K} = 10.9\ kJ/2 = 5.45\ kJ/mol$. Using Hess's Law and the equation $\Delta H = nC_p\Delta T$:

H_2O (s, 273.2 K) $\rightarrow$ H_2O (s, 258.2 K) $\Delta H_1 = 1\ mol\ (37.5\ J\ K^{-1}\ mol^{-1})\ (-15.0\ K) = -563\ J = -0.563\ kJ$

H_2O (s, 258.2 K) $\rightarrow$ H_2O (l, 258.2 K) $\Delta H_2 = 1\ mol\ (5.45\ kJ/mol) = 5.45\ kJ$

H_2O (l, 258.2 K) $\rightarrow$ H_2O (l, 273.2 K) $\Delta H_3 = 1\ mol\ (75.3\ J\ K^{-1}\ mol^{-1})\ (15.0\ K) = 1130\ J = 1.13\ kJ$

H_2O (s, 273. 2 K) $\rightarrow$ H_2O (l, 273.2 K) $\Delta H_{fus,\ 273.2} = \Delta H_1 + \Delta H_2 + \Delta H_3$

$\Delta H_{fus,\ 273.2} = -0.563\ kJ + 5.45\ kJ + 1.13\ kJ = 6.02\ kJ$; $\Delta H_{fus,\ 273.2} = 6.02\ kJ/mol$

85. Molar heat capacity of $H_2O(l) = 4.184\ J\ K^{-1}\ g^{-1}\ (18.015\ g/mol) = 75.37\ J\ K^{-1}\ mol^{-1}$
Molar heat capacity of $H_2O(g) = 2.02\ J\ K^{-1}\ g^{-1}\ (18.015\ g/mol) = 36.4\ J\ K^{-1}\ mol^{-1}$
Using Hess's Law and the equation $\Delta H = nC_p\Delta T$:

H_2O (l, 340.2 K) $\rightarrow$ H_2O (l, 373.2 K) $\Delta H_1 = 1\ mol\ (75.37\ J\ K^{-1}\ mol^{-1})(33.0\ K)(1\ kJ/1000\ J) = 2.49\ kJ$

H_2O (l, 373.2 K) $\rightarrow$ H_2O (g, 373.2 K) $\Delta H_2 = 1\ mol\ (40.66\ kJ/mol) = 40.66\ kJ$

H_2O (g, 373.2 K) $\rightarrow$ H_2O (g, 340.2 K) $\Delta H_3 = 1\ mol\ (36.4\ J\ K^{-1}\ mol^{-1})(-33.0\ K)(1\ kJ/1000\ J) = -1.20\ kJ$

H_2O (l, 340.2 K) $\rightarrow$ H_2O (g, 340.2 K) $\Delta H_{vap,\ 340.2\ K} = \Delta H_1 + \Delta H_2 + \Delta H_3 = 41.95\ kJ/mol$

87. Energy used in 8.0 hours = 40. kWh = $\frac{40.\text{ kJ h}}{\text{s}} \times \frac{3600\text{ s}}{\text{h}} = 1.4 \times 10^5$ kJ

Energy from the sun in 8.0 hours = $\frac{1.0\text{ kJ}}{\text{s m}^2} \times \frac{60\text{ s}}{\text{min}} \times \frac{60\text{ min}}{\text{h}} \times 8.0\text{ h} = 2.9 \times 10^4\text{ kJ/m}^2$

Only 13% of the sunlight is converted into electricity:

$0.13 \times (2.9 \times 10^4\text{ kJ/m}^2) \times \text{Area} = 1.4 \times 10^5\text{ kJ}$, Area = 37 m^2

Marathon Problems

89. $X \rightarrow CO_2(g) + H_2O(l) + O_2(g) + A(g)$ $\Delta H = -1893$ kJ/mol (unbalanced)

To determine X we must determine the mol of X reacted, the identity of A and the mol of A produced. For the reaction at constant P ($\Delta H = q$):

$-q_{H_2O} = q_{rxn} = -4.184\text{ J °C}^{-1}\text{ g}^{-1}\ (1.000 \times 10^4\text{ g})\ (29.52 - 25.00\text{ °C})\ (1\text{ kJ}/1000\text{ J})$

$q_{rxn} = -189.1$ kJ (carrying extra sig. figs.)

Since $\Delta H = -1893$ kJ/mol for the decomposition reaction and since only -189.1 kJ of heat were released for this reaction, then 189.1 kJ × (1 mol X/1893 kJ) = 0.100 mol X were reacted.

Molar mass of X = $\frac{22.7\text{ g X}}{0.100\text{ mol X}} = 227$ g/mol

From the problem, 0.100 mol X produced 0.300 mol CO_2, 0.250 mol H_2O and 0.025 mol O_2. Therefore, 1.00 mol X contains 3.00 mol CO_2, 2.50 mol H_2O and 0.25 mol O_2.

$$1.00\text{ mol X} = 227\text{ g} = 3.00\text{ mol } CO_2\left(\frac{44.0\text{ g}}{\text{mol}}\right) + 2.50\text{ mol } H_2O\left(\frac{18.0\text{ g}}{\text{mol}}\right) + 0.25\text{ mol } O_2\left(\frac{32.0\text{ g}}{\text{mol}}\right) + (\text{mass of A})$$

mass of A in 1.00 mol X = 227 g - 132 g - 45.0 g - 8.0 g = 42 g A

To determine A, we need the mol of A produced. The total mol of gases produced can be determined from the gas law data provided in the problem. Since $H_2O(l)$ is a product, we need to subtract P_{H_2O} from the total pressure.

$n_{total} = \frac{PV}{RT}$; $P_{total} = P_{gases} + P_{H_2O}$, $P_{gases} = 778\text{ torr} - 31\text{ torr} = 747$ torr

V = height × area; area = πr^2; $V = 59.8\text{ cm}\ (\pi)\ (8.00\text{ cm})^2\left(\frac{1\text{ L}}{1000\text{ cm}^3}\right) = 12.0$ L

T = 273.15 + 29.52 = 302.67 K

$$n_{total} = \frac{PV}{RT} = \frac{747 \text{ torr}\left(\dfrac{1 \text{ atm}}{760 \text{ torr}}\right)(12.0 \text{ L})}{\dfrac{0.08206 \text{ L atm}}{\text{K mol}}(302.67 \text{ K})} = 0.475 \text{ mol} = \text{mol } CO_2 + \text{mol } O_2 + \text{mol A}$$

mol A = 0.475 mol - 0.300 mol CO_2 - 0.025 mol O_2 = 0.150 mol A

Since 0.100 mol X reacted, then 1.00 mol X would contain 1.50 mol A which from a previous calculation represents 42 g A.

$$\text{Molar mass of A} = \frac{42 \text{ g A}}{1.50 \text{ mol A}} = 28 \text{ g/mol}$$

Since A is a gaseous element, the only element that is a gas and has this molar mass is $N_2(g)$. Thus, A = $N_2(g)$

a. Now we can determine the formula of X.

$X \rightarrow 3\ CO_2(g) + 2.5\ H_2O(l) + 0.25\ O_2(g) + 1.5\ N_2(g)$. For a balanced reaction, X = $C_3H_5N_3O_9$, which, for your information, is nitroglycerine.

b. $$w = -P\Delta V = -778 \text{ torr}\left(\frac{1 \text{ atm}}{760 \text{ torr}}\right)(12.0 \text{ L} - 0) = -12.3 \text{ L atm}$$

$$-12.3 \text{ L atm}\left(\frac{8.3145 \text{ J mol}^{-1} \text{ K}^{-1}}{0.08206 \text{ L atm mol}^{-1} \text{ K}^{-1}}\right) = -1250 \text{ J} = -1.25 \text{ kJ},\ w = -1.25 \text{ kJ}$$

c. $\Delta E = q + w$, where $q = \Delta H$ since at constant pressure. For 1 mol of X decomposed:

$w = -1.25 \text{ kJ}/0.100 \text{ mol} = -12.5 \text{ kJ/mol}$

$\Delta E = \Delta H + w = -1893 \text{ kJ/mol} + (-12.5 \text{ kJ/mol}) = -1906 \text{ kJ/mol}$

ΔH_f° for $C_3H_5N_3O_9$ can be estimated from standard enthalpies of formation data and assuming $\Delta H_{rxn} = \Delta H_{rxn}^\circ$. For the balanced reaction given in part a:

$$\Delta H_{rxn}^\circ = -1893 \text{ kJ} = [3\ \Delta H_{f,\ CO_2}^\circ + 2.5\ \Delta H_{f,\ H_2O} + 0.25\ \Delta H_{f,\ O_2} + 1.5\ \Delta H_{f,\ N_2}] - [\Delta H_{f,\ C_3H_5N_3O_9}^\circ]$$

$$-1893 \text{ kJ} = [3\ (-393.5) \text{ kJ} + 2.5\ (-286) \text{ kJ} + 0 + 0] - \Delta H_{f,\ C_3H_5N_3O_9},\ \Delta H_{f,\ C_3H_5N_3O_9} = -2.5 \text{ kJ/mol}$$

CHAPTER TEN

SPONTANEITY, ENTROPY, AND FREE ENERGY

Spontaneity and Entropy

13. We draw all of the possible arrangements of the two particles in the three levels.

2 kJ	___	___	x	___	x	xx
1 kJ	___	x	___	xx	x	___
0 kJ	xx	x	x	___	___	___
Total E =	0 kJ	1 kJ	2 kJ	2 kJ	3 kJ	4 kJ

The most likely total energy is 2 kJ.

15. Processes a, b, d, and g are spontaneous. Processes c, e, and f require an external source of energy in order to occur since they are nonspontaneous.

17. a. Positional probability increases; there is a greater volume accessible to the randomly moving gas molecules which increases disorder.

b. The positional probability doesn't change. There is no change in volume and thus, no change in the numbers of positions of the molecules.

c. Positional probability decreases because the volume decreases (P and V are inversely related).

19. There are more ways to roll a seven. We can consider all of the possible throws by constructing a table.

one die	1	2	3	4	5	6	
1	2	3	4	5	6	7	
2	3	4	5	6	7	8	
3	4	5	6	7	8	9	sum of the two dice
4	5	6	7	8	9	10	
5	6	7	8	9	10	11	
6	7	8	9	10	11	12	

There are six ways to get a seven, more than any other number. The seven is not favored by energy; rather it is favored by probability. To change the probability we would have to expend energy (do work).

Energy, Enthalpy, and Entropy Changes Involving Ideal Gases and Physical Changes

21. $15.0 \text{ g He} \times \frac{1 \text{ mol}}{4.003 \text{ g}} = 3.75 \text{ mol He}$

$q_v = \Delta E = nC_v\Delta T = 3.75 \text{ mol } (12.47 \text{ J K}^{-1} \text{ mol}^{-1})(56.5 \text{ K}) = 2640 \text{ J} = 2.64 \text{ kJ}$

23. It takes $nC_p\Delta T$ amount of energy to carry out this process. The internal energy of the system increases by $nC_v\Delta T$. So the fraction that goes into raising the internal energy is:

$$\frac{nC_v\Delta T}{nC_p\Delta T} = \frac{C_v}{C_p} = \frac{20.8}{29.1} = 0.715$$

The remainder of the energy ($nR\Delta T$) goes into expanding the gas against the constant pressure.

$100.0 \text{ g } N_2 \times \frac{1 \text{ mol}}{28.014 \text{ g}} = 3.570 \text{ mol}$

$q_v = \Delta E = nC_v\Delta T = 3.570 \text{ mol } (20.8 \text{ J mol}^{-1} \text{ K}^{-1})(60.0 \text{ K}) = 4.46 \times 10^3 \text{ J} = 4.46 \text{ kJ}$

25. The volumes for each step are:

a. $P_1 = 5.00$ atm, $n = 1.00$ mol, $T = 350.$ K; $V_1 = \frac{nRT}{P_1} = 5.74 \text{ L}$

b. $P_2 = 2.24$ atm, $V_2 = \frac{nRT}{P_2} = 12.8 \text{ L}$ c. $P_3 = 1.00$ atm, $V_3 = 28.7$ L

The process can be carried out in the following steps:

$(P_1, V_1) \rightarrow (P_2, V_1)$ $w = -P\Delta V = 0$ (constant volume process)

$(P_2, V_1) \rightarrow (P_2, V_2)$ $w = -(2.24 \text{ atm})(12.8 - 5.74 \text{ L}) = -16 \text{ L atm}$

$(P_2, V_2) \rightarrow (P_3, V_2)$ $w = 0$

$(P_3, V_2) \rightarrow (P_3, V_3)$ $w = -(1.00 \text{ atm})(28.7 - 12.8 \text{ L}) = -15.9 \text{ L atm}$

$w_{tot} = -16 - 15.9 = -32 \text{ L atm}; \; -32 \text{ L atm} \times \dfrac{101.3 \text{ J}}{1 \text{ L atm}} = -3200 \text{ J} = \text{total work}$

$w_{rev} = -nRT \ln(P_1/P_2) = -(1.00 \text{ mol})(8.3145 \text{ J mol}^{-1} \text{ K}^{-1})(350. \text{ K}) \ln(5.00/1.00) = -4680 \text{ J}$

27. a. $q_v = \Delta E = nC_v\Delta T = (1.000 \text{ mol})(28.95 \text{ J mol}^{-1} \text{ K}^{-1})(350.0 - 298.0 \text{ K})$

$q_v = 1.51 \times 10^3 \text{ J} = 1.51 \text{ kJ}$

$q_p = \Delta H = nC_p\Delta T = 1.000 (37.27)(350.0 - 298.0) = 1.94 \times 10^3 \text{ J} = 1.94 \text{ kJ}$

b. $\Delta S = S_{350} - S_{298} = nC_p \ln(T_2/T_1)$

$S_{350} - 213.64 \text{ J/K} = (1.000 \text{ mol})(37.27 \text{ J mol}^{-1} \text{ K}^{-1}) \ln(350.0/298.0)$

$S_{350} = 213.64 \text{ J/K} + 5.994 \text{ J/K} = 219.63 \text{ J/K}$ = molar entropy at 350.0 K and 1.000 atm

c. $\Delta S = nR \ln(V_2/V_1)$, $V = nRT/P$, $\Delta S = nR \ln(P_1/P_2) = S_{(350,\, 1.174)} - S_{(350,\, 1.000)}$

$\Delta S = S_{(350,\, 1.174)} - 219.63 \text{ J/K} = (1.000 \text{ mol})(8.3145 \text{ J mol}^{-1} \text{ K}^{-1}) \ln(1.000 \text{ atm}/1.174 \text{ atm})$

$\Delta S = -1.334 \text{ J/K} = S_{(350,\, 1.174)} - 219.63, \; S_{(350,\, 1.174)} = 218.30 \text{ J mol}^{-1} \text{ K}^{-1}$

29. For $A(l, 125°C) \rightarrow A(l, 75°C)$:

$\Delta S = nC_p \ln(T_2/T_1) = 1.00 \text{ mol}(75.0 \text{ J K}^{-1} \text{ mol}^{-1}) \ln(348 \text{ K}/398 \text{ K}) = -10.1 \text{ J/K}$

For $A(l, 75°C) \rightarrow A(g, 155°C)$: $\Delta S = 75.0 \text{ J K}^{-1} \text{ mol}^{-1}$

For $A(g, 155°C) \rightarrow A(g, 125°C)$:

$\Delta S = nC_p \ln(T_2/T_1) = 1.00 \text{ mol}(29.0 \text{ J K}^{-1} \text{ mol}^{-1}) \ln(398 \text{ K}/428 \text{ K}) = -2.11 \text{ J/K}$

The sum of the three step gives $A(l, 125°C) \rightarrow A(g, 125°C)$. ΔS for this process is the sum of ΔS for each of the three steps.

$\Delta S = -10.1 + 75.0 - 2.11 = 62.8 \text{ J/K}$

For a phase change, $\Delta S = \Delta H/T$. At 125°C: $\Delta H_{vap} = T\Delta S = 398 \text{ K}(62.8 \text{ J/K}) = 2.50 \times 10^4 \text{ J}$

31. Calculate the final temperature by equating heat loss to heat gain.

$(3.00 \text{ mol})(75.3 \text{ J mol}^{-1} \text{ °C}^{-1})(T_f - 0\text{°C}) = (1.00 \text{ mol})(75.3 \text{ J mol}^{-1} \text{ °C}^{-1})(100.\text{°C} - T_f)$

Solving: $T_f = 25\text{°C} = 298 \text{ K}$

Now we can calculate ΔS for the various changes using $\Delta S = nC_p \ln (T_2/T_1)$.

Heat 3 mol H_2O: $\Delta S_1 = (3.00 \text{ mol})(75.3 \text{ J mol}^{-1} \text{ K}^{-1}) \ln (298 \text{ K}/273 \text{ K}) = 19.8 \text{ J/K}$

Cool 1 mol H_2O: $\Delta S_2 = (1.00 \text{ mol})(75.3 \text{ J mol}^{-1} \text{ K}^{-1}) \ln (298/373) = -16.9 \text{ J/K}$

$\Delta S_{tot} = \Delta S_{heat} + \Delta S_{cool} = 19.8 - 16.9 = 2.9 \text{ J/K}$

Entropy and the Second Law of Thermodynamics: Free Energy

33. a. The system is the portion of the universe in which we are interested.

b. The surroundings are everything else in the universe besides the system.

c. A closed system can only exchange energy with its surroundings. Matter is not exchanged.

d. An open system can exchange both matter and energy with its surroundings.

35. No, living organisms need an outside source of matter (food) to survive.

37. a. To boil a liquid requires heat. Hence, this is an endothermic process. All endothermic processes decrease the entropy of the surroundings (ΔS_{surr} is negative).

b. This is an exothermic process. Heat is released when gas molecules slow down enough to form the solid. In exothermic processes, the entropy of the surroundings increases (ΔS_{surr} is positive).

39. a. $C_{graphite}(s)$; Diamond is a more ordered structure than graphite.

b. $C_2H_5OH(g)$; The gaseous state is more disordered than the liquid state.

c. $CO_2(g)$; The gaseous state is more disordered than the solid state.

d. $N_2O(g)$; More complicated molecule with more parts.

e. $HCl(g)$; Larger molecule, more parts (electrons), more disorder.

41. a. $2\ H_2S(g) + SO_2(g) \rightarrow 3\ S_{rhombic}(s) + 2\ H_2O(g)$; Since there are more molecules of reactant gases as compared to product molecules of gas ($\Delta n = 2 - 3 < 0$), then $\Delta S°$ will be negative.

$$\Delta S° = \Sigma n_p S°_{products} - \Sigma n_r S°_{reactants}$$

$\Delta S° = [3 \text{ mol } S_{rhombic}(s)\ (32\ J\ K^{-1}\ mol^{-1}) + 2 \text{ mol } H_2O(g)\ (189\ J\ K^{-1}\ mol^{-1})]$

$- [2 \text{ mol } H_2S(g)\ (206\ J\ K^{-1}\ mol^{-1}) + 1 \text{ mol } SO_2(g)\ (248\ J\ K^{-1}\ mol^{-1})]$

$\Delta S° = 474\ J/K - 660.\ J/K = -186\ J/K$

b. $2\ SO_3(g) \rightarrow 2\ SO_2(g) + O_2(g)$; Since Δn of gases is positive (Δn = 3-2), then $\Delta S°$ will be positive.

$\Delta S = 2 \text{ mol}(248\ J\ K^{-1}\ mol^{-1}) + 1 \text{ mol}(205\ J\ K^{-1}\ mol^{-1}) - [2 \text{ mol}(257\ J\ K^{-1}\ mol^{-1})] = 187\ J/K$

c. $Fe_2O_3(s) + 3\ H_2(g) \rightarrow 2\ Fe(s) + 3\ H_2O(g)$; Since Δn of gases = 0 (Δn = 3 - 3), then we can't easily predict if $\Delta S°$ will be positive of negative.

$\Delta S = 2 \text{ mol}(27\ J\ K^{-1}\ mol^{-1}) + 3 \text{ mol}(189\ J\ K^{-1}\ mol^{-1}) - [1 \text{ mol}(90.\ J\ K^{-1}\ mol^{-1})$
$+ 3 \text{ mol}(131\ J\ K^{-1}\ mol^{-1})]$

$\Delta S = 138\ J/K$

43. $-144\ J/K = (2 \text{ mol})\, S°_{AlBr_3} - [2(28\ J/K) + 3(152\ J/K)]$, $S°_{AlBr_3} = 184\ J\ K^{-1}\ mol^{-1}$

45. At the boiling point, $\Delta G = 0$ so $\Delta H = T\Delta S$. $T = \dfrac{\Delta H}{\Delta S} = \dfrac{58.51 \times 10^3\ J/mol}{92.92\ J\ K^{-1}\ mol^{-1}} = 629.7\ K$

47. a. $NH_3(s) \rightarrow NH_3(l)$; $\Delta G = \Delta H - T\Delta S = 5650\ J/mol - 200.\ K\ (28.9\ J\ K^{-1}\ mol^{-1})$

$\Delta G = 5650\ J/mol - 5780\ J/mol = -130\ J/mol$

Yes, NH_3 will melt since $\Delta G < 0$ at this temperature.

b. At the melting point, $\Delta G = 0$ so $T = \dfrac{\Delta H}{\Delta S} = \dfrac{5650\ J/mol}{28.9\ J\ K^{-1}\ mol^{-1}} = 196\ K$.

Free Energy and Chemical Reactions

49. a.

	$CH_4(g)$	+ $2\ O_2(g)$	→ $CO_2(g)$	+ $2\ H_2O(g)$	
$\Delta H_f°$	-75 kJ/mol	0	-393.5	-242	
$\Delta G_f°$	-51 kJ/mol	0	-394	-229	Data from Appendix 4
S°	186 $J\ K^{-1}\ mol^{-1}$	205	214	189	

$\Delta H° = \Sigma n_p \Delta H°_{f,\ products} - \Sigma n_r \Delta H°_{f,\ reactants}$; $\Delta S° = \Sigma n_p S°_{products} - \Sigma n_r S°_{reactants}$

$\Delta H° = 2\text{ mol}(-242\text{ kJ/mol}) + 1\text{ mol}(-393.5\text{ kJ/mol}) - [1\text{ mol}(-75\text{ kJ/mol})] = -803\text{ kJ}$

$\Delta S° = 2\text{ mol}(189\text{ J K}^{-1}\text{ mol}^{-1}) + 1\text{ mol}(214\text{ J K}^{-1}\text{ mol}^{-1})$

$- [1\text{ mol}(186\text{ J K}^{-1}\text{ mol}^{-1}) + 2\text{ mol}(205\text{ J K}^{-1}\text{ mol}^{-1})] = -4\text{ J/K}$

There are two ways to get $\Delta G°$. We can use $\Delta G° = \Delta H° - T\Delta S°$ (be careful of units):

$\Delta G° = \Delta H° - T\Delta S° = -803 \times 10^3\text{ J} - (298\text{ K})(-4\text{ J/K}) = -8.018 \times 10^5\text{ J} = -802\text{ kJ}$

or we can use ΔG_f° values where $\Delta G° = \Sigma n_p \Delta G^\circ_{f,\text{ products}} - \Sigma n_r \Delta G^\circ_{f,\text{ reactants}}$:

$\Delta G° = 2\text{ mol}(-229\text{ kJ/mol}) + 1\text{ mol}(-394\text{ kJ/mol}) - [1\text{ mol}(-51\text{ kJ/mol})]$

$\Delta G° = -801\text{ kJ}$ (Answers are the same within round off error.)

b. $6\ CO_2(g) + 6\ H_2O(l) \rightarrow C_6H_{12}O_6(s) + 6\ O_2(g)$

	$CO_2(g)$	$H_2O(l)$	$C_6H_{12}O_6(s)$	$O_2(g)$
ΔH_f°	-393.5 kJ/mol	-286	-1275	0
S°	214 J K^{-1} mol^{-1}	70.	212	205

$\Delta H° = -1275 - [6(-286) + 6(-393.5)] = 2802\text{ kJ}$

$\Delta S° = 6(205) + 212 - [6(214) + 6(70.)] = -262\text{ J/K}$

$\Delta G° = 2802\text{ kJ} - (298\text{ K})(-0.262\text{ kJ/K}) = 2880.\text{ kJ}$

c. $P_4O_{10}(s) + 6\ H_2O(l) \rightarrow 4\ H_3PO_4(s)$

	$P_4O_{10}(s)$	$H_2O(l)$	$H_3PO_4(s)$
ΔH_f° (kJ/mol)	-2984	-286	-1279
S° (J K^{-1} mol^{-1})	229	70.	110.

$\Delta H° = 4\text{ mol}(-1279\text{ kJ/mol}) - [1\text{ mol}(-2984\text{ kJ/mol}) + 6\text{ mol}(-286\text{ kJ/mol})] = -416\text{ kJ}$

$\Delta S° = 4(110.) - [229 + 6(70.)] = -209\text{ J/K}$

$\Delta G° = \Delta H° - T\Delta S° = -416\text{ kJ} - (298\text{ K})(-0.209\text{ kJ/K}) = -354\text{ kJ}$

d. $HCl(g) + NH_3(g) \rightarrow NH_4Cl(s)$

	$HCl(g)$	$NH_3(g)$	$NH_4Cl(s)$
ΔH_f° (kJ/mol)	-92	-46	-314
S° ($J\ K^{-1}\ mol^{-1}$)	187	193	96

$\Delta H^\circ = -314 - [-92 - 46] = -176$ kJ; $\Delta S^\circ = 96 - [187 + 193] = -284$ J/K

$\Delta G^\circ = \Delta H^\circ - T\Delta S^\circ = -176 \text{ kJ} - (298 \text{ K})(-0.284 \text{ kJ/K}) = -91 \text{ kJ}$

51. $\Delta G^\circ = -58.03 \text{ kJ} - (298 \text{ K})(-0.1766 \text{ kJ/K}) = -5.40 \text{ kJ}$

$$\Delta G^\circ = 0 = \Delta H^\circ - T\Delta S^\circ,\ T = \frac{\Delta H^\circ}{\Delta S^\circ} = \frac{-58.03 \text{ kJ}}{-0.1766 \text{ kJ/K}} = 328.6 \text{ K}$$

ΔG° is negative below 328.6 K where the favorable ΔH° term dominates.

53. a. $\Delta G^\circ = 2(-270. \text{ kJ}) - 2(-502 \text{ kJ}) = 464 \text{ kJ}$

b. **Since ΔG° is positive, then this reaction is not spontaneous at standard conditions at 298 K.**

c. $\Delta G^\circ = \Delta H^\circ - T\Delta S^\circ,\ \Delta H^\circ = \Delta G^\circ + T\Delta S^\circ = 464 \text{ kJ} + 298 \text{ K}(0.179 \text{ kJ/K}) = 517 \text{ kJ}$

We need to solve for the temperature when ΔG° = 0:

$$\Delta G^\circ = 0 = \Delta H^\circ - T\Delta S^\circ,\ T = \frac{\Delta H^\circ}{\Delta S^\circ} = \frac{517 \text{ kJ}}{0.179 \text{ kJ/K}} = 2890 \text{ K}$$

This reaction will be spontaneous at standard conditions (ΔG° < 0) when T > 2890 K. At these temperatures the favorable entropy term will dominate.

55. $CH_4(g) + CO_2(g) \rightarrow CH_3CO_2H(l)$

$\Delta H^\circ = -484 - [-75 + (-393.5)] = -16$ kJ; $\Delta S^\circ = 160. - [186 + 214] = -240.$ J/K

$\Delta G^\circ = \Delta H^\circ - T\Delta S^\circ = -16 \text{ kJ} - (298 \text{ K})(-0.240 \text{ kJ/K}) = 56 \text{ kJ}$

At standard concentrations where ΔG = ΔG°, this reaction is spontaneous only at temperatures below T = ΔH°/ΔS° = 67 K (where the favorable ΔH° term will dominate, giving a negative ΔG° value). This is not practical. Substances will be in condensed phases and rates will be very slow at this extremely low temperature.

$CH_3OH(g) + CO(g) \rightarrow CH_3CO_2H(l)$

$\Delta H^\circ = -484 - [-110.5 + (-201)] = -173$ kJ; $\Delta S^\circ = 160. - [198 + 240.] = -278$ J/K

$\Delta G° = -173 \text{ kJ} - (298 \text{ K})(-0.278 \text{ kJ/K}) = -90. \text{ kJ}$

This reaction also has a favorable enthalpy and an unfavorable entropy term. This reaction is spontaneous at temperatures below $T = \Delta H°/\Delta S° = 622 \text{ K}$ (assuming standard concentrations). The reaction of CH_3OH and CO will be preferred at standard conditions. It is spontaneous at high enough temperatures that the rates of reaction should be reasonable.

57. Enthalpy is not favorable, so ΔS must provide the driving force for the change. Thus, ΔS is positive. There is an increase in disorder, so the original enzyme has the more ordered structure.

59. Since there are more product gas molecules than reactant gas molecules ($\Delta n > 0$), then ΔS will be positive. From the signs of ΔH and ΔS, this reaction is spontaneous at all temperatures. It will cost money to heat the reaction mixture. Since there is no thermodynamic reason to do this, then the purpose of the elevated temperature must be to increase the rate of the reaction, i.e., kinetic reasons.

Free Energy: Pressure Dependence and Equilibrium

61. $\Delta G = \Delta G° + RT \ln Q = \Delta G° + RT \ln \dfrac{P_{N_2O_4}}{P^2_{NO_2}}$

$\Delta G° = 1 \text{ mol}(98 \text{ kJ/mol}) - 2 \text{ mol}(52 \text{ kJ/mol}) = -6 \text{ kJ}$

a. These are standard conditions, so $\Delta G = \Delta G°$ since $Q = 1$ and $\ln Q = 0$. Since $\Delta G°$ is negative, then the forward reaction is spontaneous. The reaction shifts right to reach equilibrium.

b. $\Delta G = -6 \times 10^3 \text{ J} + 8.3145 \text{ J K}^{-1} \text{ mol}^{-1} (298 \text{ K}) \ln \dfrac{0.50}{(0.21)^2}$

$\Delta G = -6 \times 10^3 \text{ J} + 6.0 \times 10^3 \text{ J} = 0$

Since $\Delta G = 0$, this reaction is at equilibrium (no shift).

c. $\Delta G = -6 \times 10^3 \text{ J} + 8.3145 \text{ J K}^{-1} \text{ mol}^{-1} (298 \text{ K}) \ln \dfrac{1.6}{(0.29)^2}$

$\Delta G = -6 \times 10^3 \text{ J} + 7.3 \times 10^3 \text{ J} = 1.3 \times 10^3 \text{ J} = 1 \times 10^3 \text{ J}$

Since ΔG is positive, the reverse reaction is spontaneous, so the reaction shifts to the left to reach equilibrium.

63. $\Delta H° = 2\ \Delta H°_{f,NH_3} = 2(-46) = -92 \text{ kJ};\ \Delta G° = 2\ \Delta G°_{f,NH_3} = 2(-17) = -34 \text{ kJ}$

$\Delta S° = 2(193 \text{ J/K}) - [192 \text{ J/K} + 3(131 \text{ J/K})] = -199 \text{ J/K};\ \Delta G° = -RT \ln K$

$$K = \exp\frac{-\Delta G°}{RT} = \exp\left(\frac{-(-34{,}000 \text{ J})}{(8.3145 \text{ J K}^{-1} \text{ mol}^{-1})(298 \text{ K})}\right) = e^{13.72} = 9.1 \times 10^5$$

Note: When determining exponents, we will round off after the calculation is complete. This helps eliminate excessive round off error.

a. $\Delta G = \Delta G° + RT \ln \dfrac{P^2_{NH_3}}{P_{N_2} \times P^3_{H_2}} = -34\text{ kJ} + \dfrac{(8.3145\text{ J K}^{-1}\text{ mol}^{-1})\,(298\text{ K})}{1000\text{ J/kJ}} \ln \dfrac{(50.)^2}{(200.)\,(200.)^3}$

$\Delta G = -34\text{ kJ} - 33\text{ kJ} = -67\text{ kJ}$

b. $\Delta G = -34\text{ kJ} + \dfrac{(8.3145\text{ J K}^{-1}\text{ mol}^{-1})\,(298\text{ K})}{1000\text{ J/kJ}} \ln \dfrac{(200.)^2}{(200.)\,(600.)^3}$

$\Delta G = -34\text{ kJ} - 34.4\text{ kJ} = -68\text{ kJ}$

c. Assume $\Delta H°$ and $\Delta S°$ are temperature independent.

$\Delta G°_{100} = \Delta H° - T\Delta S°,\ \ \Delta G°_{100} = -92\text{ kJ} - (100.\text{ K})(-0.199\text{ kJ/K}) = -72\text{ kJ}$

$\Delta G_{100} = \Delta G°_{100} + RT \ln Q = -72\text{ kJ} + \dfrac{(8.3145\text{ J K}^{-1}\text{ mol}^{-1})\,(100.\text{ K})}{1000\text{ J/kJ}} \ln \dfrac{(10.)^2}{(50.)\,(200.)^3}$

$\Delta G_{100} = -72\text{ kJ} - 13\text{ kJ} = -85\text{ kJ}$

d. $\Delta G°_{700} = -92\text{ kJ} - (700.\text{ K})(-0.199\text{ kJ/K}) = 47\text{ kJ}$

$\Delta G_{700} = 47\text{ kJ} + \dfrac{(8.3145\text{ J K}^{-1}\text{ mol}^{-1})\,(700.\text{ K})}{1000\text{ J/kJ}} \ln \dfrac{(10.)^2}{(50.)\,(200.)^3} = 47\text{ kJ} - 88\text{ kJ} = -41\text{ kJ}$

65. a.

	$\Delta H_f°$ (kJ/mol)	$S°$ ($\text{J K}^{-1}\text{ mol}^{-1}$)
$NH_3(g)$	-46	193
$O_2(g)$	0	205
$NO(g)$	90.	211
$H_2O(g)$	-242	189
$NO_2(g)$	34	240.
$HNO_3(l)$	-174	156
$H_2O(l)$	-286	70.

$4\,NH_3(g) + 5\,O_2(g) \rightarrow 4\,NO(g) + 6\,H_2O(g)$

$\Delta H° = 6(-242) + 4(90.) - [4(-46)] = -908\text{ kJ}$

$\Delta S° = 4(211) + 6(189) - [4(193) + 5(205)] = 181\text{ J/K}$

$\Delta G° = -908\text{ kJ} - 298\text{ K}\,(0.181\text{ kJ/K}) = -962\text{ kJ}$

$\Delta G° = -RT \ln K,\ \ \ln K = \dfrac{-\Delta G°}{RT} = \left(\dfrac{-(-962 \times 10^3\text{ J})}{8.3145\text{ J K}^{-1}\text{ mol}^{-1} \times 298\text{ K}}\right) = 388$

$\ln K = 2.303 \log K,\ \ \log K = 168,\ \ K = 10^{168}$ (an extremely large value)

$2\,NO(g) + O_2(g) \rightarrow 2\,NO_2(g)$

$\Delta H° = 2(34) - [2(90.)] = -112$ kJ; $\Delta S° = 2(240.) - [2(211) + (205)] = -147$ J/K

$\Delta G° = -112$ kJ $- (298$ K$)(-0.147$ kJ/K$) = -68$ kJ

$$K = \exp\frac{-\Delta G°}{RT} = \exp\left(\frac{-(-68{,}000\text{ J})}{8.3145\text{ J K}^{-1}\text{ mol}^{-1}\text{ (298 K)}}\right) = e^{27.44} = 8.3 \times 10^{11}$$

Note: When determining exponents, we will round off after the calculation is complete.

$3\,NO_2(g) + H_2O(l) \rightarrow 2\,HNO_3(l) + NO(g)$

$\Delta H° = 2(-174) + (90.) - [3(34) + (-286)] = -74$ kJ

$\Delta S° = 2(156) + (211) - [3(240.) + (70.)] = -267$ J/K

$\Delta G° = -74$ kJ $- (298$ K$)(-0.267$ kJ/K$) = 6$ kJ

$$K = \exp\frac{-\Delta G°}{RT} = \exp\left(\frac{-6000\text{ J}}{8.3145\text{ J K}^{-1}\text{ mol}^{-1}\text{ (298 K)}}\right) = e^{-2.4} = 9 \times 10^{-2}$$

b. $\Delta G° = -RT \ln K$; $T = 825 + 273 = 1098$ K; We must determine $\Delta G°$ at 1098 K.

$$\Delta G°_{1098} = \Delta H° - T\Delta S° = -908\text{ kJ} - (1098\text{ K})(0.181\text{ kJ/K}) = -1107\text{ kJ}$$

$$K = \exp\frac{-\Delta G°_{1098}}{RT} = \exp\left(\frac{-(-1.107 \times 10^{6}\text{ J})}{8.3145\text{ J K}^{-1}\text{ mol}^{-1}\text{ (1098 K)}}\right) = e^{121.258} = 4.589 \times 10^{52}$$

c. There is no thermodynamic reason for the elevated temperature since $\Delta H°$ is negative and $\Delta S°$ is positive. Thus, the purpose for the high temperature must be to increase the rate of the reaction.

67. $2\,SO_2(g) + O_2(g) \rightarrow 2\,SO_3(g)$; $\Delta G° = 2(-371\text{ kJ}) - [2(-300.\text{ kJ})] = -142$ kJ

$$\Delta G° = -RT \ln K,\ \ln K = \frac{-\Delta G°}{RT} = \frac{-(-142{,}000\text{ J})}{8.3145\text{ J K}^{-1}\text{mol}^{-1}\text{ (298 K)}} = 57.311,\ K = e^{57.311} = 7.76 \times 10^{24}$$

$$K = 7.76 \times 10^{24} = \frac{P^2_{SO_3}}{P^2_{SO_2} \times P_{O_2}} = \frac{(2.0)^2}{P^2_{SO_2} \times (0.50)},\ P_{SO_2} = 1.0 \times 10^{-12}\text{ atm}$$

From the negative value of $\Delta G°$, this reaction is spontaneous at standard conditions. Since there are more molecules of reactant gases than product gases, then $\Delta S°$ will be negative (unfavorable). Therefore, this reaction must be exothermic ($\Delta H° < 0$). When $\Delta H°$ and $\Delta S°$ are both negative, the reaction will be spontaneous at relatively low temperatures where the favorable $\Delta H°$ term dominates.

69. $HgbO_2 \rightarrow Hgb + O_2 \quad \Delta G° = -(-70\text{ kJ})$

$Hgb + CO \rightarrow HgbCO \quad \Delta G° = -80\text{ kJ}$

$HgbO_2 + CO \rightarrow HgbCO + O_2 \quad \Delta G° = -10\text{ kJ}$

$$\Delta G° = -RT\ln K,\quad K = \exp\left(\frac{-\Delta G°}{RT}\right) = \exp\left(\frac{-(-10\times 10^3\text{ J})}{(8.3145\text{ J K}^{-1}\text{mol}^{-1})(298\text{ K})}\right) = 60$$

71. At 25.0°C: $\Delta G° = \Delta H° - T\Delta S° = -58.03\times 10^3\text{ J/mol} - (298.2\text{ K})(-176.6\text{ J K}^{-1}\text{ mol}^{-1})$
$= -5.37\times 10^3\text{ J/mol}$

$$\Delta G° = -RT\ln K,\ \ln K = \frac{-\Delta G°}{RT} = \frac{-(-5.37\times 10^3\text{ J/mol})}{(8.3145\text{ J K}^{-1}\text{mol}^{-1})(298.2\text{ K})} = 2.166;\quad K = e^{2.166} = 8.72$$

At 100.0°C: $\Delta G° = -58.03\times 10^3\text{ J/mol} - (373.2\text{ K})(-176.6\text{ J K}^{-1}\text{ mol}^{-1}) = 7.88\times 10^3\text{ J/mol}$

$$\ln K = \frac{-(7.88\times 10^3\text{ J/mol})}{(8.3145\text{ J K}^{-1}\text{mol}^{-1})(373.2\text{ K})} = -2.540,\ K = e^{-2.540} = 0.0789$$

73. A graph of ln K vs. 1/T will yield a straight line with slope equal to $-\Delta H°/R$ and y-intercept equal to $\Delta S°/R$ (see Exercise 10.72).

a.

Temp (°C)	T(K)	1000/T (K^{-1})	K_w	ln K_w
0	273	3.66	1.14×10^{-15}	-34.408
25	298	3.36	1.00×10^{-14}	-32.236
35	308	3.25	2.09×10^{-14}	-31.499
40.	313	3.19	2.92×10^{-14}	-31.165
50.	323	3.10	5.47×10^{-14}	-30.537

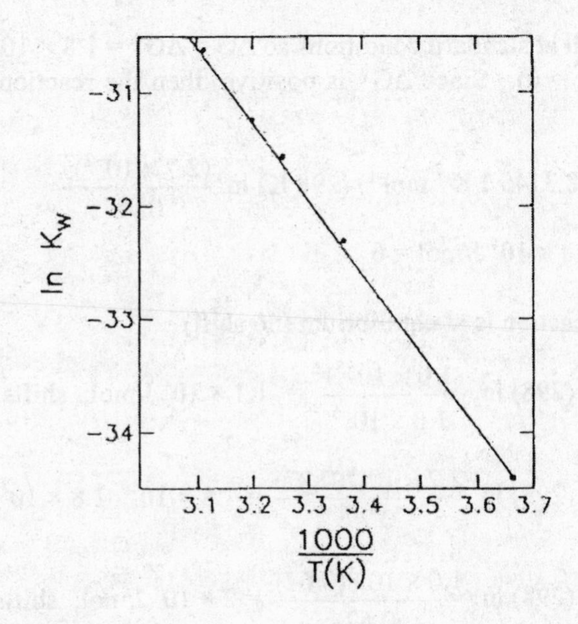

The straight line equation (from a calculator) is: $\ln K = -6.91 \times 10^3 \left(\frac{1}{T}\right) - 9.09$

$\text{Slope} = -6.91 \times 10^3\ K = \frac{-\Delta H°}{R}$

$\Delta H° = -(-6.91 \times 10^3\ K \times 8.3145\ J\ K^{-1}\ mol^{-1}) = 5.75 \times 10^4\ J/mol = 57.5\ kJ/mol$

$\text{y-intercept} = -9.09 = \frac{\Delta S°}{R}$, $\Delta S° = -9.09 \times 8.3145\ J\ K^{-1}\ mol^{-1} = -75.6\ J\ K^{-1}\ mol^{-1}$

b. From part a, $\Delta H° = 57.5$ kJ/mol and $\Delta S° = -75.6\ J\ K^{-1}\ mol^{-1}$. Assuming $\Delta H°$ and $\Delta S°$ are temperature independent:

$$\Delta G° = 57{,}500\ J/mol - 647\ K\ (-75.6\ J\ K^{-1}\ mol^{-1}) = 106{,}400\ J/mol = 106.4\ kJ/mol$$

Additional Exercises

75. No; When using $\Delta G_f°$ values in Appendix 4, we have generally specified a temperature of 25°C. Further, if gases or solutions are involved, we have specified partial pressures of 1 atm and solute concentrations of 1 molar. At other temperatures and compositions, the reaction may not be spontaneous. A negative $\Delta G°$ value means the reaction is spontaneous under standard conditions.

77. As any process occurs, ΔS_{univ} will increase; ΔS_{univ} cannot decrease. Time also goes in one direction, just as ΔS_{univ} goes in one direction.

79. The introduction of mistakes is an effect of entropy. The purpose of redundant information is to provide a control to check the "correctness" of the transmitted information.

81. $HF(aq) \rightleftharpoons H^+(aq) + F^-(aq)$; $\Delta G = \Delta G° + RT \ln \frac{[H^+][F^-]}{[HF]}$

$\Delta G° = -RT \ln K = -(8.3145\ J\ K^{-1}\ mol^{-1})\ (298\ K) \ln (7.2 \times 10^{-4}) = 1.8 \times 10^4\ J/mol$

a. The concentrations are all at standard conditions so $\Delta G = \Delta G° = 1.8 \times 10^4$ J/mol (since Q = 1.0 and ln Q = 0). Since $\Delta G°$ is positive, then the reaction shifts left to reach equilibrium.

b. $\Delta G = 1.8 \times 10^4\ J/mol + (8.3145\ J\ K^{-1}\ mol^{-1})\ (298\ K) \ln \frac{(2.7 \times 10^{-2})^2}{0.98}$

$\Delta G = 1.8 \times 10^4\ J/mol - 1.8 \times 10^4\ J/mol = 0$

Since $\Delta G = 0$, then the reaction is at equilibrium (no shift).

c. $\Delta G = 1.8 \times 10^4 + 8.3145\ (298) \ln \frac{(1.0 \times 10^{-5})^2}{1.0 \times 10^{-5}} = -1.1 \times 10^4$ J/mol; shifts right

d. $\Delta G = 1.8 \times 10^4 + 8.3145\ (298) \ln \frac{7.2 \times 10^{-4}(0.27)}{0.27} = 1.8 \times 10^4 - 1.8 \times 10^4 = 0$; at equilibrium

e. $\Delta G = 1.8 \times 10^4 + 8.3145\ (298) \ln \frac{1.0 \times 10^{-3}(0.67)}{0.52} = 2 \times 10^3$ J/mol; shifts left

83. S (monoclinic) → S (rhombic); $\Delta H° = 0 - 0.30 = -0.30$ kJ; $\Delta S° = 31.73 - 32.55 = -0.82$ J/K

At the conversion temperature: $\Delta G° = 0$ so $\Delta H° = T\Delta S°$; $T = \dfrac{\Delta H°}{\Delta S°} = \dfrac{-3.0 \times 10^2 \text{ J}}{-0.82 \text{ J/K}} = 370$ K

85. 1.00 *M* HCl → 0.100 *M* HCl $\quad \Delta G° = 0$; $\Delta G = \Delta G° + RT \ln Q = RT \ln \dfrac{[H^+][Cl^-]}{[H^+][Cl^-]}$

$$\Delta G = (8.3145 \text{ J mol}^{-1} \text{ K}^{-1})(298 \text{ K}) \ln\left(\frac{(0.100)^2}{(1.00)^2}\right) = -11{,}400 \text{ J} = -11.4 \text{ kJ}$$

The 0.100 *M* HCl is lower in free energy by 11.4 kJ.

87. $Ba(NO_3)_2(s) \rightleftharpoons Ba^{2+}(aq) + 2\ NO_3^-(aq)$ $\quad K = K_{sp}$; $\Delta G° = -561 + 2(-109) - (-797) = 18$ kJ

$$\Delta G° = -RT \ln K_{sp},\ \ln K_{sp} = \frac{-\Delta G°}{RT} = \frac{-18{,}000 \text{ J}}{8.3145 \text{ J K}^{-1} \text{ mol}^{-1} (298 \text{ K})} = -7.26,\ K_{sp} = e^{-7.26} = 7.0 \times 10^{-4}$$

89. ΔS is more favorable for reaction two than for reaction one, resulting in $K_2 > K_1$. In reaction one, seven particles in solution are forming one particle in solution. In reaction two, four particles are forming one which results in a smaller decrease in disorder than for reaction one.

91. Because of hydrogen bonding interactions, there is greater "structure" (or order) in liquid water than in most other liquids. Thus, there is a greater increase in disorder, or a greater increase in entropy when water evaporates.

93. $S = k \ln \Omega$; S has units of J K^{-1} mol^{-1} and k has units of J/K ($k = 1.38 \times 10^{-23}$ J/K)

To make units match: $S\ (\text{J K}^{-1} \text{ mol}^{-1}) = N_A k \ln \Omega$ when N_A = Avogadro's number

$$189 \text{ J K}^{-1} \text{ mol}^{-1} = 8.31 \text{ J K}^{-1} \text{ mol}^{-1} \ln \Omega_g$$
$$70.\ \text{J K}^{-1} \text{ mol}^{-1} = 8.31 \text{ J K}^{-1} \text{ mol}^{-1} \ln \Omega_l$$

Subtracting: $119 \text{ J K}^{-1} \text{ mol}^{-1} = 8.31 \text{ J K}^{-1} \text{ mol}^{-1} (\ln \Omega_g - \ln \Omega_l)$

$$14.3 = \ln(\Omega_g/\Omega_l),\quad \frac{\Omega_g}{\Omega_l} = e^{14.3} = 1.6 \times 10^6$$

95. H_2O (l, 298 K) → H_2O (g, V = 1000. L/mol); Break process into 2 steps:

Step 1: H_2O (l, 298 K) → H_2O (g, 298 K, $V = \dfrac{nR(298)}{1.00 \text{ atm}} = 24.5$ L)

$$\Delta S = S°_{H_2O(g)} - S°_{H_2O(l)} = 189 \text{ J/K} - 70.\ \text{J/K} = 119 \text{ J/K}$$

Step 2: H_2O (g, 298 K, 24.5 L) → H_2O (g, 298 K, 1000. L)

$$\Delta S = nR \ln \frac{V_2}{V_1} = (1.00 \text{ mol})(8.3145 \text{ J mol}^{-1} \text{ K}^{-1}) \ln \left(\frac{1000. \text{ L}}{24.5 \text{ L}}\right) = 30.8 \text{ J/K}$$

$$\Delta S_{tot} = 119 + 30.8 = 150. \text{ J/K}$$

$$\Delta G = 44.02 \times 10^3 \text{ J} - 298 \text{ K } (150. \text{ J/K}) = -700 \text{ J}; \text{ Spontaneous}$$

For H_2O (l, 298 K) → H_2O (g, 298 K, V = 100. L/mol):

$$\Delta S = 119 \text{ J/K} + (8.3145 \text{ J/K}) \ln \left(\frac{100. \text{ L}}{24.5 \text{ L}}\right) = 131 \text{ J/K}$$

$$\Delta G = 44.02 \times 10^3 \text{ J} - 298 \text{ K}(131 \text{ J/K}) = 5.0 \times 10^3 \text{ J}; \text{ Not spontaneous}$$

97. a. free expansion

$w = 0$ since $P_{ext} = 0$; $\Delta E = nC_v\Delta T$, since $\Delta T = 0$, $\Delta E = 0$

$\Delta E = q + w$, $q = 0$; $\Delta H = nC_p\Delta T$, since $\Delta T = 0$, $\Delta H = 0$

$$\Delta S = nR \ln \left(\frac{V_2}{V_1}\right) = (1.00 \text{ mol})\ (8.3145 \text{ J mol}^{-1} \text{ K}^{-1}) \ln \left(\frac{40.0 \text{ L}}{30.0 \text{ L}}\right) = 2.39 \text{ J/K}$$

$$\Delta G = \Delta H - T\Delta S = 0 - 300. \text{ K } (2.39 \text{ J/K}) = -717 \text{ J}$$

b. reversible expansion

$\Delta E = 0$; $\Delta H = 0$; $\Delta S = 2.39$ J/K; $\Delta G = -717$ J; These are state functions.

$$w_{rev} = -nRT \ln \left(\frac{V_2}{V_1}\right) = -(1.00 \text{ mol})(8.3145 \text{ J mol}^{-1} \text{ K}^{-1})\ (300. \text{ K}) \ln \left(\frac{40.0 \text{ L}}{30.0 \text{ L}}\right) = -718 \text{ J}$$

$\Delta E = 0 = q + w$, $q_{rev} = -w_{rev} = 718$ J

Summary:

	a) free expansion	b) reversible
q	0	718 J
w	0	-718 J
ΔE	0	0
ΔH	0	0
ΔS	2.39 J/K	2.39 J/K
ΔG	-717 J	-717 J

99. a. Isothermal: $\Delta E = 0$ and $\Delta H = 0$ if gas is ideal.

$$\Delta S = nR \ln (P_1/P_2) = (1.00 \text{ mol})(8.3145 \text{ J mol}^{-1} \text{ K}^{-1}) \ln (5.00 \text{ atm}/2.00 \text{ atm}) = 7.62 \text{ J/K}$$

$$T = \frac{PV}{nR} = \frac{5.00 \text{ atm} \times 5.00 \text{ L}}{1.00 \text{ mol} \times 0.08206 \text{ L atm K}^{-1} \text{ mol}^{-1}} = 305 \text{ K}$$

$\Delta G = \Delta H - T\Delta S = 0 - (305\text{ K})(7.62\text{ J/K}) = -2320\text{ J}$

$w = -P\Delta V = -(2.00\text{ atm})\Delta V$ where $V_f = \dfrac{nRT}{2.00\text{ atm}} = 12.5\text{ L}$

$w = -2.00\text{ atm }(12.5 - 5.00\text{ L}) \times (101.3\text{ J L}^{-1}\text{ atm}^{-1}) = -1500\text{ J}$

$\Delta E = 0 = q + w,\ q = 1500\text{ J}$

b. Second law, $\Delta S_{univ} > 0$ for spontaneous processes; $\Delta S_{univ} = \Delta S_{sys} + \Delta S_{surr} = \Delta S_{sys} - \dfrac{q_{actual}}{T}$

$\Delta S_{univ} = 7.62\text{ J/K} - \dfrac{1500\text{ J}}{305\text{ K}} = 7.62 - 4.9 = 2.7\text{ J/K}$; Thus, the process is spontaneous.

101. To calculate ΔE and ΔH, we need to determine the molar heat capacity of the ideal gas. As derived in section 10.14 of the text, for a reversible, adiabatic change ($q = 0$): $\left(\dfrac{T_2}{T_1}\right)^{C_v} = \left(\dfrac{V_1}{V_2}\right)^{R}$

$$C_v \ln\left(\frac{T_2}{T_1}\right) = R\ln\left(\frac{V_1}{V_2}\right),\quad \frac{C_v}{R} = \frac{\ln\left(\frac{V_1}{V_2}\right)}{\ln\left(\frac{T_2}{T_1}\right)}$$

Since $V_2 = 2V_1$: $\dfrac{C_v}{R} = \dfrac{\ln(1/2)}{\ln(239\text{ K}/296\text{ K})} = 3.24$

$C_v = 3.24\,(8.3145\text{ J K}^{-1}\text{ mol}^{-1}) = 26.9\text{ J K}^{-1}\text{ mol}^{-1}$

$\Delta E = nC_v\Delta T = 1.50\text{ mol }(26.9\text{ J K}^{-1}\text{ mol}^{-1})(239\text{ K} - 296\text{ K}) = -2{,}300\text{ J} = -2.3\text{ kJ}$

$\Delta H = \Delta E + nR\Delta T = -2300\text{ J} + 1.50\text{ mol }(8.3145\text{ J K}^{-1}\text{ mol}^{-1})(239\text{ K} - 296\text{ K}) = -3.0 \times 10^3\text{ J}$

$= -3.0\text{ kJ}$

Challenge Problems

103. a. $V_1 = \dfrac{nRT_1}{P_1} = \dfrac{2.00\text{ mol} \times 0.08206\text{ L atm K}^{-1}\text{mol}^{-1} \times 298\text{ K}}{2.00\text{ atm}} = 24.5\text{ L}$

For an adiabatic, reversible process, $P_1V_1^{\gamma} = P_2V_2^{\gamma}$ and $T_1V_1^{\gamma-1} = T_2V_2^{\gamma-1}$ where $\gamma = C_p/C_v$. Since argon is a monoatomic gas, then $C_p = (5/2)R$ and $C_v = (3/2)R$ so $\gamma = 5/3$.

$$V_2^{\gamma} = \frac{P_1V_1^{\gamma}}{P_2} = \frac{2.00\text{ atm}(24.5\text{ L})^{5/3}}{1.00\text{ atm}} = 413,\quad V_2 = (413)^{3/5} = 37.1\text{ L}$$

We can either use the ideal gas law or the $T_1V^{\gamma-1} = T_2V_2^{\gamma-1}$ equation to calculate the final temperature. Using the ideal gas law:

$$T_2 = \frac{P_2V_2}{nR} = \frac{1.00 \text{ atm} \times 37.1 \text{ L}}{2.00 \text{ mol} \times 0.08206 \text{ L atm K}^{-1}\text{mol}^{-1}} = 226 \text{ K}$$

b. For an adiabatic process (q = 0), $\Delta E = w = nC_v\Delta T$. For an expansion against a fixed external pressure, $w = -P\Delta V$. From the ideal gas equation (see part a), $V_1 = 24.5$ L.

$$w = -P\Delta V = nC_v\Delta T$$

$$-1.00 \text{ atm } (V_2 - 24.5 \text{ L})\left(\frac{101.3 \text{ J}}{\text{L atm}}\right) = 2.00 \text{ mol } (3/2)\left(\frac{8.3145 \text{ J}}{\text{mol K}}\right)(T_2 - 298 \text{ K})$$

We will ignore units from here. Note that both sides of the equation are in units of J.

$$-101\, V_2 + 2480 = 24.9\, T_2 - 7430$$

To solve for T_2, we need to find an expression for V_2. Using the ideal gas equation:

$$V_2 = \frac{nRT_2}{P_2} = \frac{2.00(0.08206)\, T_2}{1.00} = 0.164\, T_2; \text{ Substituting:}$$

$$-101(0.164\, T_2) + 2480 = 24.9\, T_2 - 7430, \; 41.5\, T_2 = 9910, \; T_2 = 239 \text{ K}$$

105. $K_p = P_{CO_2}$; To insure Ag_2CO_3 from decomposing, P_{CO_2} should be greater than K_p.

From Exercise 10.72, $\ln K = \frac{-\Delta H°}{RT} + \frac{\Delta S°}{R}$. For two conditions of K and T, the equation is:

$$\ln \frac{K_2}{K_1} = \frac{\Delta H°}{R}\left(\frac{1}{T_1} - \frac{1}{T_2}\right)$$

Let $T_1 = 25°C = 298$ K, $K_1 = 6.23 \times 10^{-3}$ torr; $T_2 = 110.°C = 383$ K, $K_2 = ?$

$$\ln \frac{K_2}{6.23 \times 10^{-3} \text{ torr}} = \frac{79.14 \times 10^3 \text{ J/mol}}{8.3145 \text{ J K}^{-1}\text{ mol}^{-1}}\left(\frac{1}{298 \text{ K}} - \frac{1}{383 \text{ K}}\right)$$

$$\ln \frac{K_2}{6.23 \times 10^{-3}} = 7.1, \; \frac{K_2}{6.23 \times 10^{-3}} = e^{7.1} = 1.2 \times 10^3, \; K_2 = 7.5 \text{ torr}$$

To prevent decomposition of Ag_2CO_3, the partial pressure of CO_2 should be greater than 7.5 torr.

107. $3\, O_2(g) \rightleftharpoons 2\, O_3(g)$; $\Delta H° = 2(143) = 286$ kJ; $\Delta G° = 2(163) = 326$ kJ

$$\ln K = \frac{-\Delta G°}{RT} = \frac{-326 \times 10^3 \text{ J}}{(8.3145 \text{ J K}^{-1}\text{ mol}^{-1})(298 \text{ K})} = -131.573, \; K = e^{-131.573} = 7.22 \times 10^{-58}$$

We need the value of K at 230. K. From Exercise 10.72: $\ln K = \frac{-\Delta H°}{RT} + \frac{\Delta S°}{R}$

For two sets of K and T:

$$\ln K_1 = \frac{-\Delta H°}{R}\left(\frac{1}{T_1}\right) + \frac{\Delta S°}{R};\ \ln K_2 = \frac{-\Delta H°}{R}\left(\frac{1}{T_2}\right) + \frac{\Delta S°}{R}$$

Subtracting the first expression from the second:

$$\ln K_2 - \ln K_1 = \frac{\Delta H°}{R}\left(\frac{1}{T_1} - \frac{1}{T_2}\right) \text{ or } \ln\frac{K_2}{K_1} = \frac{\Delta H°}{R}\left(\frac{1}{T_1} - \frac{1}{T_2}\right)$$

Let $K_2 = 7.22 \times 10^{-58}$, $T_2 = 298$; $K_1 = K_{230}$, $T_1 = 230.$ K; $\Delta H° = 286 \times 10^3$ J

$$\ln\frac{7.22 \times 10^{-58}}{K_{230}} = \frac{286 \times 10^3}{8.3145}\left(\frac{1}{230.} - \frac{1}{298}\right) = 34.13$$

$$\frac{7.22 \times 10^{-58}}{K_{230}} = e^{34.13} = 6.6 \times 10^{14},\ K_{230} = 1.1 \times 10^{-72}$$

$$K_{230} = 1.1 \times 10^{-72} = \frac{P_{O_3}^2}{P_{O_2}^3} = \frac{P_{O_3}^2}{(1.0 \times 10^{-3}\text{ atm})^3},\ P_{O_3} = 3.3 \times 10^{-41}\text{ atm}$$

The volume occupied by one molecule of ozone is:

$$V = \frac{nRT}{P} = \frac{(1/6.022 \times 10^{23}\text{ mol})(0.08206\text{ L atm mol}^{-1}\text{ K}^{-1})(230.\text{ K})}{(3.3 \times 10^{-41}\text{ atm})},\ V = 9.5 \times 10^{17}\text{ L}$$

Equilibrium is probably not maintained under these conditions. When only two ozone molecules are in a volume of 9.5×10^{17} L, the reaction is not at equilibrium. Under these conditions, Q > K and the reaction shifts left. But with only 2 ozone molecules in this huge volume, it is extremely unlikely that they will collide with each other. At these conditions, the concentration of ozone is not large enough to maintain equilibrium.

109. a. $\Delta G° = G_B° - G_A° = 11{,}718 - 8996 = 2722$ J

$$K = \exp\left(\frac{-\Delta G°}{RT}\right) = \exp\left(\frac{-2722\text{ J}}{(8.3145\text{ J K}^{-1}\text{ mol}^{-1})(298\text{ K})}\right) = 0.333$$

b. Since Q = 1.00 > K, reaction shifts left. Let x = atm of B(g) which reacts to reach equilibrium.

	A(g)	⇌	B(g)	$K = P_B/P_A$
Initial	1.00 atm		1.00 atm	
Equil.	1.00 + x		1.00 - x	

$$K = \frac{1.00 - x}{1.00 + x} = 0.333,\ \ 1.00 - x = 0.333 + 0.333\,x,\ \ x = 0.50 \text{ atm}$$

$P_B = 1.00 - 0.50 = 0.50$ atm; $P_A = 1.00 + 0.50 = 1.50$ atm

c. $\Delta G = \Delta G° + RT \ln Q = \Delta G° + RT \ln (P_B/P_A)$

$\Delta G = 2722 \text{ J} + (8.3145)(298) \ln (0.50/1.50) = 2722 \text{ J} - 2722 \text{ J} = 0$ (carrying extra sig. figs.)

111. Step 1: $\Delta E = 0$ and $\Delta H = 0$ since $\Delta T = 0$

$$w = -P\Delta V = -(9.87 \times 10^{-3} \text{ atm})\,\Delta V;\ \ V = \frac{nRT}{P},\ R = 0.08206 \text{ L atm mol}^{-1} \text{ K}^{-1}$$

$$\Delta V = V_f - V_i = nRT\left(\frac{1}{P_f} - \frac{1}{P_i}\right)$$

$$\Delta V = 1.00 \text{ mol } (0.08206)\ (298 \text{ K})\left(\frac{1}{9.87 \times 10^{-3} \text{ atm}} - \frac{1}{2.45 \times 10^{-2} \text{ atm}}\right)$$

$\Delta V = 1480$ L (we will carry all values to three sig. figs.)

$$w = -(9.87 \times 10^{-3} \text{ atm})(1480 \text{ L}) = -14.6 \text{ L atm } (101.3 \text{ J L}^{-1} \text{ atm}^{-1}) = -1480 \text{ J}$$

$$\Delta E = q + w = 0,\ \ q = -w = +1480 \text{ J};\ \ \Delta S = nR \ln\left(\frac{P_1}{P_2}\right)$$

$$\Delta S = 1.00 \text{ mol } (8.3145 \text{ J mol}^{-1} \text{ K}^{-1}) \ln\left(\frac{2.45 \times 10^{-2} \text{ atm}}{9.87 \times 10^{-3} \text{ atm}}\right),\ \ \Delta S = 7.56 \text{ J/K}$$

$$\Delta G = \Delta H - T\Delta S = 0 - 298 \text{ K}(7.56 \text{ J/K}) = -2250 \text{ J}$$

Step 2: $\Delta E = 0,\ \Delta H = 0$

$$w = -(4.93 \times 10^{-3} \text{ atm})\left(\frac{nRT}{4.93 \times 10^{-3}} - \frac{nRT}{9.87 \times 10^{-3}}\right) \text{L} \times \frac{101.3 \text{ J}}{\text{L atm}} = -1240 \text{ J}$$

$$q = -w = 1240 \text{ J};\ \ \Delta S = nR \ln\left(\frac{9.87 \times 10^{-3} \text{ atm}}{4.93 \times 10^{-3} \text{ atm}}\right) = 5.77 \text{ J/K}$$

$$\Delta G = 0 - 298 \text{ K}(5.77 \text{ J/K}) = -1720 \text{ J}$$

Step 3: $\Delta E = 0,\ \Delta H = 0$

$$w = -(2.45 \times 10^{-3} \text{ atm})\left(\frac{nRT}{2.45 \times 10^{-3}} - \frac{nRT}{4.93 \times 10^{-3}}\right) \text{L} \times \frac{101.3 \text{ J}}{\text{L atm}} = -1250 \text{ J}$$

$$q = -w = 1250\ \text{J};\ \ \Delta S = nR \ln\left(\frac{4.93 \times 10^{-3}\ \text{atm}}{2.45 \times 10^{-3}\ \text{atm}}\right) = 5.81\ \text{J/K}$$

$$\Delta G = 0 - 298\ \text{K}\ (5.81\ \text{J/K}) = -1730\ \text{J}$$

	q	w	ΔE	ΔS	ΔH	ΔG
Step 1	1480 J	-1480 J	0	7.56 J/K	0	-2250 J
Step 2	1240 J	-1240 J	0	5.77 J/K	0	-1720 J
Step 3	1250 J	-1250 J	0	5.81 J/K	0	-1730 J
Total	3970 J	-3970 J	0	19.14 J/K	0	-5.70×10^3 J

113. a. $\Delta G° = 2\ \text{mol}\ (-394\ \text{kJ/mol}) - 2\ \text{mol}\ (-137\ \text{kJ/mol}) = -514\ \text{kJ}$

$$K = \exp\left(\frac{-\Delta G°}{RT}\right) = \exp\left(\frac{-(-514{,}000\ \text{J})}{(8.3145\ \text{J mol}^{-1}\ \text{K}^{-1})\ (298\ \text{K})}\right) = 1.24 \times 10^{90}$$

b. $\Delta S° = 2(214\ \text{J/K}) - [2(198\ \text{J/K}) + 205\ \text{J/K}] = -173\ \text{J/K}$

$2\ CO\ (1.00\ \text{atm}) + O_2\ (1.00\ \text{atm}) \rightarrow 2\ CO_2\ (1.00\ \text{atm})$ $\quad \Delta S° = -173\ \text{J/K}$

$2\ CO_2\ (1.00\ \text{atm}) \rightarrow 2\ CO_2\ (10.0\ \text{atm})$ $\quad \Delta S = nR \ln (P_1/P_2) = -38.3\ \text{J/K}$

$2\ CO\ (10.0\ \text{atm}) \rightarrow 2\ CO\ (1.00\ \text{atm})$ $\quad \Delta S = nR \ln (P_1/P_2) = 38.3\ \text{J/K}$

$O_2\ (10.0\ \text{atm}) \rightarrow O_2\ (1.00\ \text{atm})$ $\quad \Delta S = nR \ln (P_1/P_2) = 19.1\ \text{J/K}$

$2\ CO\ (10.0\ \text{atm}) + O_2\ (10.0\ \text{atm}) \rightarrow CO_2\ (10.0\ \text{atm})$ $\quad \Delta S = -173 + 19.1 = -154\ \text{J/K}$

115.

T(°C)	T(K)	C_p(J K^{-1} mol^{-1})	C_p/T (J K^{-2} mol^{-1})
-200.	73	12	0.16
-180.	93	15	0.16
-160.	113	17	0.15
-140.	133	19	0.14
-100.	173	24	0.14
-60.	213	29	0.14
-30.	243	33	0.14
-10.	263	36	0.14
0	273	37	0.14

Total area of C_p/T vs T plot = ΔS = I + II + III (See following plot.)

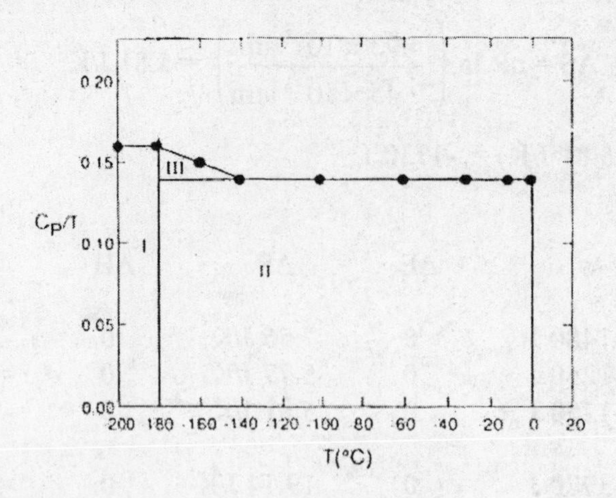

ΔS = (0.16 J K^{-2} mol^{-1})(20. K) + (0.14 J K^{-2} mol^{-1})(180. K) + 1/2(0.02 J K^{-2} mol^{-1})(40. K)

ΔS = 3.2 + 25 + 0.4 = 29 J K^{-1} mol^{-1}

117. To calculate ΔS_{sys} at 10.0°C, we need a place to start. From the data in the problem, we can calculate ΔS_{sys} at the melting point (5.5°C). For a phase change, $\Delta S_{sys} = q_{rev}/T = \Delta H/T$ where ΔH is determined at the melting point (5.5°C). Solving for ΔH at 5.5°C (using a thermochemical cycle):

C_6H_6 (l, 25.0°C) → C_6H_6 (s, 25.0°C)	ΔH = -10.04 kJ
C_6H_6 (l, 5.5°C) → C_6H_6 (l, 25.0°C)	$\Delta H = nC_p\Delta T/1000$ = 2.59 kJ
C_6H_6 (s, 25.0°C) → C_6H_6 (s, 5.5°C)	$\Delta H = nC_p\Delta T/1000$ = -1.96 kJ
C_6H_6 (l, 5.5°C) → C_6H_6 (s, 5.5°C)	ΔH = -9.41 kJ

At the melting point, $\Delta S_{sys} = \frac{\Delta H}{T} = \frac{-9.41 \times 10^3\ J}{278.7\ K}$ = -33.8 J/K

For the phase change at 10.0°C (283.2 K):

C_6H_6 (l, 278.7 K) → C_6H_6 (s, 278.7 K)	ΔS = -33.8 J/K
C_6H_6 (l, 283.2 K) → C_6H_6 (l, 278.7 K)	$\Delta S = nC_p \ln(T_2/T_1)$ = -2.130 J/K
C_6H_6 (s, 278.7 K) → C_6H_6 (s, 283.2 K)	$\Delta S = nC_p \ln(T_2/T_1)$ = 1.608 J/K
C_6H_6 (l, 283.2 K) → C_6H_6 (s, 283.2 K)	ΔS_{sys} = -34.3 J/K

To calculate ΔS_{surr}, we need ΔH at 10.0°C ($\Delta S_{surr} = \frac{-\Delta H}{T}$).

C_6H_6 (l, 25.0°C) → C_6H_6 (s, 25.0°C)	ΔH = -10.04 kJ
C_6H_6 (l, 10.0°C) → C_6H_6 (l, 25.0°C)	$\Delta H = nC_p\Delta T/1000$ = 2.00 kJ
C_6H_6 (s, 25.0°C) → C_6H_6 (s, 10.0°C)	$\Delta H = nC_p\Delta T/1000$ = -1.51 kJ
C_6H_6 (l, 10.0°C) → C_6H_6 (s, 10.0°C)	ΔH = -9.55 kJ

$$\Delta S_{surr} = \frac{-\Delta H}{T} = \frac{-(-9.55 \times 10^3\ J)}{283.2\ K} = 33.7\ J/K$$

Marathon Problems

119. a. $w = 0, q = 0, \Delta E = 0, \Delta H = 0$

b. $w = -(1.00 \text{ atm})(2.00 \text{ L}) = (101.3 \text{ J L}^{-1} \text{ atm}^{-1}) = -203 \text{ J}; \quad \Delta E = \Delta H = 0, q = 203 \text{ J}$

c. $w = -[(1.33 \text{ atm})(1.00 \text{ L}) + (1.00 \text{ atm})(1.00 \text{ L})] \times \frac{101.3 \text{ J}}{\text{L atm}} = -236 \text{ J}$

$q = 236 \text{ J}; \quad \Delta E = \Delta H = 0$

d. $w = -nRT \ln\frac{V_2}{V_1} = \left[PV \ln\left(\frac{V_2}{V_1}\right)\right] \frac{101.3 \text{ J}}{\text{L atm}} = -281 \text{ J}$

$q = 281 \text{ J}; \quad \Delta E = \Delta H = 0$

e. $w = -(2.00 \text{ atm})(-2.00 \text{ L}) \times \left(\frac{101.3 \text{ J}}{\text{L atm}}\right) = 405 \text{ J}$

$q = -405 \text{ J}; \quad \Delta E = \Delta H = 0$

f. $w = -[(1.33 \text{ atm})(-1.00 \text{ L}) + 2.00 \text{ atm } (-1.00 \text{ L})] \times \left(\frac{101.3 \text{ J}}{\text{L atm}}\right) = 337 \text{ J}$

$q = -337 \text{ J}; \quad \Delta E = \Delta H = 0$

g. $w = 281 \text{ J from } \left(w = -nRT \ln\left(\frac{V_2}{V_1}\right)\right), q = -281 \text{ J}$

Note: overall work for the 1 step process = -203 + 405 = 202 J (q = -202 J)

overall work for the 2 step process = -236 + 337 = 101 J (q = -101 J)

for the reversible process = w = -281 + 281 = 0, (q = 0)

Thus, in an overall reversible process (expan + comp), system and surroundings are unchanged.

CHAPTER ELEVEN

ELECTROCHEMISTRY

Galvanic Cells, Cell Potentials, and Standard Reduction Potentials

15. In a galvanic cell, a spontaneous reaction occurs, producing an electric current. In an electrolytic cell, electricity is used to force a reaction to occur that is not spontaneous.

17. A typical galvanic cell diagram is:

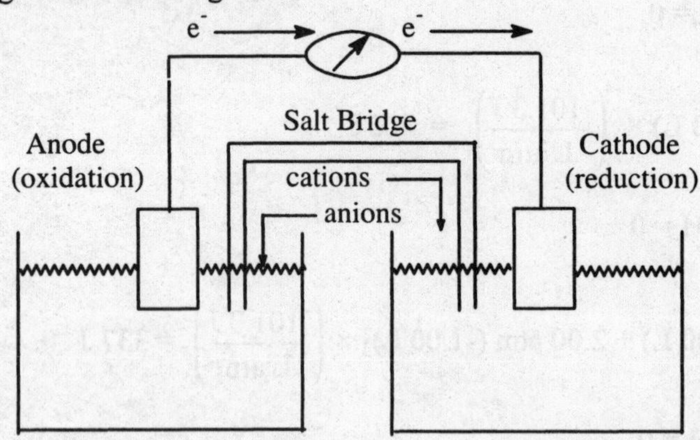

The diagram for all cells will look like this. The contents of each half-cell will be identified for each reaction, with all concentrations at 1.0 *M* and partial pressures at 1.0 atm. Note that cations always flow into the cathode compartment and anions always flow into the anode compartment. This is required to keep each compartment electrically neutral.

a. Reference Table 11.1 for standard reduction potentials. Remember that $E^\circ_{cell} = E^\circ(\text{cathode}) - E^\circ(\text{anode})$; in the Solutions Guide, we will represent E°(cathode) as E°_c and represent $-E^\circ(\text{anode})$ as $-E^\circ_a$. Also remember that standard potentials are <u>not</u> multiplied by the integer used to obtain the overall balanced equation.

$$(Cl_2 + 2\,e^- \rightarrow 2\,Cl^-) \times 3 \qquad E^\circ_c = 1.36\text{ V}$$

$$7\,H_2O + 2\,Cr^{3+} \rightarrow Cr_2O_7^{2-} + 14\,H^+ + 6\,e^- \qquad -E^\circ_a = -1.33\text{ V}$$

$$7\,H_2O(l) + 2\,Cr^{3+}(aq) + 3\,Cl_2(g) \rightarrow Cr_2O_7^{2-}(aq) + 6\,Cl^-(aq) + 14\,H^+(aq) \qquad E^\circ_{cell} = 0.03\text{ V}$$

The contents of each compartment is:

Cathode: Pt electrode; Cl_2 bubbled into solution, Cl^- in solution

Anode: Pt electrode; Cr^{3+}, H^+, and $Cr_2O_7^{2-}$ in solution

We need a nonreactive metal to use as the electrode in each case, since all of the reactants and products are in solution. Pt is the most common choice. Another possibility is graphite.

b. $Cu^{2+} + 2e^- \rightarrow Cu$ $\quad E^\circ_c = 0.34$ V

$Mg \rightarrow Mg^{2+} + 2e^-$ $\quad -E^\circ_a = 2.37$ V

$Cu^{2+}(aq) + Mg(s) \rightarrow Cu(s) + Mg^{2+}(aq)$ $\quad E^\circ_{cell} = 2.71$ V

Cathode: Cu electrode; Cu^{2+} in solution

Anode: Mg electrode; Mg^{2+} in solution

c. $5e^- + 6H^+ + IO_3^- \rightarrow 1/2\ I_2 + 3H_2O$ $\quad E^\circ_c = 1.20$ V

$(Fe^{2+} \rightarrow Fe^{3+} + e^-) \times 5$ $\quad -E^\circ_a = -0.77$ V

$6H^+ + IO_3^- + 5Fe^{2+} \rightarrow 5Fe^{3+} + 1/2\ I_2 + 3H_2O$ $\quad E^\circ_{cell} = 0.43$ V

or $12H^+(aq) + 2IO_3^-(aq) + 10Fe^{2+}(aq) \rightarrow 10Fe^{3+}(aq) + I_2(s) + 6H_2O(l)$ $\quad E^\circ_{cell} = 0.43$ V

Cathode: Pt electrode; IO_3^-, I_2 and H_2SO_4 (H^+ source) in solution.

Anode: Pt electrode; Fe^{2+} and Fe^{3+} in solution

Note: $I_2(s)$ would make a poor electrode since it sublimes.

d. $(Ag^+ + e^- \rightarrow Ag) \times 2$ $\quad E^\circ_c = 0.80$ V

$Zn \rightarrow Zn^{2+} + 2e^-$ $\quad -E^\circ_a = 0.76$ V

$Zn(s) + 2Ag^+(aq) \rightarrow 2Ag(s) + Zn^{2+}(aq)$ $\quad E^\circ_{cell} = 1.56$ V

Cathode: Ag electrode; Ag^+ in solution

Anode: Zn electrode; Zn^{2+} in solution

19. Reference Exercise 11.17 for a typical galvanic cell design. The contents of each half-cell compartment is identified below with all solute concentrations at 1.0 *M* and all gases at 1.0 atm. For each pair of half-reactions, the half-reaction with the largest standard reduction potential will be the cathode reaction and the half-reaction with the smallest reduction potential will be reversed to become the anode reaction. Only this combination gives a spontaneous overall reaction, i.e., a reaction with a positive overall standard cell potential.

a. $Cl_2 + 2e^- \rightarrow 2Cl^-$ $\quad E^\circ_c = 1.36$ V

$2Br^- \rightarrow Br_2 + 2e^-$ $\quad -E^\circ_a = -1.09$ V

$Cl_2(g) + 2Br^-(aq) \rightarrow Br_2(aq) + 2Cl^-(aq)$ $\quad E^\circ_{cell} = 0.27$ V

The contents of each compartment is:

Cathode: Pt electrode; $Cl_2(g)$ bubbled in, Cl^- in solution

Anode: Pt electrode; Br_2 and Br^- in solution

b.

$$(2\,e^- + 2\,H^+ + IO_4^- \rightarrow IO_3^- + H_2O) \times 5 \qquad E^\circ_c = 1.60\ V$$

$$(4\,H_2O + Mn^{2+} \rightarrow MnO_4^- + 8\,H^+ + 5\,e^-) \times 2 \qquad -E^\circ_a = -1.51\ V$$

$$10\,H^+ + 5\,IO_4^- + 8\,H_2O + 2\,Mn^{2+} \rightarrow 5\,IO_3^- + 5\,H_2O + 2\,MnO_4^- + 16\,H^+ \qquad E^\circ_{cell} = 0.09\ V$$

This simplifies to:

$$3\,H_2O(l) + 5\,IO_4^-(aq) + 2\,Mn^{2+}(aq) \rightarrow 5\,IO_3^-(aq) + 2\,MnO_4^-(aq) + 6\,H^+(aq) \quad E^\circ_{cell} = 0.09\ V$$

Cathode: Pt electrode; IO_4^-, IO_3^-, and H_2SO_4 (as a source of H^+) in solution

Anode: Pt electrode; Mn^{2+}, MnO_4^- and H_2SO_4 in solution

c.

$$H_2O_2 + 2\,H^+ + 2\,e^- \rightarrow 2\,H_2O \qquad E^\circ_c = 1.78\ V$$

$$H_2O_2 \rightarrow O_2 + 2\,H^+ + 2\,e^- \qquad -E^\circ_a = -0.68\ V$$

$$2\,H_2O_2(aq) \rightarrow 2\,H_2O(l) + O_2(g) \qquad E^\circ_{cell} = 1.10\ V$$

Cathode: Pt electrode; H_2O_2 and H^+ in solution

Anode: Pt electrode; $O_2(g)$ bubbled in, H_2O_2 and H^+ in solution

d.

$$(Fe^{3+} + 3\,e^- \rightarrow Fe) \times 2 \qquad E^\circ_c = -0.036\ V$$

$$(Mn \rightarrow Mn^{2+} + 2\,e^-) \times 3 \qquad -E^\circ_a = 1.18\ V$$

$$2\,Fe^{3+}(aq) + 3\,Mn(s) \rightarrow 2\,Fe(s) + 3\,Mn^{2+}(aq) \qquad E^\circ_{cell} = 1.14\ V$$

Cathode: Fe electrode; Fe^{3+} in solution; Anode: Mn electrode; Mn^{2+} in solution

21. **In standard line notation, the anode is listed first and the cathode is listed last. A double line separates the two compartments. By convention, the electrodes are on the ends with all solutes and gases towards the middle. A single line is used to indicate a phase change. We also included all concentrations.**

19a. Pt | Br^- (1.0 *M*), Br_2 (1.0 *M*) || Cl_2 (1.0 atm) | Cl^- (1.0 *M*) | Pt

19b. Pt | Mn^{2+} (1.0 *M*), MnO_4^- (1.0 *M*), H^+ (1.0 *M*) || IO_4^- (1.0 *M*), IO_3^- (1.0 *M*), H^+ (1.0 *M*) | Pt

19c. Pt | H_2O_2 (1.0 *M*), H^+ (1.0 *M*) | O_2 (1.0 atm) || H_2O_2 (1.0 *M*), H^+ (1.0 *M*)|Pt

19d. Mn | Mn^{2+} (1.0 *M*) || Fe^{3+} (1.0 *M*) | Fe

23. a. $2\ H^+ + 2\ e^- \rightarrow H_2$ $E° = 0.00$ V; $Cu \rightarrow Cu^{2+} + 2\ e^-$ $-E° = -0.34$ V

$E°_{cell} = -0.34$ V; No, H^+ cannot oxidize Cu to Cu^{2+} at standard conditions ($E°_{cell} < 0$).

b. $Fe^{3+} + e^- \rightarrow Fe^{2+}$ $E° = 0.77$ V; $2\ I^- \rightarrow I_2 + 2\ e^-$ $-E° = -0.54$ V

$E°_{cell} = 0.77 - 0.54 = 0.23$ V; Yes, Fe^{3+} can oxidize I^- to I_2.

c. $H_2 \rightarrow 2\ H^+ + 2\ e^-$ $-E° = 0.00$ V; $Ag^+ + e^- \rightarrow Ag$ $E° = 0.80$ V

$E°_{cell} = 0.80$ V; Yes, H_2 can reduce Ag^+ to Ag at standard conditions ($E°_{cell} > 0$).

d. $Fe^{2+} \rightarrow Fe^{3+} + e^-$ $-E° = -0.77$ V; $Cr^{3+} + e^- \rightarrow Cr^{2+}$ $E° = -0.50$ V

$E°_{cell} = -0.50 - 0.77 = -1.27$ V; No, Fe^{2+} cannot reduce Cr^{3+} to Cr^{2+} at standard conditions.

25. Good reducing agents are easily oxidized. The reducing agents are on the right side of the reduction half-reactions listed in Table 11.1. The best reducing agents have the most negative standard reduction potentials (E°), i.e., the best reducing agents have the most positive −E° value.

	F^-	< Cr^{3+}	< Fe^{2+}	< H_2	< Zn	< Li
−E°(V)	-2.87	-1.33	-0.77	0.00	0.76	3.05

27. a. $2\ Br^- \rightarrow Br_2 + 2\ e^-$ $-E°_a = -1.09$ V; $2\ Cl^- \rightarrow Cl_2 + 2\ e^-$ $-E°_a = -1.36$ V; $E°_c > 1.09$ V to oxidize Br^-; $E°_c < 1.36$ V to not oxidize Cl^-; $Cr_2O_7^{2-}$, O_2, MnO_2, and IO_3^- are all possible since when all of these oxidizing agents are coupled with Br^- give $E°_{cell} > 0$ and when coupled with Cl^- give $E°_{cell} < 0$ (assuming standard conditions).

b. $Mn \rightarrow Mn^{2+} + 2\ e^-$ $-E°_a = 1.18$; $Ni \rightarrow Ni^{2+} + 2\ e^-$ $-E°_a = 0.23$ V; Any oxidizing agent with $-0.23\ V > E°_c > -1.18$ V will work. $PbSO_4$, Cd^{2+}, Fe^{2+}, Cr^{3+}, Zn^{2+} and H_2O will be able to oxidize Mn but not oxidize Ni (assuming standard conditions).

29.
$$ClO^- + H_2O + 2\ e^- \rightarrow 2\ OH^- + Cl^- \qquad E° = 0.90\ V$$
$$2\ NH_3 + 2\ OH^- \rightarrow N_2H_4 + 2\ H_2O + 2\ e^- \qquad -E° = 0.10\ V$$
$$ClO^-(aq) + 2\ NH_3(aq) \rightarrow Cl^-(aq) + N_2H_4(aq) + H_2O(l) \qquad E°_{cell} = 1.00\ V$$

Since $E°_{cell}$ is positive for this reaction, then at standard conditions ClO^- can spontaneously oxidize NH_3 to the somewhat toxic N_2H_4.

31.
$$H_2O_2 + 2\ H^+ + 2\ e^- \rightarrow 2\ H_2O \qquad E°_c = 1.78\ V;\ H_2O_2 \text{ is the oxidizing agent.}$$
$$H_2O_2 \rightarrow O_2 + 2\ H^+ + 2\ e^- \qquad -E°_a = -0.68\ V;\ H_2O_2 \text{ is the reducing agent.}$$

$$H_2O_2 + 2\ H^+ + 2\ e^- \rightarrow 2\ H_2O \qquad E°_c = 1.78\ V$$
$$H_2O_2 \rightarrow O_2 + 2\ H^+ + 2\ e^- \qquad -E°_a = -0.68\ V$$
$$2\ H_2O_2(aq) \rightarrow 2\ H_2O(l) + O_2(g) \qquad E°_{cell} = 1.10\ V$$

Cell Potential, Free Energy, and Equilibrium

33. Since the cells are at standard conditions, $w_{max} = \Delta G = \Delta G° = -nFE°_{cell}$. See Exercise 17.33 for the balanced overall equations and for $E°_{cell}$.

20a. w_{max} = -(3 mol e^-)(96,485 C/mol e^-)(1.34 J/C) = -3.88×10^5 J = -388 kJ

20b. w_{max} = -(2 mol e^-)(96,485 C/mol e^-)(1.40 J/C) = -2.70×10^5 J = -270. kJ

35. Reference Exercise 11.19 for the balanced reactions and standard cell potentials. The balanced reactions are necessary to determine n, the moles of electrons transferred.

19a. $Cl_2(aq) + 2\ Br^-(aq) \rightarrow Br_2(aq) + 2\ Cl^-(aq)$ $E°_{cell}$ = 0.27 V = 0.27 J/C, n = 2 mol e^-

$\Delta G° = -nFE°_{cell}$ = -(2 mol e^-)(96,485 C/mol e^-)(0.27 J/C) = -5.2×10^4 J = -52 kJ

$$E°_{cell} = \frac{0.0591}{n} \log K,\ \log K = \frac{nE°}{0.0591} = \frac{2(0.27)}{0.0591} = 9.14,\ K = 10^{9.14} = 1.4 \times 10^9$$

19b. $\Delta G°$ = -(10 mol e^-)(96,485 C/mol e^-)(0.09 J/C) = -9×10^4 J = -90 kJ

$$\log K = \frac{10(0.09)}{0.0591} = 15.2,\ K = 10^{15.2} = 2 \times 10^{15}$$

19c. $\Delta G°$ = -(2 mol e^-)(96,485 C/mol e^-)(1.10 J/C) = -2.12×10^5 J = -212 kJ

$$\log K = \frac{2(1.10)}{0.0591} = 37.225,\ K = 1.68 \times 10^{37}$$

19d. $\Delta G°$ = -(6 mol e^-)(96,485 C/mol e^-)(1.14 J/C) = -6.60×10^5 J = -660. kJ

$$\log K = \frac{(6)(1.14)}{0.0591} = 115.736,\ K = 5.45 \times 10^{115}$$

37. a.

$(4\ H^+ + NO_3^- + 3\ e^- \rightarrow NO + 2\ H_2O) \times 2$ $E°_c$ = 0.96 V

$(Mn \rightarrow Mn^{2+} + 2\ e^-) \times 3$ $-E°_a$ = 1.18 V

$3\ Mn(s) + 8\ H^+(aq) + 2\ NO_3^-(aq) \rightarrow 2\ NO(g) + 4\ H_2O(l) + 3\ Mn^{2+}(aq)$ $E°_{cell}$ = 2.14 V

$5 \times (2\ e^- + 2\ H^+ + IO_4^- \rightarrow IO_3^- + H_2O)$ $E°_a$ = 1.60 V

$2 \times (Mn^{2+} + 4\ H_2O \rightarrow MnO_4^- + 8\ H^+ + 5\ e^-)$ $-E°_a$ = -1.51 V

$5\ IO_4^-(aq) + 2\ Mn^{2+}(aq) + 3\ H_2O(l) \rightarrow 5\ IO_3^-(aq) + 2\ MnO_4^-(aq) + 6\ H^+(aq)$ $E°_{cell}$ = 0.09 V

b. Nitric acid oxidation (see part a for $E°_{cell}$):

$\Delta G° = -nFE°_{cell}$ = -(6 mol e^-)(96,485 C/mol e^-)(2.14 J/C) = -1.24×10^6 J = -1240 kJ

$$\log K = \frac{nE°}{0.0591} = \frac{6(2.14)}{0.0591} = 217,\ K \approx 10^{217}$$

Periodate oxidation (see part a for E°_{cell}):

$$\Delta G^\circ = -(10 \text{ mol e}^-)(96{,}485 \text{ C/mol e}^-)(0.09 \text{ J/C})(1 \text{ kJ}/1000 \text{ J}) = -90 \text{ kJ}$$

$$\log K = \frac{10(0.09)}{0.0591} = 15.2, \; K = 10^{15.2} = 2 \times 10^{15}$$

39. $\Delta G^\circ = -nFE^\circ = \Delta H^\circ - T\Delta S^\circ, \; E^\circ = \frac{T\Delta S^\circ}{nF} - \frac{\Delta H^\circ}{nF}$

If we graph E° vs. T we should get a straight line (y = mx + b). The slope of the line (m) is equal to $\Delta S^\circ/nF$ and the y-intercept (b) is equal to $-\Delta H^\circ/nF$.

41. $CH_3OH(l) + 3/2\ O_2(g) \rightarrow CO_2(g) + 2\ H_2O(l)$

$$\Delta G^\circ = \Sigma n_p \Delta G^\circ_{f,\ products} - \Sigma n_r \Delta G^\circ_{f,\ reactants} = 2(-237) + (-394) - [-166] = -702 \text{ kJ}$$

The balanced half-reactions are:

$$H_2O + CH_3OH \rightarrow CO_2 + 6\ H^+ + 6\ e^- \text{ and } O_2 + 4\ H^+ + 4\ e^- \rightarrow 2\ H_2O$$

For 3/2 mol O_2, 6 moles of electrons will be transferred (n = 6).

$$\Delta G^\circ = -nFE^\circ, \; E^\circ = \frac{-\Delta G^\circ}{nF} = \frac{-(-702{,}000 \text{ J})}{(6 \text{ mol e}^-)(96{,}485 \text{ C/mol e}^-)} = 1.21 \text{ J/C} = 1.21 \text{ V}$$

For this reaction: $\Delta S^\circ = 2(70.) + 214 - [127 + 3/2(205)] = -81$ J/K

From Exercise 11.39, $E^\circ = \frac{T\Delta S^\circ}{nF} - \frac{\Delta H^\circ}{nF}$.

Since ΔS° is negative, E° will decrease with an increase in temperature.

43. a.

$Cu^+ + e^- \rightarrow Cu$ $\quad E^\circ_c = 0.52$ V

$Cu^+ \rightarrow Cu^{2+} + e^-$ $\quad -E^\circ_a = -0.16$ V

$2\ Cu^+(aq) \rightarrow Cu^{2+}(aq) + Cu(s)$ $\quad E^\circ_{cell} = 0.36$ V; Spontaneous

b.

$Fe^{2+} + 2\ e^- \rightarrow Fe$ $\quad E^\circ_c = -0.44$ V

$(Fe^{2+} \rightarrow Fe^{3+} + e^-) \times 2$ $\quad -E^\circ_a = -0.77$ V

$3\ Fe^{2+}(aq) \rightarrow 2\ Fe^{3+}(aq) + Fe(s)$ $\quad E^\circ_{cell} = -1.21$ V; Not spontaneous

c.

$HClO_2 + 2\ H^+ + 2\ e^- \rightarrow HClO + H_2O$ $\quad E^\circ_c = 1.65$ V

$HClO_2 + H_2O \rightarrow ClO_3^- + 3\ H^+ + 2\ e^-$ $\quad -E^\circ_a = -1.21$ V

$2\ HClO_2(aq) \rightarrow ClO_3^-(aq) + H^+(aq) + HClO(aq)$ $\quad E^\circ_{cell} = 0.44$ V; Spontaneous

45.
$$Al^{3+} + 3\,e^- \rightarrow Al \qquad E_c^\circ = -1.66\ V$$
$$Al + 6\,F^- \rightarrow AlF_6^{3-} + 3\,e^- \qquad E_a^\circ = 2.07\ V$$

$$Al^{3+}(aq) + 6\,F^-(aq) \rightarrow AlF_6^{3-}(aq) \qquad E_{cell}^\circ = 0.41\ V \quad K = ?$$

$$\log K = \frac{nE^\circ}{0.0591} = \frac{3(0.41)}{0.0591} = 20.81,\ K = 10^{20.81} = 6.5 \times 10^{20}$$

47.
$$CdS + 2\,e^- \rightarrow Cd + S^{2-} \qquad E_c^\circ = -1.21\ V$$
$$Cd \rightarrow Cd^{2+} + 2\,e^- \qquad -E_a^\circ = 0.402\ V$$

$$CdS(s) \rightarrow Cd^{2+}(aq) + S^{2-}(aq) \qquad E_{cell}^\circ = -0.81\ V \qquad K = K_{sp} = ?$$

For this overall reaction, $E_{cell}^\circ = \frac{0.0591}{n} \log K_{sp}$

$$\log K_{sp} = \frac{nE^\circ}{0.0591} = \frac{2(-0.81)}{0.0591} = -27.41, \quad K_{sp} = 10^{-27.41} = 3.9 \times 10^{-28}$$

Galvanic Cells: Concentration Dependence

49. a.
$$Au^{3+} + 3\,e^- \rightarrow Au \qquad E_c^\circ = 1.50\ V$$
$$(Tl \rightarrow Tl^+ + e^-) \times 3 \qquad -E_a^\circ = 0.34\ V$$

$$Au^{3+}(aq) + 3\,Tl(s) \rightarrow Au(s) + 3\,Tl^+(aq) \qquad E_{cell}^\circ = 1.84\ V$$

b. $\Delta G^\circ = -nFE_{cell}^\circ = -(3\ mol\ e^-)(96{,}485\ C/mol\ e^-)(1.84\ J/C) = -5.33 \times 10^5\ J = -533\ kJ$

$$\log K = \frac{nE^\circ}{0.0591} = \frac{3(1.84)}{0.0591} = 93.401,\ K = 10^{93.401} = 2.52 \times 10^{93}$$

c. At 25°C, $E_{cell} = E_{cell}^\circ - \frac{0.0591}{n} \log Q$ where $Q = \frac{[Tl^+]^3}{[Au^{3+}]}$

$$E_{cell} = 1.84\ V - \frac{0.0591}{3} \log \frac{[Tl^+]^3}{[Au^{3+}]} = 1.84 - \frac{0.0591}{3} \log \frac{(1.0 \times 10^{-4})^3}{1.0 \times 10^{-2}}$$

$$E_{cell} = 1.84 - (-0.20) = 2.04\ V$$

51.
$$(Pb^{2+} + 2\,e^- \rightarrow Pb) \times 3 \qquad E_c^\circ = -0.13\ V$$
$$(Al \rightarrow Al^{3+} + 3\,e^-) \times 2 \qquad -E_a^\circ = 1.66\ V$$

$$3\,Pb^{2+}(aq) + 2\,Al(s) \rightarrow 3\,Pb(s) + 2\,Al^{3+}(aq) \qquad E_{cell}^\circ = 1.53\ V$$

From the balanced reaction, when the $[Al^{3+}]$ has increased by 0.60 mol/L (Al^{3+} is a product in the spontaneous reaction), then the Pb^{2+} concentration has decreased by 3/2 (0.60 mol/L) = 0.90 *M*.

$$E_{cell} = 1.53\text{ V} - \frac{0.0591}{6}\log\frac{[Al^{3+}]^2}{[Pb^{2+}]^3} = 1.53 - \frac{0.0591}{6}\log\frac{(1.60)^2}{(0.10)^3}$$

$$E_{cell} = 1.53\text{ V} - 0.034\text{ V} = 1.50\text{ V}$$

53. a. n = 2 for this reaction (lead goes from $Pb \rightarrow Pb^{2+}$ in $PbSO_4$).

$$E_{cell} = E^\circ_{cell} - \frac{0.0591}{2}\log\left(\frac{1}{[H^+]^2[HSO_4^-]^2}\right) = 2.04\text{ V} - \frac{0.0591}{2}\log\frac{1}{(4.5)^2(4.5)^2}$$

2.04 V - (-0.077 V) = 2.12 V

b. We can calculate ΔG° from $\Delta G^\circ = \Delta H^\circ - T\Delta S^\circ$ and then E° from $\Delta G^\circ = -nFE^\circ$; or we can use the equation derived in Exercise 11.39.

$$E^\circ_{-20} = \frac{T\Delta S^\circ - \Delta H^\circ}{nF} = \frac{(253\text{ K})(263.5\text{ J/K}) + 315.9 \times 10^3\text{ J}}{(2\text{ mol e}^-)(96{,}485\text{ C/mol e}^-)} = 1.98\text{ J/C} = 1.98\text{ V}$$

c. $$E_{-20} = E^\circ_{-20} - \frac{RT}{nF}\ln Q = 1.98\text{ V} - \frac{RT}{nF}\ln\frac{1}{[H^+]^2[HSO_4^-]^2}$$

$$E_{-20} = 1.98\text{ V} - \frac{(8.3145\text{ J K}^{-1}\text{ mol}^{-1})(253\text{ K})}{(2\text{ mol e}^-)(96{,}485\text{ C/mol e}^-)}\ln\frac{1}{(4.5)^2(4.5)^2} = 1.98\text{ V} - (-0.066\text{ V}) = 2.05\text{ V}$$

d. As the temperature decreases, the cell potential decreases. Also, oil becomes more viscous at lower temperatures, which adds to the difficulty of starting an engine on a cold day. The combination of these two factors results in batteries failing more often on cold days than on warm days.

55. As is the case for all concentration cells, $E^\circ_{cell} = 0$, and the smaller ion concentration is always in the anode compartment. The general Nernst equation for the Ni | Ni^{2+} (*x M*) || Ni^{2+}(*y M*) | Ni concentration cell is:

$$E_{cell} = E^\circ_{cell} - \frac{0.0591}{n}\log Q = \frac{-0.0591}{2}\log\frac{[Ni^{2+}]_{anode}}{[Ni^{2+}]_{cathode}}$$

a. Since both compartments are at standard conditions ($[Ni^{2+}]$ = 1.0 *M*), then $E_{cell} = E^\circ_{cell} = 0$ V. No electron flow occurs.

b. Cathode = 2.0 *M* Ni^{2+}; Anode = 1.0 *M* Ni^{2+}; Electron flow is always from the anode to the cathode, so electrons flow to the right in the diagram.

$$E_{cell} = \frac{-0.0591}{2}\log\frac{[Ni^{2+}]_{anode}}{[Ni^{2+}]_{cathode}} = \frac{-0.0591}{2}\log\frac{1.0}{2.0} = 8.9 \times 10^{-3}\text{ V}$$

c. Cathode = 1.0 *M* Ni^{2+}; Anode = 0.10 *M* Ni^{2+}; Electrons flow to the left in the diagram.

$$E_{cell} = \frac{-0.0591}{2}\log\frac{0.10}{1.0} = 0.030\text{ V}$$

d. Cathode = 1.0 *M* Ni^{2+}; Anode = 4.0×10^{-5} *M* Ni^{2+}; Electrons flow to the left in the diagram.

$$E_{cell} = \frac{-0.0591}{2} \log \frac{4.0 \times 10^{-5}}{1.0} = 0.13 \text{ V}$$

e. Since both concentrations are equal, log (2.5/2.5) = log 1.0 = 0 and $E_{cell} = 0$. No electron flow occurs.

57. $Cu^{2+}(aq) + H_2(g) \rightarrow 2\,H^+(aq) + Cu(s)$ $E^\circ_{cell} = 0.34 \text{ V} - 0.00\text{V} = 0.34 \text{ V}$ and n = 2

Since $P_{H_2} = 1.0$ atm and $[H^+] = 1.0$ *M*: $E_{cell} = E^\circ_{cell} - \frac{0.0591}{2} \log \frac{1}{[Cu^{2+}]}$

a. $E_{cell} = 0.34 \text{ V} - \frac{0.0591}{2} \log \frac{1}{2.5 \times 10^{-4}} = 0.34 \text{ V} - 0.11 \text{ V} = 0.23 \text{ V}$

b. Use the K_{sp} expression to calculate the Cu^{2+} concentration in the cell.

$Cu(OH)_2(s) \rightleftharpoons Cu^{2+}(aq) + 2\,OH^-(aq)$ $K_{sp} = 1.6 \times 10^{-19} = [Cu^{2+}][OH^-]^2$

From the problem, $[OH^-] = 0.10$ *M*, so: $[Cu^{2+}] = \frac{1.6 \times 10^{-19}}{(0.10)^2} = 1.6 \times 10^{-17}$ *M*

$$E_{cell} = E^\circ_{cell} - \frac{0.0591}{2} \log \frac{1}{[Cu^{2+}]} = 0.34 \text{ V} - \frac{0.0591}{2} \log \frac{1}{1.6 \times 10^{-17}} = 0.34 - 0.50 = -0.16 \text{ V}$$

Since $E_{cell} < 0$, then the forward reaction is not spontaneous, but the reverse reaction is spontaneous. The Cu electrode becomes the anode and $E_{cell} = 0.16$ V for the reverse reaction. The cell reaction is: $2\,H^+(aq) + Cu(s) \rightarrow Cu^{2+}(aq) + H_2(g)$.

c. $0.195 \text{ V} = 0.34 \text{ V} - \frac{0.0591}{2} \log \frac{1}{[Cu^{2+}]}$, $\log \frac{1}{[Cu^{2+}]} = 4.91$, $[Cu^{2+}] = 10^{-4.91} = 1.2 \times 10^{-5}$ *M*

Note: When determining exponents, we will carry extra significant figures.

d. $E_{cell} = E^\circ_{cell} - (0.0591/2) \log (1/[Cu^{2+}]) = E^\circ_{cell} + 0.0296 \log [Cu^{2+}]$; This equation is in the form of a straight line equation, y = mx + b. A graph of E_{cell} vs. log $[Cu^{2+}]$ will yield a straight line with slope equal to 0.0296 V or 29.6 mV.

59. From Exercise 11.17a: $3\,Cl_2(g) + 2\,Cr^{3+}(aq) + 7\,H_2O(l) \rightleftharpoons 14\,H^+(aq) + Cr_2O_7^{2-}(aq) + 6\,Cl^-(aq)$
$E^\circ_{cell} = 0.03$ V

$$E_{cell} = E^\circ_{cell} - \frac{0.0591}{6} \log \frac{[Cr_2O_7^{2-}][H^+]^{14}[Cl^-]^6}{[Cr^{3+}]^2 P_{Cl_2}^3}$$

When $K_2Cr_2O_7$ and Cl^- are added to concentrated H_2SO_4, Q becomes a large number due to $[H^+]^{14}$ term. The log of a large number is positive. E_{cell} becomes negative, which means the reverse reaction becomes spontaneous. The pungent fumes were $Cl_2(g)$.

Electrolysis

61. $15\text{A} = \frac{15\ \text{C}}{\text{s}} \times \frac{60\ \text{s}}{\text{min}} \times \frac{60\ \text{min}}{\text{h}} = 5.4 \times 10^4$ **C of charge passed in 1 hour**

a. $5.4 \times 10^4\ \text{C} \times \frac{1\ \text{mol e}^-}{96{,}485\ \text{C}} \times \frac{1\ \text{mol Co}}{2\ \text{mol e}^-} \times \frac{58.9\ \text{g}}{\text{mol}} = 16\ \text{g Co}$

b. $5.4 \times 10^4\ \text{C} \times \frac{1\ \text{mol e}^-}{96{,}485\ \text{C}} \times \frac{1\ \text{mol Hf}}{4\ \text{mol e}^-} \times \frac{178.5\ \text{g}}{\text{mol}} = 25\ \text{g Hf}$

c. $2\ \text{I}^- \rightarrow \text{I}_2 + 2\text{e}^-$; $5.4 \times 10^4\ \text{C} \times \frac{1\ \text{mol e}^-}{96{,}485\ \text{C}} \times \frac{1\ \text{mol I}_2}{2\ \text{mol e}^-} \times \frac{253.8\ \text{g I}_2}{\text{mol I}_2} = 71\ \text{g I}_2$

d. **Cr is in the +6 oxidation state in CrO_3. Six mol of e^- are needed to produce 1 mol Cr from molten CrO_3.**

$$5.4 \times 10^4\ \text{C} \times \frac{1\ \text{mol e}^-}{96{,}485\ \text{C}} \times \frac{1\ \text{mol Cr}}{6\ \text{mol e}^-} \times \frac{52.0\ \text{g Cr}}{\text{mol Cr}} = 4.9\ \text{g Cr}$$

63. **First determine the species present, then reference Table 11.1 to help you identify each species as a possible oxidizing agent (species reduced) or as a possible reducing agent (species oxidized). Of all the possible oxidizing agents, the species that will be reduced at the cathode will have the most positive E°_c value; the species that will be oxidized at the anode will be the reducing agent with the most positive $-E^\circ_a$ value.**

a. **Species present: Ni^{2+} and Br^-; Ni^{2+} can be reduced to Ni and Br^- can be oxidized to Br_2 (from Table 11.1). The reactions are:**

Cathode: $Ni^{2+} + 2e^- \rightarrow Ni$ $\quad E^\circ_c = -0.23\ \text{V}$

Anode: $2\ Br^- \rightarrow Br_2 + 2\ e^-$ $\quad -E^\circ_a = -1.09\ \text{V}$

b. **Species present: Al^{3+} and F^-; Al^{3+} can be reduced and F^- can be oxidized. The reactions are:**

Cathode: $Al^{3+} + 3\ e^- \rightarrow Al$ $\quad E^\circ_c = -1.66\ \text{V}$

Anode: $2\ F^- \rightarrow F_2 + 2\ e^-$ $\quad -E^\circ_a = -2.87\ \text{V}$

c. **Species present: Mn^{2+} and I^-; Mn^{2+} can be reduced and I^- can be oxidized. The reactions are:**

Cathode: $Mn^{2+} + 2\ e^- \rightarrow Mn$ $\quad E^\circ_c = -1.18\ \text{V}$

Anode: $2\ I^- \rightarrow I_2 + 2\ e^-$ $\quad -E^\circ_a = -0.54\ \text{V}$

d. **For aqueous solutions, we must now consider H_2O as a possible oxidizing agent and a possible reducing agent. Species present: Ni^{2+}, Br^- and H_2O. Possible cathode reactions are:**

$Ni^{2+} + 2e^- \rightarrow Ni$ $\quad E^\circ_c = -0.23\ \text{V}$

$2\ H_2O + 2\ e^- \rightarrow H_2 + 2\ OH^-$ $\quad E^\circ_c = -0.83\ \text{V}$

Since it is easier to reduce Ni^{2+} than H_2O (assuming standard conditions), then Ni^{2+} will be reduced by the above cathode reaction.

Possible anode reactions are:

$$2\ Br^- \rightarrow Br_2 + 2\ e^- \qquad -E_a^\circ = -1.09\ V$$
$$2\ H_2O \rightarrow O_2 + 4\ H^+ + 4\ e^- \qquad -E_a^\circ = -1.23\ V$$

Since Br^- is easier to oxidize than H_2O (assuming standard conditions), then Br^- will be oxidized by the above anode reaction.

e. Species present: Al^{3+}, F^- and H_2O; Al^{3+} and H_2O can be reduced. The reduction potentials are $E_c^\circ = -1.66$ V for Al^{3+} and $E_c^\circ = -0.83$ V for H_2O (assuming standard conditions). H_2O should be reduced at the cathode ($2\ H_2O + 2\ e^- \rightarrow H_2 + 2\ OH^-$).

F^- and H_2O can be oxidized. The oxidation potentials are $-E_a^\circ = -2.87$ V for F^- and $-E_a^\circ = -1.23$ V for H_2O (assuming standard conditions). From the potentials, we would predict H_2O to be oxidized at the anode ($2\ H_2O \rightarrow O_2 + 4\ H^+ + 4\ e^-$).

f. Species present: Mn^{2+}, I^- and H_2O; Mn^{2+} and H_2O can be reduced. The possible cathode reactions are:

$$Mn^{2+} + 2\ e^- \rightarrow Mn \qquad E_c^\circ = -1.18\ V$$
$$2\ H_2O + 2\ e^- \rightarrow H_2 + 2\ OH^- \qquad E_c^\circ = -0.83\ V$$

Reduction of H_2O should occur at the cathode since $E_{H_2O}^\circ$ is most positive.

I^- and H_2O can be oxidized. The possible anode reactions are:

$$2\ I^- \rightarrow I_2 + 2\ e^- \qquad -E_a^\circ = -0.54\ V$$
$$2\ H_2O \rightarrow O_2 + 4\ H^+ + 4\ e^- \qquad -E_a^\circ = -1.23\ V$$

Oxidation of I^- will occur at the anode since $-E_{I^-}^\circ$ is most positive.

65. $Au^{3+} + 3\ e^- \rightarrow Au \quad E^\circ = 1.50$ V $\qquad Ni^{2+} + 2\ e^- \rightarrow Ni \quad E^\circ = -0.23$ V
$Ag^+ + e^- \rightarrow Ag \quad E^\circ = 0.80$ V $\qquad Cd^{2+} + 2\ e^- \rightarrow Cd \quad E^\circ = -0.40$ V

$2\ H_2O + 2e^- \rightarrow H_2 + 2\ OH^- \quad E^\circ = -0.83$ V

Au(s) will plate out first since it has the most positive reduction potential, followed by Ag(s), which is followed by Ni(s), and finally Cd(s) will plate out last since it has the most negative reduction potential of the metals listed.

67. To begin plating out Pd: $E_c = 0.62\ V - \frac{0.0591}{2}\log\frac{[Cl^-]^4}{[PdCl_4^{2-}]} = 0.62 - \frac{0.0591}{2}\log\frac{(1.0)^4}{0.020}$

$E_c = 0.62\ V - 0.050\ V = 0.57\ V$

When 99% of Pd has plated out, $[PdCl_4^-] = \frac{1}{100}(0.020) = 0.00020\ M$.

$$E_c = 0.62 - \frac{0.0591}{2}\log\frac{(1.0)^4}{2.0\times 10^{-4}} = 0.62\ V - 0.11V = 0.51\ V$$

To begin Pt plating: $E_c = 0.73\ V - \frac{0.0591}{2}\log\frac{(1.0)^4}{0.020} = 0.73 - 0.050 = 0.68\ V$

When 99% of Pt plated: $E_c = 0.73 - \frac{0.0591}{2}\log\frac{(1.0)^4}{2.0\times 10^{-4}} = 0.083 - 0.11 = 0.62\ V$

To begin Ir plating: $E_c = 0.77\ V - \frac{0.0591}{3}\log\frac{(1.0)^4}{0.020} = 0.77 - 0.033 = 0.74\ V$

When 99% of Ir plated: $E_c = 0.77 - \frac{0.0591}{3}\log\frac{(1.0)^4}{2.0\times 10^{-4}} = 0.77 - 0.073 = 0.70\ V$

Yes, since the range of potentials for plating out each metal do not overlap, we should be able to separate the three metals. The exact potential to apply depends on the oxidation reaction. The order of plating will be Ir(s) first, followed by Pt(s) and finally Pd(s) as the potential is gradually increased.

69. Alkaline earth metals form +2 ions, so 2 mol of e^- are transferred to form the metal, M.

$$\text{mol M} = 748\ s \times \frac{5.00\ C}{s} \times \frac{1\ \text{mol } e^-}{96{,}485\ C} \times \frac{1\ \text{mol M}}{2\ \text{mol } e^-} = 1.94\times 10^{-2}\ \text{mol M}$$

$$\text{molar mass of M} = \frac{0.471\ g\ M}{1.94\times 10^{-2}\ \text{mol M}} = 24.3\ g/mol;$$ $MgCl_2$ was electrolyzed.

71. F_2 is produced at the anode: $2\ F^- \rightarrow F_2 + 2\ e^-$

$$2.00\ h \times \frac{60\ min}{h} \times \frac{60\ s}{min} \times \frac{10.0\ C}{s} \times \frac{1\ \text{mol } e^-}{96{,}485\ C} = 0.746\ \text{mol } e^-$$

$$0.746\ \text{mol } e^- \times \frac{1\ \text{mol } F_2}{2\ \text{mol } e^-} = 0.373\ \text{mol } F_2;\quad PV = nRT,\quad V = \frac{nRT}{P}$$

$$V = \frac{(0.373\ mol)(0.08206\ L\ atm\ K^{-1} mol^{-1})(298\ K)}{1.00\ atm} = 9.12\ L\ F_2$$

K is produced at the cathode: $K^+ + e^- \rightarrow K$

$$0.746\ \text{mol } e^- \times \frac{1\ \text{mol K}}{\text{mol } e^-} \times \frac{39.10\ g\ K}{\text{mol K}} = 29.2\ g\ K$$

73. In the electrolysis of aqueous sodium chloride, H_2O is reduced in preference to Na^+ and Cl^- is oxidized in preference to H_2O. The anode reaction is $2\ Cl^- \rightarrow Cl_2 + 2\ e^-$ and the cathode reaction is $2\ H_2O + 2\ e^- \rightarrow H_2 + 2\ OH^-$. The overall reaction is $2\ H_2O(l) + 2\ Cl^-(aq) \rightarrow Cl_2(g) + H_2(g) + 2\ OH^-(aq)$.

From the 1:1 mol ratio between Cl_2 and H_2 in the overall balanced reaction, if 6.00 L of $H_2(g)$ are produced, then 6.00 L of $Cl_2(g)$ will also be produced since moles and volume of gas are directly proportional at constant T and P (see Chapter 5 of text).

Additional Exercises

75.

$$(CO + O^{2-} \rightarrow CO_2 + 2\ e^-) \times 2$$
$$O_2 + 4\ e^- \rightarrow 2\ O^{2-}$$
$$\overline{2\ CO + O_2 \rightarrow 2\ CO_2}$$

$$\Delta G = -nFE, \quad E = \frac{-\Delta G}{nF} = \frac{-(-380 \times 10^3\ J)}{(4\ mol\ e^-)(96{,}485\ C/mol\ e^-)} = 0.98\ V$$

77. $Zn \rightarrow Zn^{2+} + 2\ e^- \quad -E^\circ_a = 0.76\ V; \quad Fe \rightarrow Fe^{2+} + 2\ e^- \quad -E^\circ_a = 0.44\ V$

It is easier to oxidize Zn than Fe, so the Zn will be oxidized, protecting the iron of the *Monitor's* hull.

79. As a battery discharges, E_{cell} decreases, eventually reaching zero. A charged battery is not at equilibrium. At equilibrium $E_{cell} = 0$ and $\Delta G = 0$. We get no work out of an equilibrium system. A battery is useful to us because it can do work as it approaches equilibrium.

81.

$$O_2 + 2\ H_2O + 4\ e^- \rightarrow 4\ OH^- \qquad E^\circ_c = 0.40\ V$$
$$(H_2 + 2\ OH^- \rightarrow 2\ H_2O + 2\ e^-) \times 2 \qquad -E^\circ_a = 0.83\ V$$
$$\overline{2\ H_2(g) + O_2(g) \rightarrow 2\ H_2O(l) \qquad E^\circ_{cell} = 1.23\ V = 1.23\ J/C}$$

Since standard conditions are assumed, then $w_{max} = \Delta G^\circ$ for 2 mol H_2O produced.

$$\Delta G^\circ = -nFE^\circ_{cell} = -(4\ mol\ e^-)(96{,}485\ C/mol\ e^-)(1.23\ J/C) = -475{,}000\ J = -475\ kJ$$

For 1.00×10^3 g H_2O produced, w_{max} is:

$$1.00 \times 10^3\ g\ H_2O \times \frac{1\ mol\ H_2O}{18.02\ g\ H_2O} \times \frac{-475\ kJ}{2\ mol\ H_2O} = -13{,}200\ kJ = w_{max}$$

The work done can be no larger than the free energy change. The best that could happen is that all of the free energy released goes into doing work, but this does not occur in any real process since there is always waste energy in any real process.

83. $(Al^{3+} + 3\ e^- \rightarrow Al) \times 2$ $\qquad E^\circ_c = -1.66$ V

$(M \rightarrow M^{2+} + 2\ e^-) \times 3$ $\qquad -E^\circ_a = ?$

$3\ M(s) + 2\ Al^{3+}(aq) \rightarrow 2\ Al(s) + 3\ M^{2+}(aq)$ $\qquad E^\circ_{cell} = -E^\circ_a - 1.66$ V

$\Delta G^\circ = -nFE^\circ_{cell}$, $-411 \times 10^3\ J = -(6\ mol\ e^-)(96{,}485\ C/mol\ e^-)(E^\circ_{cell})$, $E^\circ_{cell} = 0.71$ V

$E^\circ_{cell} = -E^\circ_a - 1.66\ V = 0.71\ V$, $-E^\circ_a = 2.37$ or $E^\circ_c = -2.37$

From Table 11.1, the reduction potential for $Mg^{2+} + 2\ e^- \rightarrow Mg$ is -2.37 V which fits the data. Hence, the metal is magnesium.

85. a. $3\ e^- + 4\ H^+ + NO_3^- \rightarrow NO + 2\ H_2O$ $\qquad E^\circ = 0.96$ V

Nitric acid can oxidize Co to Co^{2+} ($E^\circ_{cell} > 0$), but is not strong enough to oxidize Co to Co^{3+} ($E^\circ_{cell} < 0$). Co^{2+} is the primary product assuming standard conditions.

b. Concentrated nitric acid is about 16 mol/L. $[H^+] = [NO_3^-] = 16\ M$; Assume $P_{NO} = 1$ atm

$$E = 0.96\ V - \frac{0.0591}{3}\log\frac{P_{NO}}{[H^+]^4\,[NO_3^-]} = 0.96 - \frac{0.0591}{3}\log\frac{1}{(16)^5} = 0.96 - (-0.12) = 1.08\ V$$

No, concentrated nitric acid will still only be able to oxidize Co to Co^{2+}.

87. a. $E_{cell} = E_{ref} + 0.05916\ pH$, $0.480\ V = 0.250\ V + 0.05916\ pH$

$$pH = \frac{0.480 - 0.250}{0.05916} = 3.888;\ \text{Uncertainty} = \pm 1\ mV = \pm 0.001\ V$$

$$pH_{max} = \frac{0.481 - 0.250}{0.05916} = 3.905;\quad pH_{min} = \frac{0.479 - 0.250}{0.05916} = 3.871$$

So if the uncertainty in potential is ± 0.001 V, then the uncertainty in pH is ± 0.017 or about ± 0.02 pH units. For this measurement, $[H^+] = 10^{-3.888} = 1.29 \times 10^{-4}\ M$. For an error of +1 mV, $[H^+] = 10^{-3.905} = 1.24 \times 10^{-4}\ M$. For an error of -1 mV, $[H^+] = 10^{-3.871} = 1.35 \times 10^{-4}\ M$. So the uncertainty in $[H^+]$ is $\pm 0.06 \times 10^{-4}\ M = \pm 6 \times 10^{-6}\ M$.

b. From part a, we will be within ± 0.02 pH units if we measure the potential to the nearest ± 0.001 V (1 mV).

89. $2\ Ag^+(aq) + Cu(s) \rightarrow Cu^{2+}(aq) + 2\ Ag(s)$ $\quad E^\circ_{cell} = 0.80 - 0.34\ V = 0.46\ V$; A galvanic cell produces a voltage as the forward reaction occurs. Any stress that increases the tendency of the forward reaction to occur will increase the cell potential, while a stress that decreases the tendency of the forward reaction to occur will decrease the cell potential.

a. Added Cu^{2+} (a product ion) will decrease the tendency of the forward reaction to occur which will decrease the cell potential.

b. Added NH_3 removes Cu^{2+} in the form of $Cu(NH_3)_4^{2+}$. As a product ion is removed, this will increase the tendency of the forward reaction to occur which will increase the cell potential.

c. Added Cl^- removes Ag^+ in the form of AgCl(s). As a reactant ion is removed, this will decrease the tendency of the forward reaction to occur which will decrease the cell potential.

d. $Q_1 = \frac{[Cu^{2+}]_o}{[Ag^+]_o^2}$; As the volume of solution is doubled, each concentration is halved.

$$Q_2 = \frac{1/2\ [Cu^{2+}]_o}{(1/2\ [Ag^+]_o)^2} = \frac{2[Cu^{2+}]_o}{[Ag^+]_o^2} = 2\ Q_1$$

The reaction quotient is doubled as the concentrations are halved. Since reactions are spontaneous when Q < K and since Q increases when the solution volume doubles, then the reaction is closer to equilibrium which will decrease the cell potential.

e. Since Ag(s) is not a reactant in this spontaneous reaction and since solids do not appear in the reaction quotient expressions, then replacing the silver electrode with a platinum electrode will have no effect on the cell potential.

Challenge Problems

91. a. $Zn(s) + Cu^{2+}(aq) \rightarrow Zn^{2+}(aq) + Cu(s)$ $E^\circ_{cell} = 1.10$ V; $E_{cell} = 1.10\ V - \frac{0.0591}{2}\log\frac{[Zn^{2+}]}{[Cu^{2+}]}$

$$E_{cell} = 1.10\ V - \frac{0.0591}{2}\log\frac{0.10}{2.50} = 1.10\ V - (-0.041\ V) = 1.14\ V$$

b. $10.0\ h \times \frac{60\ min}{h} \times \frac{60\ s}{min} \times \frac{10.0\ C}{s} \times \frac{1\ mol\ e^-}{96{,}485\ C} \times \frac{1\ mol\ Cu}{2\ mol\ e^-} = 1.87$ mol Cu produced

The Cu^{2+} concentration decreases by 1.87 mol/L and the Zn^{2+} concentration will increase by 1.87 mol/L.

$[Cu^{2+}] = 2.50 - 1.87 = 0.63\ M$; $[Zn^{2+}] = 0.10 + 1.87 = 1.97\ M$

$$E_{cell} = 1.10\ V - \frac{0.0591}{2}\log\frac{1.97}{0.63} = 1.10\ V - 0.015\ V = 1.09\ V$$

c. 1.87 mol Zn consumed $\times \frac{65.38\ g\ Zn}{mol\ Zn} = 122$ g Zn; Mass of electrode = 200. - 122 = 78 g Zn

1.87 mol Cu formed $\times \frac{63.55\ g\ Cu}{mol\ Cu} = 119$ g Cu; Mass of electrode = 200. + 119 = 319 g Cu

d. Three things could possibly cause this battery to go dead:

1. All of the Zn is consumed.
2. All of the Cu^{2+} is consumed.
3. Equilibrium is reached ($E_{cell} = 0$).

We began with 2.50 mol Cu^{2+} and 200. g Zn × 1 mol Zn/65.38 g Zn = 3.06 mol Zn. Cu^{2+} is the limiting reagent and will run out first. To react all the Cu^{2+} requires:

$$2.50 \text{ mol Cu}^{2+} \times \frac{2 \text{ mol e}^-}{\text{mol Cu}^{2+}} \times \frac{96{,}485 \text{ C}}{\text{mol e}^-} \times \frac{1 \text{ s}}{10.0 \text{ C}} \times \frac{1 \text{ h}}{3600 \text{ s}} = 13.4 \text{ h}$$

For equilibrium to be reached: $E = 0 = 1.10 \text{ V} - \frac{0.0591}{2} \log \frac{[Zn^{2+}]}{[Cu^{2+}]}$

$$\frac{[Zn^{2+}]}{[Cu^{2+}]} = K = 10^{2(1.10)/0.0591} = 1.68 \times 10^{37}$$

This is such a large equilibrium constant that virtually all of the Cu^{2+} must react to reach equilibrium. So, the battery will go dead in 13.4 hours.

93. a. $E_{meas} = E_{ref} - 0.05916 \log [F^-]$, $0.4462 = 0.2420 - 0.05916 \log [F^-]$

$\log [F^-] = -3.4517$, $[F^-] = 3.534 \times 10^{-4}\ M$

b. $pH = 9.00$; $pOH = 5.00$; $[OH^-] = 1.0 \times 10^{-5}\ M$

$0.4462 = 0.2420 - 0.05916 \log [[F^-] + 10.0(1.0 \times 10^{-5})]$

$\log ([F^-] + 1.0 \times 10^{-4}) = -3.452$, $[F^-] + 1.0 \times 10^{-4} = 3.532 \times 10^{-4}$, $[F^-] = 2.5 \times 10^{-4}\ M$

True value is 2.5×10^{-4} and by ignoring the $[OH^-]$ we would say $[F^-]$ was 3.5×10^{-4}, so:

$$\% \text{ Error} = \frac{1.0 \times 10^{-4}}{2.5 \times 10^{-4}} \times 100 = 40.\ \%$$

c. $[F^-] = 2.5 \times 10^{-4}\ M$; $\frac{[F^-]}{k[OH^-]} = 50. = \frac{2.5 \times 10^{-4}}{10.0[OH^-]}$

$[OH^-] = \frac{2.5 \times 10^{-4}}{10. \times 50.} = 5.0 \times 10^{-7}\ M$; $pOH = 6.30$; $pH = 7.70$

d. $HF \rightleftharpoons H^+ + F^-$ $K_a = \frac{[H^+][F^-]}{[HF]} = 7.2 \times 10^{-4}$; **If 99% is F^-, then $[F^-]/[HF] = 99$.**

$99[H^+] = 7.2 \times 10^{-4}$, $[H^+] = 7.3 \times 10^{-6}\ M$; $pH = 5.14$

e. **The buffer controls the pH so that there is little HF present and that there is little response to OH^-. Typically a buffer of pH = 6.0 is used.**

95. a.

$$3 \times (e^- + 2\,H^+ + NO_3^- \rightarrow NO_2 + H_2O) \qquad E^\circ_c = 0.775 \text{ V}$$
$$2\,H_2O + NO \rightarrow NO_3^- + 4\,H^+ + 3\,e^- \qquad -E^\circ_a = -0.957 \text{ V}$$
$$2\,H^+(aq) + 2\,NO_3^-(aq) + NO(g) \rightarrow 3\,NO_2(g) + H_2O(l) \qquad E^\circ_{cell} = -0.182 \text{ V} \quad K = ?$$

$$\log K = \frac{nE^\circ}{0.0591} = \frac{3(-0.182)}{0.0591} = -9.239,\ K = 10^{-9.239} = 5.77 \times 10^{-10}$$

b. Let C = concentration of $HNO_3 = [H^+] = [NO_3^-]$

$$5.77 \times 10^{-10} = \frac{P_{NO_2}^3}{P_{NO} \times [H^+]^2 \times [NO_3^-]^2} = \frac{P_{NO_2}^3}{P_{NO} \times C^4}$$

If 0.20 mol % NO_2 and P_{tot} = 1.00 atm:

$$P_{NO_2} = \frac{0.20 \text{ mol } NO_2}{100. \text{ mol total}} \times 1.00 \text{ atm} = 2.0 \times 10^{-3} \text{ atm}; \; P_{NO} = 1.00 - 0.0020 = 1.00 \text{ atm}$$

$$5.77 \times 10^{-10} = \frac{(2.0 \times 10^{-3})^3}{(1.00)\, C^4}, \; C = 1.9\, M \; HNO_3$$

97.

$$2 H^+ + 2 e^- \rightarrow H_2 \qquad E_c^\circ = 0.000 \text{ V}$$

$$Fe \rightarrow Fe^{2+} + 2e^- \qquad -E_a^\circ = -(-0.440\text{V})$$

$$2 H^+(aq) + Fe(s) \rightarrow H_2(g) + Fe^{3+}(aq) \qquad E_{cell}^\circ = 0.440 \text{ V}$$

$$E_{cell} = E_{cell}^\circ - \frac{0.0591}{n} \log Q, \text{ where n = 2 and } Q = \frac{P_{H_2} \times [Fe^{3+}]}{[H^+]^2}$$

To determine K_a for the weak acid, first use the electrochemical data to determine the H^+ concentration in the half-cell containing the weak acid.

$$0.333\text{V} = 0.440 \text{ V} - \frac{0.0591}{2} \log \frac{1.00 \text{ atm}\,(1.00 \times 10^{-3} \text{ M})}{[H^+]^2}$$

$$\frac{0.107(2)}{0.0591} = \log \frac{1.00 \times 10^{-3}}{[H^+]^2}, \quad \frac{1.00 \times 10^{-3}}{[H^+]^2} = 10^{3.621} = 4.18 \times 10^3, \quad [H^+] = 4.89 \times 10^{-4}\, M$$

Now we can solve for the K_a value of the weak acid HA through the normal set-up for a weak acid problem.

	HA	⇌	H^+	+	A^-
Initial	1.00 *M*		~0		0
Equil.	1.00 - *x*		*x*		*x*

$$K_a = \frac{[H^+][A^-]}{[HA]}$$

$$K_a = \frac{x^2}{1.00 - x} \text{ where } x = [H^+] = 4.89 \times 10^{-4}\, M, \; K_a = \frac{(4.89 \times 10^{-4})^2}{1.00 - 4.89 \times 10^{-4}} = 2.39 \times 10^{-7}$$

99. a.

$$(Ag^+ + e^- \rightarrow Ag) \times 2 \qquad E_c^\circ = 0.80 \text{ V}$$

$$Cu \rightarrow Cu^{2+} + 2 e^- \qquad -E_a^\circ = -0.34 \text{ V}$$

$$2 Ag^+(aq) + Cu(s) \rightarrow 2 Ag(s) + Cu^{2+}(aq) \qquad E_{cell}^\circ = 0.46 \text{ V}$$

$$E_{cell} = E_{cell}^\circ - \frac{0.0591}{n} \log Q \text{ where n = 2 and } Q = \frac{[Cu^{2+}]}{[Ag^+]^2}$$

To calculate E_{cell}, we need to use the K_{sp} data to determine $[Ag^+]$.

	$AgCl(s)$	$\rightleftharpoons$	$Ag^+(aq)$	+	$Cl^-(aq)$	$K_{sp} = 1.6 \times 10^{-10} = [Ag^+][Cl^-]$
Initial	s = solubility (mol/L)		0		0	
Equil.			s		s	

$K_{sp} = 1.6 \times 10^{-10} = s^2$, $s = [Ag^+] = 1.3 \times 10^{-5}$ mol/L

$$E_{cell} = 0.46\text{ V} - \frac{0.0591}{2}\log\frac{2.0}{(1.3 \times 10^{-5})^2} = 0.46\text{ V} - 0.30 = 0.16\text{ V}$$

b. $Cu^{2+}(aq) + 4\,NH_3(aq) \rightleftharpoons Cu(NH_4)_4^{2+}(aq)$ $\quad K = 1.0 \times 10^{13} = \dfrac{[Cu(NH_3)_4^{2+}]}{[Cu^{2+}][NH_3]^4}$

Since K is very large for the formation of $Cu(NH_3)_4^{2+}$, then the forward reaction is dominant. At equilibrium, essentially all of the 2.0 *M* Cu^{2+} will react to form 2.0 *M* $Cu(NH_3)_4^{2+}$. This reaction requires 8.0 *M* NH_3 to react with all of the Cu^{2+} in the balanced equation. Therefore, the mol of NH_3 added to 1.0 L solution will be larger than 8.0 mol since some NH_3 must be present at equilibrium. In order to calculate how much NH_3 is present at equilibrium, we need to use the electrochemical data to determine the Cu^{2+} concentration.

$$E_{cell} = E^\circ_{cell} - \frac{0.0591}{n}\log Q,\quad 0.52\text{ V} = 0.46\text{ V} - \frac{0.0591}{2}\log\frac{[Cu^{2+}]}{(1.3 \times 10^{-5})^2}$$

$$\log\frac{[Cu^{2+}]}{(1.3 \times 10^{-5})^2} = \frac{-0.06(2)}{0.0591} = -2.03,\quad \frac{[Cu^{2+}]}{(1.3 \times 10^{-5})^2} = 10^{-2.03} = 9.3 \times 10^{-3}$$

$[Cu^{2+}] = 1.6 \times 10^{-12} = 2 \times 10^{-12}$ *M* (We carried extra significant figures in the calculation.)

Note: Our assumption that the 2.0 *M* Cu^{2+} essentially reacts to completion is excellent as only 2×10^{-12} *M* Cu^{2+} remains after this reaction. Now we can solve for the equilibrium $[NH_3]$.

$$K = 1.0 \times 10^{13} = \frac{[Cu(NH_3)_4^{2+}]}{[Cu^{2+}][NH_3]^4} = \frac{(2.0)}{(2 \times 10^{-12})[NH_3]^4},\quad [NH_3] = 0.6\ M$$

Since 1.0 L of solution is present, then 0.6 mol NH_3 remains at equilibrium. The total mol of NH_3 added is 0.6 mol plus the 8.0 mol NH_3 necessary to form 2.0 *M* $Cu(NH_3)_4^{2+}$. Therefore, 8.0 + 0.6 = 8.6 mol NH_3 were added.

101. a. From Table 11.1: $2\,H_2O + 2\,e^- \rightarrow H_2 + 2\,OH^-$ $\quad E^\circ = -0.83$ V

$$E^\circ_{cell} = E^\circ_{H_2O} - E^\circ_{Zr} = -0.83\text{ V} + 2.36\text{ V} = 1.53\text{ V}$$

Yes, the reduction of H_2O to H_2 by Zr is spontaneous at standard conditions since $E^\circ_{cell} > 0$.

b. $(2\ H_2O + 2\ e^- \rightarrow H_2 + 2\ OH^-) \times 2$

$Zr + 4\ OH^- \rightarrow ZrO_2 \bullet H_2O + H_2O + 4\ e^-$

$3\ H_2O(l) + Zr(s) \rightarrow 2\ H_2(g) + ZrO_2 \bullet H_2O(s)$

c. $\Delta G° = -nFE° = -(4\text{ mol e}^-)(96{,}485\text{ C/mol e}^-)(1.53\text{ J/C}) = -5.90 \times 10^5\text{ J} = -590.\text{ kJ}$

$E = E° - \frac{0.0591}{n}\log Q$; At equilibrium, E = 0 and Q = K.

$E° = \frac{0.0591}{n}\log K,\ \log K = \frac{4(1.53)}{0.0591} = 104,\ K \approx 10^{104}$

d. $1.00 \times 10^3\text{ kg Zr} \times \frac{1000\text{ g}}{\text{kg}} \times \frac{1\text{ mol Zr}}{91.22\text{ g Zr}} \times \frac{2\text{ mol } H_2}{\text{mol Zr}} = 2.19 \times 10^4\text{ mol } H_2$

$2.19 \times 10^4\text{ mol } H_2 \times \frac{2.016\text{ g } H_2}{\text{mol } H_2} = 4.42 \times 10^4\text{ g } H_2$

$V = \frac{nRT}{P} = \frac{(2.19 \times 10^4\text{ mol})(0.08206\text{ L atm mol}^{-1}\text{K}^{-1})(1273\text{ K})}{1.0\text{ atm}} = 2.3 \times 10^6\text{ L } H_2$

e. Probably yes; Less radioactivity overall was released by venting the H_2 than what would have been released if the H_2 had exploded inside the reactor (as happened at Chernobyl). Neither alternative is pleasant, but venting the radioactive hydrogen is the less unpleasant of the two alternatives.

Marathon Problems

103. Begin by choosing any reduction potential as 0.00 V. For example, let's assume

$B^{2+} + 2\text{ e-} \rightarrow B\quad E° = 0.00\text{ V}$

Now, since, for example, when B/B^{2+}; E/E^{2+} are together as a cell, E = 0.81 V.

$E^{2+} + 2\ e^- \rightarrow E$ must have a potential of -0.81 V or 0.81 V.

Using this type of reasoning, we can get the table:

$B^{2+} + 2\ e^- \rightarrow B$ 0.00 V

$E^{2+} + 2\ e^- \rightarrow E$ -0.81 V

$D^{2+} + 2\ e^- \rightarrow D$ 0.19 V

$C^{2+} + 2\ e^- \rightarrow C$ -0.94 V

$A^{2+} + 2\ e^- \rightarrow A$ -0.53 V

Arrange this from largest to smallest:

$D^{2+} + 2\ e^- \rightarrow D$ 0.19 V

$B^{2+} + 2\ e^- \rightarrow B$ 0.00 V

$A^{2+} + 2\ e^- \rightarrow A$ -0.53 V

$E^{2+} + 2\ e^- \rightarrow E$ -0.81 V

$C^{2+} + 2\ e^- \rightarrow C$ -0.94 V

So, $A^{2+} + 2\ e^- \rightarrow A$ is in the middle. Let's call this 0.00 V. We get:

$D^{2+} + 2\ e^- \rightarrow D$ 0.72 V

$B^{2+} + 2\ e^- \rightarrow B$ 0.53 V

$A^{2+} + 2\ e^- \rightarrow A$ 0.00 V

$E^{2+} + 2\ e^- \rightarrow E$ -0.28 V

$C^{2+} + 2\ e^- \rightarrow C$ -0.41 V

Of course, we can also get:

$C^{2+} + 2\ e^- \rightarrow C$ 0.41 V

$E^{2+} + 2\ e^- \rightarrow E$ 0.28 V

$A^{2+} + 2\ e^- \rightarrow A$ 0.00 V

$B^{2+} + 2\ e^- \rightarrow B$ -0.53 V

$D^{2+} + 2\ e^- \rightarrow D$ -0.72 V

One way to determine which table is correct is to add metal C to a solution with D^{2+} and metal D to a solution with C^{2+}. If D comes out of solution, the first table is correct. If C comes out of solution, the second table is correct.

CHAPTER TWELVE

QUANTUM MECHANICS AND ATOMIC THEORY

Light and Matter

21. **Planck found that heated bodies only give off certain frequencies of light. Einstein's analysis of the photoelectric effect used Planck's concepts suggesting that electromagnetic radiation is "quantized".**

23. $\nu = \frac{c}{\lambda} = \frac{3.00 \times 10^8 \text{ m/s}}{1.0 \times 10^{-2} \text{ m}} = 3.0 \times 10^{10} \text{ s}^{-1}$

$E = h\nu = 6.63 \times 10^{-34} \text{ J s} \times 3.0 \times 10^{10} \text{ s}^{-1} = 2.0 \times 10^{-23}$ J/photon

$$\frac{2.0 \times 10^{-23} \text{ J}}{\text{photon}} \times \frac{6.02 \times 10^{23} \text{ photons}}{\text{mol}} = 12 \text{ J/mol}$$

25. $99.5 \text{ MHz} = 99.5 \times 10^6 \text{ Hz} = 99.5 \times 10^6 \text{ s}^{-1};\ \lambda = \frac{c}{\nu} = \frac{2.998 \times 10^8 \text{ m/s}}{99.5 \times 10^6 \text{ s}^{-1}} = 3.01 \text{ m}$

27. The energy needed to remove a single electron is:

$$\frac{279.7 \text{ kJ}}{\text{mol}} \times \frac{1 \text{ mol}}{6.0221 \times 10^{23}} = 4.645 \times 10^{-22} \text{ kJ} = 4.645 \times 10^{-19} \text{ J}$$

$$E = \frac{hc}{\lambda},\ \lambda = \frac{hc}{E} = \frac{6.6261 \times 10^{-34} \text{ J s} \times 2.9979 \times 10^8 \text{ m/s}}{4.645 \times 10^{-19} \text{ J}} = 4.277 \times 10^{-7} \text{ m} = 427.7 \text{ nm}$$

29. The energy to remove a single electron is:

$$\frac{208.4 \text{ kJ}}{\text{mol}} \times \frac{1 \text{ mol}}{6.022 \times 10^{23}} = 3.461 \times 10^{-22} \text{ kJ} = 3.461 \times 10^{-19} \text{ J} = E_w$$

Energy of 254 nm light is:

$$E = \frac{hc}{\lambda} = \frac{(6.626 \times 10^{-34} \text{ J s})(2.998 \times 10^8 \text{ m/s})}{254 \times 10^{-9} \text{ m}} = 7.82 \times 10^{-19} \text{ J}$$

$E_{photon} = E_K + E_w,\ E_K = 7.82 \times 10^{-19} \text{ J} - 3.461 \times 10^{-19} \text{ J} = 4.36 \times 10^{-19} \text{ J} =$ maximum KE

Hydrogen Atom: The Bohr Model

31. For the H-atom (Z = 1): $E_n = -2.178 \times 10^{-18}$ J/n^2; For a spectral transition, $\Delta E = E_f - E_i$:

$$\Delta E = -2.178 \times 10^{-18}\ \text{J}\left(\frac{1}{n_f^2} - \frac{1}{n_i^2}\right)$$

where n_i and n_f are the levels of the initial and final states, respectively. A positive value of ΔE always corresponds to an absorption of light and a negative value of ΔE always corresponds to an emission of light.

a. $$\Delta E = -2.178 \times 10^{-18}\ \text{J}\left(\frac{1}{2^2} - \frac{1}{3^2}\right) = -2.178 \times 10^{-18}\ \text{J}\left(\frac{1}{4} - \frac{1}{9}\right)$$

$$\Delta E = -2.178 \times 10^{-18}\ \text{J} \times (0.2500 - 0.1111) = -3.025 \times 10^{-19}\ \text{J}$$

The photon of light must have precisely this energy (3.025×10^{-19} J).

$$|\Delta E| = E_{photon} = h\nu = \frac{hc}{\lambda} \text{ or } \lambda = \frac{hc}{|\Delta E|} = \frac{6.6261 \times 10^{-34}\ \text{J s} \times 2.9979 \times 10^8\ \text{m/s}}{3.025 \times 10^{-19}\ \text{J}}$$

$$= 6.567 \times 10^{-7}\ \text{m} = 656.7\ \text{nm}$$

From Figure 12.3, this is visible electromagnetic radiation (red light).

b. $$\Delta E = -2.178 \times 10^{-18}\ \text{J}\left(\frac{1}{2^2} - \frac{1}{4^2}\right) = -4.084 \times 10^{-19}\ \text{J}$$

$$\lambda = \frac{hc}{|\Delta E|} = \frac{6.6261 \times 10^{-34}\ \text{J s} \times 2.9979 \times 10^8\ \text{m/s}}{4.084 \times 10^{-19}\ \text{J}} = 4.864 \times 10^{-7}\ \text{m} = 486.4\ \text{nm}$$

This is visible electromagnetic radiation (green-blue light).

c. $$\Delta E = -2.178 \times 10^{-18}\ \text{J}\left(\frac{1}{1^2} - \frac{1}{2^2}\right) = -1.634 \times 10^{-18}\ \text{J}$$

$$\lambda = \frac{6.6261 \times 10^{-34}\ \text{J s} \times 2.9979 \times 10^8\ \text{m/s}}{1.634 \times 10^{-18}\ \text{J}} = 1.216 \times 10^{-7}\ \text{m} = 121.6\ \text{nm}$$

This is ultraviolet electromagnetic radiation.

d. $\Delta E = -2.178 \times 10^{-18}\ J \left(\frac{1}{3^2} - \frac{1}{4^2}\right) = -1.059 \times 10^{-19}\ J$

$$\lambda = \frac{hc}{|\Delta E|} = \frac{6.6261 \times 10^{-34}\ J\ s \times 2.9979 \times 10^8\ m/s}{1.059 \times 10^{-19}\ J} = 1.876 \times 10^{-6}\ m \text{ or } 1876\ nm$$

This is infrared electromagnetic radiation.

33. a. False; It takes less energy to ionize an electron from n = 3 than from the ground state.

b. True

c. False; The energy difference between n = 3 and n = 2 is smaller than the energy difference between n = 3 and n = 1. Thus, the wavelength of light emitted is longer for the n = 3 to n = 2 electronic transition than for the n = 3 to n = 1 transition. E and λ are inversely proportional to each other ($E = hc/\lambda$).

d. True

e. False; The ground state in hydrogen is n = 1 and all other allowed energy states are called excited states; n = 2 is the first excited state and n = 3 is the second excited state.

35. $|\Delta E| = E_{photon} = \frac{hc}{\lambda} = \frac{6.6261 \times 10^{-34}\ J\ s \times 2.9979 \times 10^8\ m/s}{397.2 \times 10^{-9}\ m} = 5.001 \times 10^{-19}\ J$

$\Delta E = -5.001 \times 10^{-19}$ J since we have an emission.

$$-5.001 \times 10^{-19}\ J = E_2 - E_n = -2.178 \times 10^{-18}\ J \left(\frac{1}{2^2} - \frac{1}{n^2}\right)$$

$$0.2296 = \frac{1}{4} - \frac{1}{n^2},\quad \frac{1}{n^2} = 0.0204,\ n = 7$$

Wave Mechanics and Particle in a Box

37. a. $\lambda = \frac{h}{mv} = \frac{6.626 \times 10^{-34}\ J\ s}{(1.675 \times 10^{-27}\ kg)(0.0100 \times 2.998 \times 10^8\ m/s)} = 1.32 \times 10^{-13}\ m$

b. $\lambda = \frac{h}{mv},\ v = \frac{h}{\lambda m} = \frac{6.626 \times 10^{-34}\ J\ s}{(75 \times 10^{-12}\ m)(1.675 \times 10^{-27}\ kg)} = 5.3 \times 10^3\ m/s$

39. Only very small particles with a tiny mass exhibit wave and particle properties, e.g., an electron. Some evidence supporting the wave properties of matter are:

1) Electrons can be diffracted like light.

2) The electron microscope uses electrons in a fashion similar to the way in which light is used in a light microscope.

41. **a.** $\Delta p = m\Delta v = 9.11 \times 10^{-31}\ \text{kg} \times 0.100\ \text{m/s} = \dfrac{9.11 \times 10^{-32}\ \text{kg m}}{\text{s}}$

$$\Delta p \Delta x \geq \frac{h}{4\pi},\ \Delta x = \frac{h}{4\pi \Delta p} = \frac{6.626 \times 10^{-34}\ \text{J s}}{4 \times 3.142 \times (9.11 \times 10^{-32}\ \text{kg m/s})} = 5.79 \times 10^{-4}\ \text{m}$$

b. $\Delta x = \dfrac{h}{4\pi \Delta p} = \dfrac{6.626 \times 10^{-34}\ \text{J s}}{4 \times 3.142 \times 0.145\ \text{kg} \times 0.100\ \text{m/s}} = 3.64 \times 10^{-33}\ \text{m}$

c. **The diameter of an H atom is roughly 1.0×10^{-8} cm. The uncertainty in position is much larger than the size of the atom.**

d. **The uncertainty is insignificant compared to the size of a baseball.**

43. $E_n = \dfrac{n^2h^2}{8\ mL^2}$; $\Delta E = E_3 - E_2 = \dfrac{9\ h^2}{8\ mL^2} - \dfrac{4\ h^2}{8\ mL^2} = \dfrac{5\ h^2}{8\ mL^2}$

$$\Delta E = \frac{hc}{\lambda} = \frac{(6.626 \times 10^{-34}\ \text{J s})(2.998 \times 10^8\ \text{m/s})}{8080 \times 10^{-9}\ \text{m}} = 2.46 \times 10^{-20}\ \text{J}$$

$$\Delta E = 2.46 \times 10^{-20}\ \text{J} = \frac{5\ h^2}{8\ mL^2} = \frac{5(6.626 \times 10^{-34}\ \text{J s})^2}{8(9.109 \times 10^{-31}\ \text{kg})\ L^2},\ L = 3.50 \times 10^{-9}\ \text{m} = 3.50\ \text{nm}$$

45. $E_n = \dfrac{n^2h^2}{8\ mL^2}$; **As L increases, E_n will decrease and the spacing between energy levels will also decrease.**

47. $E_n = \dfrac{n^2h^2}{8\ mL^2}$, **n = 1 for ground state; From equation, as L increases, E_n decreases.**

Using numbers: 10^{-6} m box: $E_1 = \dfrac{h^2}{8\ m}(10^{12}\ \text{m}^{-2})$; 10^{-10} m box: $E_1 = \dfrac{h^2}{8\ m}(10^{20}\ \text{m}^{-2})$

As expected, the electron in the 10^{-6} m box has the lowest ground state energy.

Orbitals and Quantum Numbers

49. **The possible values for n, ℓ and m_ℓ are: n = 1, 2, 3, ... ; ℓ = 0, 1, 2, ... (n - 1); $m_\ell = -\ell$ ... -2, -1, 0, 1, 2, ...$+\ell$;**

1p: n = 1, ℓ = 1 is not possible; 3f: n = 3, ℓ = 3 is not possible; 2d: n = 2, ℓ = 2 is not possible; In all three incorrect cases, n = ℓ. The maximum value ℓ can have is n - 1, not n.

51. No, for n = 2, the allowed values of ℓ are 0 and 1; for n = 3 the allowed values of ℓ are 0, 1, and 2.

53. 1p, 0 electrons ($\ell \neq 1$ when $n = 1$); $6d_{x^2-y^2}$, 2 electrons (specifies one atomic orbital); 4f, 14 electrons (7 orbitals have 4f designation); $7p_y$, 2 electrons (specifies one atomic orbital); 2s, 2 electrons (specifies one atomic orbital); $n = 3$, 18 electrons (3s, 3p and 3d orbitals are possible; there are one 3s orbital, three 3p orbitals and five 3d orbitals).

55. The diagrams of the orbitals in the text only give 90% probabilities of where the electron may reside. We can never be 100% certain of the path of the electrons due to Heisenburg's uncertainty principle.

57. A nodal surface in an atomic orbital is a surface in which the probability of finding an electron is zero.

59. For $r = a_o$ and $\theta = 0°$ (Z = 1 for H):

$$\psi_{2p_z} = \frac{1}{4(2\pi)^{1/2}}\left(\frac{1}{5.29 \times 10^{-11}}\right)^{3/2} (1)\, e^{-1/2} \cos 0 = 1.57 \times 10^{14};\ \psi^2 = 2.46 \times 10^{28}$$

For $r = a_o$ and $\theta = 90°$: $\psi_{2p_z} = 0$ since $\cos 90° = 0$; $\psi^2 = 0$

Polyelectronic Atoms

61. The electrostatic energy of repulsion from Coulomb's Law will be of the form Q^2/r where Q is the charge of the electron and r is the distance between the two electrons. From the Heisenberg uncertainty principle, we cannot know precisely the path of each electron. Thus, we cannot precisely know the distance between the electrons nor the value of the electrostatic repulsions.

63. The size of the 1s orbitals would be proportional to 1/Z, that is, as Z increases, the electrons are more strongly attracted to the nucleus and will be drawn in closer. Thus the relative sizes would be:

$$H : He^{+} : Li^{2+} : C^{5+} : Fe^{25+} \rightarrow 1 : \frac{1}{2} : \frac{1}{3} : \frac{1}{6} : \frac{1}{26}$$

65. No, the spin is a convenient model. Since we cannot know the exact path of the electron, we cannot determine if it is spinning.

67. Si: $1s^2 2s^2 2p^6 3s^2 3p^2$ or $[Ne]3s^2 3p^2$; Ga: $1s^2 2s^2 2p^6 3s^2 3p^6 4s^2 3d^{10} 4p^1$ or $[Ar]4s^2 3d^{10} 4p^1$

As: $[Ar]4s^2 3d^{10} 4p^3$; Ge: $[Ar]4s^2 3d^{10} 4p^2$; Al: $[Ne]3s^2 3p^1$; Cd: $[Kr]5s^2 4d^{10}$

S: $[Ne]3s^2 3p^4$; Se: $[Ar]4s^2 3d^{10} 4p^4$

69. The following are complete electron configurations. Noble gas shorthand notation could also be used.

Sc: $1s^22s^22p^63s^23p^64s^23d^1$; Fe: $1s^22s^22p^63s^23p^64s^23d^6$

P: $1s^22s^22p^63s^23p^3$; Cs: $1s^22s^22p^63s^23p^64s^23d^{10}4p^65s^24d^{10}5p^66s^1$

Eu: $1s^22s^22p^63s^23p^64s^23d^{10}4p^65s^24d^{10}5p^66s^24f^65d^1$*

Pt: $1s^22s^22p^63s^23p^64s^23d^{10}4p^65s^24d^{10}5p^66s^24f^{14}5d^8$*

Xe: $1s^22s^22p^63s^23p^64s^23d^{10}4p^65s^24d^{10}5p^6$; Br: $1s^22s^22p^63s^23p^64s^23d^{10}4p^5$

*Note: These electron configurations were written down using only the periodic table.

Actual electron configurations are: Eu: $[Xe]6s^24f^7$ and Pt: $[Xe]6s^14f^{14}5d^9$

71. Exceptions: Cr, Cu, Nb, Mo, Tc, Ru, Rh, Pd, Ag, Pt, Au; Tc, Ru, Rh, Pd and Pt do not correspond to the supposed extra stability of half-filled and filled subshells.

73. There is a higher probability of finding the 4s electron very close to the nucleus than that for the 3d electron.

75. We get the number of unpaired electrons by examining the incompletely filled subshells. The paramagnetic substances have unpaired electrons, and the ones with no unpaired electrons are not paramagnetic (they are called diamagnetic).

Li: $1s^22s^1$ $\underbrace{\underline{\uparrow}}_{2s}$; Paramagnetic with 1 unpaired electron.

N: $1s^22s^22p^3$ $\underbrace{\underline{\uparrow}\ \underline{\uparrow}\ \underline{\uparrow}}_{2p}$; Paramagnetic with 3 unpaired electrons.

Ni: $[Ar]4s^23d^8$ $\underbrace{\underline{\uparrow\downarrow}\ \underline{\uparrow\downarrow}\ \underline{\uparrow\downarrow}\ \underline{\uparrow}\ \underline{\uparrow}}_{3d}$; Paramagnetic with 2 unpaired electrons.

Te: $[Kr]5s^24d^{10}5p^4$ $\underbrace{\underline{\uparrow\downarrow}\ \underline{\uparrow}\ \underline{\uparrow}}_{5p}$; Paramagnetic with 2 unpaired electrons.

Ba: $[Xe]6s^2$ $\underbrace{\underline{\uparrow\downarrow}}_{6s}$; Not paramagnetic since no unpaired electrons.

Hg: $[Xe]6s^24f^{14}5d^{10}$ $\underline{\uparrow\downarrow}\ \underline{\uparrow\downarrow}\ \underline{\uparrow\downarrow}\ \underline{\uparrow\downarrow}\ \underline{\uparrow\downarrow}$; Not paramagnetic since no unpaired electrons.

77. We get the number of unpaired electrons by examining the incompletely filled subshells.

O: $[He]2s^22p^4$	$2p^4$:	$\underline{\uparrow\downarrow}\ \underline{\uparrow}\ \underline{\uparrow}$	two unpaired e^-
O^+: $[He]2s^22p^3$	$2p^3$:	$\underline{\uparrow}\ \underline{\uparrow}\ \underline{\uparrow}$	three unpaired e^-
O^-: $[He]2s^22p^5$	$2p^5$:	$\underline{\uparrow\downarrow}\ \underline{\uparrow\downarrow}\ \underline{\uparrow}$	one unpaired e^-
Os: $[Xe]6s^24f^{14}5d^6$	$5d^6$:	$\underline{\uparrow\downarrow}\ \underline{\uparrow}\ \underline{\uparrow}\ \underline{\uparrow}\ \underline{\uparrow}$	four unpaired e^-

Zr: $[Kr]5s^24d^2$ $4d^2$: ↑ ↑ _ _ _ two unpaired e^-

S: $[Ne]3s^23p^4$ $3p^4$: ↑↓ ↑ ↑ two unpaired e^-

F: $[He]2s^22p^5$ $2p^5$: ↑↓ ↑↓ ↑ one unpaired e^-

Ar: $[Ne]3s^23p^6$ $3p^6$ ↑↓ ↑↓ ↑↓ zero unpaired e^-

The Periodic Table and Periodic Properties

79. In general, size decreases from left to right across a period and size increases in going down a group.

a. Be < Mg < Ca b. Xe < I < Te c. Ge < Ga < In

d. Be < Na < Rb e. Ne < Se < Sr

81. a. Ba b. K

c. O; In general, group 6A elements have a lower ionization energy than neighboring group 5A elements. This is an exception to the general ionization energy trend across the periodic table.

d. S^{2-}; This ion has the most electrons as compared to the other sulfur species present. S^{2-} has the largest amount of electron-electron repulsions which leads to S^{2-} having the largest size and smallest ionization energy.

e. Cs; This follows the general ionization energy trend.

83. The outermost electrons are the valence electrons. When atoms interact with each other, it will be the outermost electrons that are involved in these interactions. In addition, how tightly the nucleus holds these outermost electrons determines atomic size, ionization energy and other properties of atoms.

Elements in the same group have similar valence electron configurations and, as a result, have similar chemical properties.

85. As: $[Ar]4s^23d^{10}4p^3$; Se: $[Ar]4s^23d^{10}4p^4$; The general ionization energy trend predicts that Se should have a higher ionization energy than As. Se is an exception to the general ionization energy trend. There are extra electron-electron repulsions in Se because two electrons are in the same 4p orbital, resulting in a lower ionization energy for Se than predicted.

87. Size also decreases going across a period. Sc and Ti along with Y and Zr are adjacent elements. There are 14 elements (the lanthanides) between La and Hf, making Hf considerably smaller.

89. a. Uus will have 117 electrons. $[Rn]7s^25f^{14}6d^{10}7p^5$

b. It will be in the halogen family and will be most similar to astatine, At.

c. Like the other halogens: NaUus, $Mg(Uus)_2$, $C(Uus)_4$, $O(Uus)_2$

d. Like the other halogens: $UusO^-$, $UusO_2^-$, $UusO_3^-$, $UusO_4^-$

91. Electron-electron repulsions become important when we try to add electrons to an atom. From the standpoint of electron-electron repulsions, larger atoms would have more favorable (more exothermic) electron affinities. Considering only electron-nucleus attractions, smaller atoms would be expected to have the more favorable (more exothermic) EA values. These two factors are the opposite of each other. Thus, the overall variation in EA is not as great as ionization energy in which attractions to the nucleus dominate.

93. O; The electron-electron repulsions will be much more severe for $O^- + e^- \rightarrow O^{2-}$ than for $O + e^- \rightarrow O^-$.

95. a. $P(g) \rightarrow P^+(g) + e^-$; IE refers to atoms in the gas phase. b. $P(g) + e^- \rightarrow P^-(g)$

97. As successive electrons are removed, the net positive charge on the resultant ion increases. This increase in positive charge binds the remaining electrons more firmly, and the ionization energy increases.

The electron configuration for Si is $1s^22s^22p^63s^23p^2$. There is a large jump in ionization energy when going from the removal of valence electrons to the removal of core electrons. For silicon, this occurs when the fifth electron is removed since we go from the valence electrons in $n = 3$ to the core electrons in $n = 2$. There should be another big jump when the thirteenth electron is removed, i.e., when the 1s electrons are removed.

The Alkali Metals

99. Yes; the ionization energy general trend is to decrease down a group, and the atomic radius trend is to increase down a group. The data in Table 12.9 confirm both of these general trends.

101. a. $4\ Li(s) + O_2(g) \rightarrow 2\ Li_2O(s)$ b. $2\ K(s) + S(s) \rightarrow K_2S(s)$

c. $2\ Cs(s) + 2\ H_2O(l) \rightarrow 2\ CsOH(aq) + H_2(g)$ d. $2\ Na(s) + Cl_2(g) \rightarrow 2\ NaCl(s)$

103. a. Carbonate ion is CO_3^{2-}. Lithium form Li^+ ions. Thus, lithium carbonate is Li_2CO_3.

b. $\frac{1 \times 10^{-3}\ \text{mol Li}}{\text{L blood}} \times \frac{6.9\ \text{g Li}}{\text{mol Li}} = \frac{7 \times 10^{-3}\ \text{g Li}}{\text{L blood}}$

105. It should be element #119 with ground state electron configuration: $[Rn]\ 7s^25f^{14}6d^{10}7p^68s^1$

Additional Exercises

107. a. n b. n and ℓ

109. Ionization energy is for removal of the electron from the atom in the gas phase. The work function is for the removal of an electron from a solid.

$M(g) \rightarrow M^+(g) + e^-$ $\Delta H = IE$; $M(s) \rightarrow M^+(s) + e^-$ $\Delta H =$ work function

111. Yes, the maximum number of unpaired electrons in any configuration corresponds to a minimum in electron-electron repulsions.

113. Expected order from IE trend: Li < Be < B < C < N < O < F < Ne

B and O are out of order. The IE of O is lower because of the extra electron-electron repulsions present when two electrons are paired in the same orbital. B is out of order because of the smaller penetrating ability of the 2p electron in B as compared to the 2s electrons in Be.

115. $60 \times 10^6\ \text{km} \times \dfrac{1000\ \text{m}}{\text{km}} \times \dfrac{1\ \text{s}}{3.00 \times 10^8\ \text{m}} = 200\ \text{s}$ (about 3 minutes)

117. $n = 5$; $m_\ell = -4, -3, -2, -1, 0, 1, 2, 3, 4$; 18 electrons since there are 9 degenerate g orbitals.

119. a.

$$Cu^+(g) + e^- \rightarrow Cu(g) \qquad -I_1 = -746\ \text{kJ}$$
$$Cu^+(g) \rightarrow Cu^{2+}(g) + e^- \qquad I_2 = 1958\ \text{kJ}$$
$$2\,Cu^+(g) \rightarrow Cu(g) + Cu^{2+}(g) \qquad \Delta H = 1212\ \text{kJ}$$

b.

$$Na^-(g) \rightarrow Na(g) + e^- \qquad -EA = 52\ \text{kJ}$$
$$Na^+(g) + e^- \rightarrow Na(g) \qquad -I_1 = -495\ \text{kJ}$$
$$Na^-(g) + Na^+(g) \rightarrow 2\,Na(g) \qquad \Delta H = -443\ \text{kJ}$$

c.

$$Mg^{2+}(g) + e^- \rightarrow Mg^+(g) \qquad -I_2 = -1445\ \text{kJ}$$
$$K(g) \rightarrow K^+(g) + e^- \qquad I_1 = 419\ \text{kJ}$$
$$Mg^{2+}(g) + K(g) \rightarrow Mg^+(g) + K^+(g) \qquad \Delta H = -1026\ \text{kJ}$$

d.

$$Na(g) \rightarrow Na^+(g) + e^- \qquad I_1 = 495\ \text{kJ}$$
$$Cl(g) + e^- \rightarrow Cl^-(g) \qquad EA = -348.7\ \text{kJ}$$
$$Na(g) + Cl(g) \rightarrow Na^+(g) + Cl^-(g) \qquad \Delta H = 146\ \text{kJ}$$

e.

$$Mg(g) \rightarrow Mg^+(g) + e^- \qquad I_1 = 735\ \text{kJ}$$
$$F(g) + e^- \rightarrow F^-(g) \qquad EA = -327.8\ \text{kJ}$$
$$Mg(g) + F(g) \rightarrow Mg^+(g) + F^-(g) \qquad \Delta H = 407\ \text{kJ}$$

f. $Mg^+(g) \rightarrow Mg^{2+}(g) + e^-$ $\quad I_2 = 1445$ kJ

$F(g) + e^- \rightarrow F^-(g)$ $\quad EA = -327.8$ kJ

$Mg^+(g) + F(g) \rightarrow Mg^{2+}(g) + F^-(g)$ $\quad \Delta H = 1117$ kJ

g. From parts e and f we get:

$Mg(g) + F(g) \rightarrow Mg^+(g) + F^-(g)$ $\quad \Delta H = 407$ kJ

$Mg^+(g) + F(g) \rightarrow Mg^{2+}(g) + F^-(g)$ $\quad \Delta H = 1117$ kJ

$Mg(g) + 2\ F(g) \rightarrow Mg^{2+}(g) + 2\ F^-(g)$ $\quad \Delta H = 1524$ kJ

121. Valence electrons are easier to remove than inner core electrons. The large difference in energy between I_2 and I_3 indicates that this element has two valence electrons. This element is most likely an alkaline earth metal since alkaline earth metal elements all have two valence electrons.

123. a. The 4+ ion contains 20 electrons. Thus, the electrically neutral atom will contain 24 electrons. The atomic number is 24.

b. The ground state electron configuration of the ion must be: $1s^22s^22p^63s^23p^64s^03d^2$; There are 6 electrons in s orbitals.

c. 12 $\qquad$ d. 2

e. This is the isotope $^{50}_{24}Cr$. There are 26 neutrons in the nucleus.

f. 3.01×10^{23} atoms $\times \dfrac{1\ \text{mol}}{6.022 \times 10^{23}\ \text{atoms}} \times \dfrac{49.9\ \text{g}}{\text{mol}} = 24.9$ g

g. $1s^22s^22p^63s^23p^64s^13d^5$ is the ground state electron configuration for Cr. Cr is an exception to the normal filling order.

125. a. True for H only. $\qquad$ b. True for all atoms. $\qquad$ c. True for all atoms.

d. This is false for all atoms. In the presence of a magnetic field, the d orbitals are no longer degenerate.

127. None of the noble gases and no subatomic particles had been discovered when Mendeleev published his periodic table. Thus, there was no element out of place in terms of reactivity. There was no reason to predict an entire family of elements. Mendeleev ordered his table by mass; he had no way of knowing there were gaps in atomic numbers (they hadn't been invented yet).

129. At $x = 0$, the value of the square of the wave function must be zero. The particle must be inside the box. For $\psi = A \cos(Lx)$, at $x = 0$, $\cos(0) = 1$ and $\psi^2 = A^2$. This violates the boundary condition.

131. a. Since wavelength is inversely proportional to energy, then the spectral line to the right of B (at a larger wavelength) represents the lowest possible energy transition; this is n = 4 to n = 3. The B line represents the next lowest energy transition which is n = 5 to n = 3 and the A line corresponds to the n = 6 to n = 3 electronic transition.

b. Since this spectrum is for a one electron ion, then $E_n = -2.178 \times 10^{-18}\ J\ (Z^2/n^2)$. To determine ΔE and, in turn, the wavelength of spectral line A, we must determine Z, the atomic number of the one electron species. Use spectral line B data to determine Z.

$$\Delta E_{5\to3} = -2.178 \times 10^{-18}\ J\left(\frac{Z^2}{3^2} - \frac{Z^2}{5^2}\right) = -2.178 \times 10^{-18}\left(\frac{16\,Z^2}{9 \times 25}\right)$$

$$E = \frac{hc}{\lambda} = \frac{6.6261 \times 10^{-34}\ J\ s\ (2.9979 \times 10^8\ m/s)}{142.5 \times 10^{-9}\ m} = 1.394 \times 10^{-18}\ J$$

Since an emission occurs, $\Delta E_{5\to3} = -1.394 \times 10^{-18}$ J.

$$\Delta E = -1.394 \times 10^{-18}\ J = -2.178 \times 10^{-18}\ J\left(\frac{16\,Z^2}{9 \times 25}\right),\ Z^2 = 9.001,\ Z = 3;\ \text{The ion is } Li^{2+}.$$

Solving for the wavelength of line A:

$$\Delta E_{6\to3} = -2.178 \times 10^{-18}\ (3)^2\left(\frac{1}{3^2} - \frac{1}{6^2}\right) = -1.634 \times 10^{-18}\ J$$

$$\lambda = \frac{hc}{|\Delta E|} = \frac{6.6261 \times 10^{-34}\ J\ s\ (2.9979 \times 10^8\ m/s)}{1.634 \times 10^{-18}\ J} = 1.216 \times 10^{-7}\ m = 121.6\ nm$$

Challenge Problems

133. a. Since the energy levels, E_{xy}, are inversely proportional to L^2, then the $n_x = 2$, $n_y = 1$ energy level will be lower in energy than the $n_x = 1$, $n_y = 2$ energy level since $L_x > L_y$. The first three energy levels, E_{xy}, in order of increasing energy are:

$$E_{11} < E_{21} < E_{12}$$

The quantum numbers are:

ground state (E_{11}) → $n_x = 1$, $n_y = 1$
first excited state (E_{21}) → $n_x = 2$, $n_y = 1$
second excited state (E_{12}) → $n_x = 1$, $n_y = 2$

b. $E_{21} \to E_{12}$ is the transition. $E_{xy} = \dfrac{h^2}{8\,m}\left(\dfrac{n_x^2}{L_x^2} + \dfrac{n_y^2}{L_y^2}\right)$

$$E_{12} = \frac{h^2}{8\,m}\left[\frac{1^2}{(8.00 \times 10^{-9}\ m)^2} + \frac{2^2}{(5.00 \times 10^{-9}\ m)^2}\right] = \frac{1.76 \times 10^{17}\ h^2}{8\ m}$$

$$E_{21} = \frac{h^2}{8\,m}\left[\frac{2^2}{(8.00 \times 10^{-9}\,m)^2} + \frac{1^2}{(5.00 \times 10^{-9})^2}\right] = \frac{1.03 \times 10^{17}\,h^2}{8\,m}$$

$$\Delta E = E_{12} - E_{21} = \frac{1.76 \times 10^{17}\,h^2}{8\,m} - \frac{1.03 \times 10^{17}\,h^2}{8\,m} = \frac{7.3 \times 10^{16}\,h^2}{8\,m}$$

$$\Delta E = \frac{(7.3 \times 10^{16}\,m^{-2})\,(6.626 \times 10^{-34}\,J\,s)^2}{8(9.11 \times 10^{-31}\,kg)} = 4.4 \times 10^{-21}\,J$$

$$\lambda = \frac{hc}{\Delta E} = \frac{6.626 \times 10^{-34}\,J\,s\,(2.998 \times 10^8\,m/s)}{4.4 \times 10^{-21}\,J} = 4.5 \times 10^{-5}\,m$$

135. $\psi_{1s} = \frac{1}{\sqrt{\pi}}\left(\frac{Z}{a_0}\right)^{3/2} e^{-\sigma}$; $Z = 1$ for H, $\sigma = \frac{Zr}{a_0} = \frac{r}{a_0}$, $a_0 = 5.29 \times 10^{-11}$ m

$$\psi_{1s} = \frac{1}{\sqrt{\pi}}\left(\frac{1}{a_0}\right)^{3/2} \exp\left(\frac{-r}{a_0}\right)$$

Probability is proportional to ψ^2: $\psi_{1s}^2 = \frac{1}{\pi}\left(\frac{1}{a_0}\right)^3 \exp\left(\frac{-2r}{a_0}\right)$ (units of $\psi^2 = m^{-3}$)

a. ψ_{1s}^2 (at nucleus) $= \frac{1}{\pi}\left(\frac{1}{a_0}\right)^3 \exp\left[\frac{-2\,(0)}{a_0}\right] = 2.15 \times 10^{30}\,m^{-3}$

If we assume this probability is constant throughout the $1.0 \times 10^{-3}\,pm^3$ volume then the total probability, p, is $\psi_{1s}^2 \times V$.

$1.0 \times 10^{-3}\,pm^3 = (1.0 \times 10^{-3}\,pm) \times (10^{-12}\,m/pm)^3 = 1.0 \times 10^{-39}\,m^3$

total probability = p = $(2.15 \times 10^{30}\,m^{-3}) \times (1.0 \times 10^{-39}\,m^3) = 2.2 \times 10^{-9}$

b. For an electron that is 1.0×10^{-11} m from the nucleus:

$$\psi_{1s}^2 = \frac{1}{\pi}\left(\frac{1}{5.29 \times 10^{-11}}\right)^3 \exp\left[\frac{-2(1.0 \times 10^{-11})}{(5.29 \times 10^{-11})}\right] = 1.5 \times 10^{30}\,m^{-3}$$

$V = 1.0 \times 10^{-39}\,m^3$; $p = \psi_{1s}^2 \times V = 1.5 \times 10^{-9}$

c. $\psi_{1s}^2 = 2.15 \times 10^{30}\,m^{-3} \exp\left[\frac{-2(53 \times 10^{-12})}{(5.29 \times 10^{-11})}\right] = 2.9 \times 10^{29}$; $V = 1.0 \times 10^{-39}\,m^3$

$p = \psi_{1s}^2 \times V = 2.9 \times 10^{-10}$

d. $V = \frac{4}{3}\pi[(10.05 \times 10^{-12}\text{ m})^3 - (9.95 \times 10^{-12}\text{ m})^3] = 1.3 \times 10^{-34}\text{ m}^3$

We shall evaluate ψ_{1s}^2 at the middle of the shell, r = 10.00 pm, and assume ψ_{1s}^2 is constant from r = 9.95 to 10.05 pm. The concentric spheres are assumed centered about the nucleus.

$$\psi_{1s}^2 = 2.15 \times 10^{30}\text{ m}^{-3} \exp\left[\frac{-2(10.0 \times 10^{-12}\text{ m})}{(5.29 \times 10^{-11}\text{ m})}\right] = 1.47 \times 10^{30}\text{ m}^{-3}$$

$$p = (1.47 \times 10^{30}\text{ m}^{-3})(1.3 \times 10^{-34}\text{ m}^3) = 1.9 \times 10^{-4}$$

e. $V = \frac{4}{3}\pi[(52.95 \times 10^{-12}\text{ m})^3 - (52.85 \times 10^{-12}\text{ m})^3] = 4 \times 10^{-33}\text{ m}^3$

Evaluate ψ_{1s}^2 at r = 52.90 pm: $\psi_{1s}^2 = 2.15 \times 10^{30}\text{ m}^{-3}(e^{-2}) = 2.91 \times 10^{29}\text{ m}^{-3}$; $p = 1 \times 10^{-3}$

137. $E = \frac{h^2(n_x^2 + n_y^2 + n_z^2)}{8\text{ m}L^2}$; $E_{111} = \frac{3\text{ h}^2}{8\text{ m}L^2}$; $E_{112} = \frac{6\text{ h}^2}{8\text{ m}L^2}$; $\Delta E = E_{112} - E_{111} = \frac{3\text{ h}^2}{8\text{ m}L^2}$

$$\Delta E = \frac{hc}{\lambda} = \frac{(6.626 \times 10^{-34}\text{ J sec})(2.998 \times 10^8\text{ m/s})}{9.50 \times 10^{-9}\text{ m}} = 2.09 \times 10^{-17}\text{ J}$$

$$L^2 = \frac{3\text{ h}^2}{8\text{ m}\Delta E},\ L = \left(\frac{3\text{ h}^2}{8\text{ m}\Delta E}\right)^{1/2} = \left[\frac{3(6.626 \times 10^{-34}\text{ J sec})^2}{8(9.109 \times 10^{-31}\text{ kg})(2.09 \times 10^{-17}\text{ J})}\right]^{1/2}$$

$L = 9.30 \times 10^{-11}\text{ m} = 93.0\text{ pm}$

The sphere that fits in this cube will touch the cube at the center of each face. The diameter of the sphere will equal the length of the cube. So:

2 r = L and r = 46.5 pm

139. a. **1st period:** p = 1, q = 1, r = 0, s = ± 1/2 (2 elements)

2nd period: p = 2, q = 1, r = 0, s = ± 1/2 (2 elements)

3rd period: p = 3, q = 1, r = 0, s = ± 1/2 (2 elements)

p = 3, q = 3, r = -2, s = ± 1/2 (2 elements)

p = 3, q = 3, r = 0, s = ± 1/2 (2 elements)

p = 3, q = 3, r = +2, s = ± 1/2 (2 elements)

4th period: $p = 4$; q and r values are the same as with $p = 3$ (8 total elements)

1							2
3							4
5	6	7	8	9	10	11	12
13	14	15	16	17	18	19	20

b. Elements 2, 4, 12 and 20 all have filled shells and will be least reactive.

c. Draw similarities to the modern periodic table.

XY could be $X^{+}Y^{-}$, $X^{2+}Y^{2-}$ or $X^{3+}Y^{3-}$. Possible ions for each are:

X^{+} could be elements 1, 3, 5 or 13; Y^{-} could be 11 or 19.

X^{2+} could be 6 or 14; Y^{2-} could be 10 or 18.

X^{3+} could be 7 or 15; Y^{3-} could be 9 or 17.

Note: X^{4+} and Y^{4-} ions probably won't form.

XY_2 will be $X^{2+}(Y^{-})_2$; See above for possible ions.

X_2Y will be $(X^{+})_2Y^{2-}$; See above for possible ions.

XY_3 will be $X^{3+}(Y^{-})_3$; See above for possible ions.

X_2Y_3 will be $(X^{3+})_2(Y^{2-})_3$; See above for possible ions.

d. From (a), we can see that eight electrons can have $p = 3$.

e. $p = 4$, $q = 3$, $r = 2$, $s = \pm 1/2$ (2 electrons)

f. $p = 4$, $q = 3$, $r = -2$, $s = \pm 1/2$ (2)

$p = 4$, $q = 3$, $r = 0$, $s = \pm 1/2$ (2)

$p = 4$, $q = 3$, $r = +2$, $s = \pm 1/2$ (2)

A total of 6 electrons can have $p = 4$ and $q = 3$.

g. $p = 3$, $q = 0$, $r = 0$: This is not allowed; q must be odd. Zero electrons can have these quantum numbers.

h. **$p = 5$, $q = 1$, $r = 0$**

$p = 5$, $q = 3$, $r = -2, 0, +2$

$p = 5$, $q = 5$, $r = -4, -2, 0, +2, +4$

i. **$p = 6$, $q = 1$, $r = 0$, $s = \pm 1/2$ (2 electrons)**

$p = 6$, $q = 3$, $r = -2, 0, +2$; $s = \pm 1/2$ (6)

$p = 6$, $q = 5$, $r = -4, -2, 0, +2, +4$; $s = \pm 1/2$ (10)

Eighteen electrons can have $p = 6$.

CHAPTER THIRTEEN

BONDING: GENERAL CONCEPTS

Chemical Bonds and Electronegativity

11. a. Electronegativity: The ability of an atom <u>in a molecule</u> to attract electrons to itself.

    Electron affinity: The energy change for $M(g) + e^- \rightarrow M^-(g)$. EA deals with isolated atoms in the gas phase.

    b. Covalent bond: Sharing of electron pair(s); Polar covalent bond: Unequal sharing of electron pair(s).

    c. Ionic bond: Electrons are no longer shared, i.e., complete transfer of electron(s) from one atom to another.

13. Using the periodic table we expect the general trend for electronegativity to be:
    1) increase as we go from left to right across a period
    2) decrease as we go down a group

    a. C < N < O   b. Se < S < Cl   c. Sn < Ge < Si

    d. Tl < Ge < S   e. Rb < K < Na   f. Ga < B < O

15. The general trends in electronegativity used on Exercises 13.13 and 13.14 are only rules of thumb. In this exercise we use experimental values of electronegativities and can begin to see several exceptions. The order of EN using Figure 13.3 is:

    a. C (2.6) < N (3.0) < O (3.4)   same as predicted

    b. Se (2.6) = S (2.6) < Cl (3.2)   different

    c. Si (1.9) < Ge (2.0) = Sn (2.0)   different   d. Tl (2.0) = Ge (2.0) < S (2.6)   different

    e. Rb (0.8) = K (0.8) < Na (0.9)   different   f. Ga (1.8) < B (2.0) < O (3.4)   same

    Most polar bonds using actual EN values:

    a. Si-F (Ge-F predicted)   b. P-Cl (same as predicted)

c. S-F (same as predicted)

d. Ti-Cl (same as predicted)

e. C-H (Sn-H predicted)

f. Al-Br (Tl-Br predicted)

Ionic Compounds

17. a. $Cu > Cu^+ > Cu^{2+}$

b. $Pt^{2+} > Pd^{2+} > Ni^{2+}$

c. $O^{2-} > O^- > O$

d. $La^{3+} > Eu^{3+} > Gd^{3+} > Yb^{3+}$

e. $Te^{2-} > I^- > Cs^+ > Ba^{2+} > La^{3+}$

For answer a, as electrons are removed from an atom, size decreases. Answers b and d follow the radii trend. For answer c, as electrons are added to an atom, size increases. Answer e follows the trend for an isoelectronic series, i.e., the smallest ion has the most protons.

19. a. Cs_2S is composed of Cs^+ and S^{2-}. Cs^+ has the same electron configuration as Xe, and S^{2-} has the same configuration as Ar.

b. SrF_2; Sr^{2+} has the Kr electron configuration and F^- has the Ne configuration.

c. Ca_3N_2; Ca^{2+} has the Ar electron configuration and N^{3-} has the Ne configuration.

d. $AlBr_3$; Al^{3+} has the Ne electron configuration and Br^- has the Kr configuration.

21. Isoelectronic: Same number of electrons. There are two variables, number of protons and number of electrons, that will determine the size of an ion. Keeping the number of electrons constant, we only have to consider the number of protons to predict trends in size. The smallest ion has the most protons.

Se^{2-}, Br^-, Rb^+, Sr^{2+}, Y^{3+}, Zr^{4+} are some ions which are isoelectronic with Kr (36 electrons). In terms of size, the ion with the most protons will hold the electrons tightest and will be the smallest. The size trend is:

$$Zr^{4+} < Y^{3+} < Sr^{2+} < Rb^+ < Br^- < Se^{2-}$$

smallest ... largest

23. a. Al^{3+} and Cl^-; $AlCl_3$, aluminum chloride

b. Na^+ and O^{2-}; Na_2O, sodium oxide

c. Sr^{2+} and F^-; SrF_2, strontium fluoride

d. Ca^{2+} and P^{3-}; Ca_3P_2, calcium phosphide

25.

Reaction	ΔH
$Na(s) \rightarrow Na(g)$	$\Delta H = 109$ kJ (sublimation)
$Na(g) \rightarrow Na^+(g) + e^-$	$\Delta H = 495$ kJ (ionization energy)
$1/2\ Cl_2(g) \rightarrow Cl(g)$	$\Delta H = 239/2$ kJ (bond energy)
$Cl(g) + e^- \rightarrow Cl^-(g)$	$\Delta H = -349$ kJ (electron affinity)
$Na^+(g) + Cl^-(g) \rightarrow NaCl(s)$	$\Delta H = -786$ kJ (lattice energy)
$Na(s) + 1/2\ Cl_2(g) \rightarrow NaCl(s)$	$\Delta H_f^\circ = -412$ kJ/mol

27. a. From the data given, less energy is required to produce $Mg^+(g) + O^-(g)$ than to produce $Mg^{2+}(g) + O^{2-}(g)$. However, the lattice energy for $Mg^{2+}O^{2-}$ will be much more exothermic than for Mg^+O^- (due to the greater charges in $Mg^{2+}O^{2-}$). The favorable lattice energy term will dominate and $Mg^{2+}O^{2-}$ forms.

b. Mg^+ and O^- both have unpaired electrons. In Mg^{2+} and O^{2-}, there are no unpaired electrons. Hence, Mg^+O^- would be paramagnetic; $Mg^{2+}O^{2-}$ would be diamagnetic. Paramagnetism can be detected by measuring the mass of a sample in the presence and absence of a magnetic field. The apparent mass of a paramagnetic substance will be larger in a magnetic field because of the force between the unpaired electrons and the field.

29. Ca^{2+} has a greater charge than Na^+, and Se^{2-} is smaller than Te^{2-}. The effect of charge on the lattice energy is greater than the effect of size. We expect the trend from most exothermic to least exothermic to be:

$CaSe > CaTe > Na_2Se > Na_2Te$

(-2862) (-2721) (-2130) (-2095 kJ/mol) This is what we observe.

Bond Energies

31. a. $H-H + Cl-Cl \rightarrow 2\ H-Cl$

Bonds broken: Bonds formed:

1 H – H (432 kJ/mol) 2 H – Cl (427 kJ/mol)
1 Cl – Cl (239 kJ/mol)

$\Delta H = \Sigma D_{broken} - \Sigma D_{formed}$, $\Delta H = 432\ kJ + 239\ kJ - 2(427)\ kJ = -183\ kJ$

b. $N \equiv N + 3\ H-H \longrightarrow 2\ H-\underset{\displaystyle H}{\underset{|}{N}}-H$

Bonds broken: Bonds formed:

1 N ≡ N (941 kJ/mol) 6 N – H (391 kJ/mol)
3 H – H (432 kJ/mol)

$\Delta H = 941\ kJ + 3(432)\ kJ - 6(391)\ kJ = -109\ kJ$

c. Sometimes some of the bonds remain the same between reactants and products. To save time, only break and form bonds that are involved in the reaction.

$$H-C\equiv N + 2\ H-H \longrightarrow H-\overset{\displaystyle H}{\underset{\displaystyle H}{\overset{|}{\underset{|}{C}}}}-\overset{\displaystyle H}{\underset{\displaystyle H}{\overset{|}{\underset{|}{N}}}}$$

Bonds broken:

1 C ≡ N (891 kJ/mol)
2 H − H (432 kJ/mol)

Bonds formed:

1 C − N (305 kJ/mol)
2 C − H (413 kJ/mol)
2 N − H (391 kJ/mol)

ΔH = 891 kJ + 2(432 kJ) − [305 kJ + 2(413 kJ) + 2(391 kJ)] = -158 kJ

d. 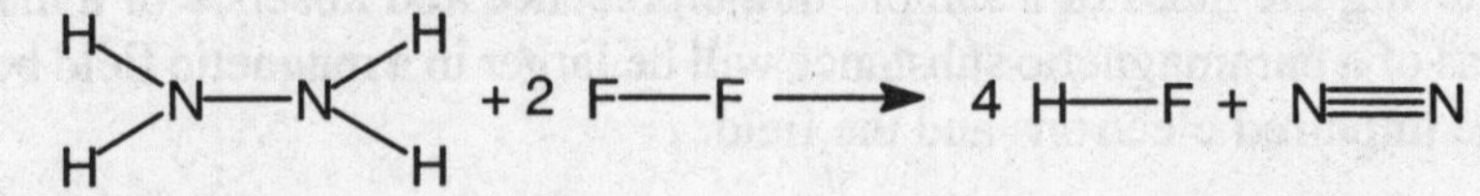 $H_2N{-}NH_2$ + 2 F—F ⟶ 4 H—F + N≡N

Bonds broken:

1 N − N (160. kJ/mol)
4 N − H (391 kJ/mol)
2 F − F (154 kJ/mol)

Bonds formed:

4 H − F (565 kJ/mol)
1 N ≡ N (941 kJ/mol)

ΔH = 160. kJ + 4(391 kJ) + 2(154 kJ) - [4(565 kJ) + 941 kJ] = -1169 kJ

33.

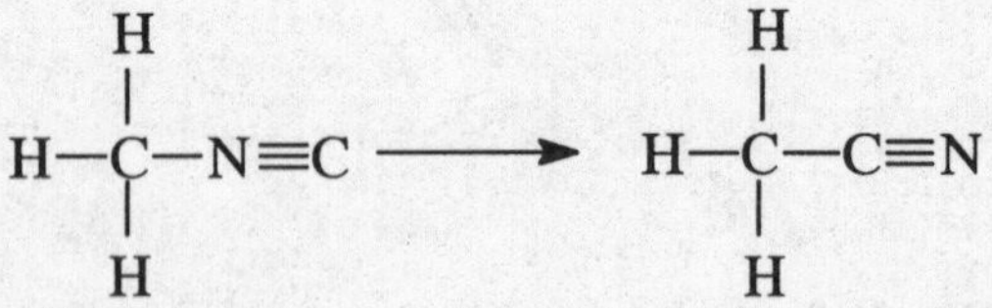

$H_3C{-}N{\equiv}C$ ⟶ $H_3C{-}C{\equiv}N$

Bonds broken: 1 C − N (305 kJ/mol) **Bonds formed: 1 C − C (347 kJ/mol)**

$\Delta H = \Sigma D_{broken} - \Sigma D_{formed}$, ΔH = 305 - 347 = -42 kJ

Note: Some bonds usually remain the same between reactants and products. To save time, only break and form bonds that are involved in the reaction.

35.

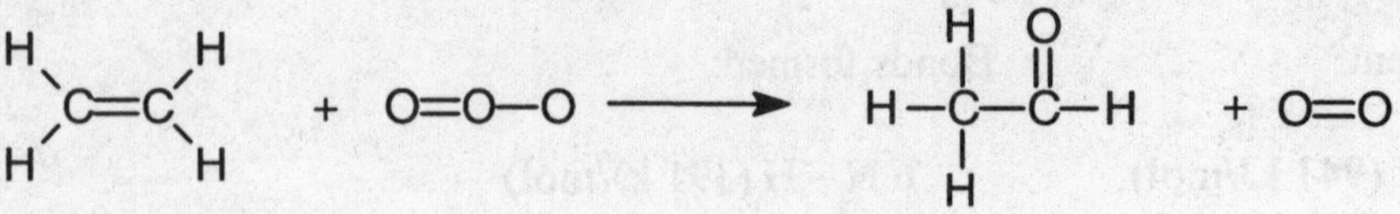

$H_2C{=}CH_2$ + O=O—O ⟶ H—C(H)(H)—C(=O)—H + O=O

Bonds broken:

1 C=C (614 kJ/mol)
1 O − O (146 kJ/mol)
1 C − H (413 kJ/mol)

Bonds formed:

1 C − C (347 kJ/mol)
1 C=O (745 kJ/mol)
1 C − H (413 kJ/mol)

ΔH = 614 + 146 + 413 - (347 + 745 + 413) = -332 kJ

37.

$$4\ \mathrm{H{-}CH_2{-}N(H){-}NH_2} + 5\ \mathrm{O_2N{-}NO_2} \longrightarrow 12\ \mathrm{H{-}O{-}H} + 9\ \mathrm{N{\equiv}N} + 4\ \mathrm{O{=}C{=}O}$$

Bonds broken:

9 N – N (160. kJ/mol)
4 N – C (305 kJ/mol)
12 C – H (413 kJ/mol)
12 N – H (391 kJ/mol)
10 N = O (607 kJ/mol)
10 N – O (201 kJ/mol)

Bonds made:

24 O – H (467 kJ/mol)
9 N ≡ N (941 kJ/mol)
8 C = O (799 kJ/mol)

$\Delta H = 9(160.) + 4(305) + 12(413) + 12(391) + 10(607) + 10(201) - [24(467) + 9(941) + 8(799)]$

$\Delta H = 20{,}388\text{ kJ} - 26{,}069\text{ kJ} = -5681\text{ kJ}$

39. Since both reactions are highly exothermic, the high temperature is not needed to provide energy. It must be necessary for some other reason. The reason is to increase the speed of the reaction. This will be discussed in Chapter 15 on kinetics.

41. a.

$HF(g) \rightarrow H(g) + F(g)$	$\Delta H = 565\text{ kJ}$
$H(g) \rightarrow H^+(g) + e^-$	$\Delta H = 1312\text{ kJ}$
$F(g) + e^- \rightarrow F^-(g)$	$\Delta H = -327.8\text{ kJ}$
$HF(g) \rightarrow H^+(g) + F^-(g)$	$\Delta H = 1549\text{ kJ}$

b.

$HCl(g) \rightarrow H(g) + Cl(g)$	$\Delta H = 427\text{ kJ}$
$H(g) \rightarrow H^+(g) + e^-$	$\Delta H = 1312\text{ kJ}$
$Cl(g) + e^- \rightarrow Cl^-(g)$	$\Delta H = -348.7\text{ kJ}$
$HCl(g) \rightarrow H^+(g) + Cl^-(g)$	$\Delta H = 1390.\text{ kJ}$

c.

$HI(g) \rightarrow H(g) + I(g)$	$\Delta H = 295\text{ kJ}$
$H(g) \rightarrow H^+(g) + e^-$	$\Delta H = 1312\text{ kJ}$
$I(g) + e^- \rightarrow I^-(g)$	$\Delta H = -295.2\text{ kJ}$
$HI(g) \rightarrow H^+(g) + I^-(g)$	$\Delta H = 1312\text{ kJ}$

d.

$H_2O(g) \rightarrow OH(g) + H(g)$	$\Delta H = 467\text{ kJ}$
$H(g) \rightarrow H^+(g) + e^-$	$\Delta H = 1312\text{ kJ}$
$OH(g) + e^- \rightarrow OH^-(g)$	$\Delta H = -180.\text{ kJ}$
$H_2O(g) \rightarrow H^+(g) + OH^-(g)$	$\Delta H = 1599\text{ kJ}$

43. $NH_3(g) \rightarrow N(g) + 3\ H(g)$

$\Delta H° = 3\ D_{NH} = 472.7\ kJ + 3(216.0\ kJ) - (-46.1\ kJ) = 1166.8\ kJ$

$$D_{NH} = \frac{1166.8\ kJ}{3\ mol\ NH\ bonds} = 388.93\ kJ/mol$$

$D_{calc} = 389$ kJ/mol as compared to 391 kJ/mol in the table. There is good agreement.

45. From Exercise 13.43, the N–H bond energy is 388.9 kJ/mol.

$H_2N{-}NH_2 \longrightarrow 2\ N(g) + 4\ H(g) \qquad \Delta H = D_{N\text{-}N} + 4\ D_{N\text{-}H} = D_{N\text{-}N} + 4(388.9)$

$\Delta H° = 2\Delta H°_{f,N} + 4\Delta H°_{f,H} - \Delta H°_{f,N_2H_4} = 2(472.7\ kJ) + 4(216.0\ kJ) - 95.4\ kJ$

$\Delta H° = 1714.0\ kJ = D_{N\text{-}N} + 4(388.9),\ D_{N\text{-}N} = 158.4\ kJ/mol$ (160. kJ/mol in Table 13.6)

Lewis Structures and Resonance

47. Drawing Lewis structures is mostly trial and error. However, the first two steps are always the same. These steps are 1) count the valence electrons available in the molecule or ion and 2) attach all atoms to each other with single bonds (called the skeletal structure). Generally, the atom listed first is assumed to be the atom in the middle (called the central atom) and all other atoms in the formulas are attached to this atom. The most notable exceptions to the rule are formulas which begin with H, e.g., H_2O, H_2CO, etc. Hydrogen can never be a central atom since this would require H to have more than two electrons.

After counting valence electrons and drawing the skeletal structure, the rest is trial and error. We place the remaining electrons around the various atoms in an attempt to satisfy the octet rule (or duet rule for H). Keep in mind that practice makes perfect. After practicing you can (and will) become very adept at drawing Lewis structures.

a. HCN has 1 + 4 + 5 = 10 valence electrons.

H—C—N   H—C≡N:

Skeletal structure   Lewis structure

Skeletal structure uses 4 e^-; 6 e^- remain

b. PH_3 has 5 + 3(1) = 8 valence electrons.

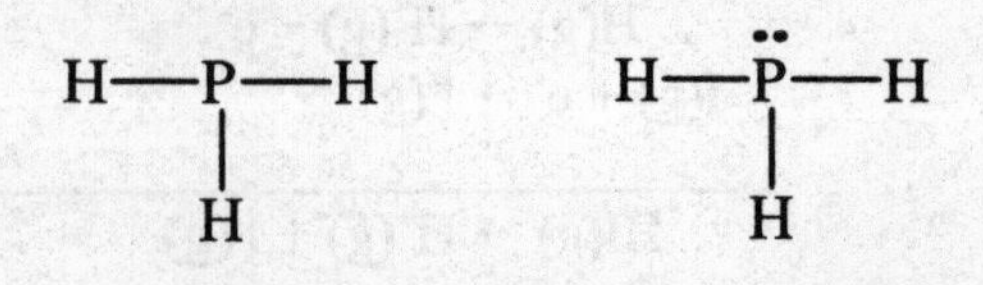

Skeletal structure   Lewis structure

Skeletal structure uses 6 e^-; 2 e^- remain

c. $CHCl_3$ has 4 + 1 + 3(7) = 26 valence electrons.

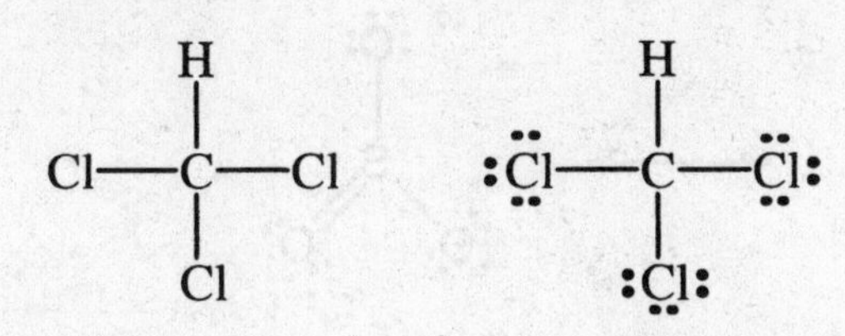

Skeletal structure   Lewis structure

Skeletal structure uses 8 e⁻; 18 e⁻ remain

d. NH_4^+ has 5 + 4(1) - 1 = 8 valence electrons.

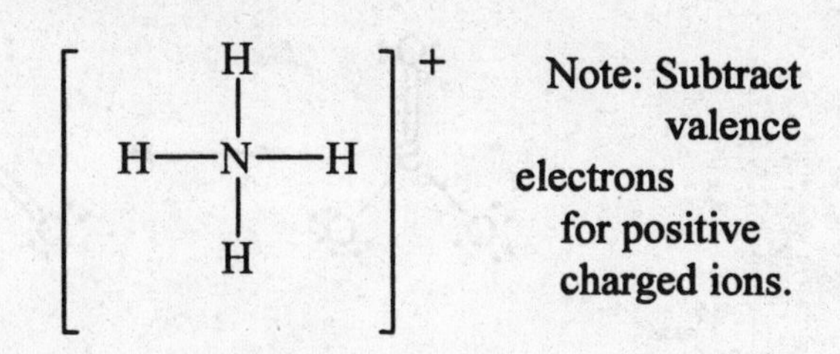

Lewis structure

e. H_2CO has 2(1) + 4 + 6 = 12 valence electrons.

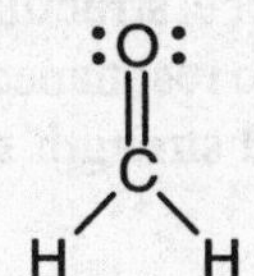

f. SeF_2 has 6 + 2(7) = 20 valence electrons.

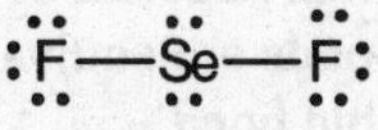

:F—Se—F:

g. CO_2 has 4 + 2(6) = 16 valence electrons.

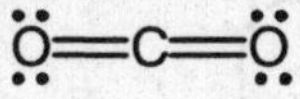

O=C=O

h. O_2 has 2(6) = 12 valence electrons.

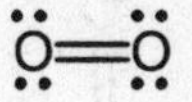

O=O

i. HBr has 1 + 7 = 8 valence electrons.

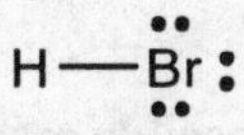

H—Br:

49. Molecules/ions that have the same number of valence electrons and the same number of atoms will have similar Lewis structures.

51. Ozone: O_3 has 3(6) = 18 valence electrons. Two resonance structures can be drawn.

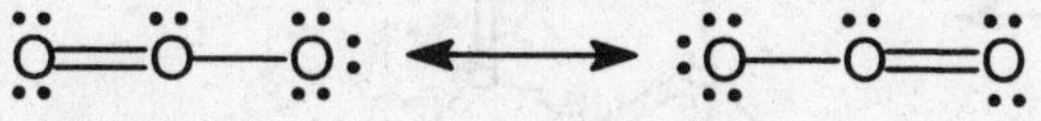

O=O—O: ⟷ :O—O=O

Sulfur dioxide: SO_2 has 6 + 2(6) = 18 valence electrons. Two resonance structures are possible.

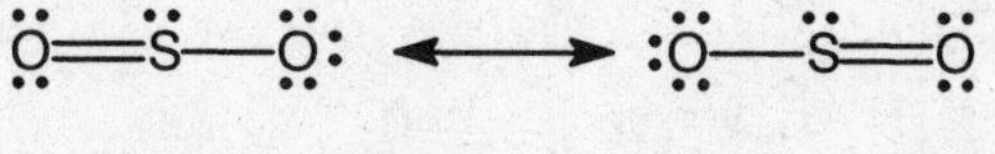

O=S—O: ⟷ :O—S=O

Sulfur trioxide: SO_3 has 6 + 3(6) = 24 valence electrons. Three resonance structures are possible.

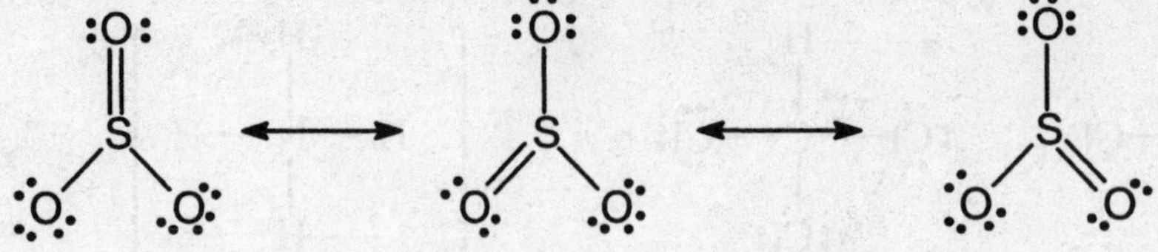

53. **CH_3NCO has 4 + 3(1) + 5 + 4 + 6 = 22 valence electrons. The order of the elements in the formula give the skeletal structure.**

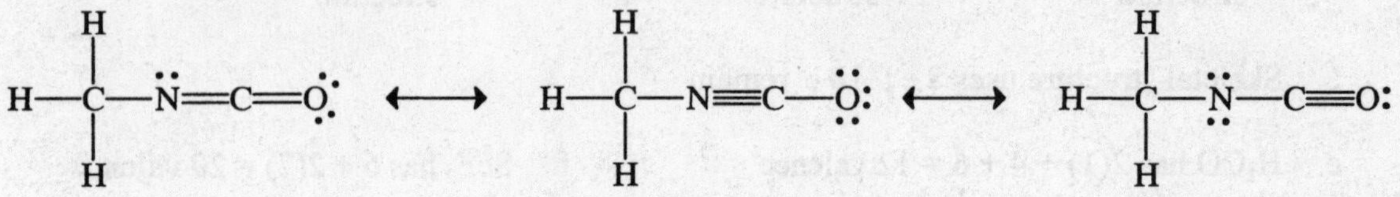

55. **Benzene has 6(4) + 6(1) = 30 valence electrons. Two resonance structures can be drawn for benzene. The actual structure of benzene is an average of these two resonance structures, that is, all carbon-carbon bonds are equivalent with a bond length and bond strength somewhere between a single and a double bond.**

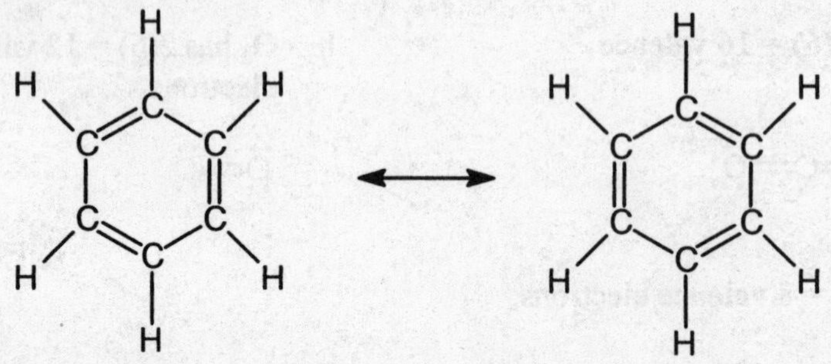

57. **Borazine ($B_3N_3H_6$) has 3(3) + 3(5) + 6(1) = 30 valence electrons. The possible resonance structures are similar to those of benzene in Exercise 13.55.**

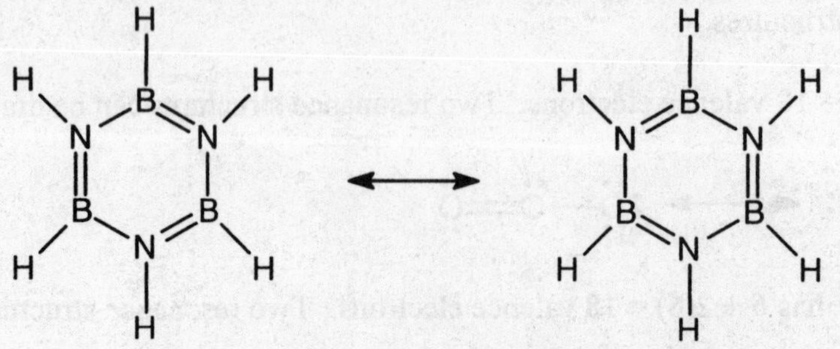

59. In each case in this problem, the octet rule cannot be satisfied for the central atom. BeH_2 and BH_3 have too few electrons around the central atom and all the others have to many electrons around the central atom. Always try to satisfy the octet rule for every atom, but when it is impossible, the central atom is the species assumed to disobey the octet rule.

PF_5, $5 + 5(7) = 40\ e^-$

BeH_2, $2 + 2(1) = 4\ e^-$

H – Be – H

BH_3, $3 + 3(1) = 6\ e^-$

Br_3^-, $3(7) + 1 = 22\ e^-$

SF_4, $6 + 4(7) = 34\ e^-$

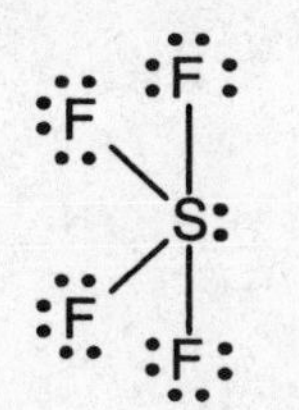

XeF_4, $8 + 4(7) = 36\ e^-$

ClF_5, $7 + 5(7) = 42\ e^-$

SF_6, $6 + 6(7) = 48\ e^-$

61. CO_3^{2-} has $4 + 3(6) + 2 = 24$ valence electrons.

HCO_3^- has 1 + 4 + 3(6) + 1 = 24 valence electrons.

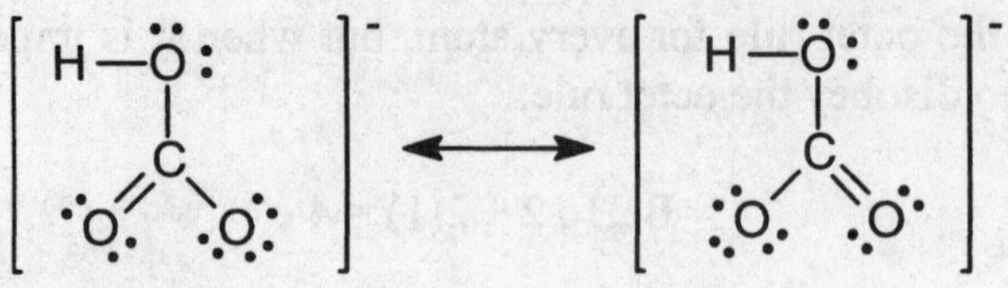

H_2CO_3 has 2(1) + 4 + 3(6) = 24 valence electrons.

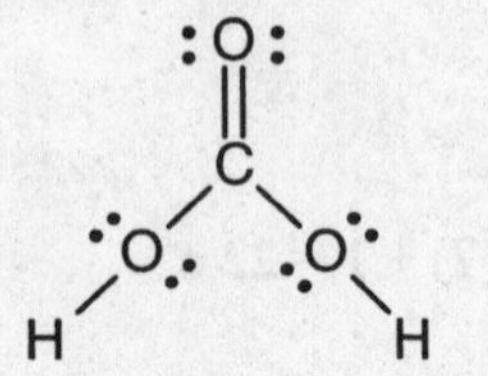

The Lewis structures for the reactants and products are:

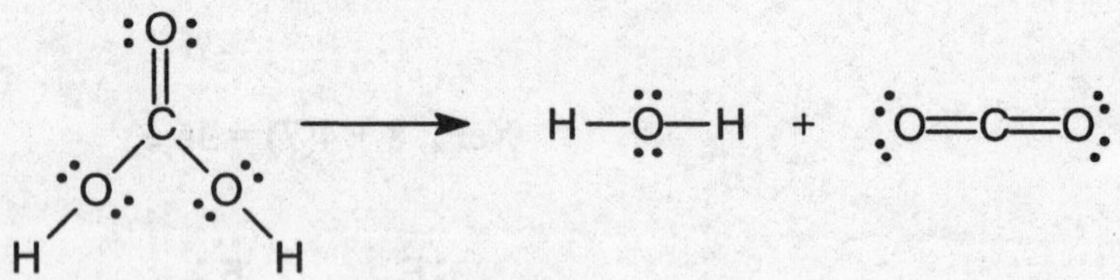

Bonds broken:	Bonds formed:
2 C – O (358 kJ/mol)	1 C = O (799 kJ/mol)
1 O – H (467 kJ/mol)	1 O – H (467 kJ/mol)

ΔH = 2(358) + 467 - [799 + 467] = -83 kJ; The carbon-oxygen double bond is stronger than two carbon-oxygen single bonds, hence CO_2 and H_2O are more stable than H_2CO_3.

63. CO_3^{2-} has 4 + 3(6) + 2 = 24 valence electrons.

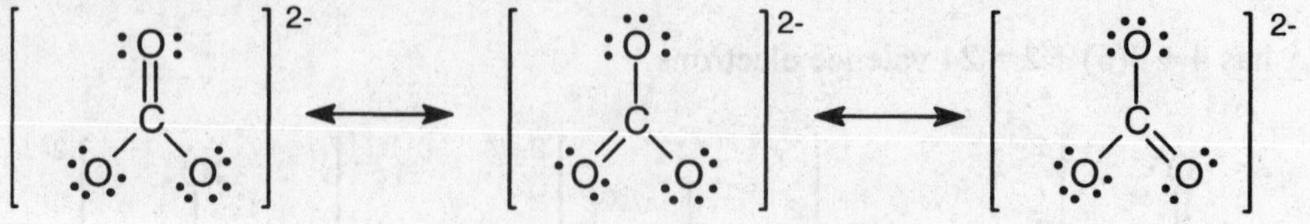

Three resonance structures can be drawn for CO_3^{2-}. The actual structure for CO_3^{2-} is an average of these three resonance structures. That is, the three C – O bond lengths are all equivalent, with a length somewhere between a single and a double bond. The actual bond length of 136 pm is consistent with this resonance view of CO_3^{2-}.

65. N_2 (10 e^-): :N≡N: Triple bond between N and N.

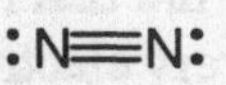

N_2F_4 (38 e^-): :F̤—N̈—N̈—F̤: with :F̤: on each N. Single bond between N and N.

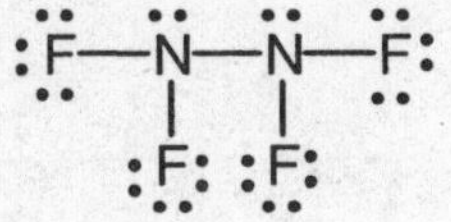

N_2F_2 (24 e^-): :F̤—N̈=N̈—F̤: Double bond between N and N.

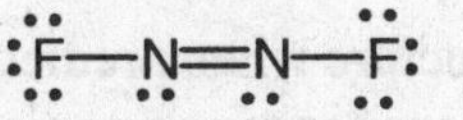

As the number of bonds increase between two atoms, bond strength increases and bond length decreases. From the Lewis structure, the shortest to longest N-N bonds is: $N_2 < N_2F_2 < N_2F_4$.

Formal Charge

67. **See Exercise 13.48a for the Lewis structures of $POCl_3$, SO_4^{2-}, ClO_4^- and PO_4^{3-}. All of these compounds/ions have similar Lewis structures to those of SO_2Cl_2 and XeO_4 shown below.**

a. $POCl_3$: P, FC = 5 - 1/2(8) = +1

b. SO_4^{2-}: S, FC = 6 - 1/2(8) = +2

c. ClO_4^-: Cl, FC = 7 - 1/2(8) = +3

d. PO_4^{3-}: P, FC = 5 - 1/2(8) = +1

e. SO_2Cl_2, 6 + 2(6) + 2(7) = 32 e^-

f. XeO_4, 8 + 4(6) = 32 e^-

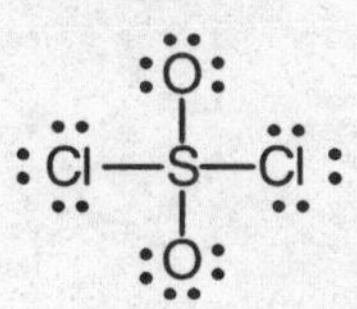

:Ö: / :C̈l—S—C̈l: / :Ö: (S bonded to two O and two Cl)

:Ö: / :Ö—Xe—Ö: / :Ö: (Xe bonded to four O)

S, FC = 6 - 1/2(8) = +2

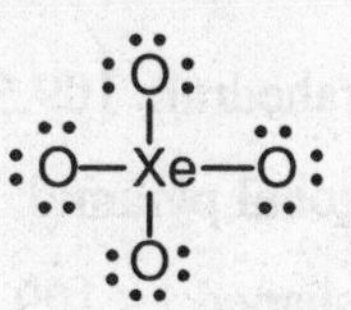

Xe, FC = 8 - 1/2(8) = +4

g. ClO_3^-, 7 + 3(6) + 1 = 26 e^-

h. NO_4^{3-}, 5 + 4(6) + 3 = 32 e^-

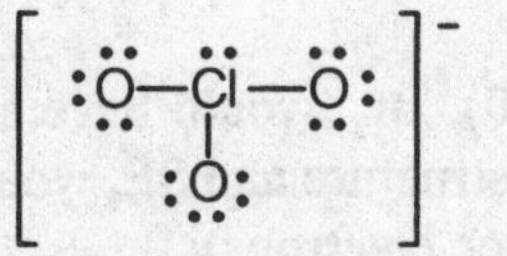

[:Ö—C̈l—Ö: with :Ö: below Cl]$^-$

[:Ö: / :Ö—N—Ö: / :Ö:]$^{3-}$

Cl, FC = 7 - 2 - 1/2(6) = +2

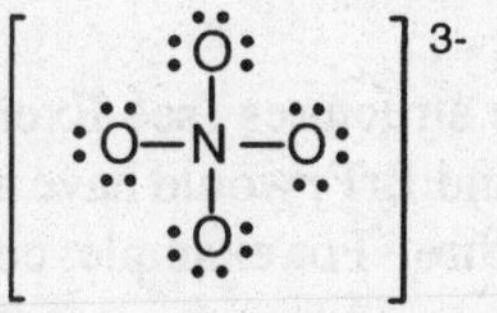

N, FC = 5 - 1/2(8) = +1

69. **O_2F_2 has 2(6) + 2(7) = 26 valence e^-. The formal charge and oxidation number of each atom is below the Lewis structure of O_2F_2.**

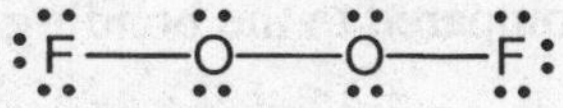

:F̤—Ö—Ö—F̤:

	F	O	O	F
Formal Charge	0	0	0	0
Oxid. Number	-1	+1	+1	-1

Oxidation numbers are more useful when accounting for the reactivity of O_2F_2. We are forced to assign +1 as the oxidation number for oxygen. Oxygen is very electronegative, and +1 is not a stable oxidation state for this element.

Molecular Structure and Polarity

71. The first step always is to draw a valid Lewis structure when predicting molecular structure. When resonance is possible, only one of the possible resonance structures is necessary to predict the correct structure since all resonance structures give the same structure. The Lewis structures are in Exercises 13.47, 13.48 and 13.50. The structures and bond angles for each follow.

13.47 a. HCN: linear, 180°

b. PH_3: trigonal pyramid, < 109.5°

c. $CHCl_3$: tetrahedral, 109.5°

d. NH_4^+: tetrahedral, 109.5°

e. H_2CO: trigonal planar, 120°

f. SeF_2: V-shaped or bent, < 109.5°

g. CO_2: linear, 180°

h and i. O_2 and HBr are both linear, but there is no bond angle in either.

Note: PH_3 and SeF_2 both have lone pairs of electrons on the central atom which result in bond angles that are something less than predicted from a tetrahedral arrangement (109.5°). However, we cannot predict the exact number. For these cases, we will just insert a less than sign to show this phenomenon.

13.48 a. All are tetrahedral; 109.5°

b. All are trigonal pyramid; < 109.5°

c. All are V-shaped; < 109.5°

13.50 a. NO_2^-: V-shaped, ≈ 120°; NO_3^-: trigonal planar, 120°

N_2O_4: trigonal planar, 120° about both N atoms

b. OCN^-, SCN^- and N_3^- are all linear with 180° bond angles.

73. From the Lewis structures (see Exercises 13.59 and 13.60), Br_3^- would have a linear molecular structure, ClF_3 and BrF_3 would have a T-shaped molecular structure and SF_4 would have a see-saw molecular structure. For example, consider ClF_3 (28 valence electrons):

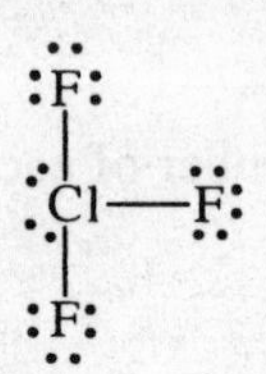

The central Cl atom is surrounded by 5 electron pairs, which requires a trigonal bipyramid geometry. Since there are 3 bonded atoms and 2 lone pairs of electrons about Cl, we describe the molecular structure of ClF_3 as T-shaped with predicted bond angles of about 90°. The actual bond angles will be slightly less than 90° due to the stronger repulsive effect of the lone pair electrons as compared to the bonding electrons.

75. a. $XeCl_2$ has 8 + 2(7) = 22 valence electrons.

There are 5 pairs of electrons about the central Xe atom. The structure will be based on a trigonal bipyramid geometry. The most stable arrangement of the atoms in $XeCl_2$ is a linear molecular structure with a 180° bond angle.

b. ICl_3 has 7 + 3(7) = 28 valence electrons.

T-shaped; The ClICl angles are ≈ 90°. Since the lone pairs will take up more space, the ClICl bond angles will probably be slightly less than 90°.

c. TeF_4 has 6 + 4(7) = 34 valence electrons.

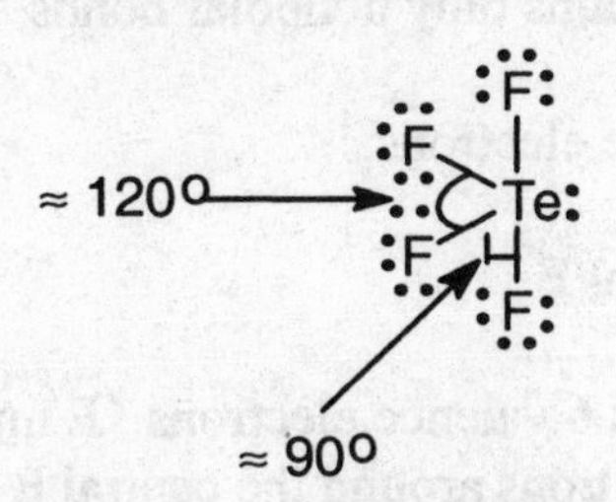

See-saw or teeter-totter or distorted tetrahedron

d. PCl_5 has 5 + 5(7) = 40 valence electrons.

Trigonal bipyramid

All of the species in this exercise have 5 pairs of electrons around the central atom. All of the structures are based on a trigonal bipyramid geometry, but only in PCl_5 are all of the pairs bonding pairs. Thus, PCl_5 is the only one we describe the molecular structure as trigonal bipyramid. Still, we had to begin with the trigonal bipyramid geometry to get to the structures (and bond angles) of the others.

77. Let us consider the molecules with three pairs of electrons around the central atom first; these molecules are SeO_3 and SeO_2 and both have a trigonal planar arrangement of electron pairs. Both these molecules have polar bonds but only SeO_2 has a dipole moment. The three bond dipoles from the three polar Se – O bonds in SeO_3 will all cancel when summed together. Hence, SeO_3 is nonpolar since the overall molecule has no resulting dipole moment. In SeO_2, the two Se – O bond dipoles do not cancel when summed together, hence SeO_2 has a dipole moment (is polar). Since O is more electronegative than Se, the negative end of the dipole moment is between the two O atoms and the positive end is around the Se atom. The arrow in the following illustration represents the overall dipole moment in SeO_2. Note that to predict polarity for SeO_2, either of the two resonance structures can be used.

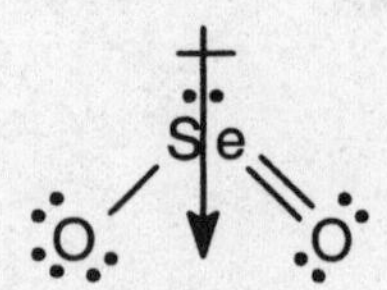

The other molecules in Exercise 13.72 (PCl_3, SCl_2, and SiF_4) have a tetrahedral arrangement of electron pairs. All have polar bonds; in SiF_4 the individual bond dipoles cancel when summed together, and in PCl_3 and SCl_2 the individual bond dipoles do not cancel. Therefore, SiF_4 has no dipole moment (is nonpolar) and PCl_3 and SCl_2 have dipole moments (are polar). For PCl_3, the negative end of the dipole moment is between the more electronegative chlorine atoms and the positive end is around P. For SCl_2, the negative end is between the more electronegative Cl atoms and the positive end of the dipole moment is around S.

79. The two usual requirements for a polar molecule are:
    1. polar bonds (bond dipoles)
    2. a structure such that the dipoles due to the polar bonds do not cancel.

In addition, some molecules that have no polar bonds, but contain unsymmetrical lone pairs are polar. For example, PH_3 is slightly polar even though it contains only nonpolar bonds.

81. The formula is EF_2O^{2-} and the Lewis structure has 28 valence electrons.

$$28 = x + 2(7) + 6 + 2, \ x = 6 \text{ valence electrons for element E}$$

Element E must belong to the group 6A elements since E has 6 valence electrons. E must also be a row 3 or heavier element since this ion has more than 8 electrons around the central E atom (row 2 elements never have more than 8 electrons around them). Some possible identities for E are S, Se and Te. The ion has a T-shaped molecular structure (see Exercise 13.73) with bond angles of $\approx 90°$.

83. Molecules which have an overall dipole moment are called polar molecules and molecules which do not have an overall dipole moment are called nonpolar molecules.

a. OCl_2, $6 + 2(7) = 20\ e^-$

V-shaped, polar; OCl_2 is polar because the two O – Cl bond dipoles don't cancel each other. The resultant dipole moment is shown in the drawing.

KrF_2, $8 + 2(7) = 22\ e^-$

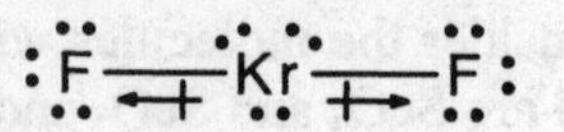

Linear, nonpolar; The molecule is nonpolar because the two Kr – F bond dipoles cancel each other.

BeH_2, $2 + 2(1) = 4\ e^-$

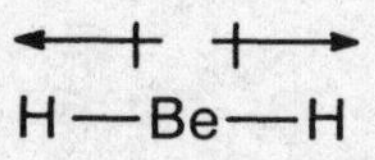

SO_2, $6 + 2(6) = 18\ e^-$

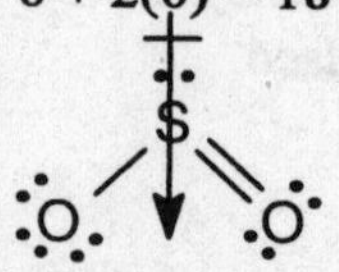

Linear, nonpolar; Be – H bond dipoles are equal and point in opposite directions. They cancel each other. BeH_2 is nonpolar.

V-shaped, polar; The S – O bond dipoles do not cancel so SO_2 is polar (has a dipole moment). Only one resonance structure is shown.

Note: All four species contain three atoms. They have different structures because the number of lone pairs of electrons around the central atom are different in each case.

b. SO_3, $6 + 3(6) = 24\ e^-$

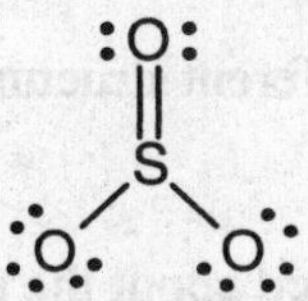

Trigonal planar, nonpolar; Bond dipoles cancel. Only one resonance structure is shown.

NF_3, $5 + 3(7) = 26\ e^-$

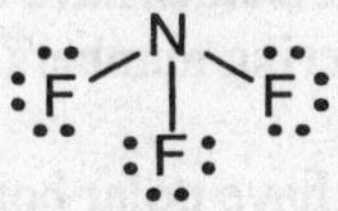

Trigonal pyramid, polar; Bond dipoles do not cancel.

IF_3 has $7 + 3(7) = 28$ valence electrons.

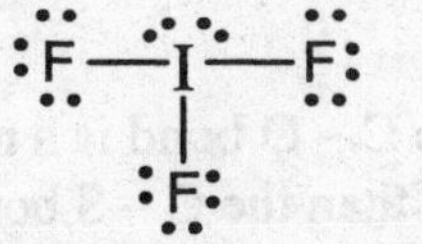

T-shaped, polar; Bond dipoles do not cancel.

Note: Each molecule has the same number of atoms, but the structures are different because of differing numbers of lone pairs around each central atom.

c. CF_4, $4 + 4(7) = 32\ e^-$

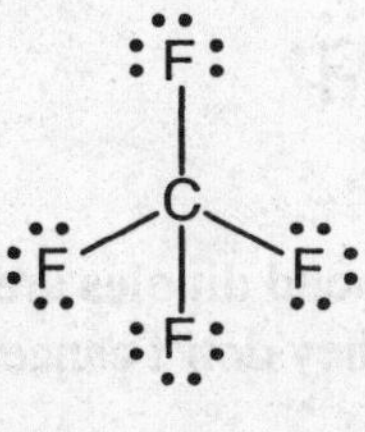

Tetrahedral, nonpolar; Bond dipoles cancel.

SeF_4, $6 + 4(7) = 34\ e^-$

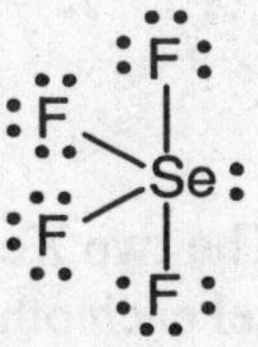

See-saw, polar; Bond dipoles do not cancel.

KrF_4, $8 + 4(7) = 36$ valence electrons

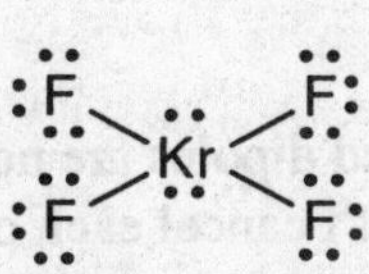

Square planar, nonpolar; Bond dipoles cancel.

Again, each molecule has the same number of atoms, but a different structure because of differing numbers of lone pairs around the central atom.

d. IF_5, $7 + 5(7) = 42$ e$^-$

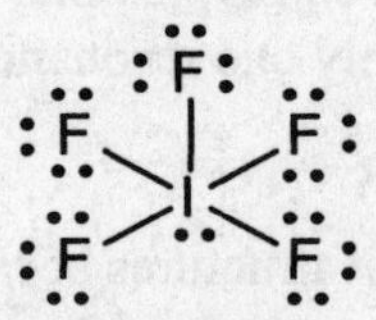

Square pyramid, polar; Bond dipoles do not cancel.

AsF_5, $5 + 5(7) = 40$ e$^-$

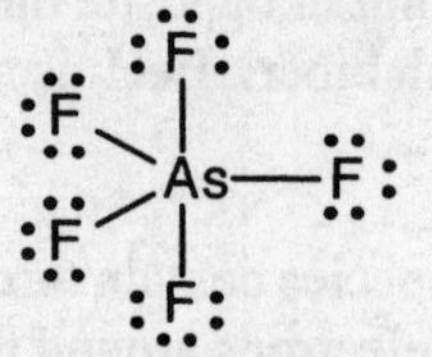

Trigonal bipyramid, nonpolar; Bond dipoles cancel.

Yet again, the molecules have the same number of atoms, but different structures because of the presence of differing numbers of lone pairs.

85. All these molecules have polar bonds that are symmetrically arranged about the central atoms. In each molecule, the individual bond dipoles cancel to give no net overall dipole moment. All these molecules are nonpolar even though they all contain polar bonds.

87. a.

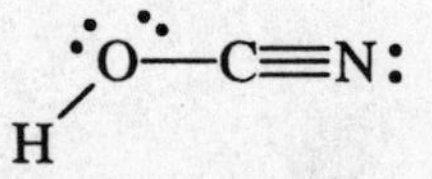

Polar; The bond dipoles do not cancel.

b.

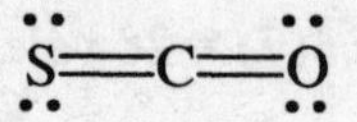

Polar; The C – O bond is a more polar bond than the C – S bond. So the two bond dipoles do not cancel each other.

c.

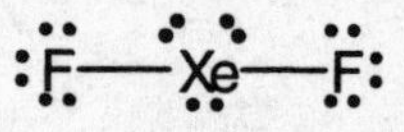

Nonpolar; The two Xe–F bond dipoles cancel each other.

d.

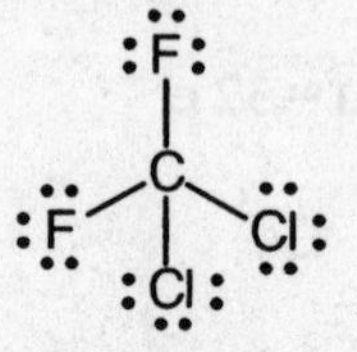

Polar; All the bond dipoles are not equivalent, so they don't cancel each other.

e.

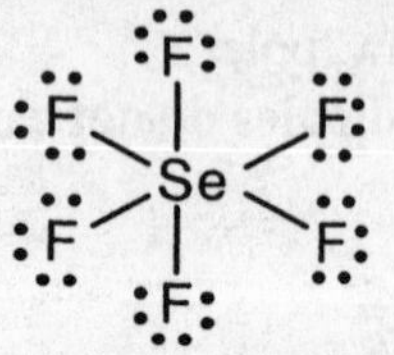

Nonpolar; The six Se – F bond dipoles cancel each other.

f.

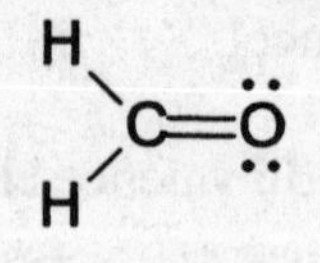

Polar; Bond dipoles are not equivalent so they don't cancel each other.

Additional Exercises

89. CS_2 has 4 + 2(6) = 16 valence electrons. C_3S_2 has 3(4) + 2(6) = 24 valence electrons.

$$:\ddot{S}=C=\ddot{S}: \quad \text{linear;} \qquad :\ddot{S}=C=C=C=\ddot{S}: \quad \text{linear}$$

91. As the halogen atoms get larger, it becomes more difficult to fit three halogen atoms around the small nitrogen atom, and the NX_3 molecule becomes less stable.

93. The stable species are:

a. NaBr: In $NaBr_2$, the sodium ion would have a + 2 charge assuming each bromine has a -1 charge. Sodium doesn't form stable Na^{2+} compounds.

b. ClO_4^-: ClO_4 has 31 valence electrons so it is impossible to satisfy the octet rule for all atoms in ClO_4. The extra electron from the -1 charge in ClO_4^- allows for complete octets for all atoms.

c. XeO_4: We can't draw a Lewis structure that obeys the octet rule for SO_4 (30 electrons), unlike with XeO_4 (32 electrons).

d. SeF_4: Both compounds require the central atom to expand its octet. O is too small and doesn't have low energy d orbitals to expand its octet (which is true for all row 2 elements).

95. a. Radius: $N^+ < N < N^-$; IE: $N^- < N < N^+$

N^+ has the fewest electrons held by the 7 protons in the nucleus while N^- has the most electrons held by the 7 protons. The 7 protons in the nucleus will hold the electrons most tightly in N^+ and least tightly in N^-. Therefore, N^+ has the smallest radius with the largest ionization energy (IE) and N^- is the largest species with the smallest IE.

b. Radius: $Cl^+ < Cl < Se < Se^-$; IE: $Se^- < Se < Cl < Cl^+$

The general trends tell us that Cl has a smaller radius than Se and a larger IE than Se. Cl^+, with fewer electron-electron repulsions than Cl, will be smaller than Cl and have a larger IE. Se^-, with more electron-electron repulsions than Se, will be larger than Se and have a smaller IE.

c. Radius: $Sr^{2+} < Rb^+ < Br^-$; IE: $Br^- < Rb^+ < Sr^{2+}$

These ions are isoelectronic. The species with the most protons (Sr^{2+}) will hold the electrons most tightly and will have the smallest radius and largest IE. The ion with the fewest protons (Br^-) will hold the electrons least tightly and will have the largest radius and smallest IE.

97. Nonmetals form covalent compounds. Nonmetals have valence electrons in the s and p orbitals. Since there are 4 total s and p orbitals, then there is room for only eight electrons (the octet rule).

99. S_2N_2 has 2(6) + 2(5) = 22 valence electrons.

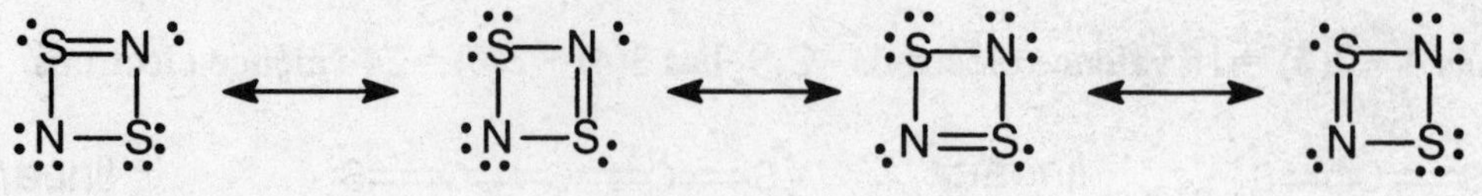

101. a. Al_2Cl_6 has 2(3) + 6(7) = 48 valence electrons.

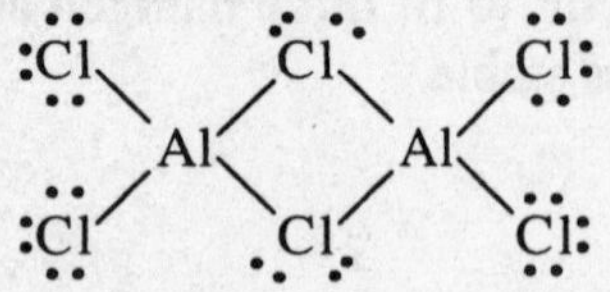

b. There are 4 pairs of electrons about each Al, we would predict the bond angles to be close to a tetrahedral angle of 109.5°.

c. nonpolar; The individual bond dipoles will cancel.

103. Yes, each structure has the same number of effective pairs around the central atom, giving the same predicted molecular structure for each compound/ion. (A multiple bond is counted as a single group of electrons.)

Challenge Problems

105. KrF_2, 8 + 2(7) = 22 e^-; From the Lewis structure, we have a trigonal bipyramid arrangement of electron pairs with a linear molecular structure.

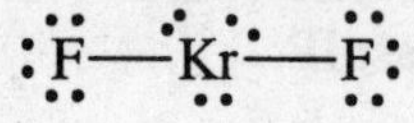

Hyperconjugation assumes that the overall bonding in KrF_2 is a combination of covalent and ionic contributions (see Section 13.13 of the text for discussion of hyperconjugation). Using hyperconjugation, two resonance structures are possible which keep the linear structure.

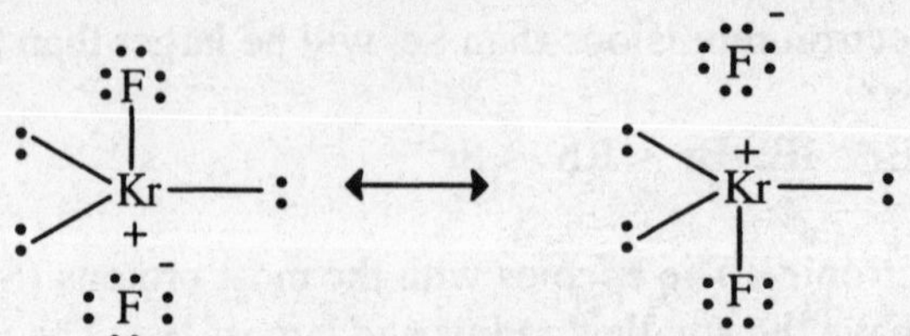

107. a. $N(NO_2)_2^-$ contains 5 + 2(5) + 4(6) + 1 = 40 valence electrons.

The most likely structures are:

There are other possible resonance structures, but these are most likely.

b. The NNN and all ONN and ONO bond angles should be about 120°.

c. $NH_4N(NO_2)_2 \rightarrow 2\ N_2 + 2\ H_2O + O_2$; Break and form all bonds.

Bonds broken:

4 N – H (391 kJ/mol)
1 N – N (160. kJ/mol)
1 N = N (418 kJ/mol)
3 N – O (201 kJ/mol)
1 N = O (607 kJ/mol)

ΣD_{broken} = 3352 kJ

Bonds formed:

2 N ≡ N (941 kJ/mol)
4 H – O (467 kJ/mol)
1 O = O (495 kJ/mol)

ΣD_{formed} = 4245 kJ

$\Delta H = \Sigma D_{broken} - \Sigma D_{formed}$ = 3352 kJ - 4245 kJ = -893 kJ

d. To estimate ΔH, we completely ignored the ionic interactions between NH_4^+ and $N(NO_2)_2^-$. In addition, we assumed the bond energies in Table 13.5 applied to the $N(NO_2)^-$ bonds in any one of the resonance structures above. This is a bad assumption since molecules that exhibit resonance generally have stronger overall bonds than predicted. All of these assumptions give an estimated ΔH value which is too negative.

109. a. i. $C_6H_6N_{12}O_{12} \rightarrow 6\ CO + 6\ N_2 + 3\ H_2O + 3/2\ O_2$

The NO_2 groups have one N – O single bond and one N = O double bond and each carbon atom has one C – H single bond. We must break and form all bonds.

Bonds broken:

3 C − C (347 kJ/mol)
6 C − H (413 kJ/mol)
12 C − N (305 kJ/mol)
6 N − N (160. kJ/mol)
6 N − O (201 kJ/mol)
6 N = O (607 kJ/mol)

ΣD_{broken} = 12,987 kJ

Bonds formed:

6 C ≡ O (1072 kJ/mol)
6 N ≡ N (941 kJ/mol)
6 H − O (467 kJ/mol)
3/2 O = O (495 kJ/mol)

ΣD_{formed} = 15,623 kJ

$\Delta H = \Sigma D_{broken} - \Sigma D_{formed}$ = 12,987 kJ - 15,623 kJ = -2636 kJ

ii. $C_6H_6N_{12}O_{12} \rightarrow 3\ CO + 3\ CO_2 + 6\ N_2 + 3\ H_2O$

Note: The bonds broken will be the same for all three reactions.

Bonds formed:

3 C ≡ O (1072 kJ/mol)
6 C = O (799 kJ/mol)
6 N ≡ N (941 kJ/mol)
6 H − O (467 kJ/mol)

ΣD_{formed} = 16,458 kJ

ΔH = 12,987 kJ - 16,458 kJ = -3471 kJ

iii. $C_6H_6N_{12}O_{12} \rightarrow 6\ CO_2 + 6\ N_2 + 3\ H_2$

Bonds formed:

12 C = O (799 kJ/mol)
6 N ≡ N (941 kJ/mol)
3 H − H (432 kJ/mol)

ΣD_{formed} = 16,530. kJ

ΔH = 12,987 kJ - 16,530. kJ = -3543 kJ

b. Reaction iii yields the most energy per mole of CL-20 so it will yield the most energy per kg.

$$\frac{-3543\ \text{kJ}}{\text{mol}} \times \frac{1\ \text{mol}}{438.23\ \text{g}} \times \frac{1000\ \text{g}}{\text{kg}} = -8085\ \text{kJ/kg}$$

Marathon Problem

111. Compound A: This compound is a strong acid (part g). HNO_3 is a strong acid and is available in concentrated solutions of 16 *M* (part c). The highest possible oxidation state of nitrogen is +5, and in HNO_3, the oxidation state of nitrogen is +5 (part b). Therefore, compound A is most likely HNO_3. The Lewis structures for HNO_3 are:

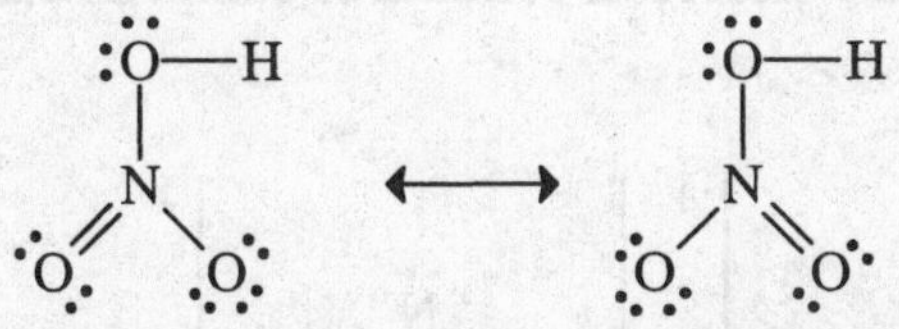

Compound B: This compound is basic (part g) and has one nitrogen (part b). The formal charge of zero (part b) tells us that there are three bonds to the nitrogen and the nitrogen has one lone pair. Assuming compound B is monobasic, then the data in part g tells us that the molar mass of B is 33.0 g/mol (21.98 mL of 1.000 *M* HCl = 0.02198 mol HCl, thus there are 0.02198 mol of B; 0.726 g/0.02198 mol = 33.0 g/mol). Because this number is rather small, it limits the possibilities. That is, there is one nitrogen, and the remainder of the atoms are O and H. Since the molar mass of B is 33.0 g/mol, then only one O oxygen atom can be present. The N and O atoms have a combined molar mass of 30.0 g/mol; the rest is made up of hydrogens (3 H atoms), giving the formula NH_3O. From the list of K_b values for weak bases in Appendix 5.3 of the text, compound B is most likely NH_2OH. The Lewis structure is:

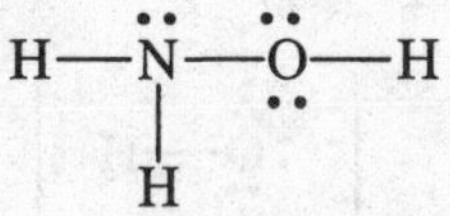

Compound C: From parts a and f and assuming compound A is HNO_3, then compound C contains the nitrate ion, NO_3^-. Because part b tells us that there are two nitrogens, the other ion needs to have one N and some H's. In addition, compound C must be a weak acid (part g), which must be due to the other ion since NO_3^- has no acidic properties. Also, the nitrogen atom in the other ion must have an oxidation state of -3 (part b) and a formal charge of +1. The ammonium ion fits the data. Thus, compound C is most likely NH_4NO_3. A Lewis structure is:

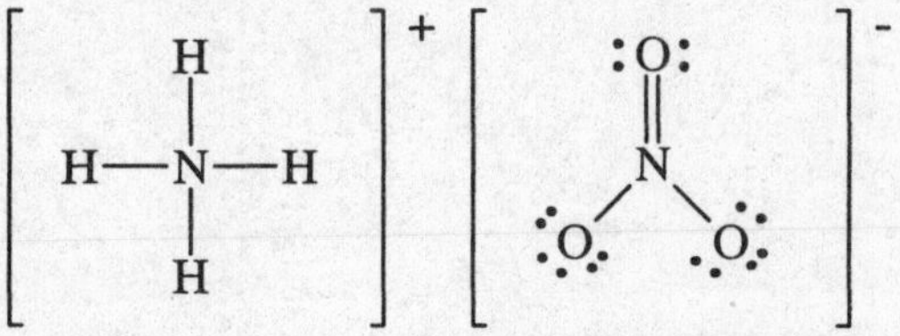

Note: Two more resonance structures can be drawn for NO_3^-.

Compound D: From part f, this compound has one less oxygen atom than compound C, thus NH_4NO_2 is a likely formula. Data from part e confirms this. Assuming 100.0 g of compound, we have:

$$43.7 \text{ g N} \times 1 \text{ mol}/14.01 \text{ g} = 3.12 \text{ mol N}$$
$$50.0 \text{ g O} \times 1 \text{ mol}/16.00 \text{ g} = 3.12 \text{ mol O}$$
$$6.3 \text{ g H} \times 1 \text{ mol}/1.008 \text{ g} = 6.25 \text{ mol H}$$

There is a 1:1:2 mole ratio of N:O:H, The empirical formula is NOH_2, which has an empirical formula mass of 32.0 g/mol.

$$\text{Molar mass} = \frac{dRT}{P} = \frac{2.86\text{ g/L }(0.08206\text{ L atm K}^{-1}\text{ mol}^{-1})(273\text{ K})}{1.00\text{ atm}} = 64.1\text{ g/mol}$$

For a correct molar mass, the molecular formula of compound D is $N_2O_2H_4$ or NH_4NO_2. A Lewis structure is:

$$\left[\begin{array}{c} H \\ | \\ H—N—H \\ | \\ H \end{array}\right]^+ \left[\; \ddot{N}\text{ bonded to }:\ddot{\underset{..}{O}}\text{ (single) and }=\ddot{O}\text{ (double)} \;\right]^-$$

Note: One more resonance structure for NO_2^- can be drawn.

<u>Compound E:</u> A basic solution (part g) which is commercially available at 15 *M* (part c) is ammonium hydroxide, NH_4OH. This is also consistent with the information given in parts b and d. The Lewis structure for NH_4OH is:

$$\left[\begin{array}{c} H \\ | \\ H—N—H \\ | \\ H \end{array}\right]^+ \left[\; :\ddot{\underset{..}{O}}—H \;\right]^-$$

CHAPTER FOURTEEN

COVALENT BONDING: ORBITALS

The Localized Electron Model and Hybrid Orbitals

9. H_2O has 2(1) + 6 = 8 valence electrons.

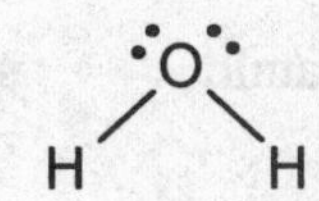

H_2O has a tetrahedral arrangement of the electron pairs about the O atom which requires sp^3 hybridization. Two of the four sp^3 hybrid orbitals are used to form bonds to the two hydrogen atoms and the other two sp^3 hybrid orbitals hold the two lone pairs on oxygen. The two O – H bonds are formed from overlap of the sp^3 hybrid orbitals from oxygen with the 1s atomic orbitals from the hydrogen atoms. Each O–H covalent bond is called a sigma (σ) bond since the shared electron pair in each bond is centered in an area on a line running between the two atoms.

11. See Exercises 13.47. 13.48 and 13.50 for the Lewis structures. To predict the hybridization, first determine the arrangement of electron pairs about each central atom using the VSEPR model, then utilize the information in Figure 14.24 of the text to deduce the hybridization required for that arrangement of electron pairs.

13.47 a. HCN; C is sp hybridized.

b. PH_3; P is sp^3 hybridized.

c. $CHCl_3$; C is sp^3 hybridized.

d. NH_4^+; N is sp^3 hybridized.

e. H_2CO; C is sp^2 hybridized.

f. SeF_2; Se is sp^3 hybridized.

g. CO_2; C is sp hybridized.

h. O_2; Each O atom is sp^2 hybridized.

i. HBr; Br is sp^3 hybridized.

13.48 a. All the central atoms are sp^3 hybridized.

b. All the central atoms are sp^3 hybridized.

c. All the central atoms are sp^3 hybridized.

13.50 a. In NO_2^-, N is sp^2 hybridized in NO_3^-, N is sp^2 hybridized, and in N_2O_4, both central N atoms are also sp^2 hybridized.

b. In OCN^- and SCN^-, the central carbon atoms in each ion are sp hybridized and in N_3^-, the central N atom is also sp hybridized.

13. a. b.

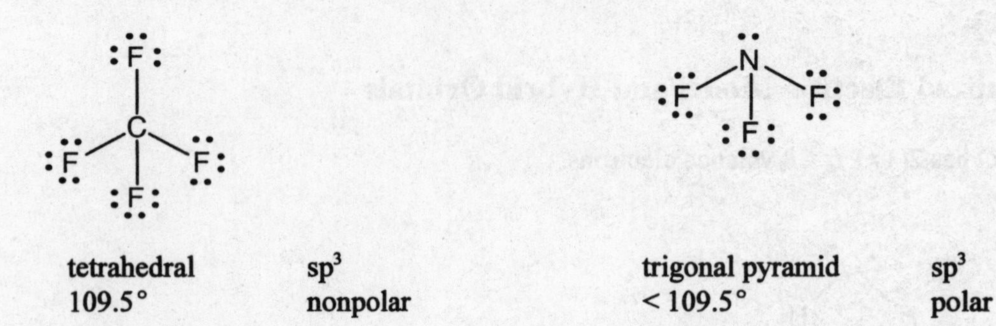

a.		b.	
tetrahedral	sp^3	trigonal pyramid	sp^3
109.5°	nonpolar	< 109.5°	polar

The angles in NF_3 should be slightly less than 109.5° because the lone pair requires more room than the bonding pairs.

c.		d.	
V-shaped	sp^3	trigonal planar	sp^2
< 109.5°	polar	120°	nonpolar

e.		f.	
linear	sp	see-saw	dsp^3
180°	nonpolar	a. ≈ 120°, b. ≈ 90°	polar

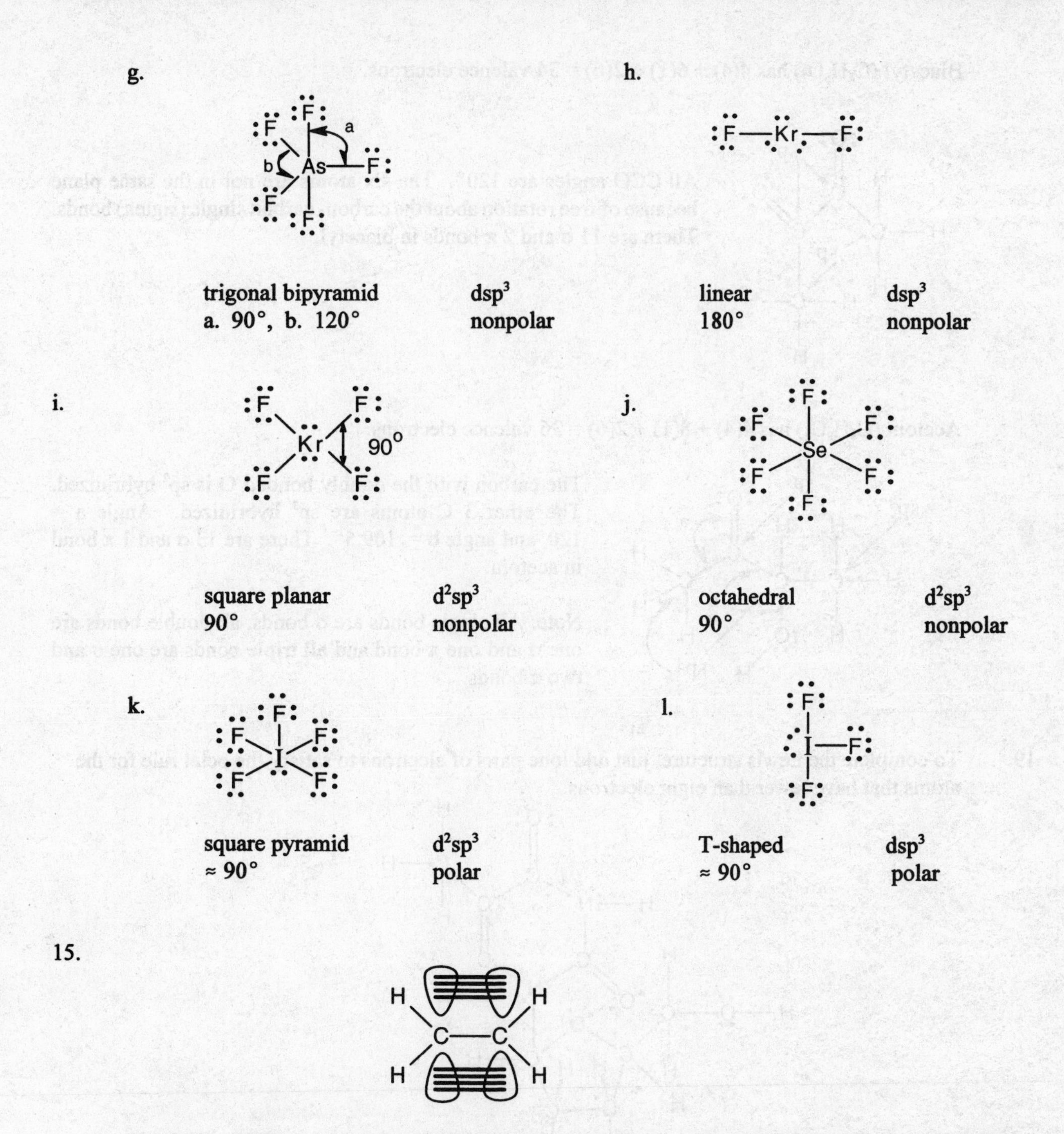

For the p-orbitals to properly line up to form the π bond, all six atoms are forced into the same plane. If the atoms are not in the same plane, then the π bond could not form since the p-orbitals would no longer be parallel to each other.

17. To complete the Lewis structures, just add lone pairs of electrons to satisfy the octet rule for the atoms with fewer than eight electrons.

Biacetyl ($C_4H_6O_2$) has 4(4) + 6(1) + 2(6) = 34 valence electrons.

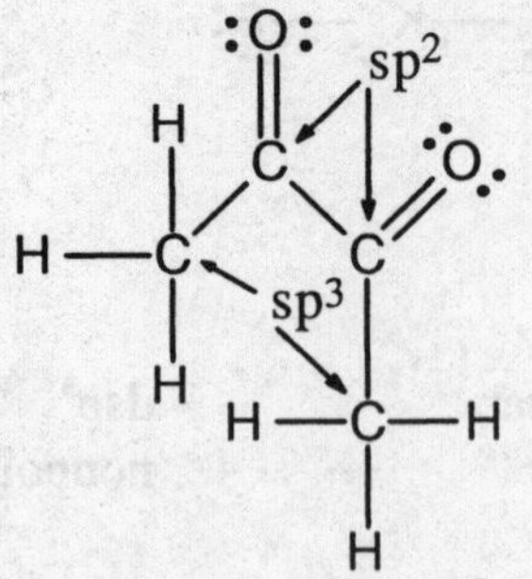

All CCO angles are 120°. The six atoms are not in the same plane because of free rotation about the carbon- carbon single (sigma) bonds. There are 11 σ and 2 π bonds in biacetyl.

Acetoin ($C_4H_8O_2$) has 4(4) + 8(1) + 2(6) = 36 valence electrons.

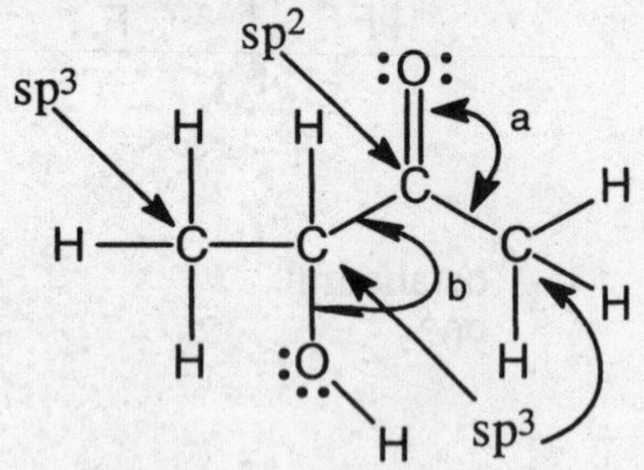

The carbon with the doubly bonded O is sp² hybridized. The other 3 C atoms are sp³ hybridized. Angle a = 120° and angle b = 109.5°. There are 13 σ and 1 π bond in acetoin.

Note: All single bonds are σ bonds, all double bonds are one σ and one π bond and all triple bonds are one σ and two π bonds.

19. To complete the Lewis structure, just add lone pairs of electrons to satisfy the octet rule for the atoms that have fewer than eight electrons.

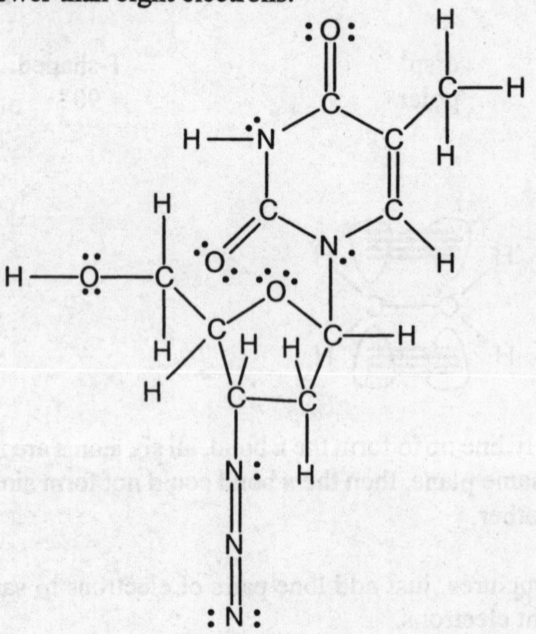

a. 6 b. 4 c. The center N in $-N=N=N$ group

d. 33 σ e. 5 π f. 180°

g. ≈ 109.5° h. sp^3

21. a. **Piperine and capsaicin are molecules classified as organic compounds, i.e., compounds based on carbon. The majority of Lewis structures for organic compounds have all atoms with zero formal charge. Therefore, carbon atoms in organic compounds will usually form four bonds, nitrogen atoms will form three bonds and complete the octet with one lone pair of electrons, and oxygen atoms will form two bonds and complete the octet with two lone pairs of electrons. Using these guidelines, the Lewis structures are:**

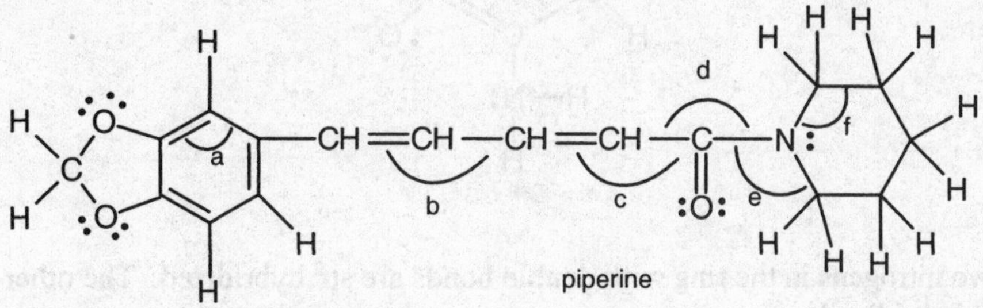

piperine

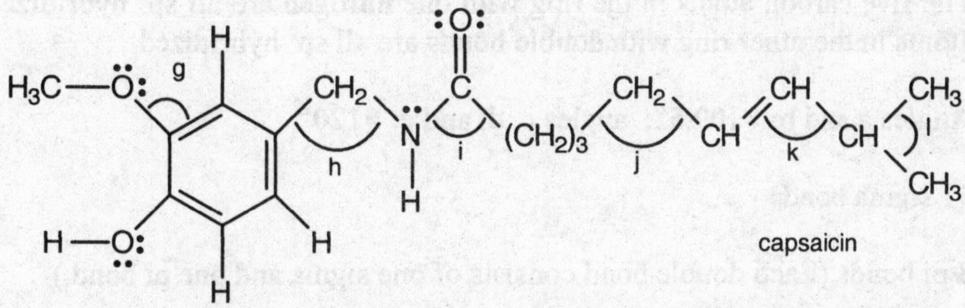

capsaicin

Note: The ring structures are all shorthand notation for rings of carbon atoms. In piperine, the first ring contains 6 carbon atoms and the second ring contains 5 carbon atoms (plus nitrogen). Also notice that CH_3, CH_2 and CH are shorthand for a carbon atoms singly bonded to hydrogen atoms.

b. **piperine: 0 sp, 11 sp^2 and 6 sp^3 carbons; capsaicin: 0 sp, 9 sp^2 and 9 sp^3 carbons**

c. **The nitrogens are sp^3 hybridized in each molecule.**

d. a. 120° e. ≈ 109.5° i. 120°
b. 120° f. 109.5° j. 109.5°
c. 120° g. 120° k. 120°
d. 120° h. 109.5° l. 109.5°

23. To complete the Lewis structure, just add lone pairs of electrons to satisfy the octet rule for the atoms with fewer than eight electrons.

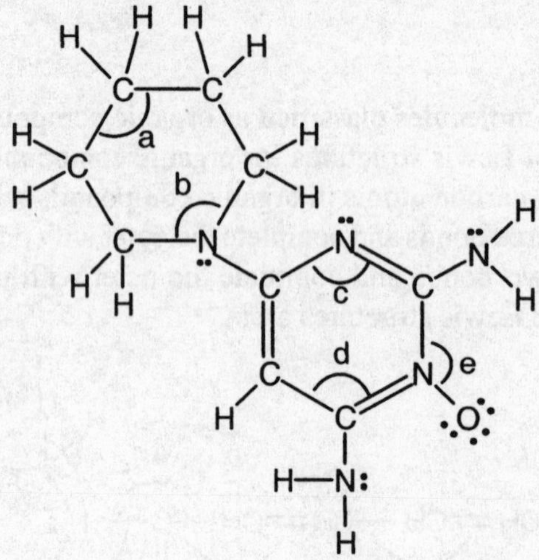

a. The two nitrogens in the ring with double bonds are sp^2 hybridized. The other three nitrogens are sp^3 hybridized.

b. The five carbon atoms in the ring with one nitrogen are all sp^3 hybridized. The four carbon atoms in the other ring with double bonds are all sp^2 hybridized.

c. Angles a and b: $\approx 109.5°$; angles c, d, and e: $\approx 120°$

d. 31 sigma bonds

e. 3 pi bonds (Each double bond consists of one sigma and one pi bond.)

The Molecular Orbital (MO) Model

25. Bond energy is directly proportional to bond order. Bond length is inversely proportional to bond order. Bond energy and bond length can be measured.

27. Paramagnetic: Unpaired electrons are present. Measure the mass of a substance in the presence and absence of a magnetic field. A substance with unpaired electrons will be attracted by the magnetic field, giving an apparent increase in mass in the presence of the field. Greater number of unpaired electrons will give greater attraction and greater observed mass increase.

29.

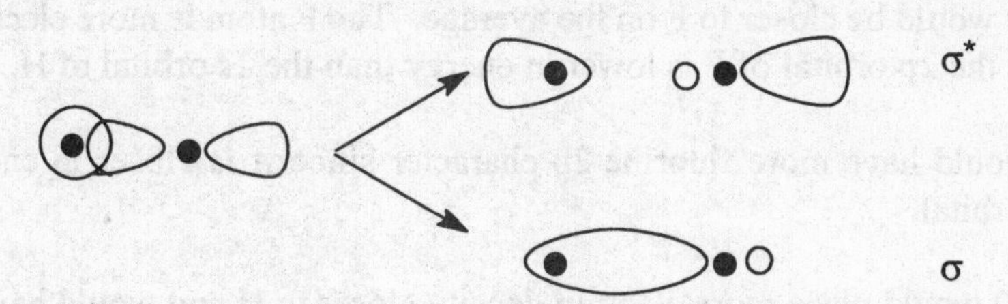

These molecular orbitals are sigma MOs since the electron density is cylindrically symmetric about the internuclear axis.

31.

a.	H_2:	$(\sigma_{1s})^2$	B.O. = (2-0)/2 = 1, diamagetic (0 unpaired e^-)
b.	B_2:	$(\sigma_{2s})^2(\sigma_{2s}*)^2(\pi_{2p})^2$	B.O. = (4-2)/2 = 1, paramagnetic (2 unpaired e^-)
c.	NO:	$(\sigma_{2s})^2(\sigma_{2s}*)^2(\pi_{2p})^4(\sigma_{2p})^2(\pi_{2p}*)^1$	B.O. = (8-3)/2 = 2.5, paramagnetic (1 unpaired e^-)
d.	CN^+:	$(\sigma_{2s})^2(\sigma_{2s}*)^2(\pi_{2p})^4$	B.O. = (6-2)/2 = 2, diamagnetic
e.	CN:	$(\sigma_{2s})^2(\sigma_{2s}*)^2(\pi_{2p})^4(\sigma_{2p})^1$	B.O. = (7-2)/2 = 2.5, paramagnetic (1 unpaired e^-)
f.	CN^-:	$(\sigma_{2s})^2(\sigma_{2s}*)^2(\pi_{2p})^4(\sigma_{2p})^2$	B.O. = 3, diamagnetic
g.	N_2:	$(\sigma_{2s})^2(\sigma_{2s}*)^2(\pi_{2p})^4(\sigma_{2p})^2$	B.O. = 3, diamagnetic
h.	N_2^+:	$(\sigma_{2s})^2(\sigma_{2s}*)^2(\pi_{2p})^4(\sigma_{2p})^1$	B.O. = 2.5, paramagnetic (1 unpaired e^-)
i.	N_2^-:	$(\sigma_{2s})^2(\sigma_{2s}*)^2(\pi_{2p})^4(\sigma_{2p})^2(\pi_{2p}*)^1$	B.O. = 2.5, paramagnetic (1 unpaired e^-)

33. CN: $(\sigma_{2s})^2(\sigma_{2s}*)^2(\pi_{2p})^4(\sigma_{2p})^1$
NO: $(\sigma_{2s})^2(\sigma_{2s}*)^2(\pi_{2p})^4(\pi_{2p})^2(\pi_{2p}*)^1$
O_2^{2+}: $(\sigma_{2s})^2(\sigma_{2s}*)^2(\sigma_{2p})^2(\pi_{2p})^4$
N_2^{2+}: $(\sigma_{2s})^2(\sigma_{2s}*)^2(\pi_{2p})^4$

If the added electron goes into a bonding orbital, the bond order would increase, making the species more stable and more likely to form. Between CN and NO, CN would most likely form CN^- since the bond order increases (unlike NO^- where the added electron goes into an antibonding orbital). Between O_2^{2+} and N_2^{2+}, N_2^+ would most likely form since the bond order increases (unlike O_2^+).

35. The electron configurations are (assuming the same orbital order as that for N_2):

CO:	$(\sigma_{2s})^2(\sigma_{2s}*)^2(\pi_{2p})^4(\sigma_{2p})^2$	B.O. = (8-2)/2 = 3; 0 unpaired e^-
CO^+:	$(\sigma_{2s})^2(\sigma_{2s}*)^2(\pi_{2p})^4(\sigma_{2p})^1$	B.O. = (7-2)/2 = 2.5; 1 unpaired e^-
CO^{2+}:	$(\sigma_{2s})^2(\sigma_{2s}*)^2(\pi_{2p})^4$	B.O. = (6-2)/2 = 2; 0 unpaired e^-

Since bond order is directly related to bond energy and, in turn, inversely related to bond length, then the bond length order should be: $CO < CO^+ < CO^{2+}$.

37. The π bonds between S atoms and between C and S atoms are not as strong. The atomic orbitals do not overlap with each other as well as the smaller atomic orbitals of C and O overlap.

39. Side to side overlap of these d-orbitals would produce a π molecular orbital. There would be no probability of finding an electron on the axis joining the two nuclei, which is characteristic of π MOs.

41. a. The electron density would be closer to F on the average. The F atom is more electronegative than the H atom and the 2p orbital of F is lower in energy than the 1s orbital of H.

b. The bonding MO would have more fluorine 2p character since it is closer in energy to the fluorine 2p atomic orbital.

c. The antibonding MO would place more electron density closer to H and would have a greater contribution from the higher energy hydrogen 1s atomic orbital.

Spectroscopy

43. reduced mass $= \mu = \dfrac{m_1 m_2}{m_1 + m_2} = \dfrac{(1.0078)(78.918)}{1.0078 + 78.918}$ amu $\times \dfrac{1.66054 \times 10^{-27}\ \text{kg}}{\text{amu}} = 1.6524 \times 10^{-27}$ kg

$$\nu_o = \frac{1}{2\pi}\sqrt{\frac{k}{\mu}} = \frac{c}{\lambda} = 2.9979 \times 10^{10}\ \text{cm s}^{-1} \times 2650.\ \text{cm}^{-1} = 7.944 \times 10^{13}\ \text{s}^{-1}$$

$$7.944 \times 10^{13}\ \text{s}^{-1} = \frac{1}{2\pi}\sqrt{\frac{k}{\mu}} = \frac{1}{2\pi}\sqrt{\frac{k}{1.6524 \times 10^{-27}\ \text{kg}}}$$

Solving for k, the force constant: $k = 411.7\ \text{kg s}^{-2} = 411.7\ \text{N m}^{-1}$

Note: 1 newton = 1 N = 1 kg m s^{-2}

45. a. $\Delta E = 2hB\,(J_i + 1) = h\nu = \dfrac{hc}{\lambda},\ \dfrac{c}{\lambda} = 2B\,(J_i + 1)$

$$\frac{c}{\lambda} = 2B\,(0 + 1) = 2B = \frac{2.998 \times 10^{8}\ \text{m s}^{-1}}{2.60 \times 10^{-3}\ \text{m}} = 1.15 \times 10^{11}\ \text{s}^{-1}$$

$$B = \frac{1.15 \times 10^{-11}\ \text{s}^{-1}}{2} = 5.75 \times 10^{10}\ \text{s}^{-1}$$

$$I = \frac{h}{8\pi^2 B} = \frac{6.626 \times 10^{-34}\ \text{J s}}{8\pi^2\,(5.75 \times 10^{10}\ \text{s}^{-1})} = 1.46 \times 10^{-46}\ \text{kg m}^2$$

$$I = \mu R_e^2,\ \mu = \frac{m_1 m_2}{m_1 + m_2} = \frac{12.000\,(15.995)}{12.000 + 15.995}\ \text{amu} \times \frac{1.66054 \times 10^{-27}\ \text{kg}}{\text{amu}} = 1.1385 \times 10^{-26}\ \text{kg}$$

$$R_e^2 = \frac{I}{\mu} = \frac{1.46 \times 10^{-46}\ \text{kg m}^2}{1.1385 \times 10^{-26}\ \text{kg}} = 1.28 \times 10^{-20}\ \text{m}^2,\ R_e = \text{bond length} = 1.13 \times 10^{-10}\ \text{m} = 113\ \text{pm}$$

b. $\nu = \dfrac{\Delta E}{h} = 2B\,(J_i + 1) = 2B\,(2 + 1) = 6B$

From part a, $B = 5.75 \times 10^{10}\ \text{s}^{-1}$, so: $\nu = 6(5.75 \times 10^{10}\ \text{s}^{-1}) = 3.45 \times 10^{11}\ \text{s}^{-1}$

47. a. **The $-CH_2$ group neighbors a $-CH_3$ group so a quartet of peaks should result (iv).**

b. **The $-CH_3$ H-atoms are separated by more than three sigma bonds from other H-atoms, so no spin-spin coupling should occur. A singlet peak should result (i).**

c. **The $-CH_2$ H-atoms neighbor two H-atoms, so a triplet peak should result (iii).**

d. **The 2 H-atoms in the $-CH_2F$ group neighbor one H-atom, so a doublet peak should result (ii).**

Additional Exercises

49. a. **XeO_3, $8 + 3(6) = 26\ e^-$**

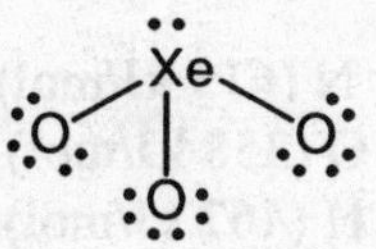

trigonal pyramid; sp^3

b. **XeO_4, $8 + 4(6) = 32\ e^-$**

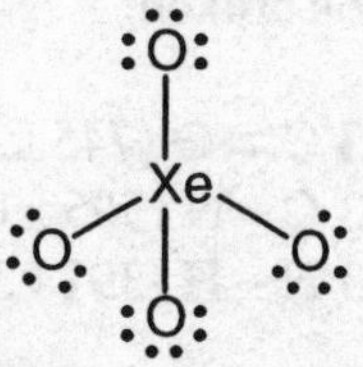

tetrahedral; sp^3

c. **$XeOF_4$, $8 + 6 + 4(7) = 42\ e^-$**

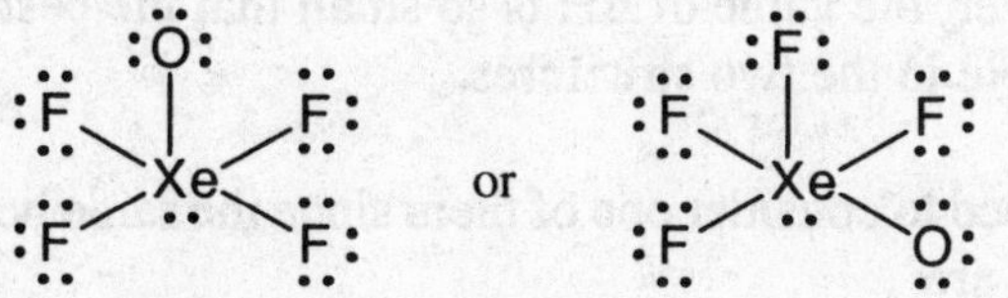

or

square pyramid; d^2sp^3

d. **$XeOF_2$, $8 + 6 + 2(7) = 28\ e^-$**

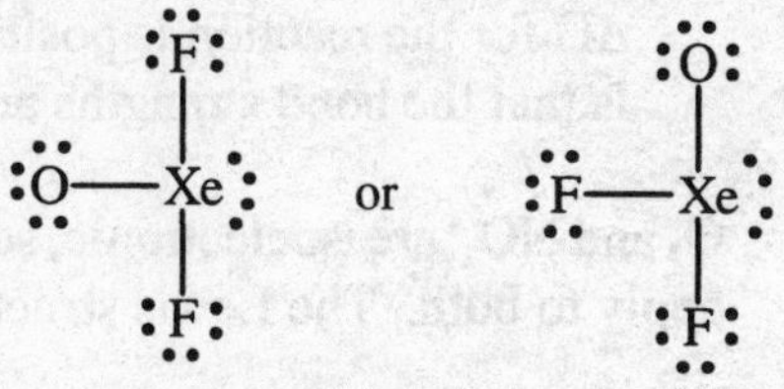

or

T-shaped; dsp^3

e. **XeO_3F_2 has $8 + 3(6) + 2(7) = 40$ valence electrons.**

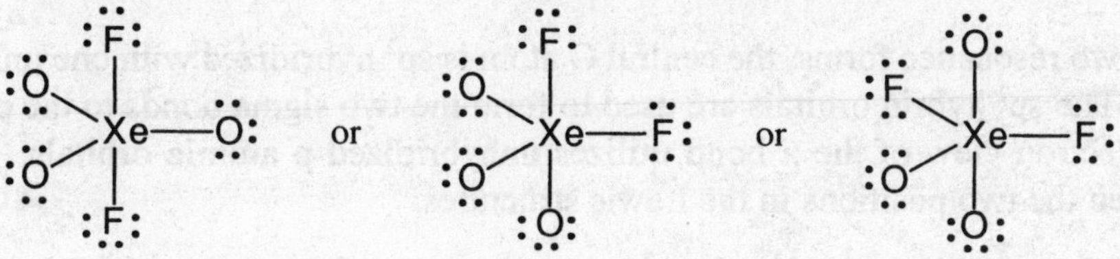

or or

trigonal bipyramid; dsp^3

51. a. **No, some atoms are in different places. Thus, these are not resonance structures; they are different compounds.**

b. **For the first Lewis structure, all nitrogens are sp^3 hybridized and all carbons are sp^2 hybridized. In the second Lewis structure, all nitrogens and carbons are sp^2 hybridized.**

c. For the reaction:

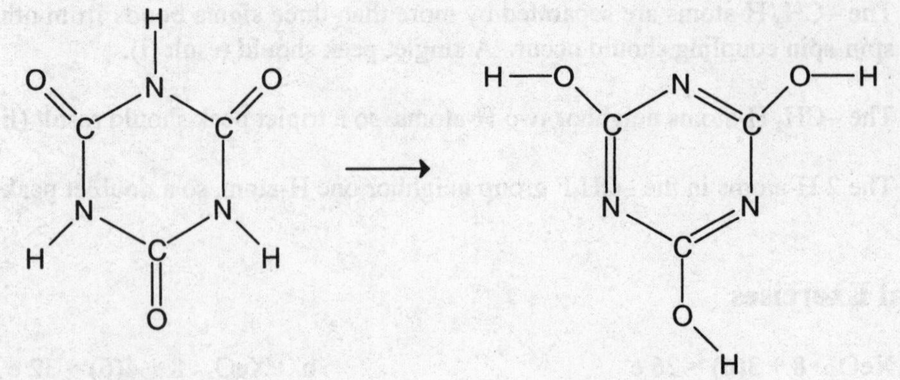

Bonds broken:

3 C = O (745 kJ/mol)
3 C − N (305 kJ/mol)
3 N − H (391 kJ/mol)

Bonds formed:

3 C = N (615 kJ/mol)
3 C − O (358 kJ/mol)
3 O − H (467 kJ/mol)

$$\Delta H = 3(745) + 3(305) + 3(391) - [3(615) + 3(358) + 3(467)]$$

$$\Delta H = 4323 \text{ kJ} - 4320 \text{ kJ} = 3 \text{ kJ}$$

The bonds are slightly stronger in the first structure with the carbon-oxygen double bonds since ΔH for the reaction is positive. However, the value of ΔH is so small that the best conclusion is that the bond strengths are comparable in the two structures.

53. O_3 and NO_2^- are isoelectronic, so we only need to consider one of them since the same bonding ideas apply to both. The Lewis structures for O_3 are:

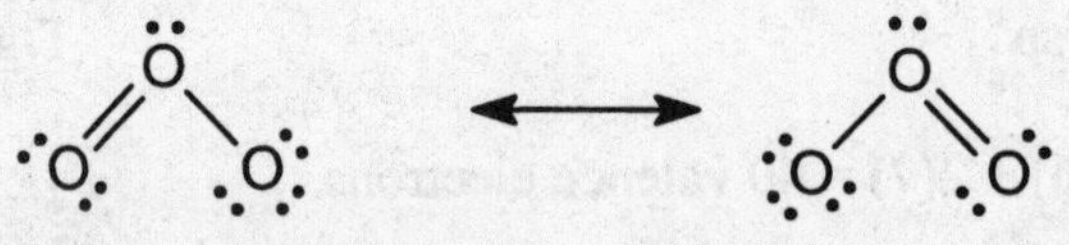

For each of the two resonance forms, the central O atom is sp^2 hybridized with one unhybridized p atomic orbital. The sp^2 hybrid orbitals are used to form the two sigma bonds to the central atom. The localized electron view of the π bond utilizes unhybridized p atomic orbitals. The π bond resonates between the two positions in the Lewis structures:

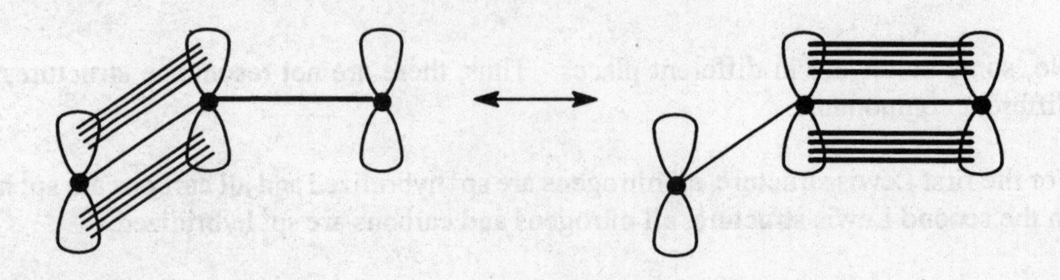

In the MO picture of the π bond, all three unhybridized p-orbitals overlap at the same time, resulting in π electrons that are delocalized over the entire surface of the molecule. This is represented as:

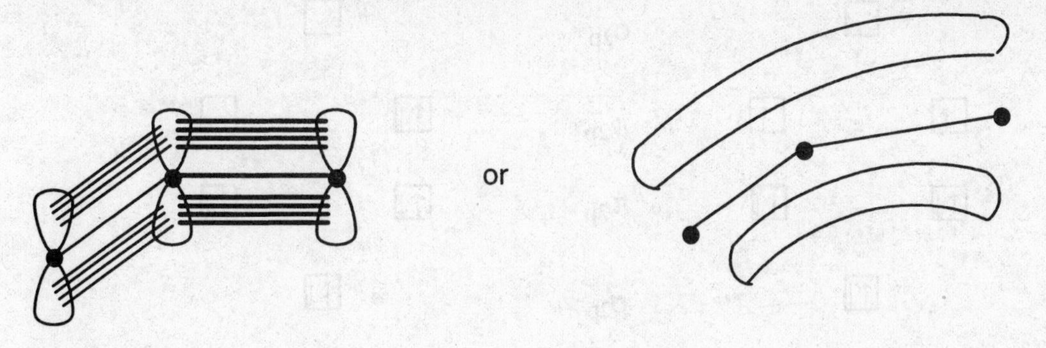

55. Lewis structures:

NO^+: $[:N \equiv O:]^+$

NO: $N=O \longleftrightarrow N=O \longleftrightarrow N=O$

NO^-: $[N=O]^-$

M.O. model (assuming the same orbital order as that for N_2):

NO^+: $(\sigma_{2s})^2(\sigma_{2s}*)^2(\pi_{2p})^4(\sigma_{2p})^2$, B.O. = 3, 0 unpaired e^- (diamagnetic)

NO: $(\sigma_{2s})^2(\sigma_{2s}*)^2(\pi_{2p})^4(\sigma_{2p})^2(\pi_{2p}*)^1$, B.O. = 2.5, 1 unpaired e^- (paramagnetic)

NO^-: $(\sigma_{2s})^2(\sigma_{2s}*)^2(\pi_{2p})^4(\sigma_{2p})^2(\pi_{2p}*)^2$ B.O. = 2, 2 unpaired e^- (paramagnetic)

From the bond orders:

bond energies: $NO^- < NO < NO^+$; bond lengths: $NO^+ < NO < NO^-$

The two models only give the same results for NO^+ (a triple bond with no unpaired electrons). Lewis structures are not adequate for NO and NO^-. The MO model gives a better representation for all three species. For NO, Lewis structures are poor for odd electron species. For NO^-, both models predict a double bond but only the MO model correctly predicts that NO^- is paramagnetic.

57. Considering only the twelve valence electrons in O_2, the MO models would be:

O₂ ground state		MO	Arrangement consistent with Lewis structure	
	☐	σ_{2p}^*	☐	
↑	↑	π_{2p}^*	↑↓	☐
↑↓	↑↓	π_{2p}	↑↓	↑↓
	↑↓	σ_{2p}	↑↓	
	↑↓	σ_{2s}^*	↑↓	
	↑↓	σ_{2s}	↑↓	

O_2 ground state

Arrangement of electrons consistent with the Lewis structure (double bond and no unpaired electrons).

It takes energy to pair electrons in the same orbital. Thus, the structure with no unpaired electrons is at a higher energy; it is an excited state.

59. a. $COCl_2$ has 4 + 6 + 2(7) = 24 valence electrons.

trigonal planar
polar
120°
sp^2

b. N_2F_2 has 2(5) + 2(7) = 24 valence electrons.

Can also be:

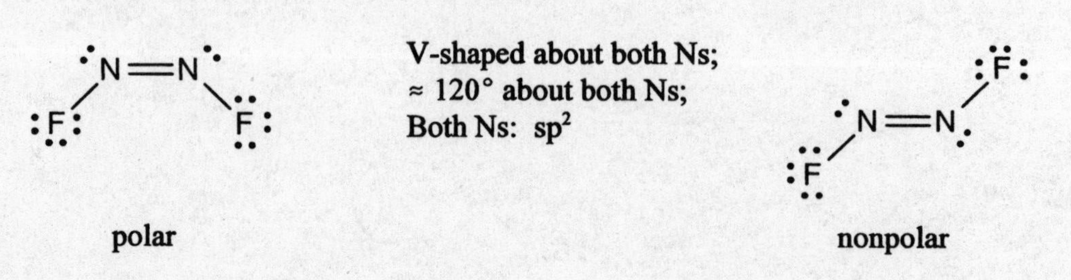

polar

nonpolar

These are distinctly different molecules.

c. COS has 4 + 6 + 6 = 16 valence electrons.

:O=C=S: linear, polar, 180°, sp

d. ICl_3 has 7 + 3(7) = 28 valence electrons.

61. a. The Lewis structures for NNO and NON are:

:N=N=O: ⟷ :N≡N—O: ⟷ :N—N≡O:

:N=O=N: ⟷ :N≡O—N: ⟷ :N—O≡N:

The NNO structure is correct. From the Lewis structures we would predict both NNO and NON to be linear. However, we would predict NNO to be polar and NON to be nonpolar. Since experiments show N_2O to be polar, then NNO is the correct structure.

b. Formal charge = number of valence electrons of atoms - [(number of lone pair electrons) + 1/2 (number of shared electrons)].

:N=N=O: ⟷ :N≡N—O: ⟷ :N—N≡O:

-1 +1 0   0 +1 -1   -2 +1 +1

The formal charges for the atoms in the various resonance structures are below each atom. The central N is sp hybridized in all of the resonance structures. We can probably ignore the 3rd resonance structure on the basis of the relatively large formal charges as compared to the first two resonance structures.

c. The sp hybrid orbitals from the center N overlap with atomic orbitals (or appropriate hybrid orbitals) from the other two atoms to form the two sigma bonds. The remaining two unhybridized p orbitals from the center N overlap with two p orbitals from the peripheral N to form the two π bonds.

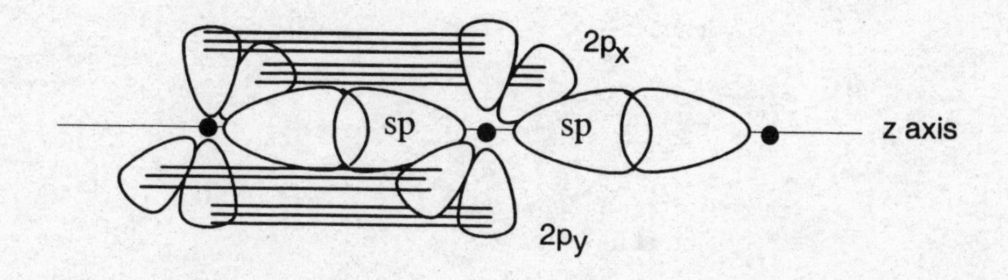

Challenge Problems

63. a. $E = \frac{hc}{\lambda} = \frac{(6.626 \times 10^{-34} \text{ J s})(2.998 \times 10^8 \text{ m/s})}{25 \times 10^{-9} \text{ m}} = 7.9 \times 10^{-18} \text{ J}$

$$7.9 \times 10^{-18} \text{ J} \times \frac{6.022 \times 10^{23}}{\text{mol}} \times \frac{1 \text{ kJ}}{1000 \text{ J}} = 4800 \text{ kJ/mol}$$

Using ΔH values from the various reactions, 25 nm light has sufficient energy to ionize N_2 and N, and to break the triple bond. Thus, N_2, N_2^+, N, and N^+ will all be present, assuming excess N_2.

b. To produce atomic nitrogen but no ions, the range of energies of the light must be from 941 kJ/mol to just below 1402 kJ/mol.

$$\frac{941 \text{ kJ}}{\text{mol}} \times \frac{1 \text{ mol}}{6.022 \times 10^{23}} \times \frac{1000 \text{ J}}{\text{kJ}} = 1.56 \times 10^{-18} \text{ J/photon}$$

$$\lambda = \frac{hc}{E} = \frac{(6.626 \times 10^{-34} \text{ J s})(2.998 \times 10^8 \text{ m/s})}{1.56 \times 10^{-18} \text{ J}} = 1.27 \times 10^{-7} \text{ m} = 127 \text{ nm}$$

$$\frac{1402 \text{ kJ}}{\text{mol}} \times \frac{1 \text{ mol}}{6.0221 \times 10^{23}} \times \frac{1000 \text{ J}}{\text{kJ}} = 2.328 \times 10^{-18} \text{ J/photon}$$

$$\lambda = \frac{hc}{E} = \frac{(6.6261 \times 10^{-34} \text{ J s})(2.9979 \times 10^8 \text{ m/s})}{2.328 \times 10^{-18} \text{ J}} = 8.533 \times 10^{-8} \text{ m} = 85.33 \text{ nm}$$

Light with wavelengths in the range of $85.33 \text{ nm} < \lambda \le 127 \text{ nm}$ will produce N but no ions.

c. N_2: $(\sigma_{2s})^2(\sigma_{2s}^*)^2(\pi_{2p})^4(\sigma_{2p})^2$; The electron removed from N_2 is in the σ_{2p} molecular orbital which is lower in energy than the 2p atomic orbital from which the electron in atomic nitrogen is removed. Since the electron removed from N_2 is lower in energy than the electron removed from N, then the ionization energy of N_2 is greater than that for N.

65. The complete Lewis structure follows. All but two of the carbon atoms are sp^3 hybridized. The two carbon atoms which contain the double bond are sp^2 hybridized (see *).

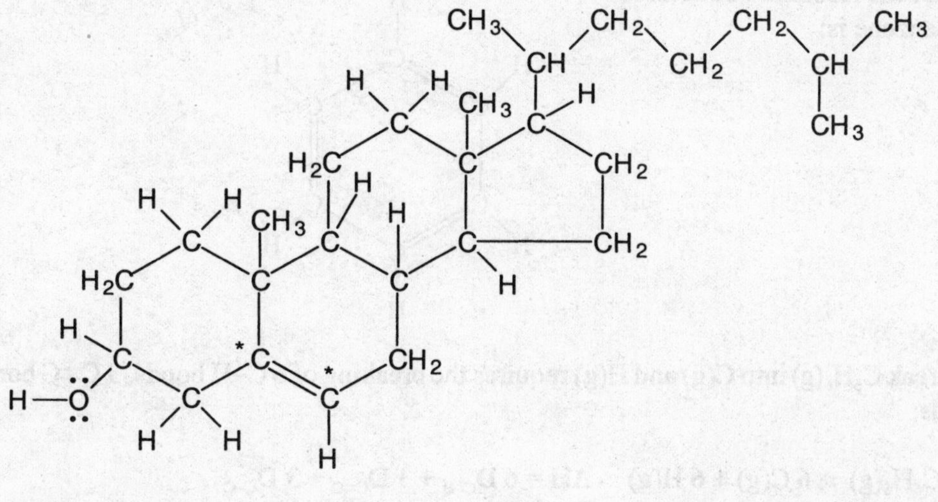

No; most of the carbons are not in the same plane since a majority of carbon atoms exhibit a tetrahedral structure. Note: CH, CH_2 and CH_3 are shorthand for carbon atoms singly bonded to hydrogen atoms.

67. a. The CO bond is polar with the negative end around the more electronegative oxygen atom. We would expect metal cations to be attracted to and to bond to the oxygen end of CO on the basis of electronegativity.

b. :C≡O:

FC (carbon) = 4 - 2 - 1/2(6) = -1

FC (oxygen) = 6 - 2 - 1/2(6) = +1

From formal charge, we would expect metal cations to bond to the carbon (with the negative formal charge).

c. In molecular orbital theory, only orbitals with proper symmetry overlap to form bonding orbitals. The metals that form bonds to CO are usually transition metals, all of which have outer electrons in the d orbitals. The only molecular orbitals of CO that have proper symmetry to overlap with d orbitals are the π_{2p}* orbitals, whose shape is similar to the d orbitals (see Figure 14.36). Since the antibonding molecular orbitals have more carbon character, one would expect the bond to form through carbon.

69. One of the resonance structures for benzene is:

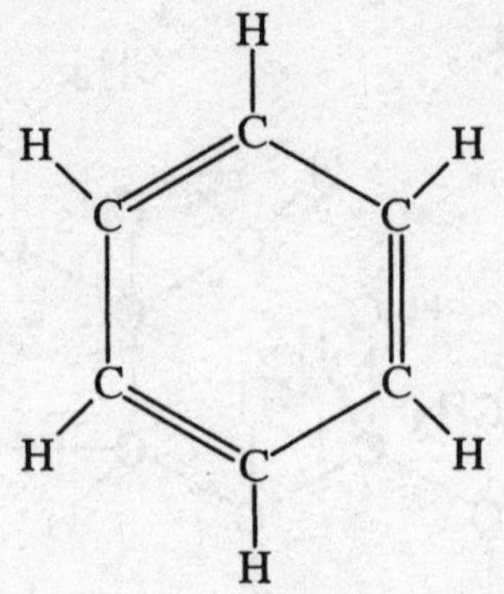

To break $C_6H_6(g)$ into C(g) and H(g) requires the breaking of 6 C–H bonds, 3 C=C bonds and 3 C–C bonds:

$$C_6H_6(g) \rightarrow 6\ C(g) + 6\ H(g) \quad \Delta H = 6\ D_{C-H} + 3\ D_{C=C} + 3\ D_{C-C}$$

$$\Delta H = 6(413\ kJ) + 3(614\ kJ) + 3(347\ kJ) = 5361\ kJ$$

The question wants ΔH_f° for $C_6H_6(g)$ which is ΔH for the reaction:

$$6\ C(s) + 3\ H_2(g) \rightarrow C_6H_6(g) \quad \Delta H = \Delta H^\circ_{f,\ C_6H_6(g)}$$

To calculate ΔH for this reaction, we will use Hess's law along with the ΔH_f° value for C(g) and the bond energy value for H_2 (D_{H_2} = 432 kJ/mol).

$$6\ C(g) + 6\ H(g) \rightarrow C_6H_6(g) \quad \Delta H_1 = -5361\ kJ$$
$$6\ C(s) \rightarrow 6\ C(g) \quad \Delta H_2 = 6(717\ kJ)$$
$$3\ H_2(g) \rightarrow 6\ H(g) \quad \Delta H_3 = 3(432\ kJ)$$

$$6\ C(s) + 3\ H_2(g) \rightarrow C_6H_6(g) \quad \Delta H = \Delta H_1 + \Delta H_2 + \Delta H_3 = 237\ kJ; \quad \Delta H^\circ_{f,\ C_6H_6(g)} = 237\ kJ/mol$$

The experimental ΔH_f° for $C_6H_6(g)$ is more stable (lower in energy) by 154 kJ as compared to ΔH_f° calculated from bond energies (83 - 237 = -154 kJ). This extra stability is related to benzene's ability to exhibit resonance. Two equivalent Lewis structures can be drawn for benzene. The π bonding system implied by each Lewis structure consists of three localized π bonds. This is not correct as all C–C bonds in benzene are equivalent. We say the π electrons in benzene are delocalized over the entire surface of C_6H_6 (see Section 14.5 of the text). The large discrepancy between ΔH_f° values is due to the delocalized π electrons, whose effect was not accounted for in the calculated ΔH_f° value. The extra stability associated with benzene can be called resonance stabilization. In general, molecules that exhibit resonance are usually more stable than predicted using bond energies.

CHAPTER FIFTEEN

CHEMICAL KINETICS

Reaction Rates

11. Using the coefficients in the balanced equation to relate the rates:

$$\frac{d[H_2]}{dt} = 3\frac{d[N_2]}{dt} \text{ and } \frac{d[NH_3]}{dt} = -2\frac{d[N_2]}{dt}; \text{ So, } \frac{1}{3}\frac{d[H_2]}{dt} = -\frac{1}{2}\frac{d[NH_3]}{dt} \text{ or } \frac{d[NH_3]}{dt} = -\frac{2}{3}\frac{d[H_2]}{dt}$$

Ammonia is produced at a rate equal to 2/3 of the rate of consumption of hydrogen.

13. Rate = $k[Cl]^{1/2}[CHCl_3]$, $\frac{mol}{L\,s} = k\left(\frac{mol}{L}\right)^{1/2}\left(\frac{mol}{L}\right)$; k must have units of $L^{1/2}\ mol^{-1/2}\ s^{-1}$.

Rate Laws from Experimental Data: Initial Rates Method

15. a. In the first two experiments, [NO] is held constant and $[Cl_2]$ is doubled. The rate also doubled. Thus, the reaction is first order with respect to Cl_2. Or mathematically: Rate = $k[NO]^x[Cl_2]^y$

$$\frac{0.36}{0.18} = \frac{k(0.10)^x(0.20)^y}{k(0.10)^x(0.10)^y} = \frac{(0.20)^y}{(0.10)^y},\ 2.0 = 2.0^y,\ y = 1$$

We can get the dependence on NO from the second and third experiments. Here, as the NO concentration doubles (Cl_2 concentration is constant), the rate increases by a factor of four. Thus, the reaction is second order with respect to NO. Or mathematically:

$$\frac{1.45}{0.36} = \frac{k(0.20)^x(0.20)}{k(0.10)^x(0.20)} = \frac{(0.20)^x}{(0.10)^x},\ 4.0 = 2.0^x,\ x = 2;\ \text{So, Rate} = k[NO]^2[Cl_2]$$

Try to examine experiments where only one concentration changes at a time. The more variables that change, the harder it is to determine the orders. Also, these types of problems can usually be solved by inspection. In general, we will solve using a mathematical approach, but keep in mind, you probably can solve for the orders by simple inspection of the data.

b. The rate constant k can be determined from the experiments. From experiment 1:

$$\frac{0.18 \text{ mol}}{\text{L min}} = k\left(\frac{0.10 \text{ mol}}{\text{L}}\right)^2\left(\frac{0.10 \text{ mol}}{\text{L}}\right), \; k = 180 \text{ L}^2 \text{ mol}^{-2} \text{ min}^{-1}$$

From the other experiments:

$k = 180 \text{ L}^2 \text{ mol}^{-2} \text{ min}^{-1}$ (2nd exp.); $k = 180 \text{ L}^2 \text{ mol}^{-2} \text{ min}^{-1}$ (3rd exp.)

The average rate constant is $k_{mean} = 1.8 \times 10^2 \text{ L}^2 \text{ mol}^{-2} \text{ min}^{-1}$

17. a. Rate = $k[NOCl]^n$; Using experiments two and three:

$$\frac{2.66 \times 10^4}{6.64 \times 10^3} = \frac{k(2.0 \times 10^{16})^n}{k(1.0 \times 10^{16})^n}, \; 4.01 = 2.0^n, \; n = 2; \; \text{Rate} = k[NOCl]^2$$

b.

$$\frac{5.98 \times 10^4 \text{ molecules}}{\text{cm}^3 \text{ s}} = k\left(\frac{3.0 \times 10^{16} \text{ molecules}}{\text{cm}^3}\right)^2, \; k = 6.6 \times 10^{-29} \text{ cm}^3/\text{molecules}^{-1}\text{s}^{-1}$$

The other three experiments give (6.7, 6.6 and 6.6) × 10^{-29} $\text{cm}^3 \text{ molecules}^{-1}\text{s}^{-1}$, respectively.

The mean value for k is $6.6 \times 10^{-29} \text{ cm}^3 \text{ molecules}^{-1}\text{s}^{-1}$.

c.

$$\frac{6.6 \times 10^{-29} \text{ cm}^3}{\text{molecules s}} \times \frac{1 \text{ L}}{1000 \text{ cm}^3} \times \frac{6.022 \times 10^{23} \text{molecules}}{\text{mol}} = \frac{4.0 \times 10^{-8} \text{ L}}{\text{mol s}}$$

19. a. Rate = $k[ClO_2]^x[OH^-]^y$; From the first two experiments:

$2.30 \times 10^{-1} = k(0.100)^x(0.100)^y$ and $5.75 \times 10^{-2} = k(0.0500)^x(0.100)^y$

Dividing the two rate laws: $4.00 = \frac{(0.100)^x}{(0.0500)^x} = 2.00^x, \; x = 2$

Comparing the second and third experiments:

$2.30 \times 10^{-1} = k(0.100)(0.100)^y$ and $1.15 \times 10^{-1} = k(0.100)(0.0500)^y$

Dividing: $2.00 = \frac{(0.100)^y}{(0.050)^y} = 2.0^y, \; y = 1$

The rate law is: Rate = $k[ClO_2]^2[OH^-]$

$2.30 \times 10^{-1} \text{ mol L}^{-1} \text{ s}^{-1} = k(0.100 \text{ mol/L})^2(0.100 \text{ mol/L}), \; k = 2.30 \times 10^2 \text{ L}^2 \text{ mol}^{-2} \text{ s}^{-1} = k_{mean}$

b. $\text{Rate} = \dfrac{2.30 \times 10^2\ \text{L}^2}{\text{mol}^2\ \text{s}} \times \left(\dfrac{0.175\ \text{mol}}{\text{L}}\right)^2 \times \dfrac{0.0844\ \text{mol}}{\text{L}} = 0.594\ \text{mol L}^{-1}\ \text{s}^{-1}$

21. Rate = $k[N_2O_5]^x$; The rate laws for the first two experiments are:

$$2.26 \times 10^{-3} = k(0.190)^x \text{ and } 8.90 \times 10^{-4} = k(0.0750)^x$$

Dividing: $2.54 = \dfrac{(0.190)^x}{(0.0750)^x} = (2.53)^x$, $x = 1$; Rate = $k[N_2O_5]$

$$k = \frac{\text{Rate}}{[N_2O_5]} = \frac{8.90 \times 10^{-4}\ \text{mol L}^{-1}\ \text{s}^{-1}}{0.0750\ \text{mol/L}} = 1.19 \times 10^{-2}\ \text{s}^{-1};\ k_{mean} = 1.19 \times 10^{-2}\ \text{s}^{-1}$$

Integrated Rate Laws

23. a. Since the ln[A] vs time plot was linear, then the reaction is first order in A. The slope of the ln[A] vs time plot equals -k. Therefore, the rate law, the integrated rate law and the rate constant value are:

$$\text{Rate} = k[A];\ \ln[A] = -kt + \ln[A]_o;\ k = 2.97 \times 10^{-2}\ \text{min}^{-1}$$

b. The half-life expression for a first order rate law is:

$$t_{1/2} = \frac{\ln 2}{k} = \frac{0.6931}{k},\ t_{1/2} = \frac{0.6931}{2.97 \times 10^{-2}\ \text{min}^{-1}} = 23.3\ \text{min}$$

c. 2.50×10^{-3} *M* is 1/8 of the original amount of A present initially, so the reaction is 87.5% complete. When a first order reaction is 87.5% complete (or 12.5% remains), then the reaction has gone through 3 half-lives:

$$100\% \xrightarrow{t_{1/2}} 50.0\% \xrightarrow{t_{1/2}} 25.0\% \xrightarrow{t_{1/2}} 12.5\%;\quad t = 3 \times t_{1/2} = 3 \times 23.3\ \text{min} = 69.9\ \text{min}$$

Or we can use the integrated rate law:

$$\ln\left(\frac{[A]}{[A]_o}\right) = -kt,\ \ln\left(\frac{2.50 \times 10^{-3}\ M}{2.00 \times 10^{-2}\ M}\right) = -(2.97 \times 10^{-2}\ \text{min}^{-1})\,t,\ t = \frac{\ln(0.125)}{-2.97 \times 10^{-2}\ \text{min}^{-1}}$$

$$= 70.0\ \text{min}$$

25. The first assumption to make is that the reaction is first order. For a first order reaction, a graph of ln $[H_2O_2]$ vs time will yield a straight line. If this plot is not linear, then the reaction is not first order and we make another assumption.

Time (s)	$[H_2O_2]$ (mol/L)	ln $[H_2O_2]$
0	1.00	0.000
120.	0.91	-0.094
300.	0.78	-0.25
600.	0.59	-0.53
1200.	0.37	-0.99
1800.	0.22	-1.51
2400.	0.13	-2.04
3000.	0.082	-2.50
3600.	0.050	-3.00

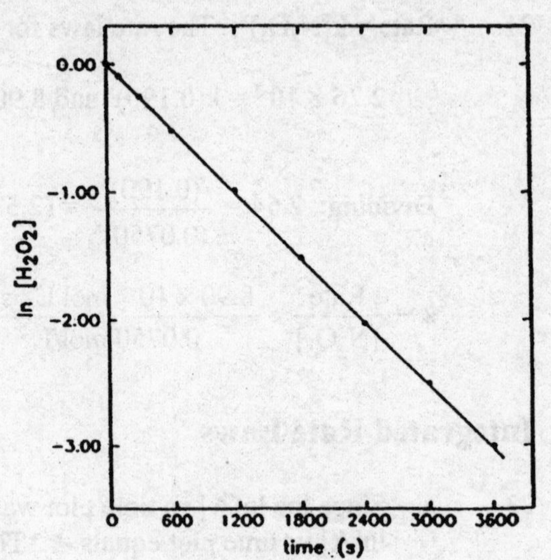

Note: We carried extra significant figures in some of the ln values in order to reduce round off error. For the plots, we will do this most of the time when the ln function is involved.

The plot of ln $[H_2O_2]$ vs. time is linear. Thus, the reaction is first order. The differential rate law and integrated rate law are: Rate = $\dfrac{-d[H_2O_2]}{dt} = k[H_2O_2]$ and $\ln [H_2O_2] = -kt + \ln [H_2O_2]_o$.

We determine the rate constant k by determining the slope of the ln $[H_2O_2]$ vs time plot (slope = -k). Using two points on the curve gives:

$$\text{slope} = -k = \frac{\Delta y}{\Delta x} = \frac{0 - (3.00)}{0 - 3600.} = -8.3 \times 10^{-4}\ s^{-1},\ k = 8.3 \times 10^{-4}\ s^{-1}$$

To determine $[H_2O_2]$ at 4000. s, use the integrated rate law where at t = 0, $[H_2O_2]_o = 1.00\ M$.

$$\ln [H_2O_2] = -kt + \ln [H_2O_2]_o \text{ or } \ln\left(\frac{[H_2O_2]}{[H_2O_2]_o}\right) = -kt$$

$$\ln\left(\frac{[H_2O_2]}{1.00}\right) = -8.3 \times 10^{-4}\ s^{-1} \times 4000.\ s,\ \ \ln [H_2O_2] = -3.3,\ [H_2O_2] = e^{-3.3} = 0.037\ M$$

27. **Assume the reaction is first order and see if the plot of ln $[NO_2]$ vs. time is linear. If this isn't linear, try the second order plot of $1/[NO_2]$ vs. time. The data and plots follow.**

Time (s)	$[NO_2]$ (M)	ln $[NO_2]$	$1/[NO_2]$ (M^{-1})
0	0.500	-0.693	2.00
1.20×10^3	0.444	-0.812	2.25
3.00×10^3	0.381	-0.965	2.62
4.50×10^3	0.340	-1.079	2.94
9.00×10^3	0.250	-1.386	4.00
1.80×10^4	0.174	-1.749	5.75

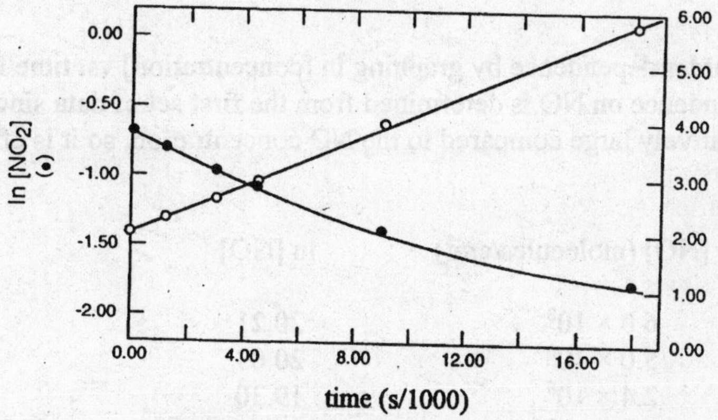

The plot of $1/[NO_2]$ vs. time is linear. The reaction is second order in NO_2. The differential rate law and integrated rate law are: Rate = $k[NO_2]^2$ and $\frac{1}{[NO_2]} = kt + \frac{1}{[NO_2]_o}$.

The slope of the plot $1/[NO_2]$ vs. t gives the value of k. Using a couple points on the plot:

$$\text{slope} = k = \frac{\Delta y}{\Delta x} = \frac{(5.75 - 2.00)\,M^{-1}}{(1.80 \times 10^4 - 0)\text{ s}} = 2.08 \times 10^{-4}\text{ L mol}^{-1}\text{ s}^{-1}$$

To determine $[NO_2]$ at 2.70×10^4 s, use the integrated rate law where $1/[NO_2]_o = 1/0.500\ M = 2.00\ M^{-1}$.

$$\frac{1}{[NO_2]} = kt + \frac{1}{[NO_2]_o},\quad \frac{1}{[NO_2]} = \frac{2.08 \times 10^{-4}\text{ L}}{\text{mol s}} \times 2.70 \times 10^4\text{ s} + 2.00\ M^{-1}$$

$$\frac{1}{[NO_2]} = 7.62,\quad [NO_2] = 0.131\ M$$

29. From the data, the pressure of C_2H_5OH decreases at a constant rate of 13 torr for every 100. s. Since the rate of disappearance of C_2H_5OH is not dependent on concentration, the reaction is zero order in C_2H_5OH.

$$k = \frac{13 \text{ torr}}{100. \text{ s}} \times \frac{1 \text{ atm}}{760 \text{ torr}} = 1.7 \times 10^{-4} \text{ atm/s}$$

The rate law and integrated rate law are:

$$\text{Rate} = k = 1.7 \times 10^{-4} \text{ atm/s}; \quad P_{C_2H_5OH} = -kt + 250. \text{ torr}\left(\frac{1 \text{ atm}}{760 \text{ torr}}\right) = -kt + 0.329 \text{ atm}$$

At 900. s: $P_{C_2H_5OH} = -1.7 \times 10^{-4} \text{ atm/s} \times 900. \text{ s} + 0.329 \text{ atm} = 0.176 \text{ atm} = 0.18 \text{ atm} = 130 \text{ torr}$

31. a. We check for first order dependence by graphing ln [concentration] vs. time for each set of data. The rate dependence on NO is determined from the first set of data since the ozone concentration is relatively large compared to the NO concentration, so it is effectively constant.

time (ms)	[NO] (molecules/cm^3)	ln [NO]
0	6.0×10^8	20.21
100.	5.0×10^8	20.03
500.	2.4×10^8	19.30
700.	1.7×10^8	18.95
1000.	9.9×10^7	18.41

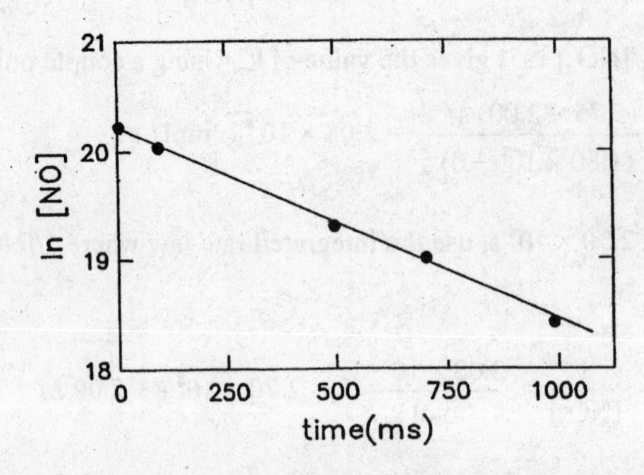

Since ln [NO] vs. t is linear, the reaction is first order with respect to NO.

We follow the same procedure for ozone using the second set of data. The data and plot are:

time (ms)	$[O_3]$ (molecules/cm^3)	ln $[O_3]$
0	1.0×10^{10}	23.03
50.	8.4×10^9	22.85
100.	7.0×10^9	22.67
200.	4.9×10^9	22.31
300.	3.4×10^9	21.95

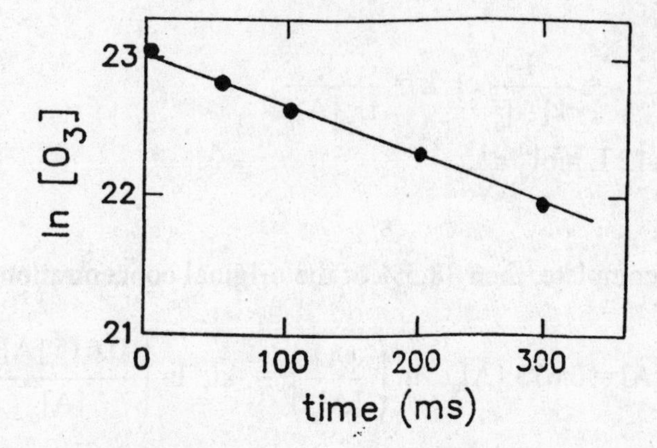

The plot of ln $[O_3]$ vs. t is linear. Hence, the reaction is first order with respect to ozone.

b. Rate = k[NO]$[O_3]$ is the overall rate law.

c. For NO experiment, Rate = k′[NO] and k′ = -(slope from graph of ln [NO] vs. t).

$$k' = \text{-slope} = -\frac{18.41 - 20.21}{(1000. - 0) \times 10^{-3}\ \text{s}} = 1.8\ \text{s}^{-1}$$

For ozone experiment, Rate = k″$[O_3]$ and k″ = -(slope from ln $[O_3]$ vs. t).

$$k'' = \text{-slope} = -\frac{(21.95 - 23.03)}{(300. - 0) \times 10^{-3}\ \text{s}} = 3.6\ \text{s}^{-1}$$

d. From NO experiment, Rate = k[NO]$[O_3]$ = k′[NO] where k′ = k$[O_3]$.

$k' = 1.8\ \text{s}^{-1} = k(1.0 \times 10^{14}\ \text{molecules/cm}^3)$, $k = 1.8 \times 10^{-14}\ \text{cm}^3\ \text{molecules}^{-1}\ \text{s}^{-1}$

We can check this from the ozone data. Rate = k″$[O_3]$ = k[NO]$[O_3]$ where k″ = k[NO].

$k'' = 3.6\ \text{s}^{-1} = k(2.0 \times 10^{14}\ \text{molecules/cm}^3)$, $k = 1.8 \times 10^{-14}\ \text{cm}^3\ \text{molecules}^{-1}\ \text{s}^{-1}$

Both values of k agree.

33. **For a first order reaction, the integrated rate law is: $\ln([A]/[A]_o) = -kt$. Solving for k:**

$$\ln\left(\frac{0.250 \text{ mol/L}}{1.00 \text{ mol/L}}\right) = -k \times 120.\ \text{s},\quad k = 0.0116\ \text{s}^{-1}$$

$$\ln\left(\frac{0.350 \text{ mol/L}}{2.00 \text{ mol/L}}\right) = -\ 0.0116\ \text{s}^{-1} \times t,\quad t = 150.\ \text{s}$$

35. **For a second order reaction: $t_{1/2} = \dfrac{1}{k[A]_o}$ or $k = \dfrac{1}{t_{1/2}[A]_o}$**

$$k = \frac{1}{143\ \text{s}(0.060\ \text{mol/L})} = 0.12\ \text{L mol}^{-1}\ \text{s}^{-1}$$

37. a. **If the reaction is 38.5% complete, then 38.5% of the original concentration is consumed, leaving 61.5%.**

$[A] = 61.5\%$ of $[A]_o$ or $[A] = 0.615\ [A]_o$; $\ln\left(\dfrac{[A]}{[A]_o}\right) = -kt$, $\ln\left(\dfrac{0.615\ [A]_o}{[A]_o}\right) = -k(480.\ \text{s})$

$\ln(0.615) = -k(480.\ \text{s})$, $-0.486 = -k(480.\ \text{s})$, $k = 1.01 \times 10^{-3}\ \text{s}^{-1}$

b. $t_{1/2} = (\ln 2)/k = 0.6931/1.01 \times 10^{-3}\ \text{s}^{-1} = 686\ \text{s}$

c. 25% complete: $[A] = 0.75\ [A]_o$; $\ln(0.75) = -1.01 \times 10^{-3}\ (t)$, $t = 280\ \text{s}$

75% complete: $[A] = 0.25\ [A]_o$; $\ln(0.25) = -1.01 \times 10^{-3}\ (t)$, $t = 1.4 \times 10^3\ \text{s}$

Or, we know it takes $2 \times t_{1/2}$ for reaction to be 75% complete. $t = 2 \times 686\ \text{s} = 1370\ \text{s}$

95% complete: $[A] = 0.05\ [A]_o$; $\ln(0.05) = -1.01 \times 10^{-3}\ (t)$, $t = 3 \times 10^3\ \text{s}$

39. a. $[A] = -kt + [A]_o$; If $k = 5.0 \times 10^{-2}\ \text{mol L}^{-1}\ \text{s}^{-1}$ and $[A]_o = 1.00 \times 10^{-3}\ M$, then:

$$[A] = -(5.0 \times 10^{-2}\ \text{mol L}^{-1}\ \text{s}^{-1})\ t + 1.00 \times 10^{-3}\ \text{mol/L}$$

b. $\dfrac{[A]_o}{2} = -(5.0 \times 10^{-2})\ t_{1/2} + [A]_o$ since at $t = t_{1/2}$, $[A] = [A]_o/2$.

$-0.50[A]_o = -(5.0 \times 10^{-2})\ t_{1/2}$, $t_{1/2} = \dfrac{0.50(1.00 \times 10^{-3})}{5.0 \times 10^{-2}} = 1.0 \times 10^{-2}\ \text{s}$; Or we can use $t_{1/2} = \dfrac{[A]_o}{2\ k}$.

c. $[A] = -kt + [A]_o = -(5.0 \times 10^{-2}\ \text{mol L}^{-1}\ \text{s}^{-1})(5.0 \times 10^{-3}\ \text{s}) + 1.00 \times 10^{-3}\ \text{mol/L} = 7.5 \times 10^{-4}\ \text{mol/L}$

$[A]_{reacted} = 1.00 \times 10^{-3}\ \text{mol/L} - 7.5 \times 10^{-4}\ \text{mol/L} = 2.5 \times 10^{-4}\ \text{mol/L}$

$[B]_{produced} = [A]_{reacted} = 2.5 \times 10^{-4}\ M$

41. a. **Since $[A]_o << [B]_o$ or $[C]_o$, then the B and C concentrations remain constant at 1.00 *M* for this experiment. So: rate = $k[A]^2[B][C] = k'[A]^2$ where $k' = k[B][C]$**

For this pseudo second order reaction:

$$\frac{1}{[A]} = k't + \frac{1}{[A]_o}, \quad \frac{1}{3.26 \times 10^{-5}\,M} = k'(3.00\text{ min}) + \frac{1}{1.00 \times 10^{-4}\,M}$$

$$k' = 6890\text{ L mol}^{-1}\text{ min}^{-1} = 115\text{ L mol}^{-1}\text{ s}^{-1}$$

$$k' = k[B][C],\ k = \frac{k'}{[B][C]} = \frac{115\text{ L mol}^{-1}\text{ s}^{-1}}{(1.00\,M)(1.00\,M)} = 115\text{ L}^3\text{ mol}^{-3}\text{ s}^{-1}$$

b. **For this pseudo second order reaction:**

$$\text{rate} = k'[A]^2,\ t_{1/2} = \frac{1}{k'[A]_o} = \frac{1}{115\text{ L mol}^{-1}\text{s}^{-1}(1.00 \times 10^{-4}\ \frac{\text{mol}}{\text{L}})} = 87.0\text{ s}$$

c. $$\frac{1}{[A]} = k't + \frac{1}{[A]_o} = 115\text{ L mol}^{-1}\text{ s}^{-1} \times 600.\text{ s} + \frac{1}{1.00 \times 10^{-4}\ \frac{\text{mol}}{\text{L}}} = 7.90 \times 10^4\text{ L/mol},$$

$$[A] = \frac{1}{7.90 \times 10^4\text{ L/mol}} = 1.27 \times 10^{-5}\ \frac{\text{mol}}{\text{L}}$$

From the stoichiometry in the balanced reaction, 1 mol of B reacts with every 3 mol of A.

amount A reacted = $1.00 \times 10^{-4}\,M - 1.27 \times 10^{-5}\,M = 8.7 \times 10^{-5}\,M$

amount B reacted = $8.7 \times 10^{-5}\text{ mol/L} \times \frac{1\text{ mol B}}{3\text{ mol A}} = 2.9 \times 10^{-5}\,M$

$[B] = 1.00\,M - 2.9 \times 10^{-5}\,M = 1.00\,M$

As we mentioned in part a, the concentration of B (and C) remain constant since the A concentration is so small.

Reaction Mechanisms

43. **For elementary reactions, the rate law can be written using the coefficients in the balanced equation to determine the orders.**

a. **Rate = $k[CH_3NC]$**

b. **Rate = $k[O_3][NO]$**

c. **Rate = $k[O_3]$**

d. **Rate = $k[O_3][O]$**

e. **Rate = $k[^{14}_{6}C]$ or Rate = kN where N = the number of $^{14}_{6}C$ atoms (convention)**

45. A mechanism consists of a series of elementary reactions where the rate law for each step can be determined using the coefficients in the balanced equations. For a plausible mechanism, the rate law derived from a mechanism must agree with the rate law determined from experiment. To derive the rate law from the mechanism, the rate of the reaction is assumed to equal the rate of the slowest step in the mechanism.

Since step 1 is the rate determining step, then the rate law for this mechanism is: Rate = $k[C_4H_9Br]$. To get the overall reaction, we sum all the individual steps of the mechanism. Summing all steps gives:

$$C_4H_9Br \rightarrow C_4H_9^+ + Br^-$$
$$C_4H_9^+ + H_2O \rightarrow C_4H_9OH_2^+$$
$$C_4H_9OH_2^+ + H_2O \rightarrow C_4H_9OH + H_3O^+$$

$$C_4H_9Br + 2\,H_2O \rightarrow C_4H_9OH + Br^- + H_3O^+$$

Intermediates in a mechanism are species that are neither reactants nor products, but that are formed and consumed during the reaction sequence. The intermediates for this mechanism are $C_4H_9^+$ and $C_4H_9OH_2^+$.

47. Let's determine the rate law for each mechanism. If the rate law derived from the mechanism is the same as the experimental rate law, then the mechanism is possible. When deriving rate laws from a mechanism, we must substitute for all intermediate concentrations.

a. Rate = $k[NO][O_2]$ <u>not possible</u>

b. Rate = $k[NO_3][NO]$ and $\dfrac{[NO_3]}{[NO][O_2]} = K_{eq} = k_1/k_{-1}$ or $[NO_3] = K_{eq}[NO][O_2]$

Rate = $kK_{eq}[NO]^2[O_2]$ <u>possible</u>

c. Rate = $k[NO]^2$ <u>not possible</u>

d. Rate = $k[N_2O_2]$ and $[N_2O_2] = K_{eq}[NO]^2$ where $K_{eq} = k_1/k_{-1}$, Rate = $kK_{eq}[NO]^2$ <u>not possible</u>

49. Rate = $k_3[Br^-][H_2BrO_3^+]$; We must substitute for the intermediate concentrations. Since steps 1 and 2 are fast equilibrium steps, then rate forward reaction = rate reverse reaction.

$k_2[HBrO_3][H^+] = k_{-2}[H_2BrO_3^+]$; $k_1[BrO_3^-][H^+] = k_{-1}[HBrO_3]$

$$[HBrO_3] = \frac{k_1}{k_{-1}}[BrO_3^-][H^+];\ [H_2BrO_3^+] = \frac{k_2}{k_{-2}}[HBrO_3][H^+] = \frac{k_2k_1}{k_{-2}k_{-1}}[BrO_3^-][H^+]^2$$

$$Rate = \frac{k_3k_2k_1}{k_{-2}k_{-1}}[Br^-][BrO_3^-][H^+]^2 = k[Br^-][BrO_3^-][H^+]^2$$

51. a. $MoCl_5^-$

b. $\text{Rate} = \dfrac{d[NO_2^-]}{dt} = k_2[NO_3^-][MoCl_5^-]$ (Only the last step contains NO_2^-.)

We use the steady-state assumption to substitute for the intermediate concentration in the rate law. The steady-state approximation assumes that the concentration of an intermediate remains constant, i.e., d[intermediate]/dt = 0. To apply the steady-state assumption, we write rate laws for all steps where the intermediate is produced and equate the sum of these rate laws to the sum of the rate laws where the intermediate is consumed. Applying the steady-state approximation to $MoCl_5^-$:

$$\frac{d[MoCl_5^-]}{dt} = 0, \text{ so } k_1[MoCl_6^{2-}] = k_{-1}[MoCl_5^-][Cl^-] + k_2[NO_3^-][MoCl_5^-]$$

$$[MoCl_5^-] = \frac{k_1[MoCl_6^{2-}]}{k_{-1}[Cl^-] + k_2[NO_3^-]}; \quad \text{Rate} = \frac{d[NO_2^-]}{dt} = \frac{k_1k_2[NO_3^-][MoCl_6^{2-}]}{k_{-1}[Cl^-] + k_2[NO_3^-]}$$

53. a. $\text{rate} = \dfrac{d[E]}{dt} = k_2[B^*]$; Assume $\dfrac{d[B^*]}{dt} = 0$, then $k_1[B]^2 = k_{-1}[B][B^*] + k_2[B^*]$

$$[B^*] = \frac{k_1[B]^2}{k_{-1}[B] + k_2}; \text{ The rate law is: Rate} = \frac{d[E]}{dt} = \frac{k_1k_2[B]^2}{k_{-1}[B] + k_2}$$

b. When $k_2 << k_{-1}[B]$, then $\text{Rate} = \dfrac{d[E]}{dt} = \dfrac{k_1k_2[B]^2}{k_{-1}[B]} = \dfrac{k_1k_2}{k_{-1}}[B]$

Reaction is first order when the rate of the second step is very slow (when k_2 is very small).

c. Collisions between B molecules only transfer energy from one B to another. This occurs at a much faster rate than the decomposition of an energetic B molecule (B*).

Temperature Dependence of Rate Constants and the Collision Model

55. In a unimolecular reaction, a single reactant molecule decomposes to product(s). In a bimolecular reaction, two molecules collide to give product(s). The probability of the simultaneous collision of three molecules with enough energy and correct orientation is very small, making termolecular steps very unlikely.

57. $k = A\exp(-E_a/RT)$ or $\ln k = \dfrac{-E_a}{RT} + \ln A$ (the Arrhenius equation)

For two conditions: $\ln\left(\dfrac{k_2}{k_1}\right) = \dfrac{E_a}{R}\left(\dfrac{1}{T_1} - \dfrac{1}{T_2}\right)$ (Assuming A is temperature independent.)

Let $k_1 = 2.0 \times 10^3\ s^{-1}$, $T_1 = 298$ K; k_2 = ?, $T_2 = 348$ K; $E_a = 15.0 \times 10^3$ J/mol

$$\ln\left(\frac{k_2}{2.0 \times 10^3\ s^{-1}}\right) = \frac{15.0 \times 10^3\ J/mol}{8.3145\ J\ K^{-1}\ mol^{-1}}\left(\frac{1}{298\ K} - \frac{1}{348\ K}\right) = 0.87$$

$$\ln\left(\frac{k_2}{2.0 \times 10^3}\right) = 0.87,\ \frac{k_2}{2.0 \times 10^3} = e^{0.87} = 2.4,\ k_2 = 2.4(2.0 \times 10^3) = 4.8 \times 10^3\ s^{-1}$$

59. $$\ln\left(\frac{k_2}{k_1}\right) = \frac{E_a}{R}\left(\frac{1}{T_1} - \frac{1}{T_2}\right);\ \frac{k_2}{k_1} = 7.00,\ T_1 = 295\ K,\ E_a = 54.0 \times 10^3\ J/mol$$

$$\ln(7.00) = \frac{54.0 \times 10^3\ J/mol}{8.3145\ J\ K^{-1} mol^{-1}}\left(\frac{1}{295\ K} - \frac{1}{T_2}\right),\ \frac{1}{295} - \frac{1}{T_2} = 3.00 \times 10^{-4}$$

$$\frac{1}{T_2} = 3.09 \times 10^{-3},\ T_2 = 324\ K = 51°C$$

61. From the Arrhenius equation in logarithmic form ($\ln k = -E_a/RT + \ln A$), a graph of ln k vs. 1/T should yield a straight line with a slope equal to $-E_a/R$ and a y-intercept equal to ln A.

a. slope $= -E_a/R$, $E_a = 1.10 \times 10^4\ K \times \frac{8.3145\ J}{K\ mol} = 9.15 \times 10^4\ J/mol = 91.5\ kJ/mol$

b. The units for A are the same as the units for k (s^{-1}).

y-intercept = ln A, $A = e^{33.5} = 3.54 \times 10^{14}\ s^{-1}$

c. $\ln k = -E_a/RT + \ln A$ or $k = A \exp(-E_a/RT)$

$$k = 3.54 \times 10^{14}\ s^{-1} \times \exp\left(\frac{-9.15 \times 10^4\ J/mol}{8.3145\ J\ K^{-1} mol^{-1} \times 298\ K}\right) = 3.24 \times 10^{-2}\ s^{-1}$$

63. The Arrhenius equation is: $k = A \exp(-E_a/RT)$ or in logarithmic form, $\ln k = -E_a/RT + \ln A$. Hence, a graph of ln k vs. 1/T should yield a straight line with a slope equal to $-E_a/R$ since the logarithmic form of the Arrhenius equation is in the form of a straight line equation, y = mx + b.

Note: We carried one extra significant figure in the following ln k values in order to reduce round off error.

T (K)	1/T (K^{-1})	k ($L\ mol^{-1}\ s^{-1}$)	ln k
195	5.13×10^{-3}	1.08×10^9	20.80
230.	4.35×10^{-3}	2.95×10^9	21.81
260.	3.85×10^{-3}	5.42×10^9	22.41
298	3.36×10^{-3}	12.0×10^9	23.21
369	2.71×10^{-3}	35.5×10^9	24.29

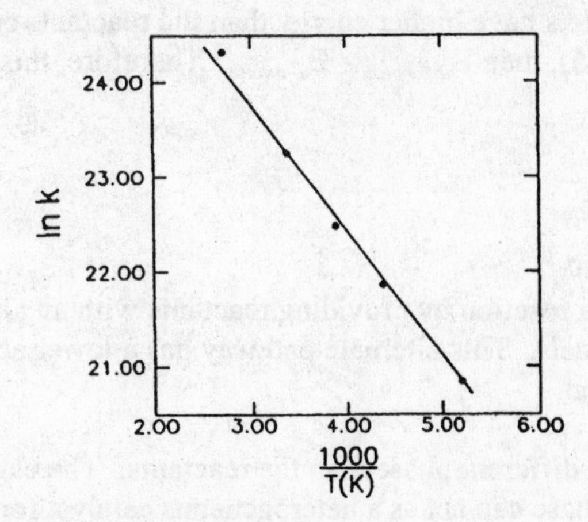

From the "eyeball" line on the graph:

$$\text{slope} = \frac{20.95 - 23.65}{5.00 \times 10^{-3} - 3.00 \times 10^{-3}} = \frac{-2.70}{2.00 \times 10^{-3}} = -1.35 \times 10^3\ \text{K} = \frac{-E_a}{R}$$

$E_a = 1.35 \times 10^3\ \text{K} \times 8.3145\ \text{J K}^{-1}\ \text{mol}^{-1} = 1.12 \times 10^4\ \text{J/mol} = 11.2\ \text{kJ/mol}$

From the best straight line (by calculator): slope = -1.43×10^3 K and E_a = 11.9 kJ/mol

65. **In the following reaction profiles, R = reactants, P = products, E_a = activation energy, ΔE = overall energy change for the reaction and RC = reaction coordinate which is the same as reaction progress.**

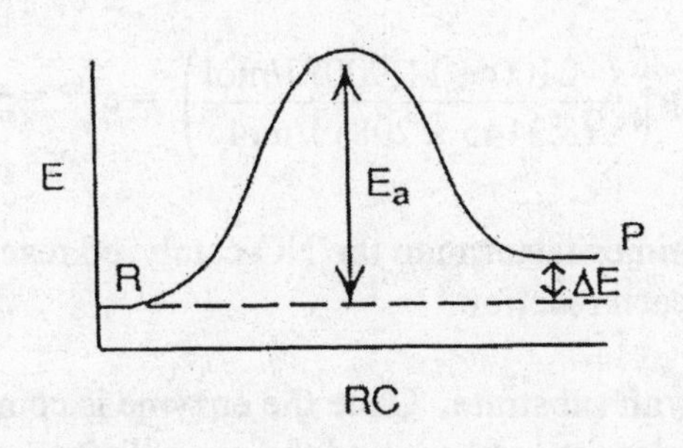

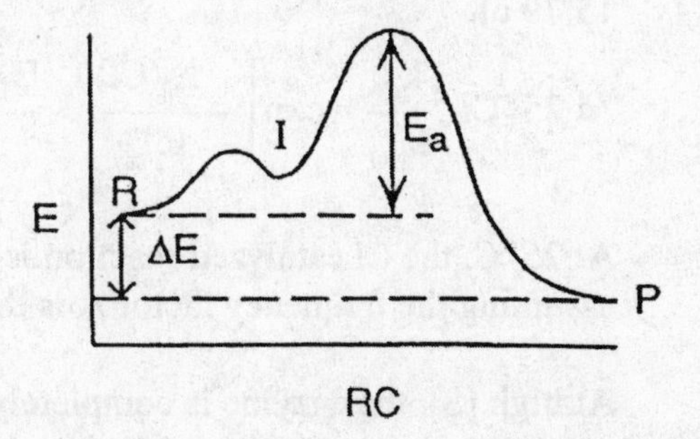

The second reaction profile represents a two step reaction since an intermediate plateau appears between the reactants and the products. This plateau (see I in plot) represents the energy of the intermediate. The general reaction mechanism for this reaction is:

$$\begin{array}{l} R \rightarrow I \\ \underline{I \rightarrow P} \\ R \rightarrow P \end{array}$$

In a mechanism, the rate of the slowest step determines the rate of the reaction. The activation energy for the slowest step will be the largest energy barrier that the reaction must overcome. Since the second hump in the diagram is at the highest energy, then the second step has the largest activation energy and will be the rate determining step (the slow step).

67. When ΔE is positive, (the products have higher energy than the reactants as represented in the energy profile for Exercise 15.65), then $E_{a,\ forward} > E_{a,\ reverse}$. Therefore, this reaction has a positive ΔE value.

Catalysts

69. A catalyst increases the rate of a reaction by providing reactants with an alternate pathway (mechanism) to convert to products. This alternate pathway has a lower activation energy, thus increasing the rate of the reaction.

A heterogeneous catalyst is in a different phase than the reactants. The catalyst is usually a solid, although a catalyst in a liquid phase can act as a heterogeneous catalyst for some gas phase reactions. Since the catalyzed reaction has a different mechanism than the uncatalyzed reaction, the catalyzed reaction most likely will have a different rate law.

71. The mechanism for the chlorine catalyzed destruction of ozone is:

$$\begin{array}{ll} O_3 + Cl \rightarrow O_2 + ClO & \text{(slow)} \\ ClO + O \rightarrow O_2 + Cl & \text{(fast)} \\ \hline O_3 + O \rightarrow 2\,O_2 & \end{array}$$

Since the chlorine atom catalyzed reaction has a lower activation energy, then the Cl catalyzed rate is faster. Hence, Cl is a more effective catalyst. Using the activation energy, we can estimate the efficiency that Cl atoms destroy ozone as compared to NO molecules (see Exercise 15.70 c).

At 25°C: $$\frac{k_{Cl}}{k_{NO}} = \exp\left(\frac{-E_a(Cl)}{RT} + \frac{E_a(NO)}{RT}\right) = \exp\left(\frac{(-2100 + 11{,}900)\ J/mol}{(8.3145 \times 298)\ J/mol}\right) = e^{3.96} = 52$$

At 25°C, the Cl catalyzed reaction is roughly 52 times faster than the NO catalyzed reaction, assuming the frequency factor A is the same for each reaction.

73. At high [S], the enzyme is completely saturated with substrate. Once the enzyme is completely saturated, the rate of decomposition of ES can no longer increase and the overall rate remains constant.

Additional Exercises

75. Rate = $k[I^-]^x[OCl^-]^y[OH^-]^z$; Comparing the first and second experiments:

$$\frac{18.7 \times 10^{-3}}{9.4 \times 10^{-3}} = \frac{k(0.0026)^x (0.012)^y (0.10)^z}{k(0.0013)^x (0.012)^y (0.10)^z},\ 2.0 = 2.0^x,\ x = 1$$

Comparing the first and third experiments:

$$\frac{9.4 \times 10^{-3}}{4.7 \times 10^{-3}} = \frac{k(0.0013)\ (0.012)^y (0.10)^z}{k(0.0013)\ (0.0060)^y (0.10)^z},\ 2.0 = 2.0^y,\ y = 1$$

Comparing the first and sixth experiments:

$$\frac{4.7 \times 10^{-3}}{9.4 \times 10^{-3}} = \frac{k(0.0013)\,(0.012)\,(0.20)^z}{k(0.0013)\,(0.012)\,(0.10)^z}, \quad 1/2 = 2.0^z, \quad z = -1$$

$$\text{Rate} = \frac{k[I^-][OCl^-]}{[OH^-]}$$; **The presence of OH^- decreases the rate of the reaction.**

For the first experiment:

$$\frac{9.4 \times 10^{-3}\ \text{mol}}{\text{L s}} = k\,\frac{(0.0013\ \text{mol/L})\,(0.012\ \text{mol/L})}{(0.10\ \text{mol/L})}, \quad k = 60.3\ \text{s}^{-1} = 60.\ \text{s}^{-1}$$

For all experiments, $k_{mean} = 60.\ s^{-1}$.

77.

Heating Time	Untreated		Deacidifying		Antioxidant	
(days)	s	ln s	s	ln s	s	ln s
0.00	100.0	4.605	100.1	4.606	114.6	4.741
1.00	67.9	4.218	60.8	4.108	65.2	4.177
2.00	38.9	3.661	26.8	3.288	28.1	3.336
3.00	16.1	2.779	-	-	11.3	2.425
6.00	6.8	1.92	-	-	-	-

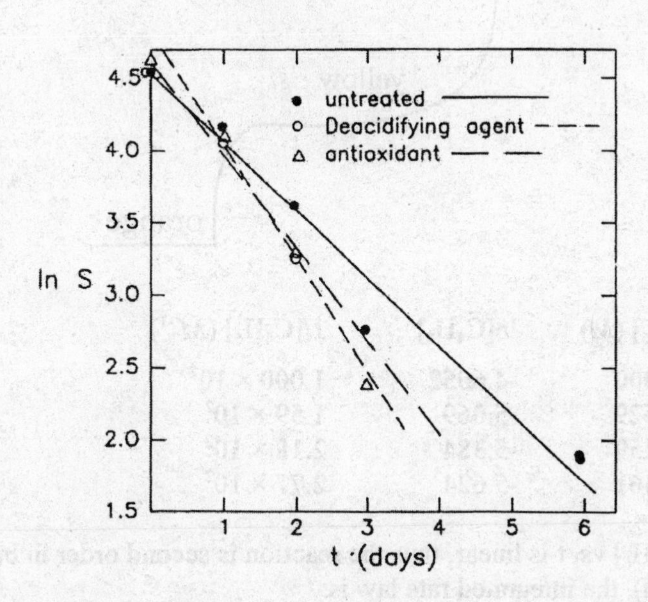

a. **We used a calculator to fit the data by least squares. The results follow.**

Untreated: $\ln s = -0.465\,t + 4.55$, $k = 0.465\ \text{day}^{-1}$

Deacidifying agent: $\ln s = -0.659\,t + 4.66$, $k = 0.659\ \text{day}^{-1}$

Antioxidant: $\ln s = -0.779\,t + 4.84$, $k = 0.779\ \text{day}^{-1}$

b. **No, the silk degrades more rapidly with the additives since k increases.**

c. $t_{1/2} = (\ln 2)/k$

Untreated: $t_{1/2}$ = 1.49 day; Deacidifying agent: $t_{1/2}$ = 1.05 day; Antioxidant: $t_{1/2}$ = 0.890 day.

79. The rate depends on the number of reactant molecules adsorbed on the surface of the catalyst. This quantity is proportional to the concentration of reactant. However, when all of the surface sites of the catalyst are occupied, the rate becomes independent of the concentration of reactant.

81. Rate = $k_2[I^-][HOCl]$; From the fast equilibrium first step:

$k_1[OCl^-] = k_{-1}[HOCl][OH^-]$, $[HOCl] = \dfrac{k_1[OCl^-]}{k_{-1}[OH^-]}$; Substituting into the rate equation:

$$\text{Rate} = \frac{k_2k_1[I^-][OCl^-]}{k_{-1}[OH^-]} = \frac{k[I^-][OCl^-]}{[OH^-]}$$

83. The yellow (rhombic crystals) form is kinetically stable. The orange (tetragonal cyrstals) form is thermodynamically stable.

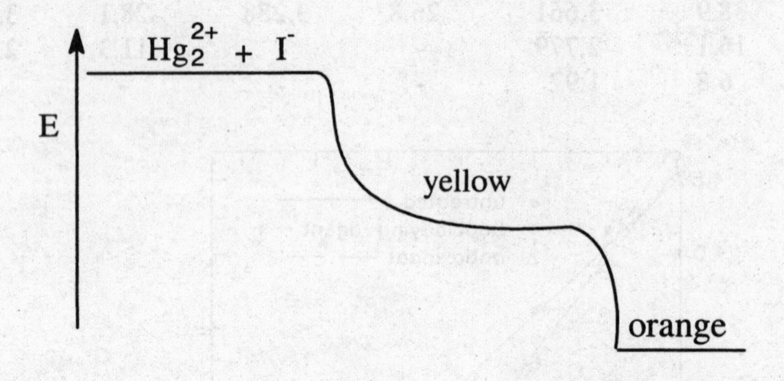

85. a.

t (s)	$[C_4H_6]$ (M)	$\ln[C_4H_6]$	$1/[C_4H_6]$ (M^{-1})
0	0.01000	-4.6052	1.000×10^2
1000.	0.00629	-5.069	1.59×10^2
2000.	0.00459	-5.384	2.18×10^2
3000.	0.00361	-5.624	2.77×10^2

The plot of $1/[C_4H_6]$ vs. t is linear, thus the reaction is second order in butadiene. From the plot (not included), the integrated rate law is:

$$\frac{1}{[C_4H_6]} = (5.90 \times 10^{-2}\ \text{L mol}^{-1}\ \text{s}^{-1})\,t + 100.0\ M^{-1}$$

b. When dimerization is 1.0% complete, 99.0% of C_4H_6 is left.

$[C_4H_6] = 0.990(0.01000) = 0.00990\ M$; $\dfrac{1}{0.00990} = 5.90 \times 10^{-2}\,t + 100.0$, $t = 17.1\ \text{s} \approx 20\ \text{s}$

c. 2.0% complete, $[C_4H_6] = 0.00980\ M$; $\frac{1}{0.00980} = 5.90 \times 10^{-2}\,t + 100.0$, $t = 34.6\ s \approx 30\ s$

d. $\frac{1}{[C_4H_6]} = kt + \frac{1}{[C_4H_6]_o}$; $[C_4H_6]_o = 0.0200\ M$; At $t = t_{1/2}$, $[C_4H_6] = 0.0100\ M$

$$\frac{1}{0.0100} = (5.90 \times 10^{-2})\,t_{1/2} + \frac{1}{0.0200},\quad t_{1/2} = 847\ s = 850\ s$$

$$\text{Or, } t_{1/2} = \frac{1}{k[A]_o} = \frac{1}{(5.90 \times 10^{-2}\ L\ mol^{-1}\ s^{-1})(2.00 \times 10^{-2}\ M)} = 847\ s$$

e. From Exercise 15.26, $k = 1.4 \times 10^{-2}\ L\ mol^{-1}\ s^{-1}$ at 500. K. From this problem, $k = 5.90 \times 10^{-2}\ L\ mol^{-1}\ s^{-1}$ at 620. K.

$$\ln\left(\frac{k_2}{k_1}\right) = \frac{E_a}{R}\left(\frac{1}{T_1} - \frac{1}{T_2}\right),\quad \ln\left(\frac{5.90 \times 10^{-2}}{1.4 \times 10^{-2}}\right) = \frac{E_a}{8.3145\ J\ K^{-1}\ mol^{-1}}\left(\frac{1}{500.\ K} - \frac{1}{620.\ K}\right)$$

$12 = E_a\,(3.9 \times 10^{-4})$, $E_a = 3.1 \times 10^4\ J/mol = 31\ kJ/mol$

87. $k = A\exp(-E_a/RT)$; $\frac{k_{cat}}{k_{uncat}} = \frac{A_{cat}\exp(-E_{a,cat}/RT)}{A_{uncat}\exp(-E_{a,uncat}/RT)} = \exp\left(\frac{-E_{a,cat} + E_{a,uncat}}{RT}\right)$

$$2.50 \times 10^3 = \frac{k_{cat}}{k_{uncat}} = \exp\left(\frac{-E_{a,cat} + 5.00 \times 10^4\ J/mol}{8.3145\ J\ K^{-1}\ mol^{-1} \times 310.\ K}\right)$$

$\ln(2.50 \times 10^3) \times 2.58 \times 10^3\ J/mol = -E_{a,cat} + 5.00 \times 10^4\ J/mol$

$E_{a,cat} = 5.00 \times 10^4\ J/mol - 2.02 \times 10^4\ J/mol = 2.98 \times 10^4\ J/mol = 29.8\ kJ/mol$

89. a.

T (K)	1/T (K^{-1})	k (min^{-1})	ln k
298.2	3.353×10^{-3}	178	5.182
293.5	3.407×10^{-3}	126	4.836
290.5	3.442×10^{-3}	100.	4.605

A plot of ln k vs. 1/T gives a straight line (plot not included). The equation for the straight line is:

$$\ln k = -6.48 \times 10^3\,(1/T) + 26.9$$

For a ln k vs. 1/T plot, the slope $= -E_a/R = -6.48 \times 10^3$ K.

$-6.48 \times 10^3\ K = -E_a/8.3145\ J\ mol^{-1}\ K^{-1}$, $E_a = 5.39 \times 10^4\ J/mol = 53.9\ kJ/mol$

b. $\ln k = -6.48 \times 10^3(1/288.2) + 26.9 = 4.42$, $k = e^{4.42} = 83\ min^{-1}$

About 83 chirps per minute per insect. Note: We carried extra sig. figs.

c. k gives the number of chirps per minute. The number or chirps in 15 sec is k/4.

T (°C)	T (°F)	k (min^{-1})	42 + 0.80 (k/4)
25.0	77.0	178	78° F
20.3	68.5	126	67°F
17.3	63.1	100.	62°F
15.0	59.0	83	59°F

The rule of thumb appears to be fairly accurate, almost ± 1°F.

91. $\text{Rate} = \frac{d[Cl_2]}{dt} = k_2[NO_2Cl][Cl]$; Assume $\frac{d[Cl]}{dt} = 0$, then:

$$k_1[NO_2Cl] = k_{-1}[NO_2][Cl] + k_2[NO_2Cl][Cl],\quad [Cl] = \frac{k_1[NO_2Cl]}{k_{-1}[NO_2] + k_2[NO_2Cl]}$$

$$\text{Rate} = \frac{d[Cl_2]}{dt} = \frac{k_1k_2[NO_2Cl]^2}{k_{-1}[NO_2] + k_2[NO_2Cl]}$$

Challenge Problems

93. a. $\text{Rate} = k[CH_3X]^x[Y]^y$; For experiment 1, [Y] is constant so $\text{Rate} = k'[CH_3X]^x$ where $k' = k(3.0\ M)^y$.

A plot (not included) of $\ln[CH_3X]$ vs t is linear (x = 1). The integrated rate law is:

$$\ln[CH_3X] = -0.93\ t - 3.99;\quad k' = 0.93\ h^{-1}$$

For Experiment 2, [Y] is again constant with $\text{Rate} = k''[CH_3X]^x$ where $k'' = k(4.5\ M)^y$. The ln plot is linear again with an integrated rate law:

$$\ln[CH_3X] = -0.93\ t - 5.40;\quad k'' = 0.93\ h^{-1}$$

Dividing the rate constant values: $\frac{k'}{k''} = \frac{0.93}{0.93} = \frac{k(3.0)^y}{k(4.5)^y}$, $1.0 = (0.67)^y$, $y = 0$

Reaction is first order in CH_3X and zero order in Y. The overall rate law is:

$$\text{Rate} = k[CH_3X] \text{ where } k = 0.93\ h^{-1} \text{ at } 25°C.$$

b. $t_{1/2} = (\ln 2)/k = 0.6931/(7.88 \times 10^8\ h^{-1}) = 8.80 \times 10^{-10}$ hour

c. $\ln\frac{k_2}{k_1} = \frac{E_a}{R}\left(\frac{1}{T_1} - \frac{1}{T_2}\right)$, $\ln\left(\frac{7.88 \times 10^8}{0.93}\right) = \frac{E_a}{8.3145\ J\ K^{-1}\ mol^{-1}}\left(\frac{1}{298\ K} - \frac{1}{358\ K}\right)$

$E_a = 3.0 \times 10^5$ J/mol $= 3.0 \times 10^2$ kJ/mol

d. From part a, the reaction is first order in CH_3X and zero order in Y. From part c, the activation energy is close to the C-X bond energy. A plausible mechanism that explains the results in parts a and c is:

$CH_3X \rightarrow CH_3 + X$ (slow)

$CH_3 + Y \rightarrow CH_3Y$ (fast)

Note: This is a possible mechanism since the derived rate law is the same as the experimental rate law.

95. Rate = $k[A]^x[B]^y[C]^z$; During the course of experiment 1, [A] and [C] are essentially constant, and Rate = $k'[B]^y$ where $k' = k[A]_o^x[C]_o^z$.

[B] (*M*)	time (s)	ln[B]	1/[B] (M^{-1})
1.0×10^{-3}	0	-6.91	1.0×10^3
2.7×10^{-4}	1.0×10^5	-8.22	3.7×10^3
1.6×10^{-4}	2.0×10^5	-8.74	6.3×10^3
1.1×10^{-4}	3.0×10^5	-9.12	9.1×10^3
8.5×10^{-5}	4.0×10^5	-9.37	12×10^3
6.9×10^{-5}	5.0×10^5	-9.58	14×10^3
5.8×10^{-5}	6.0×10^5	-9.76	17×10^3

A plot of 1/[B] vs. t is linear (plot not included). The reaction is second order in B and the integrated rate equation is:

$1/[B] = (2.7 \times 10^{-2}\ L\ mol^{-1}\ s^{-1})\,t + 1.0 \times 10^3\ M^{-1}$; $k' = 2.7 \times 10^{-2}\ L\ mol^{-1}\ s^{-1}$

For experiment 2, [B] and [C] are essentially constant and Rate = $k''[A]^x$ where $k'' = k[B]_o^y[C]_o^z = k[B]_o^2[C]_o^z$.

[A] (*M*)	time (s)	ln[A]	1/[A] (M^{-1})
1.0×10^{-2}	0	-4.61	1.0×10^2
8.9×10^{-3}	1.0	-4.72	110
7.1×10^{-3}	3.0	-4.95	140
5.5×10^{-3}	5.0	-5.20	180
3.8×10^{-3}	8.0	-5.57	260
2.9×10^{-3}	10.0	-5.84	340
2.0×10^{-3}	13.0	-6.21	5.0×10^2

A plot of ln[A] vs. t is linear. The reaction is first order in A and the integrated rate law is:

$\ln[A] = -(0.123\ s^{-1})\,t - 4.61$; $k'' = 0.123\ s^{-1}$

Note: We will carry an extra significant figure in k″.

Experiment 3: [A] and [B] are constant; Rate = $k'''[C]^z$

The plot of [C] vs. t is linear. Thus, z = 0.

The overall rate law is: Rate = $k[A][B]^2$

From Experiment 1 (to determine k):

$$k' = 2.7 \times 10^{-2}\ \text{L mol}^{-1}\ \text{s}^{-1} = k[A]_o^x\,[C]_o^z = k[A]_o = k(2.0\ M),\ \ k = 1.4 \times 10^{-2}\ \text{L}^2\ \text{mol}^{-2}\ \text{s}^{-1}$$

From Experiment 2: $k'' = 0.123\ \text{s}^{-1} = k[B]_o^2,\ \ k = \dfrac{0.123\ \text{s}^{-1}}{(3.0\ M)^2} = 1.4 \times 10^{-2}\ \text{L}^2\ \text{mol}^{-2}\ \text{s}^{-1}$

Thus, Rate = $k[A][B]^2$ and $k = 1.4 \times 10^{-2}\ \text{L}^2\ \text{mol}^{-2}\ \text{s}^{-1}$.

97. $$\text{Rate} = \frac{-d[N_2O_5]}{dt} = k_1[M][N_2O_5] - k_{-1}[NO_3][NO_2][M]$$

Assume $d[NO_3]/dt = 0$, so $k_1[N_2O_5][M] = k_{-1}[NO_3][NO_2][M] + k_2[NO_3][NO_2] + k_3[NO_3][NO]$

$$[NO_3] = \frac{k_1[N_2O_5][M]}{k_{-1}[NO_2][M] + k_2[NO_2] + k_3[NO]}$$

Assume $\dfrac{d[NO]}{dt} = 0$, so $k_2[NO_3][NO_2] = k_3[NO_3][NO]$, $[NO] = \dfrac{k_2}{k_3}[NO_2]$

Substituting: $$[NO_3] = \frac{k_1[N_2O_5][M]}{k_{-1}[NO_2][M] + k_2[NO_2] + \dfrac{k_3k_2}{k_3}[NO_2]} = \frac{k_1[N_2O_5][M]}{[NO_2](k_{-1}[M] + 2k_2)}$$

Solving for the rate law:

$$\text{Rate} = \frac{-d[N_2O_5]}{dt} = k_1[N_2O_5][M] - \frac{k_{-1}k_1[NO_2][N_2O_5][M]^2}{[NO_2](k_{-1}[M] + 2k_2)} = k_1[N_2O_5][M] - \frac{k_{-1}k_1[M]^2[N_2O_5]}{k_{-1}[M] + 2k_2}$$

$$\text{Rate} = \frac{-d[N_2O_5]}{dt} = \left(k_1 - \frac{k_{-1}k_1[M]}{k_{-1}[M] + 2k_2}\right)[N_2O_5][M];\ \text{Simplifying:}$$

$$\text{Rate} = \frac{-d[N_2O_5]}{dt} = \frac{2k_1k_2[M][N_2O_5]}{k_{-1}[M] + 2k_2}$$

99. a. [B] >> [A] so that [B] can be considered constant over the experiments. (This gives us a pseudo-order rate law equation.)

b. **Note in each case the 1/2 life doubles (in expt. 1 the first half life is 40 sec, the second is 80 sec; in expt. 2 the first half life is 20 sec, the second is 40 sec). Thus the reaction is second order in [A]. Between expt. 1 and expt. 2 we double [B] and the reaction rate doubles, thus it is first order in [B]. The overall rate law equation is rate = $k[A]^2[B]$.**

Using $t_{1/2} = \frac{1}{k[A]_o}$, we get $k = \frac{1}{(40.)(10 \times 10^{-2})} = 0.25$. But this is actually k^1 where rate = $k^1[A]^2$ and $k^1 = k[B]$.

$$k = \frac{k^1}{[B]} = \frac{0.25}{5.0} = 0.050\ L^2\ mol^{-2}\ s^{-1}$$

c. **i.** **wrong stoichiometry**

ii. **rate = $k[E][A]$**

$$k_1[A][B] = k_{-1}[E];\ [E] = \frac{k_1[A][B]}{k_{-1}}$$

$$\text{rate} = \frac{k\,k_1}{k_{-1}}[A]^2[B] \text{ could be!}$$

iii. **rate = $k[A]^2$ (no)**

So only mechanism ii is possible.

CHAPTER SIXTEEN

LIQUIDS AND SOLIDS

Intermolecular Forces and Physical Properties

11. Dipole forces are the forces that act between polar molecules. The electrostatic attraction between the positive end of one polar molecule and the negative end of another is the dipole force. Dipole forces are generally weaker than hydrogen bonding. Both of these forces are due to dipole moments in molecules. Hydrogen bonding is given a separate name from dipole forces because hydrogen bonding is a particularly strong dipole force.

    London dispersion forces can be referred to as instantaneous-induced dipole forces. As the size of the molecule increases, the strength of the London dispersion forces increases. This is because, as the electron cloud about a molecule gets larger, it is easier for the electrons to be drawn away from the nucleus. The molecule is said to be more polarizable.

13. Fusion refers to a solid converting to a liquid and vaporization refers to a liquid converting to a gas. Only a fraction of the hydrogen bonds are broken in going from the solid phase to the liquid phase. Most of the hydrogen bonds are still present in the liquid phase and must be broken during the liquid to gas phase transition. Thus, the enthalpy of vaporization is much larger than the enthalpy of fusion since more intermolecular forces are broken during the vaporization process.

15. Ionic compounds have ionic forces. Covalent compounds all have London Dispersion (LD) forces, while polar covalent compounds have dipole forces and/or hydrogen bonding forces. For hydrogen bonding (H-bonding) forces, the covalent compound must have either a N–H, O–H or F–H bond in the molecule.

    a. LD only b. dipole, LD c. H–bonding, LD

    d. ionic e. LD only (CH_4 in a nonpolar covalent compound.)

    f. dipole, LD g. ionic h. ionic

    i. LD mostly; C – F bonds are polar, but polymers like teflon are so large that LD forces are the predominant intermolecular forces.

    j. LD k. dipole, LD l. H-bonding, LD

    m. dipole, LD n. LD only

17. **a.** **$H_2NCH_2CH_2NH_2$; More extensive hydrogen bonding (H-bonding) is possible since two NH_2 groups are present.**

b. **H_2CO; H_2CO is polar while CH_3CH_3 is nonpolar. H_2CO_3 has dipole forces in addition to LD forces. CH_3CH_3 only has LD forces.**

c. **CH_3OH; CH_3OH can form relatively strong H-bonding interactions, unlike H_2CO.**

d. **HF; HF is capable of forming H-bonding interactions, HBr is not.**

19. **a.** **Neopentane is more compact than n-pentane. There is less surface area contact between neopentane molecules. This leads to weaker LD forces and a lower boiling point.**

b. **Ethanol is capable of hydrogen bonding (H-bonding), dimethyl ether is not.**

c. **HF is capable of H-bonding, HCl is not.**

d. **LiCl is ionic and HCl is a molecular solid with only dipole forces and LD forces. Ionic forces are much stronger than the intermolecular forces for molecular solids.**

e. **n-pentane is a larger molecule so it has stronger LD forces.**

f. **Dimethyl ether is polar. Dimethyl ether has dipole forces, in addition to LD forces, unlike n-propane, which only has LD forces.**

21. **Both molecules are capable of H-bonding. However, in oil of wintergreen the hydrogen bonding is <u>intra</u>molecular.**

O
CH3
C
O
H
O

------- = H-bonding

In methyl-4-hydroxybenzoate, the H-bonding is <u>inter</u>molecular, resulting in stronger intermolecular forces and a higher melting point.

23. **$C_{25}H_{52}$ has the stronger intermolecular forces because it has the higher boiling point. Even though $C_{25}H_{52}$ is nonpolar, it is so large that its London dispersion forces are much stronger than the sum of the hydrogen bonding and London dispersion interactions found in H_2O.**

25. **A single hydrogen bond in H_2O has a strength of 21 kJ/mol. Each H_2O molecule forms two H-bonds. Thus, it should take 42 kJ/mol of energy to break all of the H-bonds in water. Consider the phase transitions:**

$$\text{solid} \xrightarrow{6.0\text{ kJ}} \text{liquid} \xrightarrow{40.7\text{ kJ}} \text{vapor} \qquad \Delta H_{sub} = \Delta H_{fus} + \Delta H_{vap}$$

It takes a total of 46.7 kJ/mol to convert solid H_2O to vapor (ΔH_{sub}). This would be the amount of energy necessary to disrupt all of the intermolecular forces in ice. Thus, $(42 \div 46.7) \times 100 = 90.\%$ of the attraction in ice can be attributed to H-bonding.

Properties of Liquids

27. Critical temperature: The temperature above which a liquid cannot exist, i.e., the gas cannot be liquefied by increased pressure.

Critical pressure: The pressure that must be applied to a substance at its critical temperature to produce a liquid.

As the strength of the intermolecular forces increases, the critical temperature increases.

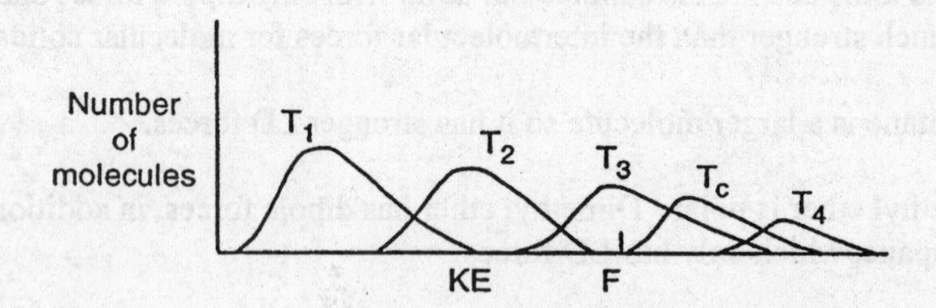

The kinetic energy distribution changes as one raises the temperature ($T_4 > T_c > T_3 > T_2 > T_1$). At the critical temperature, T_c, all molecules have kinetic energies greater than the intermolecular forces, F, and a liquid can't form. Note: The distributions above are not to scale.

29. As the strengths of intermolecular forces increase, surface tension, viscosity, and boiling point increase, while the vapor pressure decreases.

31. Water is a polar substance and wax is a nonpolar substance; they are not attracted to each other. A molecule at the surface of a drop of water is subject to attractions only by water molecules below it and to each side. The effect of this uneven pull on the surface water molecules tends to draw them into the body of the water and causes the droplet to assume the shape that has the minimum surface area, a sphere.

33. The structure of H_2O_2 is H – O – O – H, which produces greater hydrogen bonding than in water. Thus, the intermolecular forces are stronger in H_2O_2 than in H_2O resulting in a higher normal boiling point for H_2O_2 and a lower vapor pressure.

Structures and Properties of Solids

35. a. Crystalline solid: Regular, repeating structure

Amorphous solid: Irregular arrangement of atoms or molecules

b. Ionic solid: Made up of ions held together by ionic bonding.

Molecular solid: Made up of discrete covalently bonded molecules held together in the solid phase by weaker forces (LD, dipole or hydrogen bonds).

c. Molecular solid: Discrete, individual molecules

Covalent network solid: No discrete molecules; A covalent network solid is one large molecule. The interparticle forces are the covalent bonds between atoms.

d. Metallic solid: Completely delocalized electrons, conductor of electricity (ions in a sea of electrons)

Covalent network solid: Localized electrons; Insulator or semiconductor

37. No, an example is common glass, which is primarily amorphous SiO_2 (a covalent network solid) as compared to ice (a crystalline solid held together by weaker H-bonds). The interparticle forces in the amorphous solid in this case are stronger than those in the crystalline solid. Whether a solid is amorphous or crystalline depends on the long range order in the solid and not on the strengths of the interparticle forces.

39. $n\lambda = 2d \sin\theta$, $\lambda = \dfrac{2d \sin\theta}{n} = \dfrac{2 \times (201 \times 10^{-12}\ \text{m}) \sin 34.68°}{1}$, $\lambda = 2.29 \times 10^{-10}\ \text{m} = 229\ \text{pm}$

41. $n\lambda = 2d \sin\theta$, $d = \dfrac{n\lambda}{2 \sin\theta} = \dfrac{1 \times 1.54\ \text{Å}}{2 \times \sin 14.22°} = 3.13\ \text{Å} = 3.13 \times 10^{-10}\ \text{m} = 313\ \text{pm}$

43. A cubic closest packed structure has a face-centered cubic unit cell. In a face-centered cubic unit, there are:

$$8\ \text{corners} \times \frac{1/8\ \text{atom}}{\text{corner}} + 6\ \text{faces} \times \frac{1/2\ \text{atom}}{\text{face}} = 4\ \text{atoms}$$

The atoms in a face-centered cubic unit cell touch along the face diagonal of the cubic unit cell. Using the Pythagorean formula where l = length of the face diagonal and r = radius of the atom:

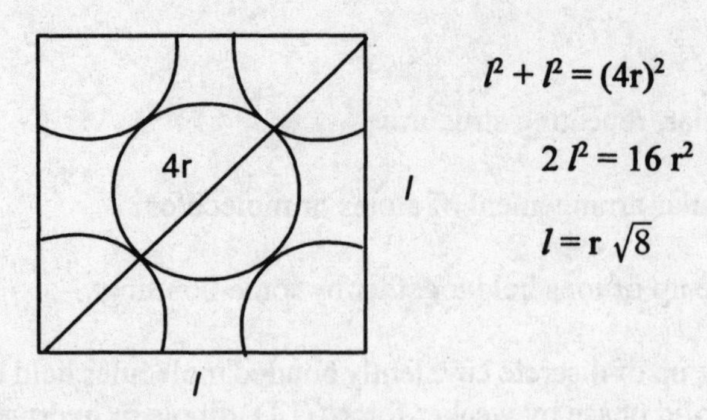

$$l^2 + l^2 = (4r)^2$$

$$2\,l^2 = 16\,r^2$$

$$l = r\sqrt{8}$$

$l = r\sqrt{8} = 197 \times 10^{-12}\text{ m} \times \sqrt{8} = 5.57 \times 10^{-10}\text{ m} = 5.57 \times 10^{-8}\text{ cm}$

Volume of a unit cell $= l^3 = (5.57 \times 10^{-8}\text{ cm})^3 = 1.73 \times 10^{-22}\text{ cm}^3$

Mass of a unit cell $= 4\text{ Ca atoms} \times \dfrac{1\text{ mol Ca}}{6.022 \times 10^{23}\text{ atoms}} \times \dfrac{40.08\text{ g Ca}}{\text{mol Ca}} = 2.662 \times 10^{-22}\text{ g Ca}$

$$\text{density} = \frac{\text{mass}}{\text{volume}} = \frac{2.662 \times 10^{-22}\text{ g}}{1.73 \times 10^{-22}\text{ cm}^3} = 1.54\text{ g/cm}^3$$

45. There are 4 Ni atoms in each unit cell: For a unit cell:

$$\text{density} = \frac{\text{mass}}{\text{volume}} = 6.84\text{ g/cm}^3 = \frac{4\text{ Ni atoms} \times \dfrac{1\text{ mol Ni}}{6.022 \times 10^{23}\text{ atoms}} \times \dfrac{58.69\text{ g Ni}}{\text{mol Ni}}}{l^3}$$

Solving: $l = 3.85 \times 10^{-8}$ cm = cube edge length

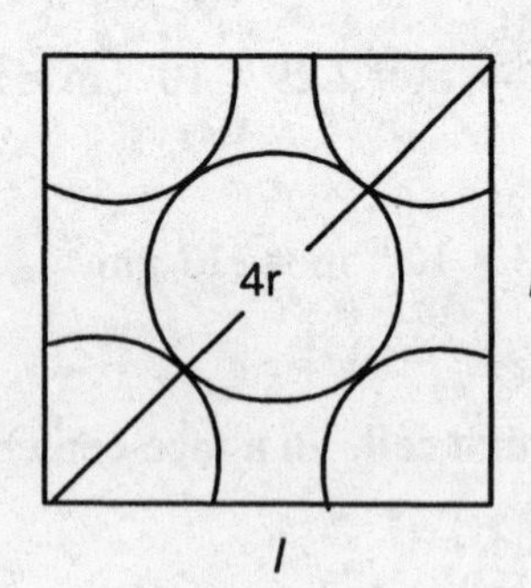

For a face centered cube:

$$(4r)^2 = l^2 + l^2 = 2\,l^2$$

$$r\sqrt{8} = l,\; r = l/\sqrt{8}$$

$$r = 3.85 \times 10^{-8}\text{ cm}/\sqrt{8}$$

$$r = 1.36 \times 10^{-8}\text{ cm} = 136\text{ pm}$$

47. For a body-centered unit cell: $8\text{ corners} \times \dfrac{1/8\text{ Ti}}{\text{corner}} + \text{Ti at body center} = 2\text{ Ti atoms}$

All body-centered unit cells have 2 atoms per unit cell. For a unit cell:

$$\text{density} = 4.50\text{ g/cm}^3 = \frac{2\text{ atoms Ti} \times \dfrac{1\text{ mol Ti}}{6.022 \times 10^{23}\text{ atoms}} \times \dfrac{47.88\text{ g Ti}}{\text{mol Ti}}}{l^3},\; l = \text{cube edge length}$$

Solving: l = edge length of unit cell = 3.28×10^{-8} cm = 328 pm

Assume Ti atoms just touch along the body diagonal of the cube, so body diagonal = 4 × radius of atoms = 4r.

The triangle we need to solve is:

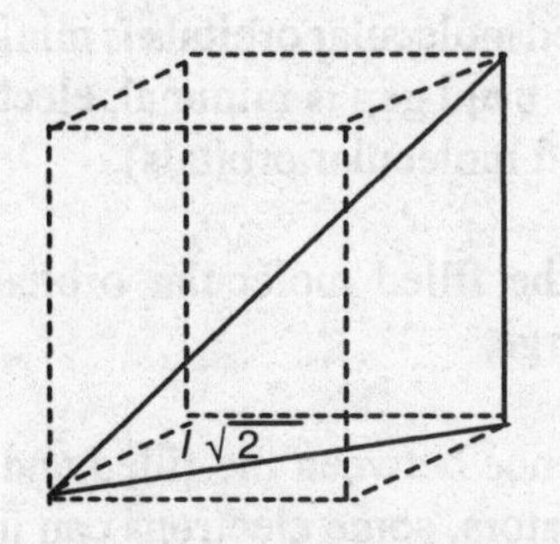

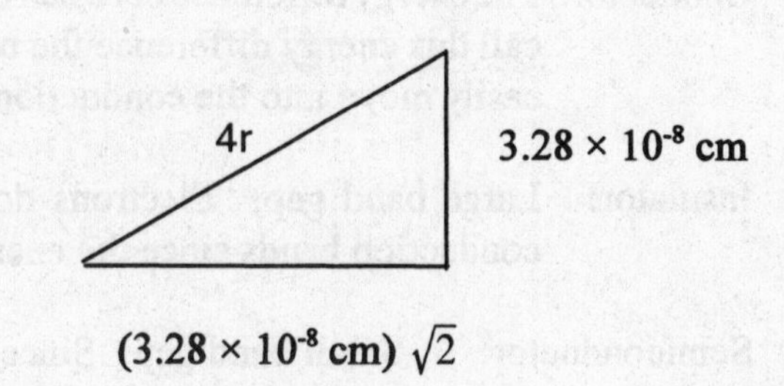

$$(4r)^2 = (3.28 \times 10^{-8}\ \text{cm})^2 + [(3.28 \times 10^{-8}\ \text{cm})\sqrt{2}\ ]^2,\ r = 1.42 \times 10^{-8}\ \text{cm} = 142\ \text{pm}$$

For a body-centered unit cell, the radius of the atom is related to the cube edge length by $4r = l\sqrt{3}$ or $l = 4r/\sqrt{3}$.

49. If a face-centered cubic structure, then 4 atoms/unit cell and from Exercise 16.43:

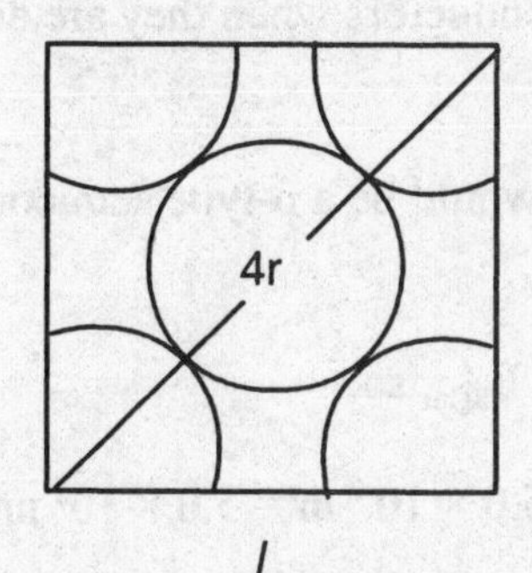

$$2\,l^2 = 16\,r^2$$

$$l = r\sqrt{8} = 144\ \text{pm}\ \sqrt{8} = 407\ \text{pm}$$

$$l = 407 \times 10^{-12}\ \text{m} = 407 \times 10^{-10}\ \text{cm}$$

$$\text{density} = \frac{4\ \text{atoms Au} \times \dfrac{1\ \text{mol Au}}{6.022 \times 10^{23}\ \text{atoms}} \times \dfrac{197.0\ \text{g Au}}{\text{mol Au}}}{(4.07 \times 10^{-8}\ \text{cm})^3} = 19.4\ \text{g/cm}^3$$

If a body-centered cubic structure, then 2 atoms/unit cell and from Exercise 16.47:

4r

l

$l\sqrt{2}$

$$16\,r^2 = l^2 + 2\,l^2$$

$$l = 4r/\sqrt{3} = 333\ \text{pm} = 333 \times 10^{-12}\ \text{m}$$

$$l = 333 \times 10^{-10}\ \text{cm} = 3.33 \times 10^{-8}\ \text{cm}$$

$$\text{density} = \frac{2 \text{ atoms Au} \times \frac{1 \text{ mol Au}}{6.022 \times 10^{23} \text{ atoms}} \times \frac{197.0 \text{ g Au}}{\text{mol Au}}}{(3.33 \times 10^{-8} \text{ cm})^3} = 17.7 \text{ g/cm}^3$$

The measured density is consistent with a face-centered cubic unit cell.

51. Conductor: The energy difference between the filled and unfilled molecular orbitals is minimal. We call this energy difference the band gap. Since the band gap is minimal, electrons can easily move into the conduction bands (the unfilled molecular orbitals).

Insulator: Large band gap; Electrons do not move from the filled molecular orbitals to the conduction bands since the energy difference is large.

Semiconductor: Small band gap; Since the energy difference between the filled and unfilled molecular orbitals is smaller than in insulators, some electrons can jump into the conduction bands. The band gap, however, is not as small as with conductors, so semiconductors have intermediate conductivity.

53. To produce an n-type semiconductor, dope Ge with a substance that has more than 4 valence electrons, e.g., a group 5A element. Phosphorus or arsenic are two substances which will produce n-type semiconductors when they are doped into germanium. To produce a p-type semiconductor, dope Ge with a substance that has fewer than 4 valence electrons, e.g., a group 3A element. Gallium or indium are two substances which will produce p-type semiconductors when they are doped into germanium.

55. In has fewer valence electrons than Se; thus Se doped with In would be a p-type semiconductor.

57. $E_{gap} = 2.5 \text{ eV} \times 1.6 \times 10^{-19} \text{ J/eV} = 4.0 \times 10^{-19} \text{ J}$; We want $E_{gap} = E_{light}$, so:

$$E_{light} = \frac{hc}{\lambda},\ \lambda = \frac{hc}{E} = \frac{(6.63 \times 10^{-34} \text{ J s})(3.00 \times 10^{8} \text{ m/s})}{4.0 \times 10^{-19} \text{ J}} = 5.0 \times 10^{-7} \text{ m} = 5.0 \times 10^{2} \text{ nm}$$

59. There is one octahedral hole per closest packed anion in a closest packed structure. If one-half of the octahedral holes are filled, then there is a 2:1 ratio of fluoride ions to cobalt ions in the crystal. The formula is CoF_2.

61. Mn ions at 8 corners: 8(1/8) = 1 Mn ion; F ions at 12 edges: 12(1/4) = 3 F ions

Formula is MnF_3. Assuming fluoride is -1 charged, then the charge on Mn is +3.

63. CsCl is a simple cubic array of Cl^- ions with Cs^+ in the middle of each unit cell. There is one Cs^+ and one Cl^- ion in each unit cell. Cs^+ and Cl^- touch along the body diagonal.

body diagonal = $2r_{Cs^+} + 2r_{Cl^-} = \sqrt{3}\,l$, l = length of cube edge

In each unit cell:

mass = 1 CsCl molecule (1 mol/6.022 × 10^{23} molecules) (168.4 g/mol) = 2.796 × 10^{-22} g

volume = l^3 = mass/density = 2.796 × 10^{-22} g/3.97 g cm^{-3} = 7.04 × 10^{-23} cm^3

$l^3 = 7.04 \times 10^{-23}$ cm^3, $l = 4.13 \times 10^{-8}$ cm = 413 pm = length of cube edge

$2r_{Cs^+} + 2r_{Cl^-} = \sqrt{3}\,l = \sqrt{3}(413 \text{ pm}) = 715 \text{ pm}$

The distance between ion centers = $r_{Cs^+} + r_{Cl^-} = 715 \text{ pm}/2 = 358 \text{ pm}$

From ionic radius: $r_{Cs^+} = 169$ pm and $r_{Cl^-} = 181$ pm; $r_{Cs^+} + r_{Cl^-} = 169 + 181 = 350.$ pm

The distance calculated from the density is 8 pm (2.3%) greater than that calculated from tables of ionic radii.

65. For a cubic hole to be filled, the cation to anion radius ratio is between $0.732 < r_+/r_- < 1.00$.

CsBr: Cs^+ radius = 169 pm, Br^- radius = 195 pm; $r_+/r_- = 169/195 = 0.867$

From the radius ratio, Cs^+ should occupy cubic holes. The structure should be the CsCl structure. The actual structure is the CsCl structure.

KF: K^+ radius = 133 pm, F^- radius = 136 pm; $r_+/r_- = 133/136 = 0.978$

Again, we would predict a structure similar to CsCl, i.e., cations in the middle of a simple cubic array of anions. The actual structure is the NaCl structure.

The radius ratio rules fail for KF. Exceptions are common for crystal structures.

67. a. $8 \text{ corners} \times \frac{1/8 \text{ Xe}}{\text{corner}} + 1 \text{ Xe inside cell} = 2 \text{ Xe}$; $8 \text{ edges} \times \frac{1/4 \text{ F}}{\text{edge}} + 2 \text{ F inside cell} = 4 \text{ F}$

Empirical formula is XeF_2. This is also the molecular formula.

b. For a unit cell:

$$\text{mass} = 2\ XeF_2 \text{ molecules} \times \frac{1 \text{ mol } XeF_2}{6.022 \times 10^{23} \text{ molecules}} \times \frac{169.3 \text{ g } XeF_2}{\text{mol } XeF_2} = 5.62 \times 10^{-22} \text{ g}$$

volume = (7.02 × 10^{-8} cm) (4.32 × 10^{-8} cm) (4.32 × 10^{-8} cm) = 1.31 × 10^{-22} cm^3

$$\text{density} = d = \frac{\text{mass}}{\text{volume}} = \frac{5.62 \times 10^{-22} \text{ g}}{1.31 \times 10^{-22} \text{ cm}^3} = 4.29 \text{ g/cm}^3$$

69. There are four sulfur ions per unit cell since the sulfur ions are cubic closest packed (face-centered cubic unit cell). Since there are one octahedral hole and two tetrahedral holes per closest packed ion, then each unit cell has 4 octahedral holes and 8 tetrahedral holes. This gives 4 (1/2) = 2 Al ions per unit cell and 8 (1/8) = 1 Zn ion per unit cell. The formula of the mineral is Al_2ZnS_4.

71. a. Y: 1 Y in center; Ba: 2 Ba in center

$$\text{Cu: } 8 \text{ corners} \times \frac{1/8 \text{ Cu}}{\text{corner}} = 1 \text{ Cu},\ 8 \text{ edges} \times \frac{1/4 \text{ Cu}}{\text{edge}} = 2 \text{ Cu},\ \text{total} = 3 \text{ Cu atoms}$$

$$\text{O: } 20 \text{ edges} \times \frac{1/4 \text{ O}}{\text{edge}} = 5 \text{ oxygen},\ 8 \text{ faces} \times \frac{1/2 \text{ O}}{\text{face}} = 4 \text{ oxygen},\ \text{total} = 9 \text{ O atoms}$$

Formula: $YBa_2Cu_3O_9$

b. The structure of this superconductor material follows the alternative perovskite structure described in Exercise 16.70 b. The $YBa_2Cu_3O_9$ structure is three of these cubic perovskite unit cells stacked on top of each other. The oxygen atoms are in the same places, Cu takes the place of Ti, two of the calcium atoms are replaced by two barium atoms, and one Ca is replaced by Y.

c. Y, Ba, and Cu are the same. Some oxygen atoms are missing.

$$12 \text{ edges} \times \frac{1/4 \text{ O}}{\text{edge}} = 3 \text{ O},\ 8 \text{ faces} \times \frac{1/2 \text{ O}}{\text{face}} = 4 \text{ O},\ \text{total} = 7 \text{ O atoms}$$

Superconductor formula is $YBa_2Cu_3O_7$.

Phase Changes and Phase Diagrams

73. Equilibrium: There is no change in composition; the vapor pressure is constant.

Dynamic: Two processes, vapor → liquid and liquid → vapor, are both occurring but with equal rates so the composition of the vapor is constant.

75. At 100.°C (373 K), the vapor pressure of H_2O is 1.00 atm = 760. torr.
For water, ΔH_{vap} = 40.7 kJ/mol.

$$\ln\left(\frac{P_1}{P_2}\right) = \frac{\Delta H_{vap}}{R}\left(\frac{1}{T_2} - \frac{1}{T_1}\right) \text{ or } \ln\left(\frac{P_2}{P_1}\right) = \frac{\Delta H_{vap}}{R}\left(\frac{1}{T_1} - \frac{1}{T_2}\right)$$

$$\ln\left(\frac{520.\text{ torr}}{760.\text{ torr}}\right) = \frac{40.7 \times 10^3 \text{ J/mol}}{8.3145 \text{ J K}^{-1}\text{mol}^{-1}}\left(\frac{1}{373 \text{ K}} - \frac{1}{T_2}\right),\ -7.75 \times 10^{-5} = \left(\frac{1}{373} - \frac{1}{T_2}\right)$$

$$-7.75 \times 10^{-5} = 2.68 \times 10^{-3} - \frac{1}{T_2},\ \frac{1}{T_2} = 2.76 \times 10^{-3},\ T_2 = \frac{1}{2.76 \times 10^{-3}} = 362 \text{ K or } 89°\text{C}$$

$$\ln\left(\frac{P_2}{1.00}\right) = \frac{40.7 \times 10^3 \text{ J/mol}}{8.3145 \text{ J K}^{-1}\text{mol}^{-1}}\left(\frac{1}{373 \text{ K}} - \frac{1}{623 \text{ K}}\right),\ \ln P_2 = 5.27,\ P_2 = e^{5.27} = 194 \text{ atm}$$

77. $\ln\left(\frac{P_1}{P_2}\right) = \frac{\Delta H_{vap}}{R}\left(\frac{1}{T_2} - \frac{1}{T_1}\right)$

P_1 = 760. torr, T_1 = 56.5°C + 273.2 = 329.7 K; P_2 = 630. torr, T_2 = ?

$$\ln\left(\frac{760.}{630.}\right) = \frac{32.0 \times 10^3 \text{ J/mol}}{8.3145 \text{ J K}^{-1}\text{mol}^{-1}}\left(\frac{1}{T_2} - \frac{1}{329.7}\right), \quad 0.188 = 3.85 \times 10^3\left(\frac{1}{T_2} - 3.033 \times 10^{-3}\right)$$

$$\frac{1}{T_2} - 3.033 \times 10^{-3} = 4.88 \times 10^{-5}, \quad \frac{1}{T_2} = 3.082 \times 10^{-3}, \quad T_2 = 324.5 \text{ K} = 51.3°\text{C}$$

$$\ln\left(\frac{630. \text{ torr}}{P_2}\right) = \frac{32.0 \times 10^3 \text{ J/mol}}{8.3145 \text{ J K}^{-1}\text{mol}^{-1}}\left(\frac{1}{298.2} - \frac{1}{324.5}\right), \quad \ln 630. - \ln P_2 = 1.05$$

$\ln P_2 = 5.40$, $P_2 = e^{5.40} = 221$ torr

79. If we graph $\ln P_{vap}$ vs 1/T, the slope of the resulting straight line will be $-\Delta H_{vap}/R$.

P_{vap}	$\ln P_{vap}$	T (Li)	1/T	T (Mg)	1/T
1 torr	0	1023 K	9.775×10^{-4} K^{-1}	893 K	11.2×10^{-4} K^{-1}
10.	2.3	1163	8.598×10^{-4}	1013	9.872×10^{-4}
100.	4.61	1353	7.391×10^{-4}	1173	8.525×10^{-4}
400.	5.99	1513	6.609×10^{-4}	1313	7.616×10^{-4}
760.	6.63	1583	6.317×10^{-4}	1383	7.231×10^{-4}

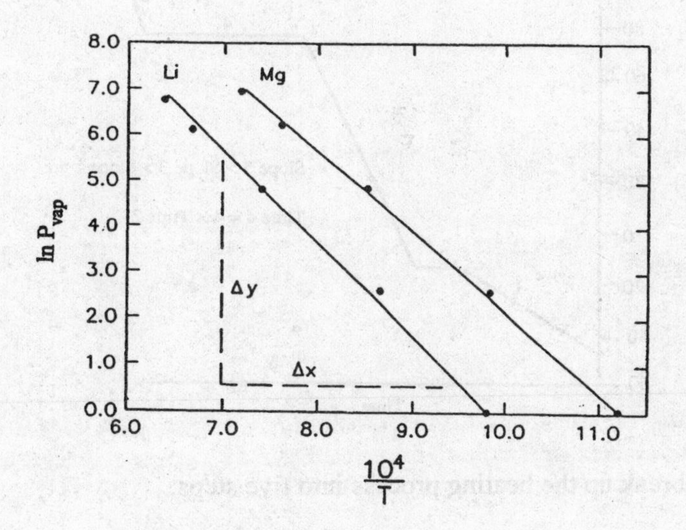

For Li:

We get the slope by taking two points (*x*, *y*) that are on the line we draw. For a line:

$$\text{slope} = \frac{\Delta y}{\Delta x} = \frac{y_2 - y_1}{x_2 - x_1}$$

or we can determine the straight line equation using a computer or calculator. The general straight line equation is y = mx + b where m = slope and b = y-intercept.

The equation of the Li line is: $\ln P_{vap} = -1.90 \times 10^4(1/T) + 18.6$, slope = -1.90×10^4 K

Slope = $-\Delta H_{vap}/R$, $\Delta H_{vap} = -\text{slope} \times R = 1.90 \times 10^4 \text{ K} \times 8.3145 \text{ J K}^{-1} \text{ mol}^{-1}$

$\Delta H_{vap} = 1.58 \times 10^5$ J/mol = 158 kJ/mol

For Mg:

The equation of the line is: $\ln P_{vap} = -1.67 \times 10^4(1/T) + 18.7$, slope = -1.67×10^4 K

$\Delta H_{vap} = -\text{slope} \times R = 1.67 \times 10^4 \text{ K} \times 8.3145 \text{ J K}^{-1} \text{ mol}^{-1}$, $\Delta H_{vap} = 1.39 \times 10^5$ J/mol = 139 kJ/mol

The bonding is stronger in Li since ΔH_{vap} is larger for Li.

81.

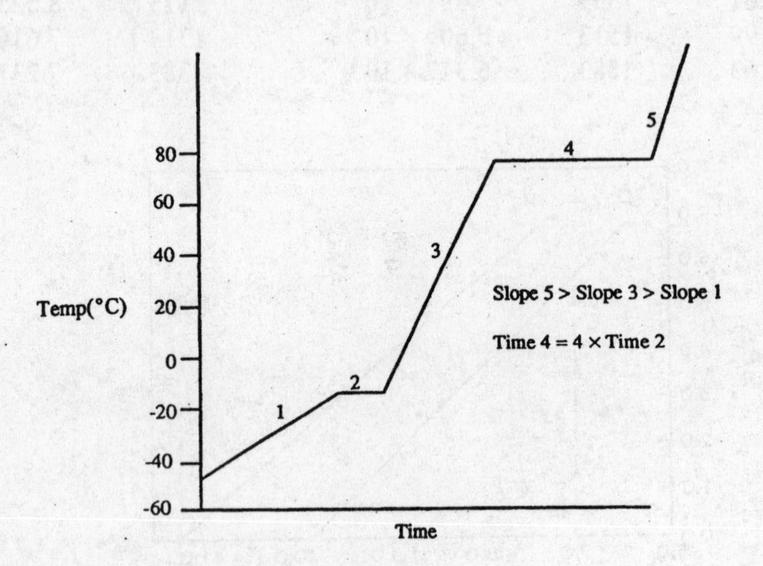

83. To calculate q_{total}, break up the heating process into five steps.

$H_2O(s, -20.°C) \rightarrow H_2O(s, 0°C)$, $\Delta T = 20.°C$ and specific heat capacity of ice = $s_{ice} = 2.1 \text{ J °C}^{-1} \text{ g}^{-1}$

$$q_1 = s_{ice} \times m \times \Delta T = \frac{2.1 \text{ J}}{\text{g °C}} \times 5.00 \times 10^2 \text{ g} \times 20.°C = 2.1 \times 10^4 \text{ J} = 21 \text{ kJ}$$

$$H_2O(s, 0°C) \rightarrow H_2O(l, 0°C),\ q_2 = 5.00 \times 10^2 \text{ g } H_2O \times \frac{1 \text{ mol}}{18.02 \text{ g}} \times \frac{6.01 \text{ kJ}}{\text{mol}} = 167 \text{ kJ}$$

$$H_2O(l, 0°C) \rightarrow H_2O(l, 100.°C),\ q_3 = \frac{4.2\ J}{g\ °C} \times 5.00 \times 10^2\ g \times 100.°C = 2.1 \times 10^5 J = 210\ kJ$$

$$H_2O(l, 100.°C) \rightarrow H_2O(g, 100.°C),\ q_4 = 5.00 \times 10^2\ g \times \frac{1\ mol}{18.02\ g} \times \frac{40.7\ kJ}{mol} = 1130\ kJ$$

$$H_2O(g, 100.°C) \rightarrow H_2O(g, 250.°C),\ q_5 = \frac{2.0\ J}{g\ °C} \times 5.00 \times 10^2\ g \times 150.°C = 1.5 \times 10^5\ J = 150\ kJ$$

$$q_{total} = q_1 + q_2 + q_3 + q_4 + q_5 = 21 + 167 + 210 + 1130 + 150 = 1680\ kJ$$

85. Heat released $= 0.250\ g\ Na \times \frac{1\ mol}{22.99\ g} \times \frac{368\ kJ}{2\ mol} = 2.00\ kJ$

To melt 50.0 g of ice requires: $50.0\ g\ ice \times \frac{1\ mol\ H_2O}{18.02\ g} \times \frac{6.01\ kJ}{mol} = 16.7\ kJ$

The reaction doesn't release enough heat to melt all of the ice. The temperature will remain at 0°C.

87. Total mass $H_2O = 18\ cubes \times \frac{30.0\ g}{cube} = 540.\ g;\ 540.\ g\ H_2O \times \frac{1\ mol\ H_2O}{18.02\ g} = 30.0\ mol\ H_2O$

Heat removed to produce ice at -5.0°C:

$$\left(\frac{4.18\ J}{g\ °C} \times 540.\ g \times 22.0\ °C\right) + \left(\frac{6.01 \times 10^3\ J}{mol} \times 30.0\ mol\right) + \left(\frac{2.08\ J}{g\ °C} \times 540.\ g \times 5.0\ °C\right)$$

$$= 4.97 \times 10^4\ J + 1.80 \times 10^5\ J + 5.6 \times 10^3\ J = 2.35 \times 10^5\ J$$

$2.35 \times 10^5\ J \times \frac{1\ g\ CF_2Cl_2}{158\ J} = 1.49 \times 10^3\ g\ CF_2Cl_2$ must be vaporized.

89. At any temperature, the plot tells us that substance A has a higher vapor pressure than substance B, with substance C having the lowest vapor pressure. Therefore, the substance with the weakest intermolecular forces is A, and the substance with the strongest intermolecular forces is C.

NH_3 can form hydrogen bonding interactions while the others cannot. Substance C is NH_3. The other two are nonpolar compounds with only London dispersion forces. Since CH_4 is smaller than SiH_4, CH_4 will have weaker LD forces and is substance A. Therefore, substance B is SiH_4.

91. A: solid; B: liquid; C: vapor

D: solid + vapor; E: solid + liquid + vapor

F: liquid + vapor; G: liquid + vapor; H: vapor

triple point: E; critical point: G

normal freezing point: temperature at which solid - liquid line is at 1.0 atm (see following plot).

normal boiling point: temperature at which liquid - vapor line is at 1.0 atm (see following plot).

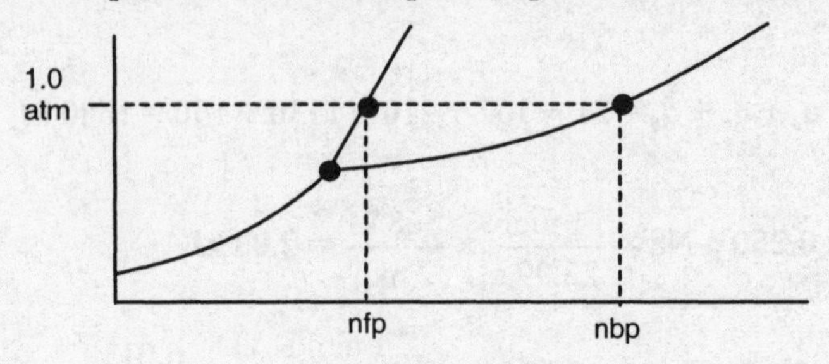

Since the solid-liquid equilibrium line has a positive slope, then the solid phase is denser than the liquid phase.

93.

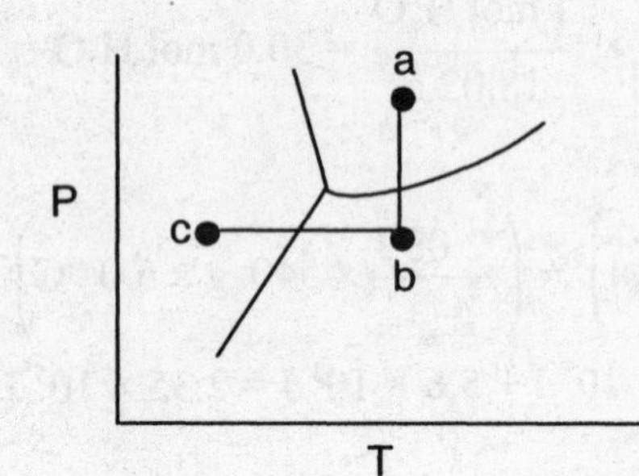

As P is lowered, we go from a to b on the phase diagram. The water boils. The evaporation of the water is endothermic and the water is cooled (b → c), forming some ice. If the pump is left on, the ice will sublime until none is left. This is the basis of freeze drying.

Additional Exercises

95. If TiO_2 conducts electricity as a liquid then it is an ionic solid; if not then TiO_2 is a network solid.

97.

H_3C—C(=O·····H—O)(—O—H·····O=)C—CH_3 ··· = H-bonding

99. B_2H_6: This compound contains only nonmetals so it is probably a molecular solid with covalent bonding. The low boiling point confirms this.

SiO_2: This is the empirical formula for quartz, which is a network solid.

CsI: This is a metal bonded to a nonmetal, which generally form ionic solids. The electrical conductivity in aqueous solution confirms this.

W: Tungsten is a metallic solid as the conductivity data confirms.

101. If we extend the liquid-vapor line of the water phase diagram to below the freezing point, we find that supercooled water will have a higher vapor pressure than ice at -10°C (see Figure 16.51 of the text). To achieve equilibrium there must be a constant vapor pressure. Over time super-cooled water will be transformed through the vapor into ice in an attempt to equilibrate the vapor pressure. Eventually there will only be ice at -10°C and its vapor at the vapor pressure given by the solid-vapor line in the phase diagram.

103.

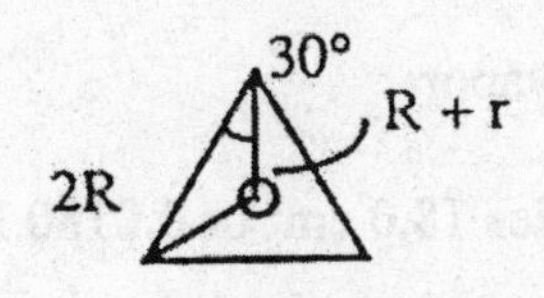

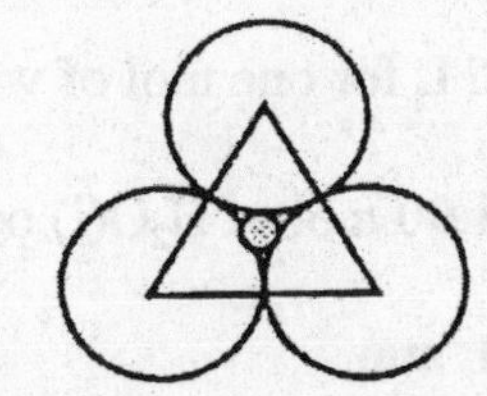

R = radius of sphere

r = radius of trigonal hole

For a right angle triangle (opposite side not drawn in):

$$\cos 30° = \frac{\text{adjacent}}{\text{hypotenuse}} = \frac{R}{R + r}, \quad 0.866 = \frac{R}{R + r}$$

$$0.866\,r = R - 0.866\,R, \quad r = \left(\frac{0.134}{0.866}\right) \times R = 0.155\,R$$

The cation must have a radius which is 0.155 times the radius of the spheres to just fit into the trigonal hole.

105. Ar is cubic closest packed. There are 4 Ar atoms per unit cell and with a face-centered unit cell, the atoms touch along the face diagonal.

face diagonal = $4r = \sqrt{2}\,l$, l = length of cube edge; $l = 4(190.\text{ pm})/\sqrt{2} = 537\text{ pm} = 5.37 \times 10^{-8}\text{ cm}$

$$\text{density} = \frac{\text{mass}}{\text{volume}} = \frac{4\text{ atoms} \times \dfrac{1\text{ mol}}{6.022 \times 10^{23}\text{ atoms}} \times \dfrac{39.95\text{ g}}{\text{mol}}}{(5.37 \times 10^{-8}\text{ cm})^3} = 1.71\text{ g/cm}^3$$

107. Out of 100.00 g: $28.31\text{ g O} \times \dfrac{1\text{ mol}}{15.999\text{ g}} = 1.769\text{ mol O}$; $71.69\text{ g Ti} \times \dfrac{1\text{ mol}}{47.88\text{ g}} = 1.497\text{ mol Ti}$

Formula is $TiO_{1.182}$ or $Ti_{0.8462}O$.
For $Ti_{0.8462}O$, let x = Ti^{2+} per mol O^{2-} and y = Ti^{3+} per mol O^{2-}. Setting up two equations and solving:

$$x + y = 0.8462 \text{ and } 2x + 3y = 2; \;\; 2x + 3(0.8462 - x) = 2$$

$$x = 0.539 \text{ mol Ti}^{2+}/\text{mol O}^{2-} \text{ and } y = 0.307 \text{ mol Ti}^{3+}/\text{mol O}^{2-}$$

$$\frac{0.539}{0.8462} \times 100 = 63.7\% \text{ of the titanium is Ti}^{2+} \text{ and } 36.3\% \text{ is Ti}^{3+}.$$

109. $24.7 \text{ g } C_6H_6 \times \frac{1 \text{ mol}}{78.11 \text{ g}} = 0.316 \text{ mol } C_6H_6$

$$P_{C_6H_6} = \frac{nRT}{V} = \frac{0.316 \text{ mol} \times \frac{0.08206 \text{ L atm}}{\text{mol K}} \times 293.2 \text{ K}}{100.0 \text{ L}} = 0.0760 \text{ atm or } 57.8 \text{ torr}$$

111. $w = -P\Delta V$; Assume const P of 1 atm.

$$V_{373} = \frac{nRT}{P} = \frac{1.00\,(0.08206)\,(373)}{1.00} = 30.6 \text{ L for one mol of water vapor}$$

Since the density of $H_2O(l)$ is 1.00 g/cm^3, 1.00 mol of $H_2O(l)$ occupies 18.0 cm^3 or 0.0180 L.

$$w = -1.00 \text{ atm } (30.6 \text{ L} - 0.0180 \text{ L}) = -30.6 \text{ L atm}$$

$$w = -30.6 \text{ L atm} \times 101.3 \text{ J L}^{-1} \text{ atm}^{-1} = -3.10 \times 10^3 \text{ J} = -3.10 \text{ kJ}$$

$$\Delta E = q + w = 41.16 \text{ kJ} - 3.10 \text{ kJ} = 38.06 \text{ kJ}$$

$$\frac{38.06}{41.16} \times 100 = 92.47\% \text{ of the energy goes to increase the internal energy of the water.}$$

The remainder of the energy (7.53%) goes to do work against the atmosphere.

113. The three sheets of B's form a cube. In this cubic unit cell there are B atoms at every face, B atoms at every edge, B atoms at every corner and one B atom in the center. A representation of this cube is:

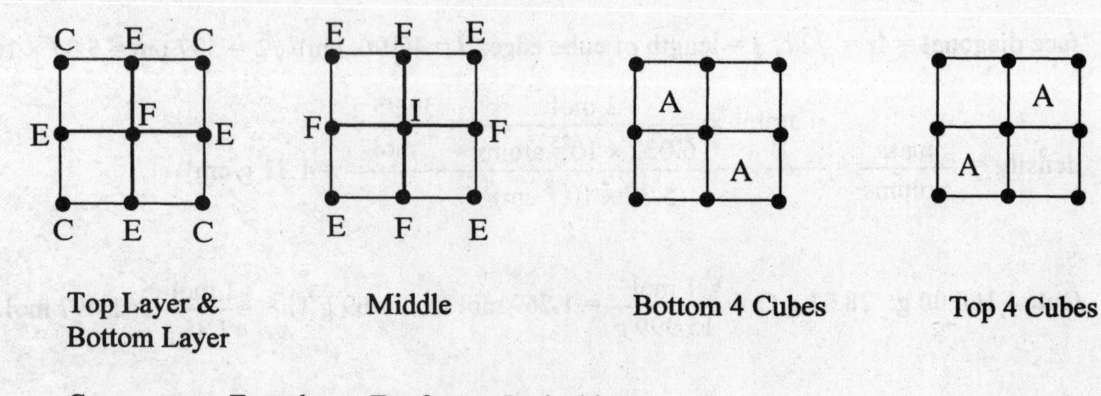

C = corner; E = edge; F = face; I = inside

Each unit cell of B atoms contains 4 A atoms. The number of B atoms in each unit cell is:

$$(8 \text{ corners} \times 1/8) + (6 \text{ faces} \times 1/2) + (12 \text{ edges} \times 1/4) + (1 \text{ middle B}) = 8 \text{ B atoms}$$

The empirical formula is AB_2.

Each A atom is in a cubic hole of B atoms so 8 B atoms surround each A atom. This will also be true in the extended lattice. The structure of B atoms in the unit cell is a cubic arrangement with B atoms at every face, edge, corner and center of the cube.

Challenge Problems

115. For water vapor at 30.0°C and 31.824 torr:

$$\text{density} = \frac{\text{P(molar mass)}}{\text{RT}} = \frac{\left(\frac{31.824 \text{ atm}}{760}\right)\left(\frac{18.015 \text{ g}}{\text{mol}}\right)}{\frac{0.08206 \text{ L atm}}{\text{mol K}} \times 303.2 \text{ K}} = 0.03032 \text{ g/L}$$

The volume of one molecule is proportional to d^3 where d is the average distance between molecules. For a large sample of molecules, the volume is still proportional to d^3. So:

$$\frac{V_{gas}}{V_{liq}} = \frac{d_{gas}^3}{d_{liq}^3}$$

If we have 0.99567 g H_2O, then $V_{liq} = 1.0000 \text{ cm}^3 = 1.0000 \times 10^{-3}$ L.

$V_{gas} = 0.99567 \text{ g} \times 1 \text{ L}/0.03032 \text{ g} = 32.84$ L

$$\frac{d_{gas}^3}{d_{liq}^3} = \frac{32.84 \text{ L}}{1.0000 \times 10^{-3} \text{ L}} = 3.284 \times 10^4, \quad \frac{d_{gas}}{d_{liq}} = (3.284 \times 10^4)^{1/3} = 32.02, \quad \frac{d_{liq}}{d_{gas}} = 0.03123$$

117. a. Structure a:

Ba: 2 Ba inside unit cell; Tl: $8 \text{ corners} \times \frac{1/8 \text{ Tl}}{\text{corner}} = 1$ Tl; Cu: $4 \text{ edges} \times \frac{1/4 \text{ Cu}}{\text{edge}} = 1$ Cu

O: $6 \text{ faces} \times \frac{1/2 \text{ O}}{\text{face}} + 8 \text{ edges} \times \frac{1/4 \text{ O}}{\text{edge}} = 5$ O

Formula = $TlBa_2CuO_5$

Structure b:

Tl and Ba are the same as in structure a.

Ca: 1 Ca inside unit cell; Cu: $8 \text{ edges} \times \frac{1/4 \text{ Cu}}{\text{edge}} = 2$ Cu

O: $10 \text{ faces} \times \frac{1/2 \text{ O}}{\text{face}} + 8 \text{ edges} \times \frac{1/4 \text{ O}}{\text{edge}} = 7$ O

Formula = $TlBa_2CaCu_2O_7$

Structure c:

Tl and Ba are the same and two Ca atoms are located inside the unit cell.

Cu: 12 edges × $\frac{1/4\ Cu}{edge}$ = 3 Cu; O: 14 faces × $\frac{1/2\ O}{face}$ + 8 edges × $\frac{1/4\ O}{edge}$ = 9 O

Formula: $TlBa_2Ca_2Cu_3O_9$

Structure d: Following similar calculations, formula = $TlBa_2Ca_3Cu_4O_{11}$

b. Structure a has one planar sheet of Cu and O atoms and the number increases by one for each of the remaining structures. The order of superconductivity temperature from lowest to highest temperature is: a < b < c < d.

c. $TlBa_2CuO_5$: 3 + 2(2) + x + 5(-2) = 0, x = +3
Only Cu^{3+} is present in each formula unit.

$TlBa_2CaCu_2O_7$: 3 + 2(2) + 2 + 2(x) + 7(-2) = 0, x = +5/2
Each formula unit contains 1 Cu^{2+} and 1 Cu^{3+}.

$TlBa_2Ca_2Cu_3O_9$: 3 + 2(2) + 2(2) + 3(x) + 9(-2) = 0, x = +7/3
Each formula unit contains 2 Cu^{2+} and 1 Cu^{3+}.

$TlBa_2Ca_3Cu_4O_{11}$: 3 + 2(2) + 3(2) + 4(x) + 11(-2) = 0, x = +9/4
Each formula unit contains 3 Cu^{2+} and 1 Cu^{3+}.

d. This superconductor material achieves variable copper oxidation states by varying the numbers of Ca, Cu and O in each unit cell. The mixtures of copper oxidation states is discussed in part c. The superconductor material in Exercise 16.71 achieves variable copper oxidation states by omitting oxygen at various sites in the lattice.

119. For an octahedral hole; the geometry is:

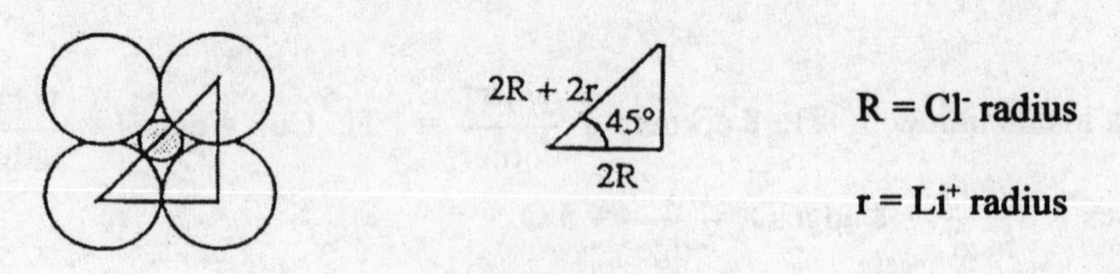

From the diagram: $\cos 45° = \frac{adjacent}{hypotenuse} = \frac{2R}{2R + 2r} = \frac{R}{R + r}$, $0.707 = \frac{R}{R + r}$

R = 0.707 (R + r), r = 0.414 R

LiCl unit cell:

length of cube edge = 2R + 2r = 514 pm

2R + 2(0.414 R) = 514 pm, R = 182 pm = Cl^- radius

2(182 pm) + 2r = 514 pm, r = 75 pm = Li^+ radius

From Figure 13.7, the Li^+ radius is 60 pm and the Cl^- radius is 181 pm. The Li^+ ion is much smaller than calculated. This probably means that the ions are not actually in contact with each other. The octahedral holes are larger than the Li^+ ion.

121. Use radius ratios and the information in Table 16.6 of the text to determine the type of structure. In the NaCl type cubic unit cell, the cations occupy octahedral holes; in the CsCl type cubic unit cell, the cations occupy cubic holes; in the Zn S type cubic unit cell, the cations occupy the tetrahedral holes. To determine the fraction of holes filled, the stoichiometry given in the formula will determine this. And finally, to determine the density, use the geometry relationships unique to each structure to determine the edge length (assuming the cations and anions touch); then go on to estimate the density.

For SnO_2: $\frac{r_+}{r_-} = \frac{71\text{ pm}}{140.\text{ pm}} = 0.51$

Radius ratios predict that octahedral holes will be filled. Therefore, we predict the NaCl type unit cell for SnO_2. Since there is 1 octahedral hole per closest packed anion (O^{2-}), then one-half of the octahedral holes will be filled by the Sn^{4+} cations. This is required by the 1:2 mol ratio of cations to anions in the SnO_2 formula.

To estimate the density in an NaCl type unit cell, the cations and anions are assumed to touch along the cube edge. Since the unit cell contains four O^{2-} ions, then two SnO_2 formula units are contained per unit cell.

ℓ = cube edge length = $2r_{Sn^{4+}} + 2r_{O^{2-}}$ = 2(71 pm) + 2(140. pm) = 422 pm

$$\text{density} = \frac{\text{mass}}{\text{volume}} = \frac{2\ SnO_2\text{ units} \times \frac{1\text{ mol } SnO_2}{6.022 \times 10^{23}\text{ units}} \times \frac{150.7\text{ g } SnO_2}{\text{mol } SnO_2}}{(4.22 \times 10^{-8}\text{ cm})^3} = 6.66\text{ g/cm}^3$$

For AlP: $\frac{r_+}{r_-} = \frac{50.\text{ pm}}{212\text{ pm}} = 0.24$

Radius ratios predict tetrahedral holes for the Al^{3+} cations, which occurs in a ZnS type unit cell. Since there are two tetrahedral holes per closest packed anion (P^{3-}), then Al^{3+} ions occupy one-half of the tetrahedral holes. This is required to give the 1:1 formula stoichiometry in AlP.

In a ZnS type unit cell, it is the body diagonal of the unit cell that allows determination of the cube edge length (ℓ). From Figures 16.38 and 16.39 of the text, the body diagonal of a cube that encloses each tetrahedral hole has a length equal to $2r_+ + 2r_-$. From Figure 16.41a, each AlP unit cell consists of eight of these smaller cubes. The length of the body diagonal of the unit cell is equal to $4r_+ + 4r_-$. This relationship, along with realizing that each unit cell contains four P^{3-} ions so four AlP formula units are contained per unit cell, allows determination of the density.

$$\text{body diagonal} = \sqrt{3}\,\ell = 4r_{Al^{3+}} + 4r_{P^{3-}},\ \ell = \frac{4(50.\text{ pm}) + 4(212\text{ pm})}{\sqrt{3}} = 605\text{ pm}$$

$$\text{density} = \frac{\text{mass}}{\text{volume}} = \frac{4\text{ AlP units} \times \dfrac{1\text{ mol AlP}}{6.022 \times 10^{23}\text{ units}} \times \dfrac{57.95\text{ g AlP}}{1\text{ mol AlP}}}{(6.05 \times 10^{-8}\text{ cm})^3} = 1.74\text{ g/cm}^3$$

For BaO: $\dfrac{r_+}{r_-} = \dfrac{135\text{ pm}}{140.\text{ pm}} = 0.964$

Radius ratios predicts that Ba^{2+} ions occupy cubic holes. This is seen in CsCl type unit cells. Since there is one cubic hole per simple cubic packing of anions (O^{2-}), then all the cubic holes will be filled by Ba^{2+} ions. This is required to give the required 1:1 formula stoichiometry in BaO.

To estimate the density, the ions in a CsCl type unit cell are assumed to just touch along the body diagonal of the unit cell so that body diagonal = $2r_+ + 2r_-$. Since the unit cell contains one O^{2-} ion, then one BaO formula unit is contained per unit cell.

$$\text{body diagonal} = \sqrt{3}\,\ell = 2r_{Ba^{2+}} + 2r_{O^{2-}},\ \ell = \frac{2(135\text{ pm}) + 2(140.\text{ pm})}{\sqrt{3}} = 318\text{ pm}$$

$$\text{density} = \frac{\text{mass}}{\text{volume}} = \frac{1\text{ BaO unit} \times \dfrac{1\text{ mol BaO}}{6.022 \times 10^{23}\text{ units}} \times \dfrac{153.3\text{ g BaO}}{1\text{ mol BaO}}}{(3.18 \times 10^{-8}\text{ cm})^3} = 7.92\text{ g/cm}^3$$

CHAPTER SEVENTEEN

PROPERTIES OF SOLUTIONS

13. $0.250\ \text{L} \times \dfrac{0.100\ \text{mol}}{\text{L}} \times \dfrac{134.0\ \text{g}}{\text{mol}} = 3.35\ \text{g}\ Na_2C_2O_4$

15. a. $Ca(NO_3)_2(s) \rightarrow Ca^{2+}(aq) + 2\ NO_3^-(aq)$; $M_{Ca^{2+}} = \dfrac{1.06 \times 10^{-3}\ \text{mol}}{\text{L}}$; $M_{NO_3^-} = \dfrac{2.12 \times 10^{-3}\ \text{mol}}{\text{L}}$

b. $1.0 \times 10^{-3}\ \text{L} \times \dfrac{1.06 \times 10^{-3}\ \text{mol}\ Ca^{2+}}{\text{L}} \times \dfrac{40.08\ \text{g}\ Ca^{2+}}{\text{mol}} = 4.2 \times 10^{-5}\ \text{g}\ Ca^{2+}$

c. $1.0 \times 10^{-6}\ \text{L} \times \dfrac{2.12 \times 10^{-3}\ \text{mol}\ NO_3^-}{\text{L}} \times \dfrac{6.02 \times 10^{23}\ NO_3^-\ \text{ions}}{\text{mol}\ NO_3^-} = 1.3 \times 10^{15}\ NO_3^-\ \text{ions}$

17. a. $HNO_3(l) \rightarrow H^+(aq) + NO_3^-(aq)$

b. $Na_2SO_4(s) \rightarrow 2\ Na^+(aq) + SO_4^{2-}(aq)$

c. $Al(NO_3)_3(s) \rightarrow Al^{3+}(aq) + 3\ NO_3^-(aq)$

d. $SrBr_2(s) \rightarrow Sr^{2+}(aq) + 2\ Br^-(aq)$

e. $KClO_4(s) \rightarrow K^+(aq) + ClO_4^-(aq)$

f. $NH_4Br(s) \rightarrow NH_4^+(aq) + Br^-(aq)$

g. $NH_4NO_3(s) \rightarrow NH_4^+(aq) + NO_3^-(aq)$

h. $CuSO_4(s) \rightarrow Cu^{2+}(aq) + SO_4^{2-}(aq)$

i. $NaOH(s) \rightarrow Na^+(aq) + OH^-(aq)$

Concentration of Solutions

19. $\text{molality} = \dfrac{40.0\ \text{g EG}}{60.0\ \text{g}\ H_2O} \times \dfrac{1000\ \text{g}}{\text{kg}} \times \dfrac{1\ \text{mol EG}}{62.07\ \text{g}} = 10.7\ \text{mol/kg}$

$\text{molarity} = \dfrac{40.0\ \text{g EG}}{100.0\ \text{g solution}} \times \dfrac{1.05\ \text{g}}{\text{cm}^3} \times \dfrac{1000\ \text{cm}^3}{\text{L}} \times \dfrac{1\ \text{mol}}{62.07\ \text{g}} = 6.77\ \text{mol/L}$

$40.0\ \text{g EG} \times \dfrac{1\ \text{mol}}{62.07\ \text{g}} = 0.644\ \text{mol EG}$; $60.0\ \text{g}\ H_2O \times \dfrac{1\ \text{mol}}{18.02\ \text{g}} = 3.33\ \text{mol}\ H_2O$

$\chi_{EG} = \dfrac{0.644}{3.33 + 0.644} = 0.162 = \text{mole fraction ethylene glycol}$

21. $25 \text{ mL } C_5H_{12} \times \frac{0.63 \text{ g}}{\text{mL}} = 16 \text{ g } C_5H_{12}$; $25 \text{ mL} \times \frac{0.63}{\text{mL}} \times \frac{1 \text{ mol}}{72.15 \text{ g}} = 0.22 \text{ mol } C_5H_{12}$

$45 \text{ mL } C_6H_{14} \times \frac{0.66 \text{ g}}{\text{mL}} = 30.\text{ g } C_6H_{14}$; $45 \text{ mL} \times \frac{0.66 \text{ g}}{\text{mL}} \times \frac{1 \text{ mol}}{86.17 \text{ g}} = 0.34 \text{ mol } C_6H_{14}$

$$\text{mass \% pentane} = \frac{\text{mass pentane}}{\text{total mass}} \times 100 = \frac{16 \text{ g}}{16 \text{ g} + 30.\text{ g}} \times 100 = 35\%$$

$$\chi_{pentane} = \frac{\text{mol pentane}}{\text{total mol}} = \frac{0.22 \text{ mol}}{0.22 \text{ mol} + 0.34 \text{ mol}} = 0.39$$

$$\text{molality} = \frac{\text{mol pentane}}{\text{kg hexane}} = \frac{0.22 \text{ mol}}{0.030 \text{ kg}} = 7.3 \text{ mol/kg}$$

$$\text{molarity} = \frac{\text{mol pentane}}{\text{L solution}} = \frac{0.22 \text{ mol}}{25 \text{ mL} + 45 \text{ mL}} \times \frac{1000 \text{ mL}}{1 \text{ L}} = 3.1 \text{ mol/L}$$

23. If we have 1.00 L of solution:

$$1.37 \text{ mol citric acid} \times \frac{192.1 \text{ g}}{\text{mol}} = 263 \text{ g citric acid}$$

$$1.00 \times 10^3 \text{ mL solution} \times \frac{1.10 \text{ g}}{\text{mL}} = 1.10 \times 10^3 \text{ g solution}$$

$$\text{mass \% of citric acid} = \frac{263 \text{ g}}{1.10 \times 10^3 \text{ g}} \times 100 = 23.9\%$$

In 1.00 L of solution, we have 263 g citric acid and $(1.10 \times 10^3 - 263) = 840$ g of H_2O.

$$\text{molality} = \frac{1.37 \text{ mol citric acid}}{0.84 \text{ kg } H_2O} = 1.6 \text{ mol/kg}$$

$840 \text{ g } H_2O \times \frac{1 \text{ mol}}{18.0 \text{ g}} = 47 \text{ mol } H_2O$; $\chi_{\text{citric acid}} = \frac{1.37}{47 + 1.37} = 0.028$

25. $0.40 \text{ mol } CH_3COCH_3 \times \frac{58.1 \text{ g}}{\text{mol}} = 23 \text{ g acetone}$; $0.60 \text{ mol } C_2H_5OH \times \frac{46.1 \text{ g}}{\text{mol}} = 28 \text{ g ethanol}$

$$\text{mass \% acetone} = \frac{23 \text{ g}}{23 \text{ g} + 28 \text{ g}} \times 100 = 45\%$$

Thermodynamics of Solutions and Solubility

27. The nature of the interparticle forces. Polar solutes and ionic solutes dissolve in polar solvents and nonpolar solutes dissolve in nonpolar solvents.

29. Using Hess's law:

$$NaI(s) \rightarrow Na^+(g) + I^-(g) \qquad \Delta H = -\Delta H_{LE} = -(-686 \text{ kJ/mol})$$
$$Na^+(g) + I^-(g) \rightarrow Na^+(aq) + I^-(aq) \qquad \Delta H = \Delta H_{hyd} = -694 \text{ kJ/mol}$$

$$NaI(s) \rightarrow Na^+(aq) + I^-(aq) \qquad \Delta H_{soln} = -8 \text{ kJ/mol}$$

ΔH_{soln} refers to the heat released or gained when a solute dissolves in a solvent. Here, an ionic compound dissolves in water.

31. Both $Al(OH)_3$ and NaOH are ionic compounds. Since the lattice energy is proportional to the charge of the ions, then the lattice energy of aluminum hydroxide is greater than that of sodium hydroxide. The attraction of water molecules for Al^{3+} and OH^- cannot overcome the larger lattice energy and $Al(OH)_3$ is insoluble. For NaOH, the favorable hydration energy is large enough to overcome the smaller lattice energy and NaOH is soluble.

33. Water is a polar molecule capable of hydrogen bonding. Polar molecules, especially molecules capable of hydrogen bonding, and ions can be hydrated. For covalent compounds, as polarity increases, the attraction to water (hydration) increases. For ionic compounds, as the charge of the ions increase and/or the size of the ions decrease, the attraction to water (hydration) increases.

a. CH_3CH_2OH; CH_3CH_2OH is polar while $CH_3CH_2CH_3$ is nonpolar.

b. $CHCl_3$; $CHCl_3$ is polar while CCl_4 is nonpolar.

c. CH_3CH_2OH; CH_3CH_2OH is much more polar than $CH_3(CH_2)_{14}CH_2OH$.

35. As the length of the hydrocarbon chain increases, the solubility decreases. The –OH end of the alcohols can hydrogen bond with water. The hydrocarbon chain, however, is basically nonpolar and interacts poorly with water. As the hydrocarbon chain gets longer, a greater portion of the molecule cannot interact with the water molecules and the solubility decreases, i.e., the effect of the –OH group decreases as the alcohols get larger.

37. hydrophobic: water hating; hydrophilic: water loving

39. $P_{gas} = kC$, $0.790 \text{ atm} = k \times \dfrac{8.21 \times 10^{-4} \text{ mol}}{L}$, $k = 962 \text{ L atm/mol}$

$P_{gas} = kC$, $1.10 \text{ atm} = \dfrac{962 \text{ L atm}}{\text{mol}} \times C$, $C = 1.14 \times 10^{-3} \text{ mol/L}$

41. As the temperature increases the gas molecules will have a greater average kinetic energy. A greater fraction of the gas molecules in solution will have a kinetic energy greater than the attractive forces between the gas molecules and the solvent molecules. More gas molecules are able to escape to the vapor phase and the solubility of the gas decreases.

Vapor Pressures of Solution

43. $P_{C_2H_5OH} = \chi_{C_2H_5OH} P^\circ_{C_2H_5OH}$; $\chi_{C_2H_5OH} = \dfrac{\text{mol } C_2H_5OH \text{ solution}}{\text{total mol in solution}}$

$$53.6 \text{ g } C_3H_8O_3 \times \frac{1 \text{ mol } C_3H_8O_3}{92.09 \text{ g}} = 0.582 \text{ mol } C_3H_8O_3$$

$$133.7 \text{ g } C_2H_5OH \times \frac{1 \text{ mol } C_2H_5OH}{46.07 \text{ g}} = 2.90 \text{ mol } C_2H_5OH; \text{ total mol} = 0.582 + 2.90 = 3.48 \text{ mol}$$

$$113 \text{ torr} = \frac{2.90 \text{ mol}}{3.48 \text{ mol}} \times P^\circ_{C_2H_5OH},\ P^\circ_{C_2H_5OH} = 136 \text{ torr}$$

45. Solution d (methanol/water); Methanol is more volatile than water, which will increase the total vapor pressure to a value greater than the vapor pressure of pure water at this temperature.

47. $P_B = \chi_B P^\circ_B$, $\chi_B = P_B/P^\circ_B = 0.900 \text{ atm}/0.930 \text{ atm} = 0.968$

$$0.968 = \frac{\text{mol benzene}}{\text{total mol}};\ \text{mol benzene} = 78.11 \text{ g } C_6H_6 \times \frac{1 \text{ mol}}{78.11} = 1.000 \text{ mol}$$

Let x = mol solute, then: $\chi_B = 0.968 = \dfrac{1.000 \text{ mol}}{1.000 + x}$, $0.968 + 0.968\,x = 1.000$, $x = 0.033$ mol

$$\text{molar mass} = \frac{10.0 \text{ g}}{0.033 \text{ mol}} = 303 \text{ g/mol} \approx 3.0 \times 10^2 \text{ g/mol}$$

49. a. $25 \text{ mL } C_5H_{12} \times \dfrac{0.63 \text{ g}}{\text{mL}} \times \dfrac{1 \text{ mol}}{72.15 \text{ g}} = 0.22 \text{ mol } C_5H_{12}$

$45 \text{ mL } C_6H_{14} \times \dfrac{0.66 \text{ g}}{\text{mL}} \times \dfrac{1 \text{ mol}}{86.17 \text{ g}} = 0.34 \text{ mol } C_6H_{14}$; total mol = 0.22 + 0.34 = 0.56 mol

$$\chi^L_{pen} = \frac{\text{mol pentane in solution}}{\text{total mol in solution}} = \frac{0.22 \text{ mol}}{0.56 \text{ mol}} = 0.39,\ \chi^L_{hex} = 1.00 - 0.39 = 0.61$$

$$P_{pen} = \chi^L_{pen} P^\circ_{pen} = 0.39(511 \text{ torr}) = 2.0 \times 10^2 \text{ torr};\ P_{hex} = 0.61(150. \text{ torr}) = 92 \text{ torr}$$

$$P_{total} = P_{pen} + P_{hex} = 2.0 \times 10^2 + 92 = 292 \text{ torr} = 290 \text{ torr}$$

b. From Chapter 5 on gases, the partial pressure of a gas is proportional to the number of moles of gas present. For the vapor phase:

$$\chi^V_{pen} = \frac{\text{mol pentane in vapor}}{\text{total mol vapor}} = \frac{P_{pen}}{P_{total}} = \frac{2.0 \times 10^2 \text{ torr}}{290 \text{ torr}} = 0.69$$

Note: In the Solutions Guide, we added V or L to the mole fraction symbol to emphasize which value we are solving. If the L or V is omitted, then the liquid phase is assumed.

51. $P_{total} = P_{meth} + P_{prop}$, 174 torr = χ^L_{meth}(303 torr) + χ^L_{prop}(44.6 torr); $\chi^L_{prop} = 1.000 - \chi^L_{meth}$

$174 = 303\ \chi^L_{meth} + (1.000 - \chi^L_{meth})\ 44.6,\quad \frac{129}{258} = \chi^L_{meth} = 0.500;\ \ \chi^L_{prop} = 1.000 - 0.500 = 0.500$

53. $50.0\ g\ CH_3COCH_3 \times \frac{1\ mol}{58.08\ g} = 0.861\ mol\ acetone$

$50.0\ g\ CH_3OH \times \frac{1\ mol}{32.04\ g} = 1.56\ mol\ methanol$

$\chi^L_{acetone} = \frac{0.861}{0.861 + 1.56} = 0.356;\ \ \chi^L_{methanol} = 1.000 - \chi^L_{acetone} = 0.644$

$P_{total} = P_{methanol} + P_{acetone} = 0.644(143\ torr) + 0.356(271\ torr) = 92.1\ torr + 96.5\ torr = 188.6\ torr$

Since partial pressures are proportional to the moles of gas present, then in the vapor phase:

$\chi^V_{acetone} = \frac{P_{acetone}}{P_{total}} = \frac{96.5\ torr}{188.6\ torr} = 0.512;\ \ \chi^V_{methanol} = 1.000 - 0.512 = 0.488$

The actual vapor pressure of the solution (161 torr) is less than the calculated pressure assuming ideal behavior (188.6 torr). Therefore, the solution exhibits negative deviations from Raoult's law. This occurs when the solute-solvent interactions are stronger than in pure solute and pure solvent.

55. No, the solution is not ideal. For an ideal solution, the strength of intermolecular forces in the solution are the same as in pure solute and pure solvent. This results in $\Delta H_{soln} = 0$ for an ideal solution. ΔH_{soln} for methanol/water is not zero. Since $\Delta H_{soln} < 0$, then this solution exhibits negative deviation from Raoult's law.

57. Solutions of A and B have vapor pressures less than ideal (see Figure 17.11 of the text), so this plot shows negative deviations from Rault's law. Negative deviations occur when the intermolecular forces are stronger in solution than in pure solvent and solute. This results in an exothermic enthalpy of solution. The only statement that is false is e. A substance boils when the vapor pressure equals the external pressure. Since $X_b = 0.6$ has a lower vapor pressure at the temperature of the plot than either pure A or pure B, then one would expect this solution to require the highest temperature in order for the vapor pressure to reach the external pressure. Therefore, the solution with $X_B = 0.6$ will have a higher boiling point than either pure A or pure B. (Note that since $P°_B > P°_A$, then B is more volatile than A.)

Colligative Properties

59. $molality = m = \frac{50.0\ g\ C_2H_6O_2}{50.0\ g\ H_2O} \times \frac{1000\ g}{kg} \times \frac{1\ mol}{62.07\ g} = 16.1\ mol/kg$

$\Delta T_f = K_f m = 1.86°C/molal \times 16.1\ molal = 29.9°C;\ \ T_f = 0.0°C - 29.9°C = -29.9°C$

$\Delta T_b = K_b m = 0.51°C/molal \times 16.1\ molal = 8.2°C;\ \ T_b = 100.0°C + 8.2°C = 108.2°C$

61. $m = \dfrac{24.0\text{ g} \times \dfrac{1\text{ mol}}{58.0\text{ g}}}{0.600\text{ kg}} = 0.690\text{ mol/kg}$; $\Delta T_b = K_b m = 0.51\,°\text{C kg/mol} \times 0.690\text{ mol/kg} = 0.35\,°\text{C}$

$T_b = 99.725\,°\text{C} + 0.35\,°\text{C} = 100.08\,°\text{C}$

63. $\Delta T_b = 77.85\,°\text{C} - 76.50\,°\text{C} = 1.35\,°\text{C}$; $m = \dfrac{\Delta T_b}{K_b} = \dfrac{1.35\,°\text{C}}{5.03\,°\text{C kg/mol}} = 0.268\text{ mol/kg}$

$$\text{mol biomolecule} = 0.0150\text{ kg solvent} \times \frac{0.268\text{ mol hydrocarbon}}{\text{kg solvent}} = 4.02 \times 10^{-3}\text{ mol}$$

From the problem, 2.00 g biomolecule was used which must contain 4.02×10^{-3} mol biomolecule. The molar mass of the biomolecule is:

$$\frac{2.00\text{ g}}{4.02 \times 10^{-3}\text{ mol}} = 498\text{ g/mol}$$

65. $\Delta T_f = K_f m$, $m = \dfrac{\Delta T_f}{K_f} = \dfrac{0.240\,°\text{C}}{4.70\,°\text{C kg/mol}} = \dfrac{5.11 \times 10^{-2}\text{ mol biomolecule}}{\text{kg solvent}}$

The mol of biomolecule present is:

$$0.0150\text{ kg solvent} \times \frac{5.11 \times 10^{-2}\text{ mol biomolecule}}{\text{kg solvent}} = 7.67 \times 10^{-4}\text{ mol biomolecule}$$

From the problem, 0.350 g biomolecule was used which must contain 7.67×10^{-4} mol biomolecule. The molar mass of the biomolecule is:

$$\text{molar mass} = \frac{0.350\text{ g}}{7.67 \times 10^{-4}\text{ mol}} = 456\text{ g/mol}$$

67. $M = \dfrac{1.0\text{ g}}{\text{L}} \times \dfrac{1\text{ mol}}{9.0 \times 10^4\text{ g}} = 1.1 \times 10^{-5}\text{ mol/L}$; $\pi = MRT$

At 298 K: $\pi = \dfrac{1.1 \times 10^{-5}\text{ mol}}{\text{L}} \times \dfrac{0.08206\text{ L atm}}{\text{mol K}} \times 298\text{ K} \times \dfrac{760\text{ torr}}{\text{atm}}$, $\pi = 0.20\text{ torr}$

Since $d = 1.0\text{ g/cm}^3$, then 1.0 L solution has a mass of 1.0 kg. Since only 1.0 g of protein is present per liter solution, then 1.0 kg of H_2O is present and molality equals molarity.

$$\Delta T_f = K_f m = \frac{1.86\,°\text{C}}{\text{molal}} \times 1.1 \times 10^{-5}\text{ molal} = 2.0 \times 10^{-5}\,°\text{C}$$

69. $\pi = MRT = \dfrac{0.1\text{ mol}}{\text{L}} \times \dfrac{0.08206\text{ L atm}}{\text{mol K}} \times 298\text{ K} = 2.45\text{ atm} \approx 2\text{ atm}$

$$\pi = 2\text{ atm} \times \frac{760\text{ mm Hg}}{\text{atm}} \approx 2000\text{ mm} \approx 2\text{ m}$$

The osmotic pressure would support a mercury column of ≈ 2 m. The height of a fluid column in a tree will be higher because Hg is more dense than the fluid in a tree. If we assume the fluid in a tree is mostly H_2O, then the fluid has a density of 1.0 g/cm^3. The density of Hg is 13.6 g/cm^3.

Height of fluid ≈ 2 m × 13.6 ≈ 30 m

71. With addition of salt or sugar, the osmotic pressure inside the fruit cells (and bacteria) is less than outside the cell. Water will leave the cells which will dehydrate any bacteria present, causing them to die.

Properties of Electrolyte Solutions

73. 19.6 torr = χ_{H_2O} (23.8 torr), χ_{H_2O} = 0.824; χ_{solute} = 1.000 - 0.824 = 0.176

0.176 is the mol fraction of all the solute particles present. Since NaCl dissolves to produce two ions in solution (Na^+ and Cl^-), 0.176 is the mole fraction of Na^+ and Cl^- ions present. The mole fraction of NaCl is 1/2 (0.176) = 0.0880 = χ_{NaCl}.

At 45°C, P_{H_2O} = 0.824 (71.9 torr) = 59.2 torr

75. $Na_3PO_4(s) \rightarrow 3\ Na^+(aq) + PO_4^{3-}(aq)$, i = 4.0; $CaBr_2(s) \rightarrow Ca^{2+}(aq) + 2\ Br^-(aq)$, i = 3.0

$KCl(s) \rightarrow K^+(aq) + Cl^-(aq)$, i = 2.0.

The effective particle concentrations of the solutions are:

4.0(0.010 molal) = 0.040 molal for Na_3PO_4 solution; 3.0(0.020 molal) = 0.060 molal for $CaBr_2$ solution; 2.0(0.020 molal) = 0.040 molal for KCl solution; slightly greater than 0.020 molal for HF solution since HF only partially dissociates in water (it is a weak acid).

a. The 0.010 *m* Na_3PO_4 solution and the 0.020 *m* KCl solution both have effective particle concentrations of 0.040 *m* (assuming complete dissociation), so both of these solutions should have the same boiling point as the 0.040 *m* $C_6H_{12}O_6$ solution (a nonelectrolyte).

b. $P = \chi P°$; As the solute concentration decreases, the solvent's vapor pressure increases since χ increases. Therefore, the 0.020 *m* HF solution will have the highest vapor pressure since it has the smallest effective particle concentration.

c. $\Delta T = K_f m$; The 0.020 *m* $CaBr_2$ solution has the largest effective particle concentration so it will have the largest freezing point depression (largest ΔT).

77. Ion pairing can occur, resulting in fewer particles than expected. This results in smaller freezing point depressions and smaller boiling point elevations ($\Delta T = Km$). Ion pairing will increase as the concentration of electrolyte increases.

79. $NaCl(s) \rightarrow Na^+(aq) + Cl^-(aq)$, $i = 2.0$

$$\pi = iMRT = 2.0 \times \frac{0.10 \text{ mol}}{\text{L}} \times \frac{0.08206 \text{ L atm}}{\text{mol K}} \times 293 \text{ K} = 4.8 \text{ atm}$$

A pressure greater than 4.8 atm should be applied to insure purification by reverse osmosis.

81. a. $MgCl_2$, i(observed) = 2.7

$\Delta T_f = iK_f m = 2.7 \times 1.86°\text{C/molal} \times 0.050 \text{ molal} = 0.25°\text{C}$; $T_f = -0.25°\text{C}$

$\Delta T_b = iK_b m = 2.7 \times 0.51°\text{C/molal} \times 0.050 \text{ molal} = 0.069°\text{C}$; $T_b = 100.069°\text{C}$

b. $FeCl_3$, i(observed) = 3.4

$\Delta T_f = iK_f m = 3.4 \times 1.86\ °\text{C/molal} \times 0.050 \text{ molal} = 0.32°\text{C}$; $T_f = -0.32°\text{C}$

$\Delta T_b = iK_b m = 3.4 \times 0.51°\text{C/molal} \times 0.050 \text{ molal} = 0.087°\text{C}$; $T_b = 100.087°\text{C}$

83. $\pi = iMRT = 3.0 \times 0.50 \text{ mol/L} \times 0.08206 \text{ L atm K}^{-1} \text{ mol}^{-1} \times 298 \text{ K} = 37 \text{ atm}$

Because of ion pairing in solution, we would expect i to be less than 3.0 which results in fewer solute particles in solution which results in a lower osmotic pressure than calculated above.

Additional Exercises

85. **Both solutions and colloids have suspended particles in some medium. The major difference between the two is the size of the particles. A colloid is a suspension of relatively large particles as compared to a solution. Because of this, colloids will scatter light while solutions will not. The scattering of light by a colloidal suspension is called the Tyndall effect.**

87.

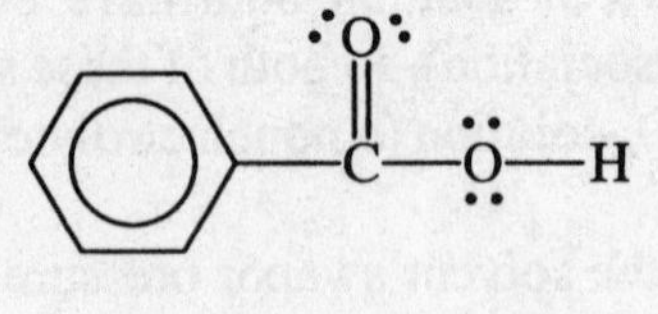

Benzoic acid is capable of hydrogen bonding, but a significant part of benzoic acid is the nonpolar benzene ring. In benzene, a hydrogen bonded dimer forms.

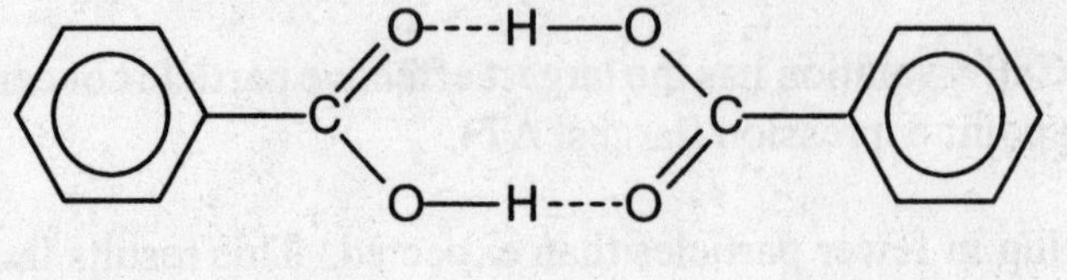

The dimer is relatively nonpolar and thus more soluble in benzene than in water. Since benzoic acid forms dimers in benzene, the effective solute particle concentration will be less than 1.0 molal. Therefore, the freezing point depression would be less than 5.12°C ($\Delta T_f = K_f m$).

89. a. $NH_4NO_3(s) \rightarrow NH_4^+(aq) + NO_3^-(aq) \quad \Delta H_{soln} = ?$

Heat gain by dissolution process = heat loss by solution; We will keep all quantities positive in order to avoid sign errors. Since the temperature of the water decreased, then the dissolution of NH_4NO_3 is endothermic (ΔH is positive). Mass of solution = 1.60 + 75.0 = 76.6 g

$$\text{heat loss by solution} = \frac{4.18\ J}{g\ °C} \times 76.6\ g \times (25.00°C - 23.34°C) = 532\ J$$

$$\Delta H_{soln} = \frac{532\ J}{1.60\ g\ NH_4NO_3} \times \frac{80.05\ g\ NH_4NO_3}{mol\ NH_4NO_3} = 2.66 \times 10^4\ J/mol = 26.6\ kJ/mol$$

b. We will use Hess's law to solve for the lattice energy. The lattice energy equation is:

$NH_4^+(g) + NO_3^-(g) \rightarrow NH_4NO_3(s) \quad \Delta H = \text{lattice energy}$

$$NH_4^+(g) + NO_3^-(g) \rightarrow NH_4^+(aq) + NO_3^-(aq) \quad \Delta H = \Delta H_{hyd} = -630.\ kJ/mol$$
$$NH_4^+(aq) + NO_3^-(aq) \rightarrow NH_4NO_3(s) \quad \Delta H = -\Delta H_{soln} = -26.6\ kJ/mol$$

$$NH_4^+(g) + NO_3^-(g) \rightarrow NH_4NO_3(s) \quad \Delta H = \Delta H_{hyd} - \Delta H_{soln} = -657\ kJ/mol$$

91. $\chi^V_{pen} = 0.15 = \frac{P_{pen}}{P_{total}}$; $P_{pen} = \chi^L_{pen} P°_{pen} = \chi^L_{pen}(511\ torr)$; $P_{total} = P_{pen} + P_{hex} = \chi^L_{pen}(511) + \chi^L_{hex}(150.)$

Since $\chi^L_{hex} = 1.000 - \chi^L_{pen}$, then: $P_{total} = \chi^L_{pen}(511) + (1.000 - \chi^L_{pen})(150.) = 150. + 361\ \chi^L_{pen}$

$$\chi^V_{pen} = \frac{P_{pen}}{P_{total}},\ 0.15 = \frac{\chi^L_{pen}(511)}{150. + 361\ \chi^L_{pen}},\ 0.15\ (150. + 361\ \chi^L_{pen}) = 511\ \chi^L_{pen}$$

$$23 + 54\ \chi^L_{pen} = 511\ \chi^L_{pen},\ \chi^L_{pen} = \frac{23}{457} = 0.050$$

93. $$14.22\ mg\ CO_2 \times \frac{12.011\ mg\ C}{44.009\ mg\ CO_2} = 3.881\ mg\ C;\ \%\ C = \frac{3.881\ mg}{4.80\ mg} \times 100 = 80.9\%\ C$$

$$1.66\ mg\ H_2O \times \frac{2.016\ mg\ H}{18.02\ mg\ H_2O} = 0.186\ mg\ H;\ \%\ H = \frac{0.186\ mg}{4.80\ mg} \times 100 = 3.88\%\ H$$

$\%\ O = 100.00 - (80.9 + 3.88) = 15.2\%\ O$

Out of 100.00 g:

$$80.9\text{ g C} \times \frac{1\text{ mol}}{12.01\text{ g}} = 6.74\text{ mol C}; \quad \frac{6.74}{0.950} = 7.09 \approx 7$$

$$3.88\text{ g H} \times \frac{1\text{ mol}}{1.008\text{ g}} = 3.85\text{ mol H}; \quad \frac{3.85}{0.950} = 4.05 \approx 4$$

$$15.2\text{ g O} \times \frac{1\text{ mol}}{16.00\text{ g}} = 0.950\text{ mol O}; \quad \frac{0.950}{0.950} = 1.00$$

Therefore, the empirical formula is C_7H_4O.

$$\Delta T_f = K_f m, \quad m = \frac{\Delta T_f}{K_f} = \frac{22.3°C}{40.°C/\text{molal}} = 0.56\text{ molal}$$

$$\text{mol anthraquinone} = 0.0114\text{ kg camphor} \times \frac{0.56\text{ mol anthraquinone}}{\text{kg camphor}} = 6.4 \times 10^{-3}\text{ mol}$$

$$\text{molar mass} = \frac{1.32\text{ g}}{6.4 \times 10^{-3}\text{ mol}} = 210\text{ g/mol}$$

The empirical mass of C_7H_4O is: $7(12) + 4(1) + 16 \approx 104$ g/mol. Since the molar mass is twice the empirical mass, then the molecular formula is $C_{14}H_8O_2$.

95. a. $$m = \frac{\Delta T_f}{K_f} = \frac{1.32°C}{5.12°C\text{ kg/mol}} = 0.258\text{ mol/kg}$$

$$\text{mol unknown} = 0.01560\text{ kg} \times \frac{0.258\text{ mol unknown}}{\text{kg}} = 4.02 \times 10^{-3}\text{ mol}$$

$$\text{molar mass of unknown} = \frac{1.22\text{ g}}{4.02 \times 10^{-3}\text{ mol}} = 303\text{ g/mol}$$

$$\text{Uncertainty in temperature} = \frac{0.04}{1.32} \times 100 = 3\%; \text{ A 3\% uncertainty in 303 g/mol} = 9\text{ g/mol.}$$

So, molar mass = 303 ± 9 g/mol.

b. **No, codeine could not be eliminated since its molar mass is in the possible range including the uncertainty.**

c. **We would like the uncertainty to be ± 1 g/mol. We need the freezing point depression to be about 10 times what it was in this problem. Two possibilities are:**

1. **make the solution ten times more concentrated (may be solubility problem) or**
2. **use a solvent with a larger K_f value, e.g., camphor**

97. $M_3X_2(s) \rightarrow 3\ M^{2+}(aq) + 2\ X^{3-}(aq) \qquad K_{sp} = [M^{2+}]^3[X^{3-}]^2$

Initial	s = solubility (mol/L)	0	0
Equil.		3s	2s

$K_{sp} = (3s)^3(2s)^2 = 108\ s^5$; Total ion concentration = 3s + 2s = 5s

$$\pi = iMRT,\ iM = \text{total ion concentration} = \frac{\pi}{RT} = \frac{2.64 \times 10^{-2}\ \text{atm}}{0.08206\ \text{L atm K}^{-1}\ \text{mol}^{-1} \times 298\ \text{K}}$$

$$= 1.08 \times 10^{-3}\ \text{mol/L}$$

$5s = 1.08 \times 10^{-3}$ mol/L, $s = 2.16 \times 10^{-4}$ mol/L; $K_{sp} = 108\ s^5 = 108(2.16 \times 10^{-4})^5 = 5.08 \times 10^{-17}$

99. $$m = \frac{0.100\ \text{g} \times \dfrac{1\ \text{mol}}{100.0\ \text{g}}}{0.5000\ \text{kg}} = 2.00 \times 10^{-3}\ \text{mol/kg} \approx 2.00 \times 10^{-3}\ \text{mol/L} \quad \text{(dilute solution)}$$

$\Delta T_f = iK_f m$, $0.0056°C = i(1.86°C/\text{molal})(2.00 \times 10^{-3}\ \text{molal})$, $i = 1.5$

If i = 1.0, % dissociation = 0% and if i = 2.0, % dissociation = 100%. Since i = 1.5, then the weak acid is 50.% dissociated.

$$HA \rightleftharpoons H^+ + A^- \qquad K_a = \frac{[H^+][A^-]}{[HA]}$$

Since the weak acid is 50.% dissociated, then:

$[H^+] = [A^-] = [HA]_o \times 0.50 = 2.00 \times 10^{-3}\ M \times 0.50 = 1.0 \times 10^{-3}\ M$

$[HA] = [HA]_o$ - amount HA reacted $= 2.00 \times 10^{-3}\ M - 1.0 \times 10^{-3}\ M = 1.0 \times 10^{-3}\ M$

$$K_a = \frac{[H^+][A^-]}{[HA]} = \frac{(1.0 \times 10^{-3})(1.0 \times 10^{-3})}{1.0 \times 10^{-3}} = 1.0 \times 10^{-3}$$

101. $$iM = \frac{\pi}{RT} = \frac{0.3950\ \text{atm}}{0.08206\ \text{L atm mol}^{-1}\ \text{K}^{-1}\ (298.2\ \text{K})} = 0.01614\ \text{mol/L} = \text{total ion concentration}$$

$0.01614\ \text{mol/L} = M_{Mg^{2+}} + M_{Na^+} + M_{Cl^-}$; $M_{Cl^-} = 2\ M_{Mg^{2+}} + M_{Na^+}$ (charge balance)

Combining: $0.01614 = 3\ M_{Mg^{2+}} + 2\ M_{Na^+}$

Let x = mass $MgCl_2$ and y = mass NaCl, then $x + y = 0.5000$ g.

$$M_{Mg^{2+}} = \frac{x}{95.218} \text{ and } M_{Na^+} = \frac{y}{58.443} \qquad \text{(Since V = 1.000 L.)}$$

Total ion concentration $= \frac{3x}{95.218} + \frac{2y}{58.443} = 0.01614$ mol/L; Rearranging: $3x + 3.2585y = 1.537$

Solving by simultaneous equations:

$$\begin{array}{rl} 3x + 3.2585\,y &= 1.537 \\ -3(x + y) &= -3(0.5000) \\ \hline 0.2585\,y &= 0.037, \quad y = 0.14 \text{ g NaCl} \end{array}$$

mass $MgCl_2$ = 0.5000 g - 0.14 g = 0.36 g; mass % $MgCl_2 = \frac{0.36 \text{ g}}{0.5000 \text{ g}} \times 100 = 72\%$

Challenge Problems

103. From the problem, $\chi^L_{C_6H_6} = \chi^L_{CCl_4} = 0.500$. We need the pure vapor pressures (P°) in order to calculate the vapor pressure of the solution. Using the thermodynamic data:

$$C_6H_6(l) \rightleftharpoons C_6Hg(g) \quad K = P_{C_6H_6} = P^\circ_{C_6H_6} \text{ at } 25°C$$

$$\Delta G^\circ_{rxn} = \Delta G^\circ_{f,\,C_6H_6(g)} - \Delta G^\circ_{f,\,C_6H_6(l)} = 129.66 \text{ kJ/mol} - 124.50 \text{ kJ/mol} = 5.16 \text{ kJ/mol}$$

$$\Delta G^\circ = -RT\ln K,\ \ln K = \frac{-\Delta G^\circ}{RT} = \frac{-5.16 \times 10^3 \text{ J/mol}}{(8.3145 \text{ J K}^{-1} \text{ mol}^{-1})(298 \text{ K})} = -2.08$$

$$K = P^\circ_{C_6H_6} = e^{-2.08} = 0.125 \text{ atm}$$

For CCl_4: $\Delta G^\circ_{rxn} = \Delta G^\circ_{f,\,CCl_4(g)} - \Delta G^\circ_{f,\,CCl_4(l)} = -60.59 \text{ kJ/mol} - (-65.21 \text{ kJ/mol}) = 4.62 \text{ kJ/mol}$

$$K = P^\circ_{CCl_4} = \exp\left(\frac{-\Delta G^\circ}{RT}\right) = \exp\left(\frac{-4620 \text{ J/mol}}{8.3145 \text{ J K}^{-1} \text{ mol}^{-1} \times 298 \text{ K}}\right) = 0.155 \text{ atm}$$

$$P_{C_6H_6} = X^L_{C_6H_6} P^\circ_{C_6H_6} = 0.500\ (0.125 \text{ atm}) = 0.0625 \text{ atm}; \quad P_{CCl_4} = 0.500\ (0.155 \text{ atm}) = 0.0775 \text{ atm}$$

$$X^V_{C_6H_6} = \frac{P_{C_6H_6}}{P_{tot}} = \frac{0.0625 \text{ atm}}{0.0625 \text{ atm} + 0.0775 \text{ atm}} = \frac{0.0625}{0.1400} = 0.446; \quad X^V_{CCl_4} = 1.000 - 0.446 = 0.554$$

105. a. Assuming $MgCO_3(s)$ does not dissociate, the solute concentration in water is:

$$\frac{560\ \mu g\ MgCO_3(s)}{mL} = \frac{560\ mg}{L} = \frac{560 \times 10^{-3}\ g}{L} \times \frac{1\ mol\ MgCO_3}{84.32\ g} = 6.6 \times 10^{-3}\ mol\ MgCO_3/L$$

An applied pressure of 8.0 atm will purify water up to a solute concentration of:

$$M = \frac{\pi}{RT} = \frac{8.0\ atm}{0.08206\ L\ atm\ K^{-1}\ mol^{-1} \times 300.\ K} = \frac{0.32\ mol}{L}$$

When the concentration of $MgCO_3(s)$ reaches 0.32 mol/L, the reverse osmosis unit can no longer purify the water. Let V = volume (L) of water remaining after purifying 45 L of H_2O. When V + 45 L of water has been processed, the moles of solute particles will equal:

$$6.6 \times 10^{-3}\ mol/L \times (45\ L + V) = 0.32\ mol/L \times V$$

Solving: $0.30 = (0.32 - 0.0066) \times V$, $V = 0.96\ L$

The minimum total volume of water that must be processed is 45 L + 0.96 L = 46 L.

Note: If $MgCO_3$ does dissociate into Mg^{2+} and CO_3^{2-} ions, then the solute concentration increases to $1.3 \times 10^{-2}\ M$ and at least 47 L of water must be processed.

b. No; A reverse osmosis system that applies 8.0 atm can only purify water with a solute concentration less than 0.32 mol/L. Salt water has a solute concentration of $2(0.60\ M) = 1.2$ mol/L ions. The solute concentration of salt water is much too high for this reverse osmosis unit to work.

107. a. Assuming no ion association between $SO_4^{2-}(aq)$ and $Fe^{3+}(aq)$, then i = 5 for $Fe_2(SO_4)_3$.

$$\pi = iMRT = 5(0.0500\ mol/L)(0.08206\ L\ atm\ K^{-1}\ mol^{-1})(298\ K) = 6.11\ atm$$

b. $Fe_2(SO_4)_3(aq) \rightarrow 2\ Fe^{3+}(aq) + 3\ SO_4^{2-}(aq)$

Under ideal circumstances, 2/5 of π calculated above results from Fe^{3+} and 3/5 results from SO_4^{2-}. The contribution to π from SO_4^{2-} is $3/5 \times 6.11$ atm = 3.67 atm. Since SO_4^{2-} is assumed unchanged in solution, then the SO_4^{2-} contribution in the actual solution will also be 3.67 atm. The contribution to the actual π from the $Fe(H_2O)_6^{3+}$ dissociation reaction is 6.73 - 3.67 = 3.06 atm.

The initial concentration of $Fe(H_2O)_6^{2+}$ is 2(0.0500) = 0.100 M. The set-up for the weak acid problem is:

$$Fe(H_2O)_6^{3+} \rightleftharpoons H^+ + Fe(OH)(H_2O)_5^{2+} \qquad K_a = \frac{[H^+][Fe(OH)(H_2O)_5^{2+}]}{[Fe(H_2O)_6^{3+}]}$$

	$Fe(H_2O)_6^{3+}$	H^+	$Fe(OH)(H_2O)_5^{2+}$
Initial	0.100 M	~0	0
	x mol/L of $Fe(H_2O)_6^{3+}$ reacts to reach equilibrium		
Equil.	$0.100 - x$	x	x

$$\pi = iMRT;\ \text{Total ion concentration} = iM = \frac{\pi}{RT} = \frac{3.06\ \text{atm}}{0.08206\ \text{L atm K}^{-1}\ \text{mol}^{-1}\ (298\ \text{K})} = 0.125\ M$$

$0.125\ M = 0.100 - x + x + x = 0.100 + x,\ x = 0.025\ M$

$$K_a = \frac{[H^+][Fe(OH)(H_2O)_5^{2+}]}{[Fe(H_2O)_6^{3+}]} = \frac{x^2}{0.100 - x} = \frac{(0.025)^2}{(0.100 - 0.025)} = \frac{(0.025)^2}{0.075},\ K_a = 8.3 \times 10^{-3}$$

109. a. $\pi = iMRT,\ iM = \dfrac{\pi}{RT} = \dfrac{7.83\ \text{atm}}{0.08206\ \text{L atm K}^{-1}\ \text{mol}^{-1} \times 298\ \text{K}} = 0.320\ \text{mol/L}$

Assuming 1.000 L of solution:

total mol solute particles = mol Na^+ + mol Cl^- + mol NaCl = 0.320 mol

mass solution = 1000. mL $\times \dfrac{1.071\ \text{g}}{\text{mL}}$ = 1071 g solution

mass NaCl in solution = 0.0100 × 1071 g = 10.7 g NaCl

mol NaCl added to solution = 10.7 g $\times \dfrac{1\ \text{mol}}{58.44\ \text{g}}$ = 0.183 mol NaCl

Some of this NaCl dissociates into Na^+ and Cl^- (two mol ions per mol NaCl) and some remains undissociated. Let x = mol undissociated NaCl = mol ion pairs.

mol solute particles = 0.320 mol = 2(0.183 - x) + x

0.320 = 0.366 - x, x = 0.046 mol ion pairs

fraction of ion pairs = $\dfrac{0.046}{0.183}$ = 0.25, or 25%

b. $\Delta T = K_f m$ where K_f = 1.86 °C kg/mol; From part a, 1.000 L of solution contains 0.320 mol of solute particles. To calculate the molality of the solution, we need the kg of solvent present in 1.000 L solution.

mass of 1.000 L solution = 1071 g; mass of NaCl = 10.7 g

mass of solvent in 1.000 L solution = 1071 g - 10.7 g = 1060. g

ΔT = 1.86 °C kg/mol $\times \dfrac{0.320\ \text{mol}}{1.060\ \text{kg}}$ = 0.562 °C

Assuming water freezes at 0.000 °C, then T_f = -0.562 °C.

Marathon Problem

111. a. **From part a information we can calculate the molar mass of Na_nA and deduce the formula.**

$$\text{mol } Na_nA = \text{mol reducing agent} = 0.01526 \text{ L} \times \frac{0.02313 \text{ mol}}{\text{L}} = 3.530 \times 10^{-4} \text{ mol } Na_nA$$

$$\text{molar mass of } Na_nA = \frac{30.0 \times 10^{-3} \text{ g}}{3.530 \times 10^{-4} \text{ mol}} = 85.0 \text{ g/mol}$$

To deduce the formula, we will assume various charges and numbers of oxygens present in the oxyanion, then use the periodic table to see if an element fits the molar mass data.

Assuming n = 1 so the formula is NaA. The molar mass of the oxyanion, A^-, is 85.0 - 23.0 = 62.0 g/mol. The oxyanion part of the formula could be EO^- or EO_2^- or EO_3^- where E is some element. If EO^-, then the molar mass of E is 62.0 - 16.0 = 46.0 g/mol; no element has this molar mass. If EO_2^-, molar mass of E = 62.0 - 32.0 = 30.0 g/mol. Phosphorus is close, but PO_2^- anions are not common. If EO_3^-, molar mass of E = 62.0 - 48.0 = 14.0. Nitrogen has this molar mass and NO_3^- anions are very common. Therefore, NO_3^- is a possible formula for A^-.

Next, we assume Na_2A and Na_3A formulas and go through the same procedure as above. In all cases, no element in the periodic table fits the data. Therefore, we assume the oxyanion is NO_3^- = A^-.

b. **The crystal data in part b allows determination of the metal, M, in the formula. See Exercise 16.47 for a review of relationships in body-centered cubic cells. In a bcc unit cell, there are 2 atoms per unit cell and the body diagonal of the cubic cell is related to the radius of the metal by the equation $4r = \sqrt{3}\ell$ where ℓ = cubic edge length.**

$$\ell = \frac{4r}{\sqrt{3}} = \frac{4(1.984 \times 10^{-8} \text{ cm})}{\sqrt{3}} = 4.582 \times 10^{-8} \text{ cm}$$

$$\text{volume of unit cell} = \ell^3 = (4.582 \times 10^{-8})^3 = 9.620 \times 10^{-23} \text{ cm}^3$$

$$\text{mass of M in a unit cell} = 9.620 \times 10^{-23} \text{ cm}^3 \times \frac{5.243 \text{ g}}{\text{cm}^3} = 5.044 \times 10^{-22} \text{ g M}$$

$$\text{mol M in a unit cell} = 2 \text{ atoms} \times \frac{1 \text{ mol}}{6.022 \times 10^{23}} = 3.321 \times 10^{-24} \text{ mol M}$$

$$\text{molar mass of M} = \frac{5.044 \times 10^{-22} \text{ g M}}{3.321 \times 10^{-24} \text{ mol M}} = 151.9 \text{ g/mol}$$

From the periodic table, M is europium, Eu. Given the charge of Eu is +3, then the formula of the salt is $Eu(NO_3)_3 \cdot zH_2O$.

c. **Part c data allows determination of the molar mass of $Eu(NO_3)_3 \cdot zH_2O$, from which we can determine z, the number of waters of hydration.**

$$\pi = iMRT,\ iM = \frac{\pi}{RT} = \frac{558\ \text{torr}\left(\frac{1\ \text{atm}}{760\ \text{torr}}\right)}{0.08206\ \text{L atm K}^{-1}\ \text{mol}^{-1}\ (298\ \text{K})} = 0.0300\ \text{mol/L}$$

The total molarity of solute particles present is 0.0300 *M*. The solute particles are Eu^{3+} and NO_3^- ions (the waters of hydration are not solute particles). Since each mol of $Eu(NO_3)_3{\bullet}zH_2O$ dissolves to form 4 ions (Eu^{3+} + 3 NO_3^-), then the molarity of $Eu(NO_3)_3{\bullet}zH_2O$ is 0.0300/4 = 0.00750 *M*.

$$\text{mol Eu(NO}_3)_3{\bullet}\text{zH}_2\text{O} = 0.01000\ \text{L} \times \frac{0.00750\ \text{mol}}{\text{L}} = 7.50 \times 10^{-5}\ \text{mol}$$

$$\text{molar mass of Eu(NO}_3)_3{\bullet}\text{zH}_2\text{O} = \frac{33.45 \times 10^{-3}\ \text{g}}{7.50 \times 10^{-5}\ \text{mol}} = 446\ \text{g/mol}$$

$$446\ \text{g/mol} = 152.0 + 3(62.0) + z(18.0),\ \ z(18.0) = 108,\ \ z = 6.00$$

The formula for the strong electrolyte is $Eu(NO_3)_3{\bullet}6H_2O$.

CHAPTER EIGHTEEN

THE REPRESENTATIVE ELEMENTS: GROUPS 1A THROUGH 4A

Group 1A Elements

1. The gravity of the earth is not strong enough to keep the light H_2 molecules in the atmosphere.

3. a. $\Delta H° = 2(-46 \text{ kJ}) = -92 \text{ kJ}$; $\Delta S° = 2(193 \text{ J/K}) - [3(131 \text{ J/K}) + 192 \text{ J/K}] = -199 \text{ J/K}$;

   $\Delta G° = \Delta H° - T\Delta S° = -92 \text{ kJ} - 298 \text{ K}(-0.199 \text{ kJ/K}) = -33 \text{ kJ}$

   b. Since $\Delta G°$ is negative, then this reaction is spontaneous at standard conditions.

   c. $\Delta G° = 0$ when $T = \dfrac{\Delta H°}{\Delta S°} = \dfrac{-92 \text{ kJ}}{-0.199 \text{ kJ/K}} = 460 \text{ K}$

   At $T < 460$ K and standard pressures, the favorable $\Delta H°$ term dominates and the reaction is spontaneous ($\Delta G° < 0$).

5. Ionic, covalent, and metallic (or interstitial); The ionic and covalent hydrides are true compounds obeying the law of definite proportions and differ from each other in the type of bonding. The interstitial hydrides are more like solid solutions of hydrogen with a transition metal, and do not obey the law of definite proportions.

7. a. $2 Rb(s) + 2 H_2O(l) \rightarrow 2 RbOH(aq) + H_2(g)$
   b. $Na_2O_2(s) + 2 H_2O(l) \rightarrow 2 NaOH(aq) + H_2O_2(aq)$
   c. $LiH(s) + H_2O(l) \rightarrow H_2(g) + LiOH(aq)$
   d. $2 KO_2(s) + 2H_2O(l) \rightarrow 2 KOH(aq) + O_2(g) + H_2O_2(aq)$

9. Hydrogen forms many compounds in which the oxidation state is +1, as do the Group 1A elements. For example, H_2SO_4 and HCl as compared to Na_2SO_4 and NaCl. On the other hand, hydrogen forms diatomic H_2 molecules and is a nonmetal, while the Group 1A elements are metals. Hydrogen also forms compounds with a -1 oxidation state, which is not characteristic of Group 1A metals, e.g., NaH.

11. $4 Li(s) + O_2(g) \rightarrow 2 Li_2O(s)$

    $16 Li(s) + S_8(s) \rightarrow 8 Li_2S(s)$; $2 Li(s) + Cl_2(g) \rightarrow 2 LiCl(s)$

    $12 Li(s) + P_4(s) \rightarrow 4 Li_3P(s)$; $2 Li(s) + H_2(g) \rightarrow 2 LiH(s)$

    $2 Li(s) + 2 H_2O(l) \rightarrow 2 LiOH(aq) + H_2(g)$; $2 Li(s) + 2 HCl(aq) \rightarrow 2 LiCl(aq) + H_2(g)$

13. $4\ KO_2(s) + 2\ CO_2(g) \rightarrow 2\ K_2CO_3(s) + 3\ O_2(g)$; Potassium superoxide can react with exhaled CO_2 to produce O_2, which then can be inhaled.

Group 2A Elements

15. Group IA and IIA metals are all easily oxidized. They must be produced in the absence of materials (H_2O, O_2) that are capable of oxidizing them.

17. The acidity decreases. Solutions of Be^{2+} are acidic, while solutions of the other M^{2+} ions are neutral.

19. One would predict that $BeCl_2(NH_3)_2$ would form in excess ammonia. $BeCl_2(NH_3)_2$ has 2 + 2(7) + 2(5) + 6(1) = 32 valence electrons. A structure for this molecule can be drawn that obeys the octet rule for all atoms and has zero formal charge on all atoms except for Be and the N atoms. This is not the case for $BeCl_2NH_3$ (see Exercise 18.18). In $BeCl_2NH_3$, the "best" Lewis structure has only 6 electrons around Be.

```
        H    :Cl:    H
        |     |      |
   H —— N —— Be —— N —— H
        |     |      |
        H    :Cl:    H
```

21. $2\ Sr(s) + O_2(g) \rightarrow 2\ SrO(s)$; $8\ Sr(s) + S_8(s) \rightarrow 8\ SrS(s)$

$Sr(s) + Cl_2(g) \rightarrow SrCl_2(s)$; $6\ Sr(s) + P_4(s) \rightarrow 2\ Sr_3P_2(s)$

$Sr(s) + H_2(g) \rightarrow SrH_2(s)$; $Sr(s) + 2\ H_2O(l) \rightarrow Sr(OH)_2(aq) + H_2(g)$

$Sr(s) + 2\ HCl(aq) \rightarrow SrCl_2(aq) + H_2(g)$

23. $CaCO_3(s) + H_2SO_4(aq) \rightarrow CaSO_4(aq) + H_2O(l) + CO_2(g)$

25. $$\frac{1\ \text{mg}\ F^-}{L} \times \frac{1\ \text{g}}{1000\ \text{mg}} \times \frac{1\ \text{mol}\ F^-}{19.0\ \text{g}\ F^-} = 5.3 \times 10^{-5}\ M\ F^- = 5 \times 10^{-5}\ M\ F^-$$

$CaF_2(s) \rightleftharpoons Ca^{2+}(aq) + 2\ F^-(aq)$ $K_{sp} = [Ca^{2+}][F^-]^2 = 4.0 \times 10^{-11}$; Precipitation will occur when $Q > K_{sp}$. Lets calculate $[Ca^{2+}]$ so that $Q = K_{sp}$.

$Q = 4.0 \times 10^{-11} = [Ca^{2+}]_o[F^-]_o^2 = [Ca^{2+}]_o(5 \times 10^{-5})^2$, $[Ca^{2+}]_o = 2 \times 10^{-2}\ M$

$CaF_2(s)$ will precipitate when $[Ca^{2+}]_o > 2 \times 10^{-2}\ M$. Therefore, hard water should have a calcium ion concentration of less than $2 \times 10^{-2}\ M$ to avoid precipitate formation.

Group 3A Elements

27. a. AlN b. GaF_3 c. Ga_2S_3

29. Element 113: [Rn] $7s^25f^{14}6d^{10}7p^1$; Element 113 would fall below Tl in the periodic table. Like Tl, we would expect element 113 to form +1 and +3 oxidation states in its compounds.

31. $Ga_2O_3(s) + 6\ H^+(aq) \rightarrow 2\ Ga^{3+}(aq) + 3\ H_2O(l)$

$Ga_2O_3(s) + 2\ OH^-(aq) + 3\ H_2O(l) \rightarrow 2\ Ga(OH)_4^-(aq)$

$In_2O_3(s) + 6\ H^+(aq) \rightarrow 2\ In^{3+}(aq) + 3\ H_2O(l)$; $In_2O_3(s) + OH^-(aq) \rightarrow$ no reaction

33. Group 3A elements have one fewer valence electron than Si or Ge. A p-type semiconductor would form.

35. $2\ Ga(s) + 3\ F_2(g) \rightarrow 2\ GaF_3(s)$; $4\ Ga(s) + 3\ O_2(g) \rightarrow 2\ Ga_2O_3(s)$
$16\ Ga(s) + 3\ S_8(s) \rightarrow 8\ Ga_2S_3(s)$

$2\ Ga(s) + 6\ HCl(aq) \rightarrow 2\ GaCl_3(aq) + 3\ H_2(g)$

37. Tl_2O_3, thallium(III) oxide; Tl_2O, thallium(I) oxide; $InCl_3$, indium(III) chloride; InCl, indium(I) chloride

Group 4A Elements

39. Compounds containing Si – Si single and multiple bonds are rare, unlike compounds of carbon. The bond strengths of the Si – Si and C – C single bonds are similar. The difference in bonding properties must be for other reasons. One reason is that silicon does not form strong π bonds, unlike carbon. Another reason is that silicon forms particularly strong sigma bonds to oxygen, resulting in compounds with Si – O bonds instead of Si – Si bonds.

41. CS_2 has 4 + 2(6) = 16 valence electrons. C_3S_2 has 3(4) + 2(6) = 24 valence electrons.

S=C=S linear; S=C=C=C=S linear

43. White tin is stable at normal temperatures. Gray tin is stable at temperatures below 13.2°C. Thus for the phase change: Sn(gray) → Sn(white), ΔG is negative at T > 13.2°C and ΔG is positive at T < 13.2°C. This is only possible if ΔH is positive and ΔS is positive. Thus, gray tin has the more ordered structure.

45. $Sn(s) + 2F_2(g) \rightarrow SnF_4(s)$, tin(IV) fluoride; $Sn(s) + F_2(g) \rightarrow SnF_2(s)$, tin(II) fluoride

47. In graphite planes of carbon atoms slide easily along each other. In addition, graphite is not volatile. The lubricant will not be lost when used in a high vacuum environment.

49. Pb_3O_4: We assign -2 for the oxidation state of O. The sum of the oxidation states of Pb must be +8. We get this if two of the lead atoms are Pb(II) and one is Pb(IV). Therefore, the mole ratio of lead(II) to lead(IV) is 2:1.

51. Lead is very toxic. As the temperature of the water increases, the solubility of lead will increase. Drinking hot tap water from pipes containing lead solder could result in higher lead concentrations in the body.

53. $Pb(NO_3)_2(aq) + H_3AsO_4(aq) \rightarrow PbHAsO_4(s) + 2\ HNO_3(aq)$

Note: The insecticide used is $PbHAsO_4$ and is commonly called lead arsenate. This is not the correct name, however. Correctly, lead arsenate would be $Pb_3(AsO_4)_2$ and $PbHAsO_4$ should be named lead hydrogen arsenate.

Additional Exercises

55. Na^+ can oxidize Na^- to Na. The purpose of the cryptand is to encapsulate the Na^+ ion so that it does not come in contact with the Na^- ion and oxidize it to sodium metal.

57. a. $2\ Na + 2\ NH_3 \rightarrow 2\ NaNH_2 + H_2$

b. $$\text{mass percent} = \frac{251.4\text{ g}}{1000.\text{ g} + 251.4\text{ g}} \times 100 = 20.09\ \%$$

$$\text{mol Na} = 251.4 \times \frac{1\text{ mol}}{22.99\text{ g}} = 10.94\text{ mol Na};\quad \text{mol NH}_3 = 1000.\text{ g} \times \frac{1\text{ mol}}{17.031\text{ g}} = 58.72\text{ mol NH}_3$$

$$\text{mole fraction} = \chi_{Na} = \frac{10.94\text{ mol}}{10.94\text{ mol} + 58.72\text{ mol}} = 0.1570$$

$$\text{molality} = \frac{251.4\text{ g Na}}{\text{kg}} \times \frac{1\text{ mol Na}}{22.99\text{ g Na}} = 10.94\text{ mol/kg}$$

59.
$$2\ H_2O + 2\ e^- \rightarrow H_2 + 2\ OH^-$$
$$Be + 4\ OH^- \rightarrow Be(OH)_4^{2-} + 2\ e^-$$

$$Be(s) + 2\ H_2O(l) + 2\ OH^-(aq) \rightarrow Be(OH)_4^{2-}(aq) + H_2(g)$$

H_2O is the oxidizing agent. Be is the reducing agent.

61. Strontium and calcium are both alkaline earth metals so both have similar chemical properties. Since milk is a good source of calcium, strontium could replace some calcium in milk without much difficulty.

63. $Tl^{3+} + 2\,e^- \rightarrow Tl^+$ $\quad E^\circ_c = 1.25$ V

$3\,I^- \rightarrow I_3^- + 2\,e^-$ $\quad -E^\circ_a = -0.55$ V

$Tl^{3+} + 3\,I^- \rightarrow Tl^+ + I_3^-$ $\quad E^\circ_{cell} = 0.70$ V (Spontaneous since $E^\circ_{cell} > 0$.)

In solution, Tl^{3+} can oxidize I^- to I_3^-. Thus, we expect TlI_3 to be thallium(I) triiodide.

65. The π electrons are free to move in graphite, thus giving it a greater conductivity (lower resistance). The electrons have the greatest mobility within sheets of carbon atoms, resulting in a lower resistance in the basal plane. Electrons in diamond are not mobile (high resistance). The structure of diamond is uniform in all directions; thus, there is no directional dependence of the resistivity.

67. Carbon cannot form the fifth bond necessary for the transition state since carbon doesn't have low energy d orbitals available to expand its octet of valence electrons.

69. For 589.0 nm: $\nu = \frac{c}{\lambda} = \frac{2.9979 \times 10^8 \text{ m/s}}{589.0 \times 10^{-9} \text{ m}} = 5.090 \times 10^{14} \text{ s}^{-1}$

$E = h\nu = 6.6261 \times 10^{-34} \text{ J s} \times 5.090 \times 10^{14} \text{ s}^{-1} = 3.373 \times 10^{-19} \text{ J}$

For 589.6 nm: $\nu = c/\lambda = 5.085 \times 10^{14} \text{ s}^{-1}$; $E = h\nu = 3.369 \times 10^{-19}$ J

The energies in kJ/mol are:

$$3.373 \times 10^{-19} \text{ J} \times \frac{1 \text{ kJ}}{1000 \text{ J}} \times \frac{6.0221 \times 10^{23}}{\text{mol}} = 203.1 \text{ kJ/mol}$$

$$3.369 \times 10^{-19} \text{ J} \times \frac{1 \text{ kJ}}{1000 \text{ J}} \times \frac{6.0221 \times 10^{23}}{\text{mol}} = 202.9 \text{ kJ/mol}$$

Challenge Problems

71. $Pb^{2+} + H_2EDTA^{2-} \rightleftharpoons PbEDTA^{2-} + 2\,H^+$

	Pb^{2+}	H_2EDTA^{2-}		$PbEDTA^{2-}$	$2\,H^+$	
Before	0.0010 *M*	0.050 *M*		0	1.0×10^{-6} *M*	(buffer, $[H^+]$ constant)
Change	-0.0010	-0.0010	→	+0.0010	No change	Reacts completely
After	0	0.049		0.0010	1.0×10^{-6}	New initial
	x mol/L $PbEDTA^{2-}$ dissociates to reach equilibrium					
Change	+*x*	+*x*	←	-*x*	---	
Equil.	*x*	0.049 + *x*		0.0010 - *x*	1.0×10^{-6}	(buffer)

$$K = 1.0 \times 10^{23} = \frac{[PbEDTA^{2-}][H^+]^2}{[Pb^{2+}][H_2EDTA^{2-}]} = \frac{(0.0010 - x)(1.0 \times 10^{-6})^2}{(x)(0.049 + x)}$$

$$1.0 \times 10^{23} \approx \frac{(0.0010)(1.0 \times 10^{-12})}{x(0.049)}, \; x = [Pb^{2+}] = 2.0 \times 10^{-37} \, M \quad \text{Assumptions good.}$$

73. $SiCl_4(l) + 2\ H_2O(l) \rightarrow SiO_2(s) + 4\ H^+(aq) + 4\ Cl^-(aq)$

$\Delta H° = -911 + 4(0) + 4(-167) - [-687 + 2(-286)] = -320.\ kJ$

$\Delta S° = 42 + 4(0) + 4(57) - [240. + 2(70.)] = -110.\ J/K$; $\Delta G° = \Delta H° - T\Delta S°$

$\Delta G° = 0$ when $T = \Delta H°/\Delta S° = -320. \times 10^3\ J/(-110.\ J/K) = 2910\ K$

Due to the favorable $\Delta H°$ term, this reaction at standard conditions is spontaneous at temperatures below 2910 K.

The corresponding reaction for CCl_4 is:

$CCl_4(l) + 2\ H_2O(l) \rightarrow CO_2(g) + 4\ H^+(aq) + 4\ Cl^-(aq)$

$\Delta H° = -393.5 + 4(0) + 4(-167) - [-135 + 2(-286)] = -355\ kJ$

$\Delta S° = 214 + 4(0) + 4(57) - [216 + 2(70.)] = 86\ J/K$

Thermodynamics predicts that this reaction (at standard conditions) would be spontaneous at any temperature.

The answer for the reactivity difference must lie with kinetics. $SiCl_4$ reacts because an activated complex can form by a water molecule attaching to silicon in $SiCl_4$. The activated complex requires silicon to form a fifth bond (See Exercise 18.67). Silicon has low energy 3d orbitals available to expand the octet. Carbon will not break the octet rule, therefore, CCl_4 does not form this activated complex. CCl_4 and H_2O require a different pathway to get to products. The different pathway has a much higher activation energy and, in turn, the reaction is much slower.

75. a. K^+ (blood) $\rightleftharpoons$ K^+ (muscle) $\Delta G° = 0$; $\Delta G = RT \ln\left(\frac{[K^+]_m}{[K^+]_b}\right)$; $\Delta G = w_{max}$

$$\Delta G = \frac{8.3145\ J}{K\ mol}(310.\ K)\ln\left(\frac{0.15}{0.0050}\right),\ \Delta G = 8.8 \times 10^3\ J/mol = 8.8\ kJ/mol$$

At least 8.8 kJ of work must be applied to transport 1 mol K^+.

b. Other ions will have to be transported to maintain electroneutrality. Either anions must be transported into the cells, or cations (Na^+) in the cell must be transported to the blood. The latter is what happens: $[Na^+]$ in blood is greater than $[Na^+]$ in cells as a result of this pumping.

c. $\Delta G° = -RT \ln K = -(8.3145\ J\ K^{-1}\ mol^{-1})(310.\ K) \ln 1.7 \times 10^5 = -3.1 \times 10^4\ J/mol = -31\ kJ/mol$

The hydrolysis of ATP (at standard conditions) provides 31 kJ/mol of energy to do work. We must add 8.8 kJ of work to transport 1.0 mol of K^+.

$$8.8\ kJ \times \frac{1\ mol\ ATP}{31\ kJ} = 0.28\ mol\ ATP\ \text{must be hydrolyzed.}$$

CHAPTER NINETEEN

THE REPRESENTATIVE ELEMENTS: GROUPS 5A THROUGH 8A

Group 5A Elements

1. NO_4^{3-}

Both NO_4^{3-} and PO_4^{3-} have 32 valence electrons so both have similar Lewis structures. From the Lewis structure for NO_4^{3-}, the central N atom has a tetrahedral arrangement of electron pairs. N is small. There is probably not enough room for all 4 oxygen atoms around N. P is larger, thus, PO_4^{3-} is stable.

PO_3^-

PO_3^- and NO_3^- both have 24 valence electrons so both have similar Lewis structures. From the Lewis structure, PO_3^- has a trigonal arrangement of electron pairs about the central P atom (two single bonds and one double bond). P = O bonds are not particularly stable, while N = O bonds are stable. Thus, NO_3^- is stable.

3. $2\ NaN_3(s) \rightarrow 2\ Na(s) + 3\ N_2(g)$

$$n_{N_2} = \frac{PV}{RT} = \frac{1.00\ \text{atm} \times 70.0\ \text{L}}{\frac{0.08206\ \text{L atm}}{\text{mol K}} \times 273\ \text{K}} = 3.12\ \text{mol}\ N_2 \text{ needed to fill air bag.}$$

$$\text{mass } NaN_3 \text{ reacted} = 3.12\ \text{mol}\ N_2 \times \frac{2\ \text{mol}\ NaN_3}{3\ \text{mol}\ N_2} \times \frac{65.02\ \text{g}\ NaN_3}{\text{mol}\ NaN_3} = 135\ \text{g}\ NaN_3$$

5. a. NO_2, $5 + 2(6) = 17\ e^-$

N_2O_4, $2(5) + 4(6) = 34\ e^-$

plus other resonance structures

plus other resonance structures

b. BF_3, $3 + 3(7) = 24\ e^-$

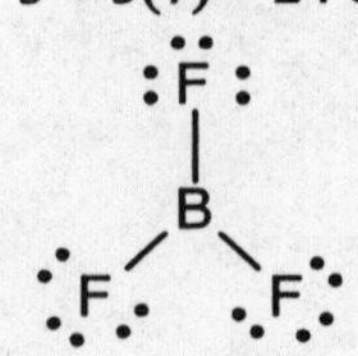

NH_3, $5 + 3(1) = 8\ e^-$

BF_3NH_3, $24 + 8 = 32\ e^-$

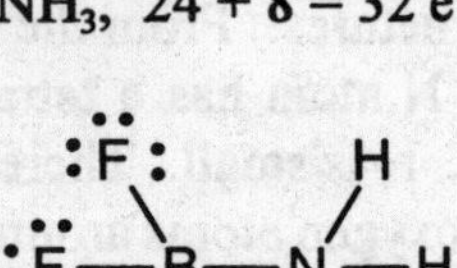

In reaction a, NO_2 has an odd number of electrons so it is impossible to satisfy the octet rule. By dimerizing to form N_2O_4, the odd electron on two NO_2 molecules can pair up giving a species whose Lewis structure can satisfy the octet rule. In general odd electron species are very reactive. In reaction b, BF_3 can be considered electron deficient; boron has only six electrons around it. By forming BF_3NH_3, the boron atom satisfies the octet rule by accepting a lone pair of electrons from NH_3 to form a fourth bond.

7. Unbalanced equation:

$$CaF_2{\bullet}3Ca_3(PO_4)_2(s) + H_2SO_4(aq) \rightarrow H_3PO_4(aq) + HF(aq) + CaSO_4{\bullet}2H_2O(s)$$

Balancing Ca^{2+}, F^-, and PO_4^{3-}:

$$CaF_2{\bullet}3Ca_3(PO_4)_2(s) + H_2SO_4(aq) \rightarrow 6\ H_3PO_4(aq) + 2\ HF(aq) + 10\ CaSO_4{\bullet}2H_2O(s)$$

On the righthand side, there are 20 extra hydrogen atoms, 10 extra sulfates, and 20 extra water molecules. We can balance the hydrogen and sulfate with 10 sulfuric acid molecules. The extra waters came from the water in the sulfuric acid solution. The balanced equation is:

$$CaF_2{\bullet}3Ca_3(PO_4)_2(s) + 10\ H_2SO_4(aq) + 20\ H_2O(l) \rightarrow 6\ H_3PO_4(aq) + 2\ HF(aq) + 10\ CaSO_4{\bullet}2H_2O(s)$$

9. OCN^- has $6 + 4 + 5 + 1 = 16$ valence electrons.

$[O{=}C{=}N]^- \longleftrightarrow [O{-}C{\equiv}N]^- \longleftrightarrow [O{\equiv}C{-}N]^-$

Formal charge: 0 0 -1 | -1 0 0 | +1 0 -2

Only the first two resonance structures should be important. The third places a positive formal charge on the most electronegative atom in the ion and a -2 formal charge on N.

CNO^-:

$$[\ddot{:}C{=}N{=}\ddot{O}:]^- \longleftrightarrow [:C{\equiv}N{-}\ddot{\underset{..}{O}}:]^- \longleftrightarrow [:\ddot{\underset{..}{C}}{-}N{\equiv}O:]^-$$

Formal charge	-2	+1	0	-1	+1	-1	-3	+1	+1

All of the resonance structures for fulminate (CNO^-) involve greater formal charges than in cynate (OCN^-), making fulminate more reactive (less stable).

11. a. $NH_4NO_3(s) \xrightarrow{heat} N_2O(g) + 2\ H_2O(g)$

b. $2\ N_2O_5(g) \rightarrow 4\ NO_2(g) + O_2(g)$

c. $2\ K_3P(s) + 6\ H_2O(l) \rightarrow 2\ PH_3(g) + 6\ KOH(aq)$

d. $PBr_3(l) + 3\ H_2O(l) \rightarrow H_3PO_3(aq) + 3\ HBr(aq)$

e. $2\ NH_3(aq) + NaOCl(aq) \rightarrow N_2H_4(aq) + NaCl(aq) + H_2O(l)$

13. **Production of bismuth:**

$$2\ Bi_2S_3(s) + 9\ O_2(g) \rightarrow 2\ Bi_2O_3(s) + 6\ SO_2(g); \quad 2\ Bi_2O_3(s) + 3\ C(s) \rightarrow 4\ Bi(s) + 3\ CO_2(g)$$

Production of antimony:

$$2\ Sb_2S_3(s) + 9\ O_2(g) \rightarrow 2\ Sb_2O_3(s) + 6\ SO_2(g); \quad 2\ Sb_2O_3(s) + 3\ C(s) \rightarrow 4\ Sb(s) + 3\ CO_2(g)$$

15. **TSP = Na_3PO_4; PO_4^{3-} is the conjugate base of the weak acid HPO_4^{2-} ($K_a = 4.8 \times 10^{-13}$). All conjugate bases of weak acids are effective bases ($K_b = K_w/K_a = 1.0 \times 10^{-14}/4.8 \times 10^{-13} = 2.1 \times 10^{-2}$). The weak base reaction of PO_4^{3-} with H_2O is: $PO_4^{3-}(aq) + H_2O(l) \rightleftharpoons HPO_4^{2-}(aq) + OH^-(aq)$**

17. **White phosphorus consists of discrete tetrahedral P_4 molecules. The bond angles in the P_4 tetrahedrons are only 60°, which makes P_4 very reactive especially towards oxygen. Red and black phosphorus are covalent network solids. In red phosphorus the P_4 tetrahedra are bonded to each other in chains making them less reactive than white phosphorus. They need a source of energy to react with oxygen, such as when one strikes a match. Black phosphorus is crystalline with the P atoms tightly bonded to each other in the crystal and are fairly unreactive towards oxygen.**

19. $4\ As(s) + 3\ O_2(g) \rightarrow As_4O_6(s)$; $4\ As(s) + 5\ O_2(g) \rightarrow As_4O_{10}(s)$

$As_4O_6(s) + 6\ H_2O(l) \rightarrow 4\ H_3AsO_3(aq)$; $As_4O_{10}(s) + 6\ H_2O(l) \rightarrow 4\ H_3AsO_4(aq)$

21. $1/2\ N_2(g) + 1/2\ O_2(g) \rightarrow NO(g)$ $\Delta G° = \Delta G°_{f,\,NO} = 87$ kJ/mol; By definition, $\Delta G°_f$ for a compound equals the free energy change that would accompany the formation of 1 mol of that compound from its elements in their standard states. NO (and some other oxides of nitrogen) have weaker bonds as compared to the triple bond of N_2 and the double bond of O_2. Because of this, NO (and some other oxides of nitrogen) have higher (positive) standard free energies of formation as compared to the relatively stable N_2 and O_2 molecules.

23. $AsCl_4^+$, $5 + 4(7) - 1 = 32\ e^-$ $AsCl_6^-$, $5 + 6(7) + 1 = 48\ e^-$

$$[AsCl_4]^+ \qquad [AsCl_6]^-$$

The reaction is a Lewis acid-base reaction. A chloride ion acts as a Lewis base when it is transferred from one $AsCl_5$ to another. Arsenic is the Lewis acid (electron pair acceptor).

25. The pollution provides nitrogen and phosphorus nutrients so the algae can grow. The algae consume oxygen, causing fish to die.

27. $5\ N_2O_4(l) + 4\ N_2H_3CH_3(l) \rightarrow 12\ H_2O(g) + 9\ N_2(g) + 4\ CO_2(g)$

$$\Delta H° = \left[12\text{ mol}\left(\frac{-242\text{ kJ}}{\text{mol}}\right) + 4\text{ mol}\left(\frac{-393.5\text{ kJ}}{\text{mol}}\right)\right]$$

$$- \left[5\text{ mol}\left(\frac{-20.\text{ kJ}}{\text{mol}}\right) + 4\text{ mol}\left(\frac{54\text{ kJ}}{\text{mol}}\right)\right] = -4594\text{ kJ}$$

Group 6A Elements

29. In the upper atmosphere, O_3 acts as a filter for UV radiation:

$$O_3 \xrightarrow{h\nu} O_2 + O \quad \text{(See Exercise 19.28.)}$$

O_3 is also a powerful oxidizing agent. It irritates the lungs and eyes, and at high concentration it is toxic. The smell of a "fresh spring day" is O_3 formed during lightning discharges. Toxic materials don't necessarily smell bad. For example, HCN smells like almonds.

31. a. As we go down the family, K_a increases. This is consistent with the bond to hydrogen getting weaker.

b. Po is below Te, so the K_a should be larger. The K_a for H_2Po should be on the order of 10^{-2}.

33. In the presence of S^{2-}, sulfur forms polysulfide ions, S_n^{2-}, which are soluble like most species with charges, e.g., $S_8 + S^{2-} \rightleftharpoons S_9^{2-}$. Nitric acid oxidizes S^{2-} to S, which then causes sulfur to precipitate out of solution.

35. a. $2\ SO_2(g) + O_2(g) \rightarrow 2\ SO_3(g)$

b. $SO_3(g) + H_2O(l) \rightarrow H_2SO_4(aq)$

c. $2\ Na_2S_2O_3(aq) + I_2(aq) \rightarrow Na_2S_4O_6(aq) + 2\ NaI(aq)$

d. $Cu(s) + 2\ H_2SO_4(aq) \rightarrow CuSO_4(aq) + 2\ H_2O(l) + SO_2(aq)$

37. $H_2SeO_4(aq) + 3\ SO_2(g) \rightarrow Se(s) + 3\ SO_3(g) + H_2O(l)$

39. This element is in the oxygen family as all oxygen family members have ns^2np^4 valence electron configurations.

a. As with all elements of the oxygen family, this element has 6 valence electrons.

b. The nonmetals in the oxygen family are O, S, Se and Te, which are all possible identities for the element.

c. Ions in the oxygen family are -2 charged in ionic compounds. Li_2X would be the formula between Li^+ and X^{2-} ions.

d. In general, radii increases from right to left across the periodic table and increases going down a family. From this trend, the radius of the unknown element must be smaller than the Ba radius.

e. The ionization energy trend is the opposite of the radii trend indicated in the previous answer. From this trend, the unknown element will have a smaller ionization energy than fluorine.

Group 7A Elements

41. O_2F_2 has $2(6) + 2(7) = 26$ valence e^-.

:F—O—O—F:

Formal Charge	0	0	0	0
Oxidation State	-1	+1	+1	-1

Oxidation states are more useful. We are forced to assign +1 as the oxidation state for oxygen. Oxygen is very electronegative and +1 is not a stable oxidation state for this element.

43. Fluorine is the most reactive of the halogens because it is the most electronegative atom and the bond in the F_2 molecule is very weak.

45. $ClO^- + H_2O + 2\ e^- \rightarrow 2\ OH^- + Cl^-$ $\qquad E^\circ_c = 0.90\ V$

$2\ NH_3 + 2\ OH^- \rightarrow N_2H_4 + 2\ H_2O + 2\ e^-$ $\qquad -E^\circ_a = 0.10\ V$

$ClO^-(aq) + 2\ NH_3(aq) \rightarrow Cl^-(aq) + N_2H_4(aq) + H_2O(l)$ $\qquad E^\circ_{cell} = 1.00\ V$

Since E°_{cell} is positive for this reaction, then at standard conditions ClO^- can spontaneously oxidize NH_3 to the somewhat toxic N_2H_4.

47. a. $BaCl_2(s) + H_2SO_4(aq) \rightarrow BaSO_4(s) + 2\ HCl(g)$

b. $BrF(s) + H_2O(l) \rightarrow HF(aq) + HOBr(aq)$

c. $SiO_2(s) + 4\ HF(aq) \rightarrow SiF_4(g) + 2\ H_2O(l)$

Group 8A Elements

49. Helium is unreactive and doesn't combine with any other elements. It is a very light gas and would easily escape the earth's gravitational pull as the planet was formed.

51. $10.0\ m \times 10.0\ m \times 10.0\ m = 1.00 \times 10^3\ m^3$; From Table 19.12, volume % Ar = 0.9%.

$$1.00 \times 10^3\ m^3 \times \left(\frac{10\ dm}{m}\right)^3 \times \frac{1\ L}{dm^3} \times \frac{0.9\ L\ Ar}{100\ L\ air} = 9 \times 10^3\ L \text{ of Ar in the room}$$

$$PV = nRT,\ n = \frac{PV}{RT} = \frac{(1.0\ atm)\ (9 \times 10^3\ L)}{(0.08206\ L\ atm\ K^{-1} mol^{-1})\ (298\ K)} = 4 \times 10^2\ mol\ Ar$$

$$4 \times 10^2\ mol\ Ar \times \frac{39.95\ g}{mol} = 2 \times 10^4\ g \text{ Ar in the room}$$

$$4 \times 10^2\ mol\ Ar \times \frac{6.022 \times 10^{23}\ atoms}{mol} = 2 \times 10^{26} \text{ atoms Ar in the room}$$

$$\text{A 2 L breath contains: } 2\ L\ air \times \frac{0.9\ L\ Ar}{100\ L\ air} = 2 \times 10^{-2}\ L\ Ar$$

$$n = \frac{PV}{RT} = \frac{(1.0\ atm)\ (2 \times 10^{-2}\ L)}{(0.08206\ L\ atm\ K^{-1} mol^{-1})\ (298\ K)} = 8 \times 10^{-4}\ mol\ Ar$$

$$8 \times 10^{-4}\ mol\ Ar \times \frac{6.022 \times 10^{23}\ atoms}{mol} = 5 \times 10^{20} \text{ atoms of Ar in a 2 L breath}$$

Since Ar and Rn are both noble gases, then both species will be relatively unreactive. However, all nuclei of Rn are radioactive, unlike most nuclei of Ar. The radioactive decay products of Rn can cause biological damage when inhaled.

53. XeF_2 can react with oxygen to produce explosive xenon oxides and oxyfluorides and react with water to form HF as well.

Additional Exercises

55. a. The Lewis structures for NNO and NON are (16 valence electrons each):

$:\!N{=}N{=}O\!: \longleftrightarrow :\!N{\equiv}N{-}\ddot{O}\!: \longleftrightarrow :\!\ddot{N}{-}N{\equiv}O\!:$

$:\!N{=}O{=}N\!: \longleftrightarrow :\!N{\equiv}O{-}\ddot{N}\!: \longleftrightarrow :\!\ddot{N}{-}O{\equiv}N\!:$

The NNO structure is correct. From the Lewis structures we would predict both NNO and NON to be linear. However, we would predict NNO to be polar and NON to be nonpolar. Since experiments show N_2O to be polar, then NNO is the correct structure.

b. Formal charge = number of valence electrons of atoms - [(number of lone pair electrons) + 1/2 (number of shared electrons)].

$:\!N{=}N{=}O\!: \longleftrightarrow :\!N{\equiv}N{-}\ddot{O}\!: \longleftrightarrow :\!\ddot{N}{-}N{\equiv}O\!:$

-1 +1 0   0 +1 -1   -2 +1 +1

The formal charges for the atoms in the various resonance structures appear below each atom. The central N is sp hybridized in all of the resonance structures. We can probably ignore the third resonance structure on the basis of the relatively large formal charges on the various atoms in N_2O as compared to the first two resonance structures.

c. The sp hybrid orbitals on the center N overlap with atomic orbitals (or hybrid orbitals) on the other two atoms to form the two sigma bonds. The remaining two unhybridized p orbitals on the center N overlap with two p orbitals on the peripheral N to form the two π bonds.

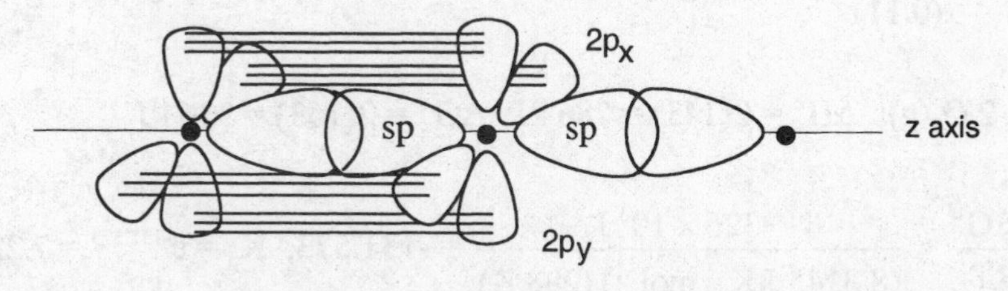

57. Xe has one more valence electron than I. Thus, the isoelectric species will have I plus one extra electron substituted for Xe, giving a species with a net minus one charge.

a. IO_4^- b. IO_3^- c. IF_2^- d. IF_4^- e. IF_6^- f. IOF_3^-

59. $$1.0 \times 10^4 \text{ kg waste} \times \frac{3.0 \text{ kg } NH_4^+}{100 \text{ kg waste}} \times \frac{1000 \text{ g}}{\text{kg}} \times \frac{1 \text{ mol } NH_4^+}{18.04 \text{ g } NH_4^+} \times \frac{1 \text{ mol } C_5H_7O_2N}{55 \text{ mol } NH_4^+}$$

$$\times \frac{113.12 \text{ g } C_5H_7O_2N}{\text{mol } C_5H_7O_2N} = 3.4 \times 10^4 \text{ g tissue if all } NH_4^+ \text{ converted}$$

Since only 95% of the NH_4^+ ions react:

mass of tissue = (0.95) (3.4×10^4 g) = 3.2×10^4 g or 32 kg bacterial tissue

Challenge Problems

61. $Mg^{2+} + P_3O_{10}^{5-} \rightleftharpoons MgP_3O_{10}^{3-}$ pK = -8.60; $[Mg^{2+}]_o = \frac{50. \times 10^{-3}\ g}{L} \times \frac{1\ mol}{24.3\ g} = 2.1 \times 10^{-3}\ M$

$$[P_3O_{10}^{5-}]_o = \frac{40.\ g\ Na_5P_3O_{10}}{L} \times \frac{1\ mol}{367.9\ g} = 0.11\ M$$

Assume the reaction goes to completion since K is large ($K = 10^{8.60} = 4.0 \times 10^8$). Then solve the back equilibrium problem to determine the small amount of Mg^{2+} present.

	Mg^{2+}	+ $P_3O_{10}^{5-}$	$\rightleftharpoons$	$MgP_3O_{10}^{3-}$	
Before	$2.1 \times 10^{-3}\ M$	$0.11\ M$		0	
Change	-2.1×10^{-3}	-2.1×10^{-3}	$\rightarrow$	$+2.1 \times 10^{-3}$	React completely
After	0	0.11		2.1×10^{-3}	New initial
	x mol/L $MgP_3O_{10}^{3-}$ dissociates to reach equilibrium				
Change	$+x$	$+x$	$\leftarrow$	$-x$	
Equil.	x	$0.11 + x$		$2.1 \times 10^{-3} - x$	

$$K = 4.0 \times 10^8 = \frac{[MgP_3O_{10}^{3-}]}{[Mg^{2+}][P_3O_{10}^{5-}]} = \frac{2.1 \times 10^{-3} - x}{x(0.11 + x)} \quad \text{(assume } x << 2.1 \times 10^{-3}\text{)}$$

$$4.0 \times 10^8 \approx \frac{2.1 \times 10^{-3}}{x(0.11)},\ x = [Mg^{2+}] = 4.8 \times 10^{-11}\ M;\ \text{Assumptions good.}$$

63. $3\ O_2(g) \rightleftharpoons 2\ O_3(g)$; $\Delta H° = 2(143) = 286$ kJ; $\Delta G° = 2(163) = 326$ kJ

$$\ln K_p = \frac{-\Delta G°}{RT} = \frac{-326 \times 10^3\ J}{(8.3145\ J\ K^{-1}\ mol^{-1})(298\ K)} = -131.573,\ K_p = e^{-131.573} = 7.22 \times 10^{-58}$$

Note: We carried extra significant figures for the K_p calculation.

We need the value of K at 230. K. From Section 10.11 of the text: $\ln K = \frac{-\Delta H°}{RT} + \frac{\Delta S°}{R}$

For two sets of K and T:

$$\ln K_1 = \frac{-\Delta H°}{R}\left(\frac{1}{T_1}\right) + \frac{\Delta S°}{R};\ \ln K_2 = \frac{-\Delta H°}{R}\left(\frac{1}{T_2}\right) + \frac{\Delta S°}{R}$$

Subtracting the first expression from the second:

$$\ln K_2 - \ln K_1 = \frac{\Delta H°}{R}\left(\frac{1}{T_1} - \frac{1}{T_2}\right) \text{ or } \ln\frac{K_2}{K_1} = \frac{\Delta H°}{R}\left(\frac{1}{T_1} - \frac{1}{T_2}\right)$$

Let $K_2 = 7.22 \times 10^{-58}$, $T_2 = 298$; $K_1 = K_{230}$, $T_1 = 230.$ K; $\Delta H° = 286 \times 10^3$ J

$$\ln\frac{7.22 \times 10^{-58}}{K_{230}} = \frac{286 \times 10^3}{8.3145}\left(\frac{1}{230.} - \frac{1}{298}\right) = 34.13 \quad \textbf{(Carrying extra sig. figs.)}$$

$$\frac{7.22 \times 10^{-58}}{K_{230}} = e^{34.13} = 6.6 \times 10^{14},\ K_{230} = 1.1 \times 10^{-72}$$

$$K_{230} = 1.1 \times 10^{-72} = \frac{P_{O_3}^2}{P_{O_2}^3} = \frac{P_{O_3}^2}{(1.0 \times 10^{-3}\ \text{atm})^3},\ P_{O_3} = 3.3 \times 10^{-41}\ \text{atm}$$

The volume occupied by one molecule of ozone is:

$$V = \frac{nRT}{P} = \frac{(1/6.022 \times 10^{23}\ \text{mol})(0.08206\ \text{L atm mol}^{-1}\ \text{K}^{-1})(230.\ \text{K})}{(3.3 \times 10^{-41}\ \text{atm})},\ V = 9.5 \times 10^{17}\ \text{L}$$

Equilibrium is probably not maintained under these conditions. When only two ozone molecules are in a volume of 9.5×10^{17} L, the reaction is not at equilibrium. Under these conditions, Q > K and the reaction shifts left. But with only 2 ozone molecules in this huge volume, it is extremely unlikely that they will collide with each other. At these conditions, the concentration of ozone is not large enough to maintain equilibrium.

65. a. NO is the catalyst. NO is present in the first step of the mechanism on the reactant side, but it is not a reactant since it is regenerated in the second step and does not appear in the overall balanced equation.

b. NO_2 is an intermediate. Intermediates also never appear in the overall balanced equation. In a mechanism, intermediates always appear first on the product side while catalysts always appear first on the reactant side.

c. $k = A\exp(-E_a/RT)$; $\dfrac{k_{cat}}{k_{un}} = \dfrac{A\exp[-E_a(cat)/RT]}{A\exp[-E_a(un)/RT]} = \exp\left(\dfrac{E_a(un) - E_a(cat)}{RT}\right)$

$$\frac{k_{cat}}{k_{un}} = \exp\left(\frac{2100\ \text{J/mol}}{8.3145\ \text{J K}^{-1}\ \text{mol}^{-1} \times 298\ \text{K}}\right) = e^{0.85} = 2.3$$

The catalyzed reaction is approximately 2.3 times faster than the uncatalyzed reaction at 25 °C.

d. **The mechanism for the chlorine catalyzed destruction of ozone is:**

$$\begin{array}{ll} O_3 + Cl \rightarrow O_2 + ClO & \text{slow} \\ ClO + O \rightarrow O_2 + Cl & \text{fast} \\ \hline O_3 + O \rightarrow 2\,O_2 & \end{array}$$

e. **Since the chlorine atom catalyzed reaction has a lower activation energy, then the Cl catalyzed rate is faster. Hence, Cl is a more effective catalyst. Using the activation energy, we can estimate the efficiency that Cl atoms destroy ozone as compared to NO molecules.**

$$\text{At } 25°C\text{:} \quad \frac{k_{Cl}}{k_{NO}} = \exp\left(\frac{-E_a(Cl)}{RT} + \frac{E_a(NO)}{RT}\right) = \exp\left(\frac{(-2100 + 11{,}900)\ \text{J/mol}}{(8.3145 \times 298)\ \text{J/mol}}\right) = e^{3.96} = 52$$

At 25°C, the Cl catalyzed reaction is roughly 52 times faster (more efficient) than the NO catalyzed reaction, assuming the frequency factor A is the same for each reaction and assuming similar rate laws.

CHAPTER TWENTY

TRANSITION METALS AND COORDINATION CHEMISTRY

Transition Metals

7. Cr and Cu are exceptions to the normal filling order of electrons.

a. Cr: $[Ar]4s^1 3d^5$ b. Cu: $[Ar]4s^1 3d^{10}$ c. V: $[Ar]4s^2 3d^3$

Cr^{2+}: $[Ar]3d^4$ Cu^+: $[Ar]3d^{10}$ V^{2+}: $[Ar]3d^3$

Cr^{3+}: $[Ar]3d^3$ Cu^{2+}: $[Ar]3d^9$ V^{3+}: $[Ar]3d^2$

9. Most transition metals have unfilled d orbitals, which creates a large number of valence electrons that can be removed. Stable ions of the representative metals are determined by how many s and p valence electrons can be removed. In general, representative metals lose all of the s and p valence electrons to form their stable ions. Transition metals generally lose the s electron(s) to form +1 and +2 ions, but they can also lose some (or all) of the d electrons to form other oxidation states as well.

11. a. molybdenum(IV) sulfide; molybdenum(VI) oxide

b. MoS_2, +4; MoO_3, +6; $(NH_4)_2Mo_2O_7$, +6; $(NH_4)_6Mo_7O_{24} \bullet 4\ H_2O$, +6

13. TiF_4: ionic compound containing Ti^{4+} ions and F^- ions. $TiCl_4$, $TiBr_4$, and TiI_4: covalent compounds containing discrete, tetrahedral TiX_4 molecules. As these covalent molecules get larger, the bp and mp increase because the London dispersion forces increase. TiF_4 has the highest bp since the interparticle forces are usually stronger in ionic compounds as compared to covalent compounds.

15. $H^+ + OH^- \rightarrow H_2O$; Sodium hydroxide (NaOH) will react with the H^+ on the product side of the reaction. This effectively removes H^+ from the equilibrium, which will shift the reaction to the right to produce more H^+ and CrO_4^{2-}. Since more CrO_4^{2-} is produced, the solution turns yellow.

Coordination Compounds

17. a. ligand: Species that donates a pair of electrons to form a covalent bond to a metal ion. Ligands act as Lewis bases (electron pair donors).

b. chelate: Ligand that can form more than one bond to a metal ion.

c. bidentate: Ligand that forms two bonds to a metal ion.

d. complex ion: Metal ion plus ligands.

19. a. pentaamminechlororuthenium(III) ion b. hexacyanoferrate(II) ion

c. tris(ethylenediamine)manganese(II) ion d. pentaamminenitrocobalt(III) ion

21. a. $K_2[CoCl_4]$ b. $[Pt(H_2O)(CO)_3]Br_2$

c. $Na_3[Fe(CN)_2(C_2O_4)_2]$ d. $[Cr(NH_3)_3Cl(NH_2CH_2CH_2NH_2)]I_2$

23. $BaCl_2$ gives no precipitate so SO_4^{2-} must be in the coordination sphere. A precipitate with $AgNO_3$ means that the Cl^- is not in the coordination sphere. Since there are only four ammonia molecules in the coordination sphere, then the SO_4^{2-} must be acting as a bidentate ligand (assuming an octahedral complex ion). The structure is:

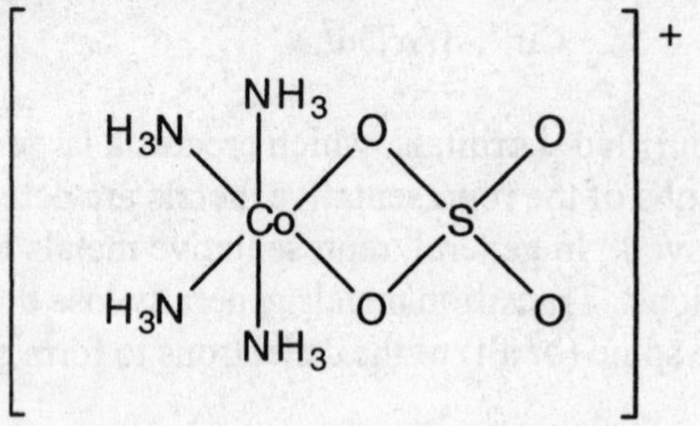

NH_3
H_3N O O
Co S Cl^-
H_3N O O
NH_3
+

25. $Hg^{2+}(aq) + 2\ I^-(aq) \rightarrow HgI_2(s)$ (orange precipitate)

$HgI_2(s) + 2\ I^-(aq) \rightarrow HgI_4^{2-}(aq)$ (soluble complex ion)

27. a. isomers: Species with the same formulas but different properties. See text for examples of the following types of isomers.

b. structural isomers: Isomers that have one or more bonds that are different.

c. stereoisomers: Isomers that contain the same bonds but differ in how the atoms are arranged in space.

d. coordination isomers: Structural isomers that differ in the atoms that make up the complex ion.

e. linkage isomers: Structural isomers that differ in how one or more ligands are attached to the transition metal.

f. geometric isomers: (cis - trans isomerism) Stereoisomers that differ in the positions of atoms with respect to a rigid ring, bond, or each other.

g. optical isomers: **Stereoisomers that are nonsuperimposable mirror images of each other; that is, they are different in the same way that our left and right hands are different.**

29. a.

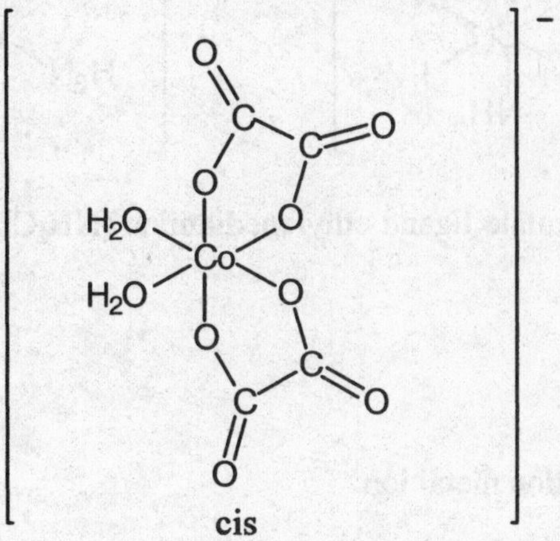

$[Co(C_2O_4)_2(H_2O)_2]^-$ structure: O, C, C, O, O, O, H_2O, H_2O, Co, O, O, C, C, O, O

cis

$[Co(C_2O_4)_2(H_2O)_2]^-$ structure: O, O, OH_2, O, O, C, C, Co, C, C, O, O, OH_2, O, O

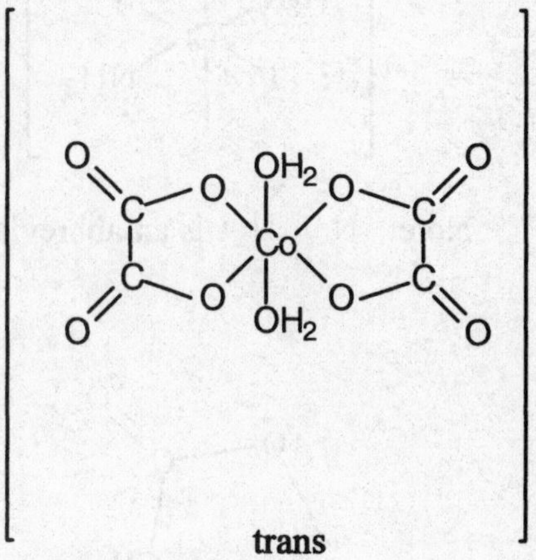

trans

Note: $C_2O_4^{2-}$ is a bidentate ligand. Bidentate ligands bond to the metal at two positions which are 90° apart from each other in octahedral complexes. Bidentate ligands do not bond to the metal at positions 180° apart.

b.

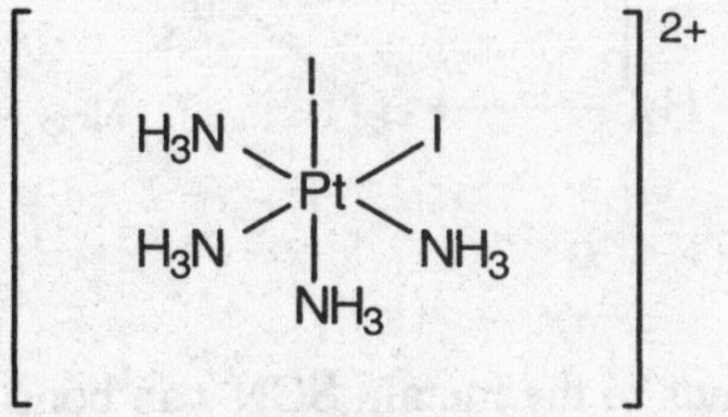

$[Pt(NH_3)_4I_2]^{2+}$ structure: I, H_3N, I, Pt, H_3N, NH_3, NH_3

cis

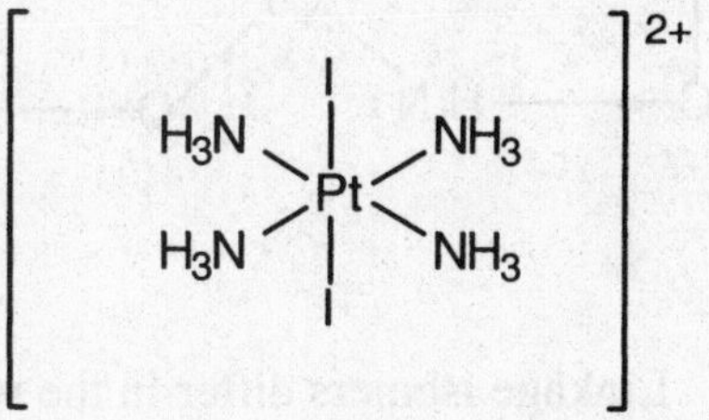

$[Pt(NH_3)_4I_2]^{2+}$ structure: I, H_3N, NH_3, Pt, H_3N, NH_3, I

trans

c.

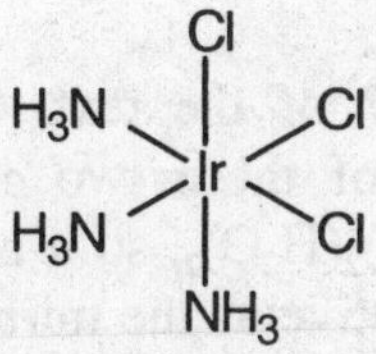

Structure: Cl, H_3N, Cl, Ir, H_3N, Cl, NH_3

cis

Structure: Cl, H_3N, Cl, Ir, H_3N, NH_3, Cl

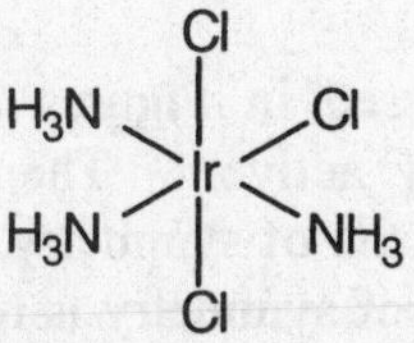

trans

d.

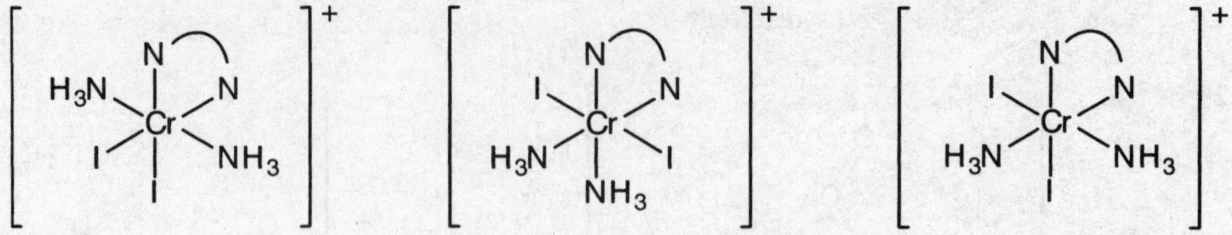

$$\left[\text{Cr}(\text{en})(\text{NH}_3)_2\text{I}_2\right]^+ \qquad \left[\text{Cr}(\text{en})(\text{NH}_3)_2\text{I}_2\right]^+ \qquad \left[\text{Cr}(\text{en})(\text{NH}_3)_2\text{I}_2\right]^+$$

Note: N⌒N **is an abbreviation for the bidentate ligand ethylenediamine ($NH_2CH_2CH_2NH_2$).**

31.

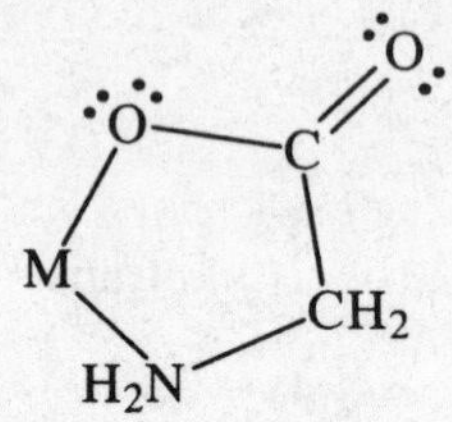

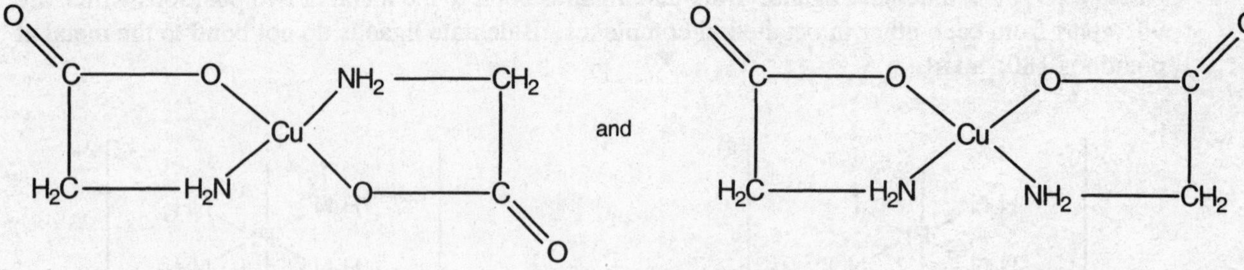

M = transition metal ion

and

33. **Linkage isomers differ in the way that the ligand bonds to the metal. SCN^- can bond through the sulfur or through the nitrogen atom. NO_2^- can bond through the nitrogen or through the oxygen atom. OCN^- can bond through the oxygen or through the nitrogen atom. N_3^-, $NH_2CH_2CH_2NH_2$ and I^- are not capable of linkage isomerism.**

35. **Similar to the molecules discussed in Figures 20.16 and 20.17 of the text, $Cr(acac)_3$ and cis-$Cr(acac)_2(H_2O)_2$ are optically active. The mirror images of these two complexes are nonsuperimposable. There is a plane of symmetry in trans-$Cr(acac)_2(H_2O)_2$, so it is not optically active. A molecule with a plane of symmetry is never optically active as the mirror images are always superimposable. A plane of symmetry is a plane through a molecule where one side reflects on the other side of the molecule.**

Bonding, Color, and Magnetism in Coordination Compounds

37. a. **Ligand that will give complex ions with the maximum number of unpaired electrons.**

b. **Ligand that will give complex ions with the minimum number of unpaired electrons.**

c. **Complex with a minimum number of unpaired electrons (low-spin = strong-field).**

d. **Complex with a maximum number of unpaired electrons (high-spin = weak-field).**

39. **Sc^{3+} has no electrons in d orbitals. Ti^{3+} and V^{3+} have d electrons present. The color of transition metal complexes results from electron transfer between split d orbitals. If no d electrons are present, no electron transfer can occur and the compounds are not colored.**

41. a. **Fe^{2+}: [Ar]$3d^6$**

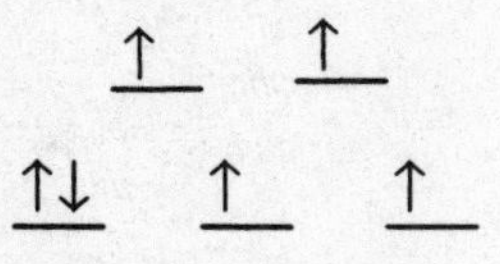

↑ ↑

↑↓ ↑ ↑

High spin, small Δ

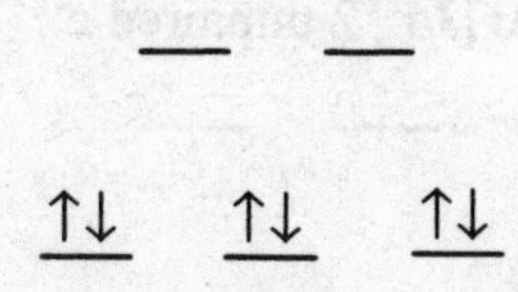

↑↓ ↑↓ ↑↓

Low spin, large Δ

b. **Fe^{3+}: [Ar]$3d^5$**

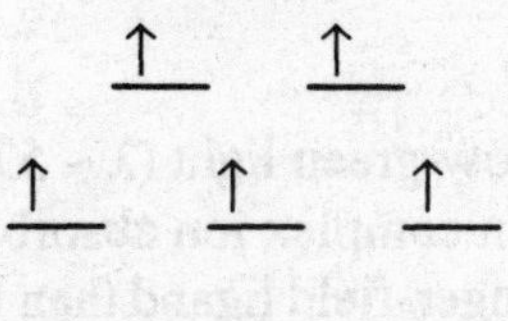

↑ ↑

↑ ↑ ↑

High spin, small Δ

c. **Ni^{2+}: [Ar]$3d^8$**

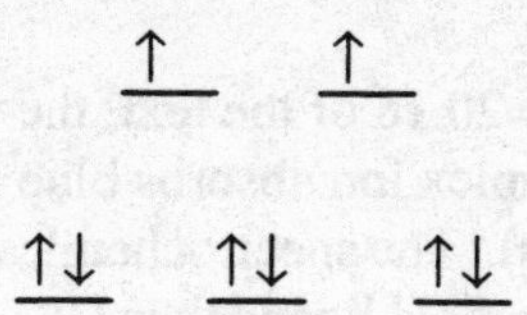

↑ ↑

↑↓ ↑↓ ↑↓

d. **Zn^{2+}: [Ar]$3d^{10}$**

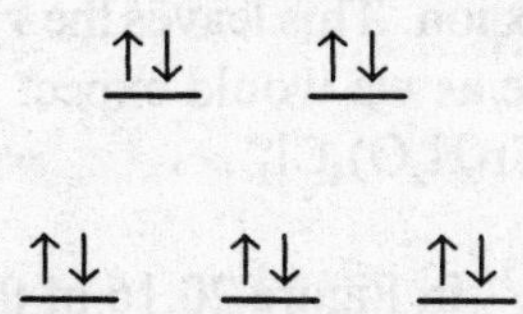

↑↓ ↑↓

↑↓ ↑↓ ↑↓

e. **Co^{2+}: [Ar]$3d^7$**

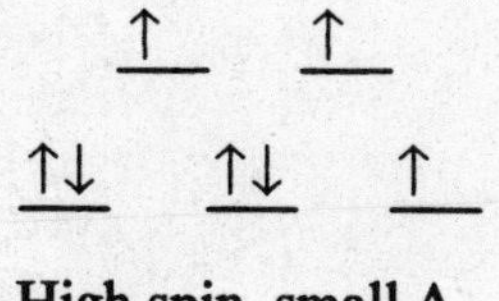

↑ ↑

↑↓ ↑↓ ↑

High spin, small Δ

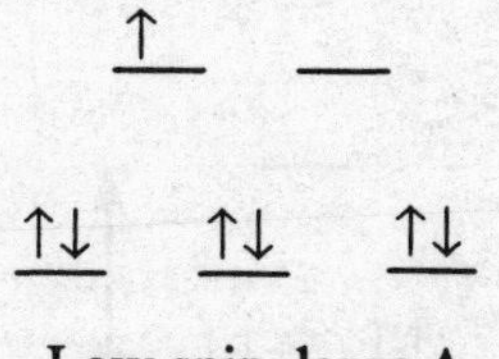

↑

↑↓ ↑↓ ↑↓

Low spin, large Δ

43. **To determine the crystal field diagrams, you need to determine the oxidation state of the transition metal which can only be determined if you know the charges of the ligands (see Table 20.13). The electron configurations and the crystal field diagrams follow.**

a. Ru^{2+}: [Kr]$4d^6$, no unpaired e^-

___ ___

↑↓ ↑↓ ↑↓

Low spin, large Δ

b. Ni^{2+}: [Ar]$3d^8$, 2 unpaired e^-

↑ ↑

↑↓ ↑↓ ↑↓

c. V^{3+}: [Ar]$3d^2$, 2 unpaired e^-

___ ___

↑ ↑ ___

Note: Ni^{2+} must have 2 unpaired electrons, whether high-spin or low-spin, and V^{3+} must have 2 unpaired electrons, whether high-spin or low-spin.

45. From Table 20.16 of the text, the violet complex ion absorbs yellow-green light ($\lambda \sim 570$ nm), the yellow complex ion absorbs blue light ($\lambda \sim 450$ nm), and the green complex ion absorbs red light ($\lambda \sim 650$ nm). The spectrochemical series shows that NH_3 is a stronger-field ligand than H_2O which is a stronger-field ligand than Cl^-. Therefore, $Cr(NH_3)_6^{3+}$ will have the largest d-orbital splitting and will absorb the lowest wavelength electromagnetic radiation ($\lambda \sim 450$ nm) since energy and wavelength are inversely related ($\lambda = hc/E$). Thus, the yellow solution contains the $Cr(NH_3)_6^{3+}$ complex ion. Similarly, we would expect the $Cr(H_2O)_4Cl_2^+$ complex ion to have the smallest d-orbital splitting since it contains the weakest-field ligands. The green solution with the longest wavelength of absorbed light contains the $Cr(H_2O)_4Cl_2^+$ complex ion. This leaves the violet solution, which contains the $Cr(H_2O)_6^{3+}$ complex ion. This makes sense as we would expect $Cr(H_2O)_6^{3+}$ to absorb light of a wavelength between that of $Cr(NH_3)_6^{3+}$ and $Cr(H_2O)_4Cl_2^+$.

47. a. $Ru(phen)_3^{2+}$ exhibits optical isomerism [similar to $Co(en)_3^{3+}$ in Figure 20.16 of the text].

b. Ru^{2+}: [Kr]$4d^6$; Since there are no unpaired electrons, then Ru^{2+} is a strong-field (low-spin) case.

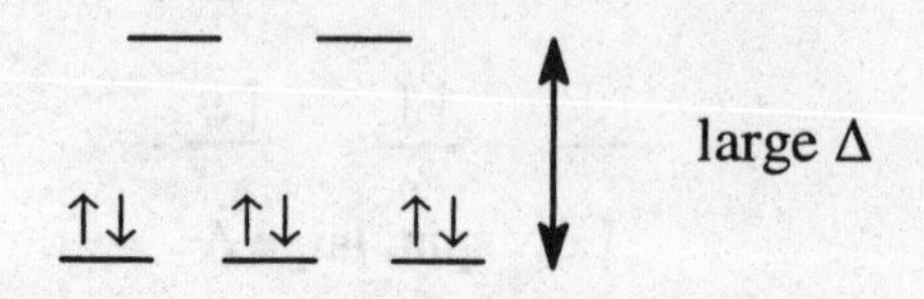

49. $CoBr_6^{4-}$ has an octahedral structure and $CoBr_4^{2-}$ has a tetrahedral structure (as do most Co^{2+} complexes with four ligands). Coordination complexes absorb electromagnetic radiation (EMR) of energy equal to the energy difference between the split d-orbitals. Since the tetra-hedral d-orbital splitting is less than one-half of the octahedral d-orbital splitting, then tetrahedral complexes will absorb lower energy EMR which corresponds to longer wavelength EMR ($E = hc/\lambda$). Therefore, $CoBr_6^{2-}$ will absorb EMR having a wavelength shorter than 3.4×10^{-6} m.

Additional Exercises

51. **$Ni(CO)_4$ is composed of 4 CO molecules and Ni. Thus, nickel has an oxidation state of zero.**

53. i. $0.0203 \text{ g } CrO_3 \times \frac{52.00 \text{ g Cr}}{100.0 \text{ g } CrO_3} = 0.0106 \text{ g Cr}; \ \% \text{ Cr} = \frac{0.0106 \text{ g}}{0.105 \text{ g}} \times 100 = 10.1\% \text{ Cr}$

ii. $32.93 \text{ mL HCl} \times \frac{0.100 \text{ mmol HCl}}{\text{mL}} \times \frac{1 \text{ mmol } NH_3}{\text{mmol HCl}} \times \frac{17.03 \text{ mg } NH_3}{\text{mmol}} = 56.1 \text{ mg } NH_3$

$\% \ NH_3 = \frac{56.1 \text{ mg}}{341 \text{ mg}} \times 100 = 16.5\% \ NH_3$

iii. **73.53% + 16.5% + 10.1% = 100.1%; The compound must be composed of only Cr, NH_3, and I.**

Out of 100.00 of compound:

$10.1 \text{ g Cr} \times \frac{1 \text{ mol}}{52.00 \text{ g}} = 0.194 \qquad \frac{0.194}{0.194} = 1.00$

$16.5 \text{ g } NH_3 \times \frac{1 \text{ mol}}{17.03 \text{ g}} = 0.969 \qquad \frac{0.969}{0.194} = 4.99$

$73.53 \text{ g I} \times \frac{1 \text{ mol}}{126.9 \text{ g}} = 0.5794 \qquad \frac{0.5794}{0.194} = 2.99$

$Cr(NH_3)_5I_3$ is the empirical formula. Cr(III) forms octahedral complexes. So compound A is made of the octahedral $[Cr(NH_3)_5I]^{2+}$ complex ion and two I^- ions as counter ions; the formula is $[Cr(NH_3)_5I]I_2$. Lets check this proposed formula using the freezing point data.

iv. **$\Delta T_f = iK_f m$; For $[Cr(NH_3)_5I]I_2$, i = 3.0 (assuming complete dissociation).**

$\text{molality} = m = \frac{0.601 \text{ g complex}}{1.000 \times 10^{-2} \text{ kg } H_2O} \times \frac{1 \text{ mol complex}}{517.9 \text{ g complex}} = 0.116 \text{ mol/kg}$

$\Delta T_f = 3.0 \times 1.86°\text{C kg/mol} \times 0.116 \text{ mol/kg} = 0.65°\text{C}$

Since ΔT_f is close to the measured value, then this is consistent with the formula $[Cr(NH_3)_5I]I_2$.

55. **M = metal ion**

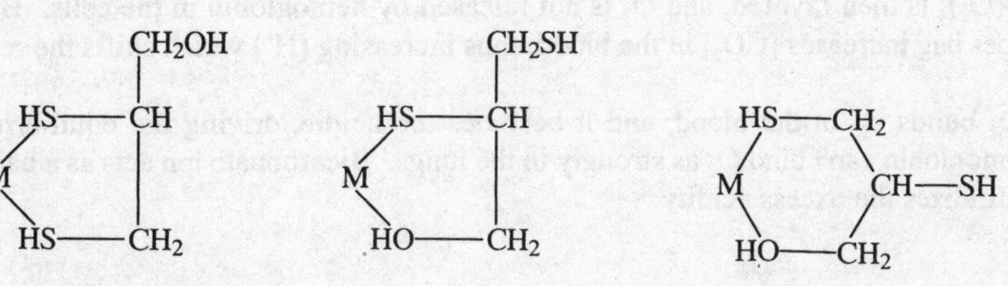

CH_2OH attached to CH; M bonded to HS—CH and HS—CH_2	CH_2SH attached to CH; M bonded to HS—CH and HO—CH_2	M bonded to HS—CH_2 and HO—CH_2; CH—SH between them

57. No; In all three cases, six bonds are formed between Ni^{2+} and nitrogen, so ΔH values should be similar. ΔS° for formation of the complex ion is most negative for 6 NH_3 molecules reacting with a metal ion (7 independent species become 1). For penten reacting with a metal ion, 2 independent species become 1, so ΔS° is least negative for this reaction as compared to the other reactions. Thus, the chelate effect occurs because the more bonds a chelating agent can form to the metal, the more favorable ΔS° is for the formation of the complex ion and the larger the formation constant.

59. $$\overset{II}{(H_2O)_5Cr} - Cl - \overset{III}{Co(NH_3)_5} \rightarrow \overset{III}{(H_2O)_5Cr} - Cl - \overset{II}{Co(NH_3)_5} \rightarrow Cr(H_2O)_5Cl^{2+} + \text{Co(II) complex}$$

Yes, this is consistent. After the oxidation, the ligands on Cr(III) won't exchange. Since Cl^- is in the coordination sphere, then it must have formed a bond to Cr(II) before the electron transfer occurred (as proposed through the formation of the intermediate).

61. a.

	$Fe(H_2O)_6^{3+}$ + H_2O	⇌	$Fe(H_2O)_5(OH)^{2+}$	+	H_3O^+
Initial	0.10 *M*		0		~0
Equil.	0.10 - *x*		*x*		*x*

$$K_a = \frac{[Fe(H_2O)_5(OH)^{2+}][H_3O^+]}{[Fe(H_2O)_6^{3+}]} = 6.0 \times 10^{-3} = \frac{x^2}{0.10 - x} \approx \frac{x^2}{0.10}$$

$x = 2.4 \times 10^{-2}$; Assumption is poor (x is 24% of 0.10). Using successive approximations:

$$\frac{x^2}{0.10 - 0.024} = 6.0 \times 10^{-3},\ x = 0.021$$

$$\frac{x^2}{0.10 - 0.021} = 6.0 \times 10^{-3},\ x = 0.022; \quad \frac{x^2}{0.10 - 0.022} = 6.0 \times 10^{-3},\ x = 0.022$$

$x = [H^+] = 0.022\ M$; pH = 1.66

b. Because of the lower charge, Fe^{2+}(aq) will not be as strong an acid as Fe^{3+}(aq). A solution of iron(II) nitrate will be less acidic (have a higher pH) than a solution with the same concentration of iron(III) nitrate.

63. a. In the lungs, there is a lot of O_2, and the equilibrium favors $Hb(O_2)_4$. In the cells, there is a deficiency of O_2, and the equilibrium favors HbH_4^{4+}.

b. CO_2 is a weak acid, $CO_2 + H_2O \rightleftharpoons HCO_3^- + H^+$. Removing CO_2 essentially decreases H^+. $Hb(O_2)_4$ is then favored, and O_2 is not released by hemoglobin in the cells. Breathing into a paper bag increases $[CO_2]$ in the blood, thus increasing $[H^+]$ which shifts the reaction left.

c. CO_2 builds up in the blood, and it becomes too acidic, driving the equilibrium to the left. Hemoglobin can't bind O_2 as strongly in the lungs. Bicarbonate ion acts as a base in water and neutralizes the excess acidity.

Challenge Problems

65.

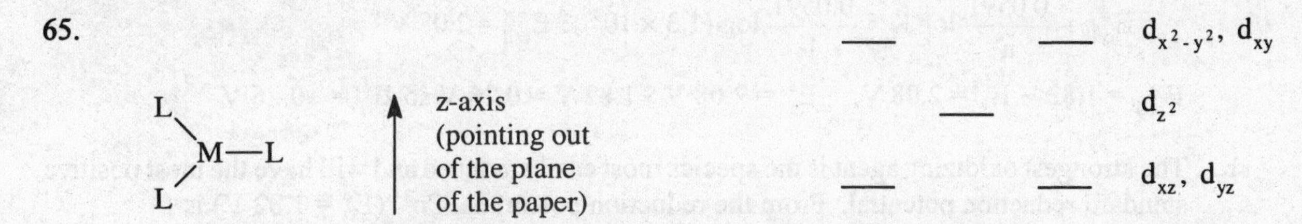

The $d_{x^2-y^2}$ and d_{xy} orbitals are in the plane of the three ligands and should be destabilized the most. The amount of destabilization should be about equal when all the possible interactions are considered. The d_{z^2} orbital has some electron density in the xy plane (the doughnut) and should be destabilized a lesser amount as compared to the $d_{x^2-y^2}$ and d_{xy} orbitals. The d_{xz} and d_{yz} orbitals have no electron density in the plane and should be lowest in energy.

67.

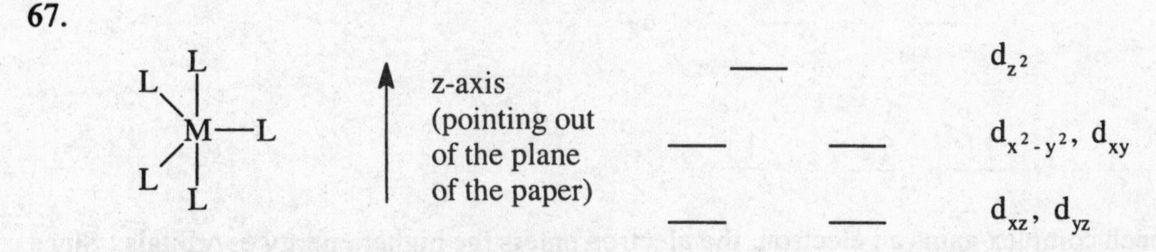

The d_{z^2} orbital will be destabilized much more than in the trigonal planar case (see Exercise 20.65). The d_{z^2} orbital has electron density on the z-axis directed at the two axial ligands. The $d_{x^2-y^2}$ and d_{xy} orbitals are in the plane of the three trigonal planar ligands and should be destabilized a lesser amount as compared to the d_{z^2} orbital; only a portion of the electron density in the $d_{x^2-y^2}$ and d_{xy} orbitals is directed at the ligands. The d_{xz} and d_{yz} orbitals will be destabilized the least since the electron density is directed between the ligands.

69. a. **Consider the following electrochemical cell:**

$$Co^{3+} + e^- \rightarrow Co^{2+} \qquad E^\circ_c = 1.82\ V$$
$$Co(en)_3{}^{2+} \rightarrow Co(en)_3{}^{3+} + e^- \qquad -E^\circ_a = ?$$
$$\overline{Co^{3+} + Co(en)_3{}^{2+} \rightarrow Co^{2+} + Co(en)_3{}^{3+} \qquad E^\circ_{cell} = 1.82 - E^\circ_a}$$

The equilibrium constant for this overall reaction is:

$$Co^{3+} + 3\ en \rightarrow Co(en)_3{}^{3+} \qquad K_1 = 2.0 \times 10^{47}$$
$$Co(en)_3{}^{2+} \rightarrow Co^{2+} + 3\ en \qquad K_2 = 1/1.5 \times 10^{12}$$
$$\overline{Co^{3+} + Co(en)_3{}^{2+} \rightarrow Co(en)_3{}^{3+} + Co^{2+} \qquad K = K_1K_2 = \frac{2.0 \times 10^{47}}{1.5 \times 10^{12}} = 1.3 \times 10^{35}}$$

From the Nernst equation for the overall reaction:

$$E^\circ_{cell} = \frac{0.0591}{n}\log K = \frac{0.0591}{1}\log(1.3\times10^{35}),\ E^\circ_{cell} = 2.08\text{ V}$$

$$E^\circ_{cell} = 1.82 - E^\circ_a = 2.08\text{ V},\ -E^\circ_a = 2.08\text{ V} - 1.82\text{ V} = 0.26\text{ V so } E^\circ_c = -0.26\text{ V}$$

b. **The strongest oxidizing agent is the species most easily reduced and will have the most positive standard reduction potential. From the reduction potentials, Co^{3+} ($E^\circ = 1.82$ V) is a much stronger oxidizing agent than $Co(en)_3^{3+}$ ($E^\circ = -0.26$ V).**

c. **In aqueous solution, Co^{3+} forms the hydrated transition metal complex, $Co(H_2O)_6^{3+}$. In both complexes, $Co(H_2O)_6^{3+}$ and $Co(en)_3^{3+}$, cobalt exists as Co^{3+} which has 6 d electrons. Assuming a strong-field case for each complex ion, then the d-orbital splitting diagram for each is:**

___ ___ e_g

↑↓ ↑↓ ↑↓ t_{2g}

When each complex gains an electron, the electron enters the higher energy e_g orbitals. Since en is a stronger field ligand than H_2O, then the d-orbital splitting is larger for $Co(en)_3^{3+}$ and it takes more energy to add an electron to $Co(en)_3^{3+}$ than to $Co(H_2O)_6^{3+}$. Therefore, it is more favorable for $Co(H_2O)_6^{3+}$ to gain an electron than for $Co(en)_3^{3+}$ to gain an electron.

71. a. $AgBr(s) \rightleftharpoons Ag^+ + Br^-$ $K_{sp} = [Ag^+][Br^-] = 5.0\times10^{-13}$

	AgBr(s)	Ag^+	Br^-
Initial	s = solubility (mol/L)	0	0
Equil.		s	s

$K_{sp} = 5.0\times10^{-13} = s^2$, $s = 7.1\times10^{-7}$ mol/L

b.

$$AgBr(s) \rightleftharpoons Ag^+ + Br^- \qquad K_{sp} = 5.0\times10^{-13}$$
$$Ag^+ + 2\,NH_3 \rightleftharpoons Ag(NH_3)_2^+ \qquad K_f = 1.7\times10^{7}$$

$$AgBr(s) + 2\,NH_3(aq) \rightleftharpoons Ag(NH_3)_2^+(aq) + Br^-(aq) \qquad K = K_{sp}\times K_f = 8.5\times10^{-6}$$

	AgBr(s) +	$2\,NH_3$	$\rightleftharpoons$ $Ag(NH_3)_2^+$ +	Br^-
Initial		3.0 *M*	0	0
	s mol/L of AgBr(s) dissolves to reach equilibrium = molar solubility			
Equil.		3.0 - 2s	s	s

$$K = \frac{[Ag(NH_3)_2^+][Br^-]}{[NH_3]^2} = \frac{s^2}{(3.0-2s)^2} = 8.5\times10^{-6} \approx \frac{s^2}{(3.0)^2},\ s = 8.7\times10^{-3}\text{ mol/L}$$

Assumption good.

c. The presence of NH_3 increases the solubility of AgBr. Added NH_3 removes Ag^+ from solution by forming the complex ion $Ag(NH_3)_2^+$. As Ag^+ is removed, more AgBr(s) will dissolve to replenish the Ag^+ concentration.

d. $$\text{mass AgBr} = 0.2500 \text{ L} \times \frac{8.7 \times 10^{-3} \text{ mol AgBr}}{\text{L}} \times \frac{187.8 \text{ g AgBr}}{\text{mol AgBr}} = 0.41 \text{ g AgBr}$$

e. Added HNO_3 will have no effect on the AgBr(s) solubility in pure water. Neither H^+ nor NO_3^- react with Ag^+ or Br^- ions. Br^- is the conjugate base of the strong acid HBr, so it is a terrible base. Added H^+ will not react with Br^- to any great extent. However, added HNO_3 will reduce the solubility of AgBr(s) in the ammonia solution. NH_3 is a weak base ($K_b = 1.8 \times 10^{-5}$). Added H^+ will react with NH_3 to form NH_4^+. As NH_3 is removed, a smaller amount of the $Ag(NH_3)_2^+$ complex ion will form resulting in a smaller amount of AgBr(s) which will dissolve.

CHAPTER TWENTY-ONE

THE NUCLEUS: A CHEMIST'S VIEW

Radioactive Decay and Nuclear Transformations

1. **All nuclear reactions must be charge balanced and mass balanced. To charge balance, balance the sum of the atomic numbers on each side of the reaction and to mass balance, balance the sum of the mass numbers on each side of the reaction.**

   a. ${}^{51}_{24}\text{Cr} + {}^{0}_{-1}\text{e} \rightarrow {}^{51}_{23}\text{V}$ b. ${}^{131}_{53}\text{I} \rightarrow {}^{0}_{-1}\text{e} + {}^{131}_{54}\text{Xe}$

   c. ${}^{32}_{15}\text{P} \rightarrow {}^{0}_{-1}\text{e} + {}^{32}_{16}\text{S}$ d. ${}^{235}_{92}\text{U} \rightarrow {}^{4}_{2}\text{He} + {}^{231}_{90}\text{Th}$

3. a. ${}^{68}_{31}\text{Ga} + {}^{0}_{-1}\text{e} \rightarrow {}^{68}_{30}\text{Zn}$ b. ${}^{62}_{29}\text{Cu} \rightarrow {}^{0}_{+1}\text{e} + {}^{62}_{28}\text{Ni}$

   c. ${}^{212}_{87}\text{Fr} \rightarrow {}^{4}_{2}\text{He} + {}^{208}_{85}\text{At}$ d. ${}^{129}_{51}\text{Sb} \rightarrow {}^{0}_{-1}\text{e} + {}^{129}_{52}\text{Te}$

5. ${}^{247}_{97}\text{Bk} \rightarrow {}^{207}_{82}\text{Pb} + ?\ {}^{4}_{2}\text{He} + ?\ {}^{0}_{-1}\text{e}$; **The change in mass number (247 - 207 = 40) is due exclusively to the alpha particles. A change in mass number of 40 requires 10 ${}^{4}_{2}$He particles to be produced. The atomic number only changes by 97 - 82 = 15. The 10 alpha particles change the atomic number by 20, so 5 ${}^{0}_{-1}$e (5 beta particles) are produced in the decay series of ^{247}Bk to ^{207}Pb.**

7. **Reference Table 21.2 of the text for potential radioactive decay processes. ^{17}F and ^{18}F contain too many protons or too few neutrons. Electron capture or positron production are both possible decay mechanisms that increase the neutron to proton ratio. Alpha particle production also increases the neutron to proton ratio, but it is not likely for these light nuclei. ^{21}F contains too many neutrons or too few protons. Beta particle production lowers the neutron to proton ratio, so we expect ^{21}F to be a β-emitter.**

9. a. ${}^{249}_{98}\text{Cf} + {}^{18}_{8}\text{O} \rightarrow {}^{263}_{106}\text{Sg} + 4\ {}^{1}_{0}\text{n}$ b. ${}^{259}_{104}\text{Rf}$; ${}^{263}_{106}\text{Sg} \rightarrow {}^{4}_{2}\text{He} + {}^{259}_{104}\text{Rf}$

Kinetics of Radioactive Decay

11. $$k = \frac{\ln 2}{t_{1/2}} = \frac{0.69315}{432.2\ \text{yr}} \times \frac{1\ \text{yr}}{365\ \text{d}} \times \frac{1\ \text{d}}{24\ \text{h}} \times \frac{1\ \text{hr}}{3600\ \text{s}} = 5.086 \times 10^{-11}\ \text{s}^{-1}$$

$$\text{Rate} = kN = 5.086 \times 10^{-11}\ \text{s}^{-1} \times 5.00\ \text{g} \times \frac{1\ \text{mol}}{241\ \text{g}} \times \frac{6.022 \times 10^{23}\ \text{nuclei}}{\text{mol}} = 6.35 \times 10^{11}\ \text{decays/s}$$

6.35×10^{11} alpha particles are emitted each second from a 5.00 g ^{241}Am sample.

13. a. $$k = \frac{\ln 2}{t_{1/2}} = \frac{0.6931}{12.8\,\text{d}} \times \frac{1\,\text{d}}{24\,\text{h}} \times \frac{1\,\text{h}}{3600\,\text{s}} = 6.27 \times 10^{-7}\ \text{s}^{-1}$$

b. $$\text{Rate} = kN = 6.27 \times 10^{-7}\ \text{s}^{-1} \times \left(28.0 \times 10^{-3}\ \text{g} \times \frac{1\ \text{mol}}{64.0\ \text{g}} \times \frac{6.022 \times 10^{23}\,\text{nuclei}}{\text{mol}}\right)$$

Rate = 1.65×10^{14} decays /s

c. **25% of the ^{64}Cu will remain after 2 half-lives (100% decays to 50% after one half-life which decays to 25% after a second half-life). Hence, 2(12.8 days) = 25.6 days is the time frame for the experiment.**

15. $$t = 59.0\ \text{yr};\ k = \frac{\ln 2}{t_{1/2}};\ \ln\left(\frac{N}{N_o}\right) = -kt = \frac{-0.6931 \times 59.0\ \text{yr}}{28.8\ \text{yr}} = -1.42,\ \left(\frac{N}{N_o}\right) = e^{-1.42} = 0.242$$

24.2% of the ^{90}Sr remains as of July 16, 2004.

17. $$175\ \text{mg}\ Na_3{}^{32}PO_4 \times \frac{32.0\ \text{mg}\ ^{32}P}{165.0\ \text{mg}\ Na_3{}^{32}PO_4} = 33.9\ \text{mg}\ ^{32}P;\ k = \frac{\ln 2}{t_{1/2}}$$

$$\ln\left(\frac{N}{N_o}\right) = -kt = \frac{-0.6931\ t}{t_{1/2}},\ \ln\left(\frac{m}{33.9\ \text{mg}}\right) = \frac{-0.6931\ (35.0\ \text{d})}{14.3\ \text{d}};$$ **Carrying extra sig. figs.:**

$$\ln(m) = -1.696 + 3.523 = 1.827,\ m = e^{1.827} = 6.22\ \text{mg}\ ^{32}P\ \text{remains}$$

19. $$t_{1/2} = 5730\ \text{yr};\ k = (\ln 2)/t_{1/2};\ \ln(N/N_o) = -kt;\ \ln\frac{15.1}{15.3} = \frac{-(\ln 2)\ t}{5730\ \text{yr}},\ t = 109\ \text{yr}$$

No; From ^{14}C dating, the painting was produced (at the earliest) during the late 1800s.

21. $$\ln\left(\frac{N}{N_o}\right) = -kt = \frac{-(\ln 2)\ t}{12.3\ \text{yr}},\ \ln\left(\frac{0.17 \times N_o}{N_o}\right) = -5.64 \times 10^{-2}\ t,\ t = 31.4\ \text{yr}$$

It takes 31.4 yr for the tritium to decay to 17% of the original amount. Hence, the watch stopped fluorescing enough to be read in 1975 (1944 + 31.4).

23. Assuming 1.000 g ^{238}U present in a sample, then 0.688 g ^{206}Pb is present. Since 1 mol ^{206}Pb is produced per mol ^{238}U decayed, then:

$$^{238}\text{U decayed} = 0.688\ \text{g Pb} \times \frac{1\ \text{mol Pb}}{206\ \text{g Pb}} \times \frac{1\ \text{mol U}}{\text{mol Pb}} \times \frac{238\ \text{g U}}{\text{mol U}} = 0.795\ \text{g}\ ^{238}\text{U}$$

Original mass ^{238}U present = 1.000 g + 0.795 g = 1.795 g ^{238}U

$$\ln\left(\frac{N}{N_o}\right) = -kt = \frac{-(\ln 2)\,t}{t_{1/2}},\ \ln\left(\frac{1.000\ \text{g}}{1.795\ \text{g}}\right) = \frac{-0.693\ (t)}{4.5 \times 10^9\ \text{yr}},\quad t = 3.8 \times 10^9\ \text{yr}$$

Energy Changes in Nuclear Reactions

25. $$\Delta E = \Delta mc^2,\ \Delta m = \frac{\Delta E}{c^2} = \frac{3.9 \times 10^{23}\ \text{kg m}^2/\text{s}^2}{(3.00 \times 10^8\ \text{m/s})^2} = 4.3 \times 10^6\ \text{kg}$$

The sun loses 4.3×10^6 kg of mass each second. Note: 1 J = 1 kg m^2/s^2

27. From the table at the back of the text, the mass of a proton = 1.00728 amu, the mass of a neutron = 1.00866 amu, and the mass of an electron = 5.486×10^{-4} amu.

Mass of nucleus = mass of atom - mass of electrons = 55.9349 - 26(0.0005486) = 55.9206 amu

$$26\,{}^{1}_{1}\text{H} + 30\,{}^{1}_{0}\text{n} \rightarrow {}^{56}_{26}\text{Fe};\ \Delta m = 55.9206\ \text{amu} - [26(1.00728) + 30(1.00866)]\ \text{amu} = -0.5285\ \text{amu}$$

$$\Delta E = \Delta mc^2 = -0.5285\ \text{amu} \times \frac{1.6605 \times 10^{-27}\ \text{kg}}{\text{amu}} \times (2.9979 \times 10^8\ \text{m/s})^2 = -7.887 \times 10^{-11}\ \text{J}$$

$$\frac{\text{binding energy}}{\text{nucleon}} = \frac{7.887 \times 10^{-11}\ \text{J}}{56\ \text{nucleons}} = 1.408 \times 10^{-12}\ \text{J/nucleon}$$

29. Let m_e = mass of electron; For ^{12}C (6e, 6p, 6n): mass defect = Δm = mass of ^{12}C nucleus -[mass of 6 protons + mass of 6 neutrons]. Note: Atomic masses given include the mass of the electrons.

$$\Delta m = 12.00000\ \text{amu} - 6\,m_e - [6(1.00782 - m_e) + 6(1.00866)];\ \text{Mass of electrons cancel.}$$

$$\Delta m = 12.00000 - [6(1.00782) + 6(1.00866)] = -0.09888\ \text{amu}$$

$$\Delta E = \Delta mc^2 = -0.09888\ \text{amu} \times \frac{1.6605 \times 10^{-27}\ \text{kg}}{\text{amu}} \times (2.9979 \times 10^8\ \text{m/s})^2 = -1.476 \times 10^{-11}\ \text{J}$$

$$\frac{\text{BE}}{\text{nucleon}} = \frac{1.476 \times 10^{-11}\ \text{J}}{12\ \text{nucleons}} = 1.230 \times 10^{-12}\ \text{J/nucleon}$$

For ^{235}U (92e, 92p, 143n):

$$\Delta m = 235.0439 - 92\ m_e - [92(1.00782 - m_e) + 143(1.00866)] = -1.9139\ \text{amu}$$

$$\Delta E = \Delta mc^2 = -1.9139\ \text{amu} \times \frac{1.66054 \times 10^{-27}\ \text{kg}}{\text{amu}} \times (2.99792 \times 10^8\ \text{m/s})^2 = -2.8563 \times 10^{-10}\ \text{J}$$

$$\frac{\text{BE}}{\text{nucleon}} = \frac{2.8563 \times 10^{-10}\ \text{J}}{235\ \text{nucleons}} = 1.2154 \times 10^{-12}\ \text{J/nucleon}$$

Since ^{26}Fe is the most stable know nucleus, then the binding energy per nucleon for ^{56}Fe (1.408×10^{-12} J/nucleon) will be larger than that of ^{12}C or ^{235}U (see Figure 21.9 of the text).

31. $^{2}_{1}\text{H} + ^{3}_{1}\text{H} \rightarrow ^{4}_{2}\text{He} + ^{1}_{0}\text{n}$; Mass of electrons cancel when determining Δm for this nuclear reaction.

$$\Delta m = [4.00260 + 1.00866 - (2.01410 + 3.01605)]\ \text{amu} = -1.889 \times 10^{-2}\ \text{amu}$$

For the production of one mol of $^{4}_{2}$He: $\Delta m = -1.889 \times 10^{-2}\ \text{g} = -1.889 \times 10^{-5}\ \text{kg}$

$$\Delta E = \Delta mc^2 = -1.889 \times 10^{-5}\ \text{kg} \times (2.9979 \times 10^8\ \text{m/s})^2 = -1.698 \times 10^{12}\ \text{J/mol}$$

For 1 nuclei of $^{4}_{2}$He:

$$\frac{-1.698 \times 10^{12}\ \text{J}}{\text{mol}} \times \frac{1\ \text{mol}}{6.0221 \times 10^{23}\ \text{nuclei}} = -2.820 \times 10^{-12}\ \text{J/nuclei}$$

Detection, Uses, and Health Effects of Radiation

33. The Geiger-Müller tube has a certain response time. After the gas in the tube ionizes to produce a "count," some time must elapse for the gas to return to an electrically neutral state. The response of the tube levels out because at high activities, radioactive particles are entering the tube faster than the tube can respond to them.

35. Fission: Splitting of a heavy nucleus into two (or more) lighter nuclei.

Fusion: Combining two light nuclei to form a heavier nucleus.

The maximum binding energy per nucleon occurs at Fe. Nuclei smaller than Fe become more stable by fusing to form heavier nuclei closer in mass to Fe. Nuclei larger than Fe form more stable nuclei by splitting to form lighter nuclei closer in mass to Fe.

37. moderator: Slows the neutrons to increase the efficiency of the fission reaction.

control rods: Absorbs neutrons to slow or halt the fission reaction.

39. Even though gamma rays penetrate human tissue very deeply, they are very small and cause only occasional ionization of biomolecules. Alpha particles, because they are much more massive, are very effective at causing ionization of biomolecules and produce a dense trail of damage once they get inside an organism.

41. Magnetic fields are needed to contain fusion reactions. Superconductors allow the production of very strong magnetic fields by their ability to carry large electric currents. Stronger magnetic fields should be more capable of containing fusion reactions.

43. A nonradioactive substance can be put in equilibrium with a radioactive substance. The two materials can then be checked to see whether all the radioactivity remains in the original material or if it has been scrambled by the equilibrium.

45. All evolved oxygen in O_2 comes from water and not from carbon dioxide.

47. (i) and (ii) mean that Pu is not a significant threat outside the body. Our skin is sufficient to keep out the α particles. If Pu gets inside the body, it is easily oxidized to Pu^{4+} (iv), which is chemically similar to Fe^{3+} (iii). Thus, Pu^{4+} will concentrate in tissues where Fe^{3+} is found, including the bone marrow where red blood cells are produced. Once inside the body, α particles can cause considerable damage.

Additional Exercises

49. $$20{,}000 \text{ ton TNT} \times \frac{4 \times 10^9 \text{ J}}{\text{ton TNT}} \times \frac{1 \text{ mol } ^{235}\text{U}}{2 \times 10^{13} \text{ J}} \times \frac{235 \text{ g } ^{235}\text{U}}{\text{mol } ^{235}\text{U}} = 940 \text{ g } ^{235}\text{U} \approx 900 \text{ g } ^{235}\text{U}$$

This assumes that all of the ^{235}U undergoes fission.

51. The only product in the fast equilibrium step is assumed to be $N^{16}O\,^{18}O_2$, where N is the central atom. However, this is a reversible reaction where $N^{16}O\,^{18}O_2$ will decompose to NO and O_2. Since any two oxygen atoms can leave $N^{16}O\,^{18}O_2$ to form O_2, then we would expect (at equilibrium) 1/3 of the NO present in this fast equilibrium step to be $N^{16}O$ and 2/3 to be $N^{18}O$. In the second step (the slow step), the intermediate $N^{16}O\,^{18}O_2$ reacts with the scrambled NO to form the NO_2 product, where N is the central atom in NO_2. Any one of the three oxygen atoms can be transferred from $N^{16}O\,^{18}O_2$ to NO when the NO_2 product is formed. The distribution of ^{18}O in the product can best be determined by forming a probability table.

	$N^{16}O$ (1/3)	$N^{18}O$ (2/3)
^{16}O (1/3) from $N^{16}O\,^{18}O_2$	$N^{16}O_2$ (1/9)	$N^{18}O^{16}O$ (2/9)
^{18}O (2/3) from $N^{16}O\,^{18}O_2$	$N^{16}O^{18}O$ (2/9)	$N^{18}O_2$ (4/9)

From the probability table, 1/9 of the NO_2 is $N^{16}O_2$, 4/9 of the NO_2 is $N^{18}O_2$ and 4/9 of the NO_2 is $N^{16}O^{18}O$ (2/9 + 2/9 = 4/9). Note: $N^{16}O^{18}O$ is the same as $N^{18}O^{16}O$. In addition, $N^{16}O^{18}O_2$ is not the only NO_3 intermediate formed; $N^{16}O_2\,^{18}O$ and $N^{18}O_3$ can also form in the fast equilibrium first step.

However, the distribution of ^{18}O in the NO_2 product is the same as calculated above, even when these other NO_3 intermediates are considered.

53. Assuming that the radionuclide is long lived enough such that no significant decay occurs during the time of the experiment, the total counts of radioactivity injected are:

$$0.10 \text{ mL} \times \frac{5.0 \times 10^3 \text{ cpm}}{\text{mL}} = 5.0 \times 10^2 \text{ cpm}$$

Assuming that the total activity is uniformly distributed only in the rats blood, the blood volume is:

$$V \times \frac{48 \text{ cpm}}{\text{mL}} = 5.0 \times 10^2 \text{ cpm}, \ V = 10.4 \text{ mL} = 10. \text{ mL}$$

Challenge Problems

55. $$\text{mol I} = \frac{33 \text{ counts}}{\text{min}} \times \frac{1 \text{ mol I} \bullet \text{min}}{5.0 \times 10^{11} \text{ counts}} = 6.6 \times 10^{-11} \text{ mol I}$$

$$[I^-] = \frac{6.6 \times 10^{-11} \text{ mol I}^-}{0.150 \text{ L}} = 4.4 \times 10^{-10} \text{ mol/L}$$

	$Hg_2I_2(s)$ →	$Hg_2^{2+}(aq)$ +	$2\ I^-(aq)$	$K_{sp} = [Hg_2^{2+}][I^-]^2$
Initial	s = solubility (mol/L)	0	0	
Equil.		s	$2s$	

From the problem, $2s = 4.4 \times 10^{-10}$ mol/L, $s = 2.2 \times 10^{-10}$ mol/L

$K_{sp} = (s)(2s)^2 = (2.2 \times 10^{-10})(4.4 \times 10^{-10})^2 = 4.3 \times 10^{-29}$

57. a. For a gas, $u_{avg} = \sqrt{8\,RT/\pi M}$ where M is the molar mass in kg. From the equation, the lighter the gas molecule, the faster the average velocity. Therefore, $^{235}UF_6$ will have the greater average velocity at a certain temperature since it is the lighter molecule.

b. From Graham's law (see Section 5.7 of the text):

$$\frac{\text{diffusion rate for } ^{235}UF_6}{\text{diffusion rate for } ^{238}UF_6} = \sqrt{\frac{M(^{238}UF_6)}{M(^{235}UF_6)}} = \sqrt{\frac{352.05 \text{ g/mol}}{349.03 \text{ g/mol}}} = 1.0043$$

Each diffusion step increases the $^{235}UF_6$ concentration by a factor of 1.0043. To determine the number of steps to get to the desired 3.00% ^{235}U, we use the following formula:

$$\frac{0.700\ ^{235}UF_6}{99.3\ ^{238}UF_6} \times (1.0043)^N = \frac{3.00\ ^{235}UF_6}{97.0\ ^{238}UF_6}$$

original ratio **final ratio**

where N represents the number of steps required.

Solving (and carrying extra sig. figs.):

$$(1.0043)^N = \frac{297.9}{67.9} = 4.387,\ N \log 1.0043 = \log 4.387$$

$$N = \frac{0.6422}{1.863 \times 10^{-3}} = 345 \text{ steps}$$

Thus, 345 steps are required to obtain the desired enrichment.

c. $$\frac{^{235}UF_6}{^{238}UF_6} \times (1.0043)^{100} = \frac{1526}{1.000 \times 10^5 - 1526},\quad \frac{^{235}UF_6}{^{238}UF_6} \times 1.5358 = \frac{1526}{98500}$$

original ratio **final ratio**

$$\frac{^{235}UF_6}{^{238}UF_6} = 1.01 \times 10^{-2} = \text{initial } ^{235}U \text{ to } ^{238}U \text{ atom ratio}$$

CHAPTER TWENTY-TWO

ORGANIC CHEMISTRY

Hydrocarbons

1. **A difficult task in this problem is recognizing different compounds from compounds that differ by rotations about one or more C–C bonds (called conformations). The best way to distinguish different compounds from conformations is to name them. Different name = different compound; same name = same compound so it is not an isomer, but instead, is a conformation.**

a.

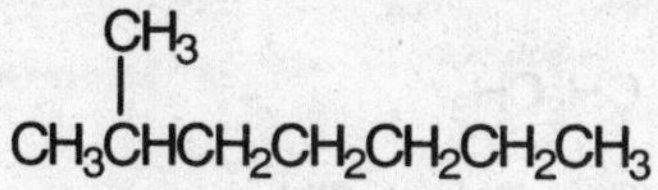

$CH_3CH(CH_3)CH_2CH_2CH_2CH_2CH_3$

2-methylheptane

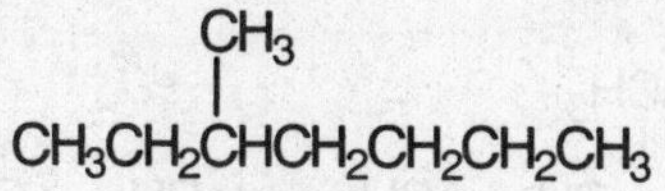

$CH_3CH_2CH(CH_3)CH_2CH_2CH_2CH_3$

3-methylheptane

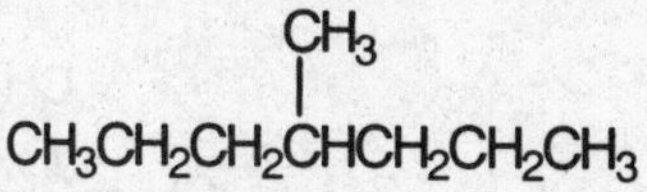

$CH_3CH_2CH_2CH(CH_3)CH_2CH_2CH_3$

4-methylheptane

b.

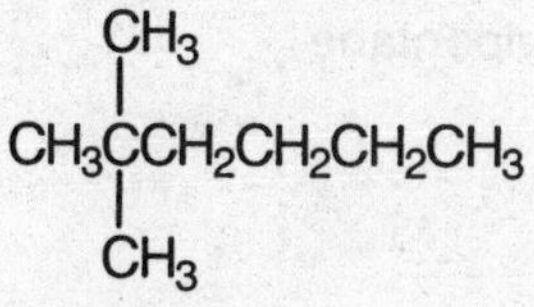

$CH_3C(CH_3)_2CH_2CH_2CH_2CH_3$

2,2-dimethylhexane

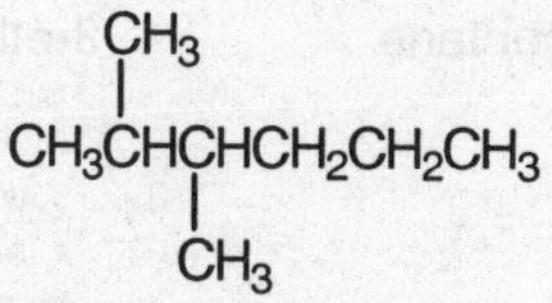

$CH_3CH(CH_3)CH(CH_3)CH_2CH_2CH_3$

2,3-dimethylhexane

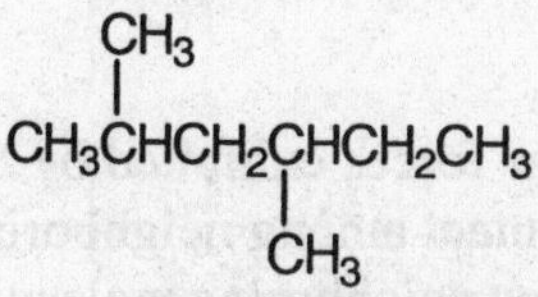

$CH_3CH(CH_3)CH_2CH(CH_3)CH_2CH_3$

2,4-dimethylhexane

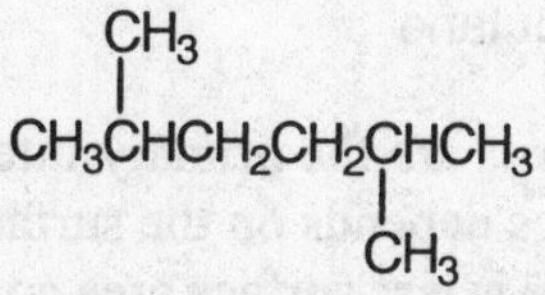

$CH_3CH(CH_3)CH_2CH_2CH(CH_3)CH_3$

2,5-dimethylhexane

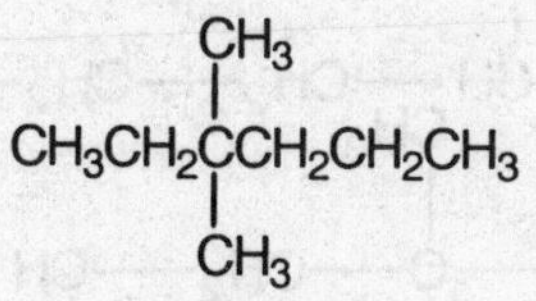

$CH_3CH_2C(CH_3)_2CH_2CH_2CH_3$

3,3-dimethylhexane

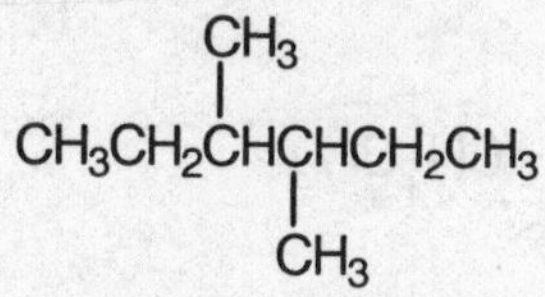

$CH_3CH_2CH(CH_3)CH(CH_3)CH_2CH_3$

3,4-dimethylhexane

$CH_3CH_2CH(CH_2CH_3)CH_2CH_2CH_3$

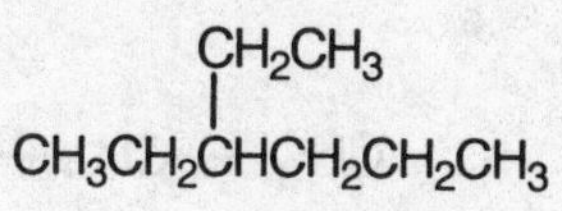

3-ethylhexane

c.

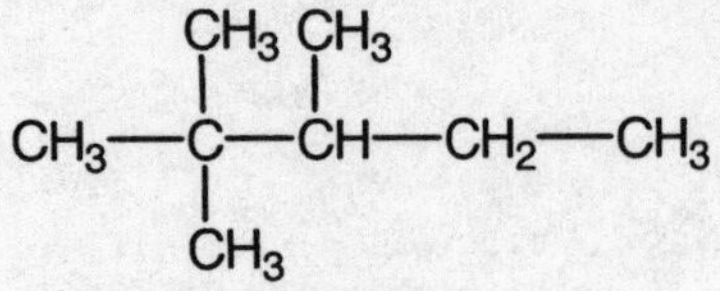

CH_3 CH_3

$CH_3-C-CH-CH_2-CH_3$

CH_3

2,2,3-trimethylpentane

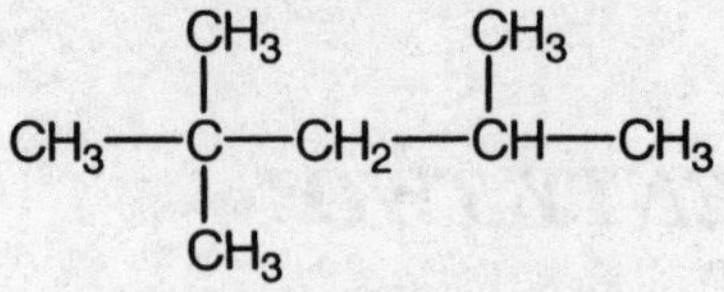

CH_3 CH_3

$CH_3-C-CH_2-CH-CH_3$

CH_3

2,2,4-trimethylpentane

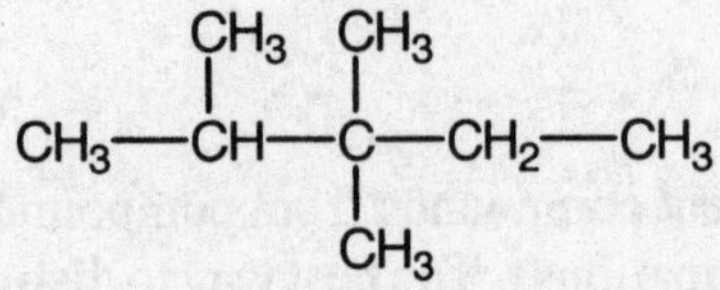

CH_3 CH_3

$CH_3-CH-C-CH_2-CH_3$

CH_3

2,3,3-trimethylpentane

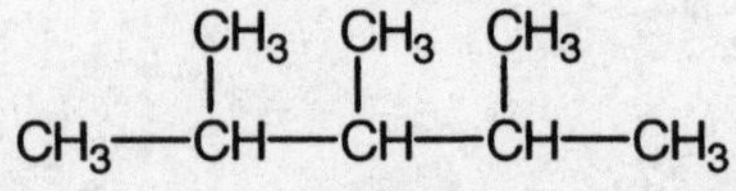

CH_3 CH_3 CH_3

$CH_3-CH-CH-CH-CH_3$

2,3,4-trimethylpentane

CH_3 CH_2CH_3

$CH_3-CH-CH-CH_2-CH_3$

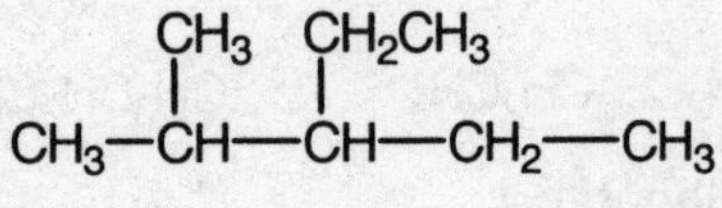

3-ethyl-2-methylpentane

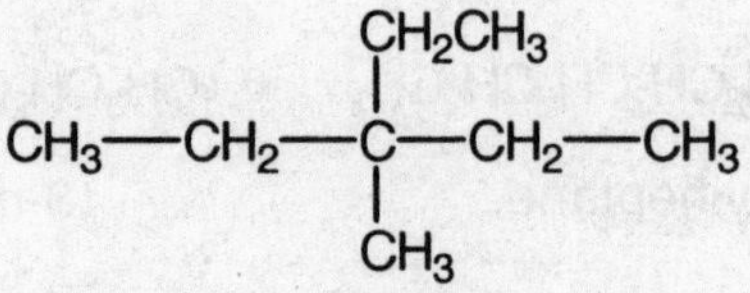

CH_2CH_3

$CH_3-CH_2-C-CH_2-CH_3$

CH_3

3-ethyl-3-methylpentane

d.

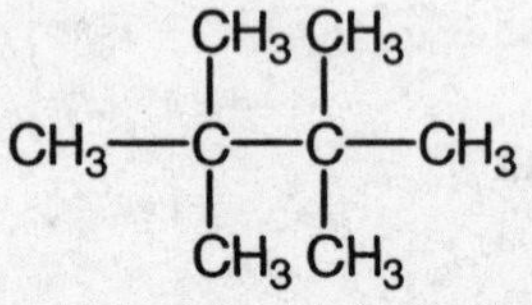

CH_3 CH_3

$CH_3-C-C-CH_3$

CH_3 CH_3

2,2,3,3-tetramethylbutane

3. **London dispersion (LD) forces are the primary intermolecular forces exhibited by hydrocarbons. The strength of the LD forces depends on the surface area contact among neighboring molecules. As branching increases, there is less surface area contact among neighboring molecules, leading to weaker LD forces and lower boiling points.**

5. a.

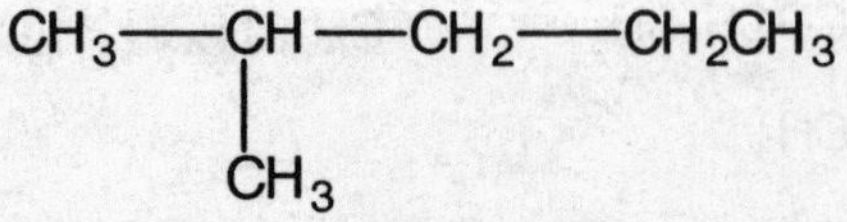

$CH_3-CH-CH_2-CH_2CH_3$

CH_3

b.

CH_3

$CH_3-CH-CH_2-CH_2-CH_3$

CH_3

$CH_3-C-CH_2-CH-CH_3$

CH_3 CH_3

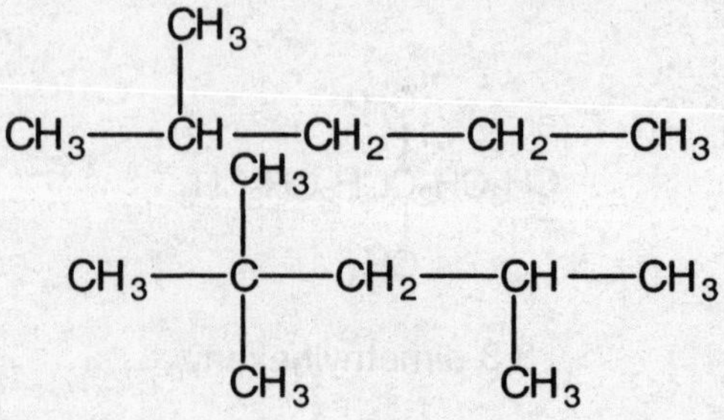

c.

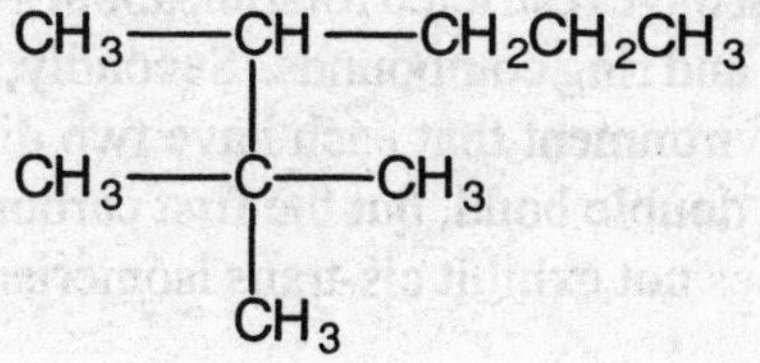

```
CH3—CH—CH2CH2CH3
     |
CH3—C—CH3
     |
    CH3
```

d. The longest chain is 6 carbons long.

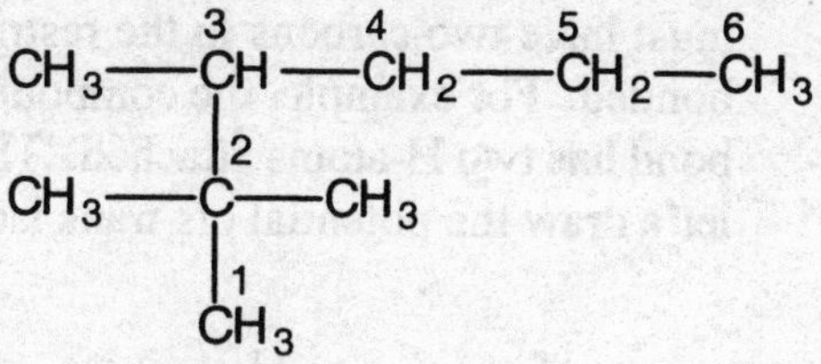

```
      3     4      5      6
CH3—CH—CH2—CH2—CH3
     |2
CH3—C—CH3
     |1
    CH3
```

2,2,3-trimethylhexane

7. **a. 2,2,4-trimethylhexane b. 5-methylnonane c. 2,2,4,4-tetramethylpentane**

d. 3-ethyl-3-methyloctane

Note: For alkanes always identify the longest carbon chain for the base name first, then number the carbons to give the lowest overall numbers for the substituent groups.

9. **a. 1-butene**
b. 2-methyl-2-butene
c. 2,5-dimethyl-3-heptene
d. 2,3-dimethyl-1-pentene
e. 1-ethyl-3-methylcyclopentene (double bond assumed between C_1 and C_2)
f. 4-ethyl-3-methylcyclopentene
g. 4-methyl-2-pentyne

Note: Multiple bonds are assigned the lowest number possible.

11. **a. 1,3-dichlorobutane**
b. 1,1,1-trichlorobutane
c. 2,3-dichloro-2,4-dimethylhexane
d. 1,2-difluoroethane
e. 3-iodo-1-butene
f. 2-bromotoluene (or 1-bromo-2-methylbenzene)
g. 1-bromo-2-methylcyclohexane
h. 4-bromo-3-methylcyclohexene (double bond assumed between C_1 and C_2)

13. **isopropylbenzene or 2-phenylpropane**

Isomerism

15. **structural isomers: Same formula but different bonding, either in the kinds of bonds present or the way in which the bonds connect atoms to each other.**

geometrical isomers: Same formula and same bonds but differ in the arrangement of atoms in space about a rigid bond or ring.

17. To exhibit cis-trans isomerism, a compound must first have restricted rotation about a carbon-carbon bond. This occurs in compounds with double bonds and ring compounds. Secondly, the compound must have two carbons in the restricted rotation environment that each have two different groups bonded. For example, the compound in 22.9a has a double bond, but the first carbon in the double bond has two H-atoms attached. This compound does not exhibit cis-trans isomerism. To see this, let's draw the potential cis-trans isomers:

These are the same compounds; they only differ by a simply rotation of the molecule. Therefore, they are not isomers of each other, but instead they are the same compound. The only compounds that fulfill the restricted rotation requirement **and** have two different groups attached to carbons in the restricted rotation are compounds c and f. The cis-trans isomerism for these follows.

c.

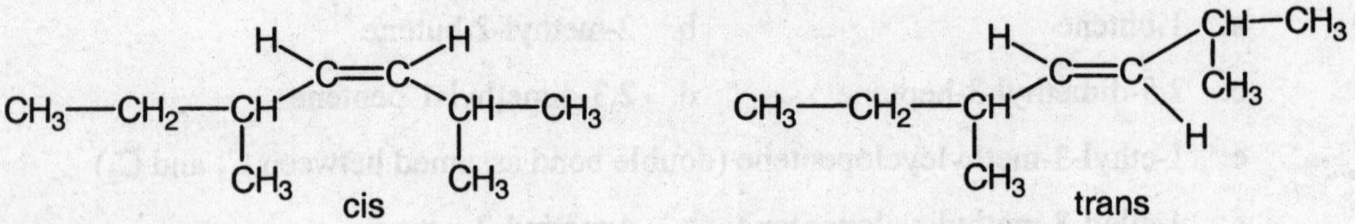

cis trans

f.

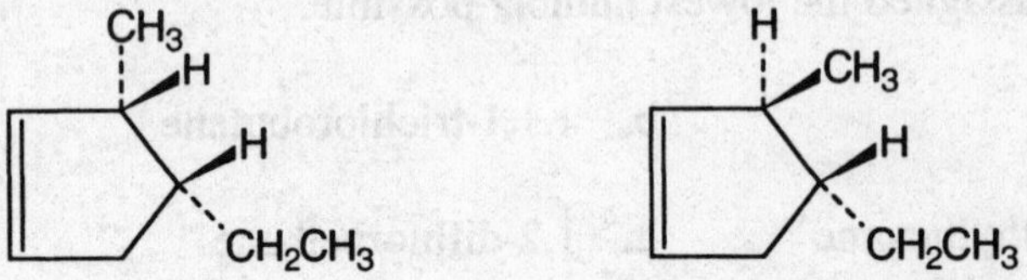

cis trans

= out of plane of paper; ----- = into plane of paper

19. The cis isomer has the CH_3 groups on the same side of the ring. The trans isomer has the CH_3 groups on opposite sides of the ring.

cis trans

21. **To help distinguish the different isomers, we will name them.**

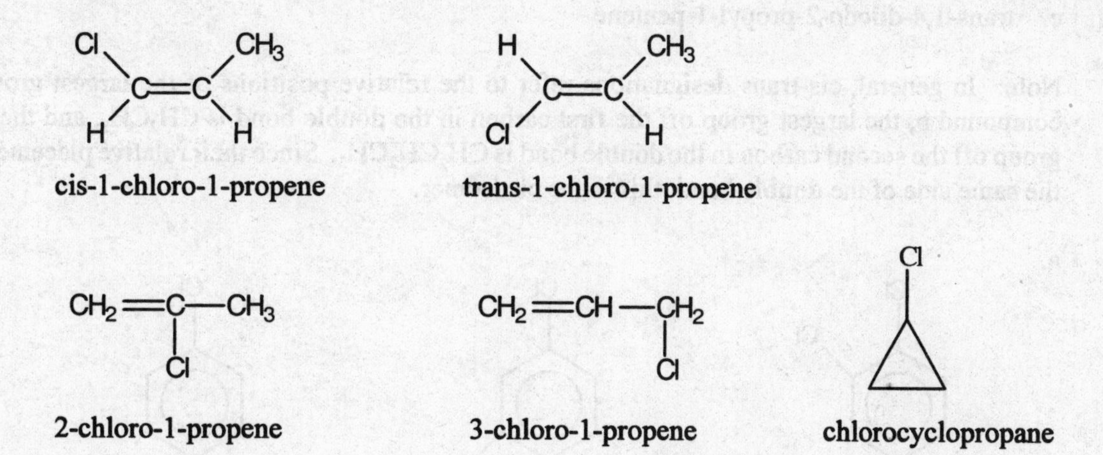

23. **C_5H_{10} has the general formula for alkenes, C_nH_{2n}. To distinguish the different isomers from each other, we will name them. Each isomer must have a different name.**

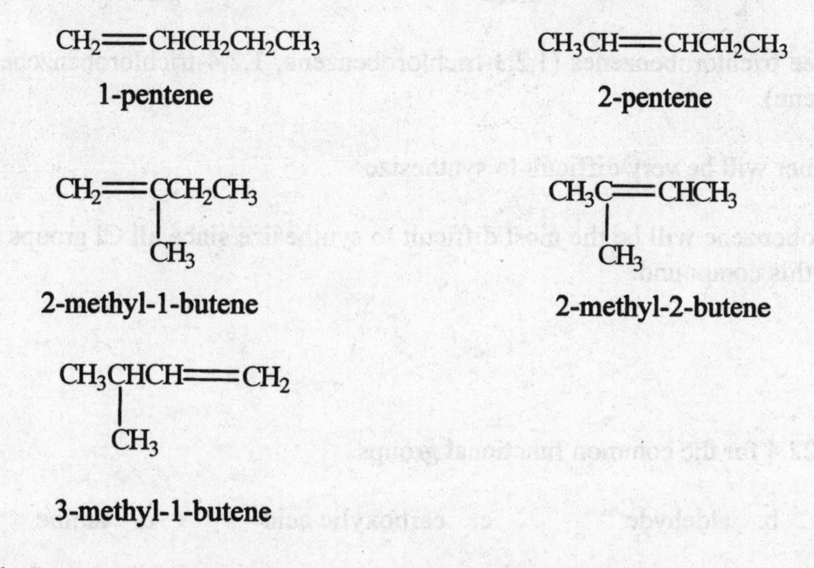

Only 2-pentene exhibits cis-trans isomerism. The isomers are:

H_3C, CH_2CH_3, C=C, H, H — **cis**

H, CH_2CH_3, C=C, H_3C, H — **trans**

The other isomers of C_5H_{10} do not contain carbons in the double bond that each contain two different groups attached.

25. a. cis-1-bromo-1-propene b. cis-4-ethyl-3-methyl-3-heptene

c. trans-1,4-diiodo-2-propyl-1-pentene

Note: In general, cis-trans designations refer to the relative positions of the largest groups. In compound b, the largest group off the first carbon in the double bond is CH_2CH_3, and the largest group off the second carbon in the double bond is $CH_2CH_2CH_3$. Since their relative placement is on the same side of the double bond, this is the cis isomer.

27. a.

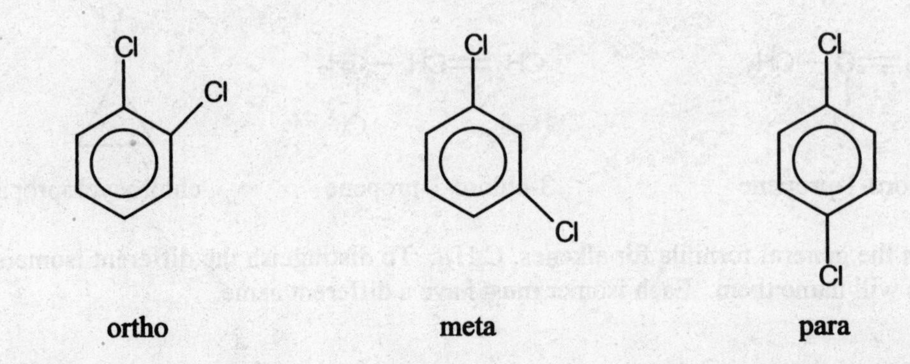

b. There are three trichlorobenzenes (1,2,3-trichlorobenzene, 1,2,4-trichlorobenzene and 1,3,5-trichlorobenzene).

c. The meta isomer will be very difficult to synthesize.

d. 1,3,5-trichlorobenzene will be the most difficult to synthesize since all Cl groups are meta to each other in this compound.

Functional Groups

29. Reference Table 22.4 for the common functional groups.

a. ketone b. aldehyde c. carboxylic acid d. amine

31. a.

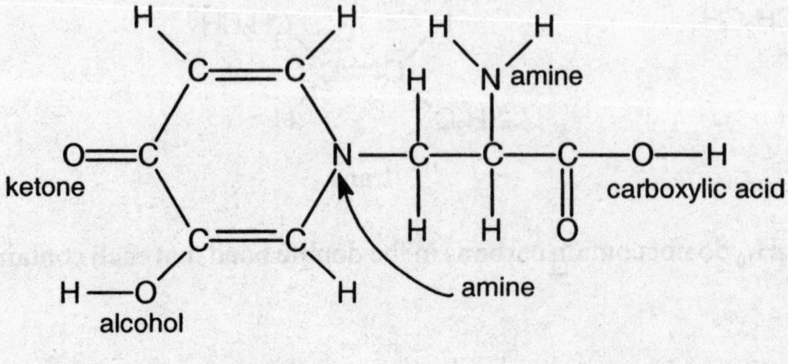

amine
ketone
carboxylic acid
amine
alcohol

b. 5 carbons in ring and the carbon in $-CO_2H$: sp^2; the other two carbons: sp^3

c. 24 sigma bonds; 4 pi bonds

33. a. 3-chloro-1-butanol: Since the carbon containing the OH group is bonded to just 1 other carbon (1 R group), then this is a primary alcohol.

b. 3-methyl-3-hexanol; Since the carbon containing the OH group is bonded to three other carbons (3 R groups), then this is a tertrary alcohol

c. 2-methylcyclopentanol; Secondary alcohol (2 R groups bonded to carbon containing the OH group); Note: In ring compounds, the alcohol group is assumed to be bonded to C_1 so the number designation is commonly omitted for the alcohol group.

35.

OH
|
$CH_3CH_2CH_2CH_2CH_2$

1-pentanol

OH
|
$CH_3CH_2CH_2CHCH_3$

2-pentanol

OH
|
$CH_3CH_2CHCH_2CH_3$

3-pentanol

OH
|
$CH_3CH_2CHCH_2$
|
CH_3

2-methyl-1-butanol

OH
|
$CH_3CHCH_2CH_2$
|
CH_3

3-methyl-1-butanol

OH
|
$CH_3CH_2CCH_3$
|
CH_3

2-methyl-2-butanol

OH
|
$CH_3CHCHCH_3$
|
CH_3

3-methyl-2-butanol

CH_3 OH
| |
$CH_3—C—CH_2$
|
CH_3

2,2-dimethyl-1-propanol

There are six isomeric ethers with formula $C_5H_{12}O$. The structures follow.

$CH_3—O—CH_2CH_2CH_2CH_3$

CH_3
|
$CH_3—O—CHCH_2CH_3$

CH_3
|
$CH_3—O—CH_2CHCH_3$

CH_3
|
$CH_3—O—C—CH_3$
|
CH_3

$CH_3CH_2—O—CH_2CH_2CH_3$

CH_3
|
$CH_3CH_2—O—CH$
|
CH_3

37. a. **4,5-dichloro-3-hexanone** b. **2,3-dimethylpentanal**

c. **3-methylbenzaldehyde or m-methylbenzaldehyde**

39. a. trans-2-butene: $CH_3(H)C{=}C(H)CH_3$ (trans), formula = C_4H_8

(cyclobutane, C_4H_8: each ring carbon bonded to two H) or (methylcyclopropane: ring with H, H; H, CH_3; H, H)

b. propanoic acid: $CH_3CH_2\overset{O}{\overset{\|}{C}}{-}OH$, formula = $C_3H_6O_2$

$CH_3\overset{O}{\overset{\|}{C}}{-}O{-}CH_3$ or $H\overset{O}{\overset{\|}{C}}{-}O{-}CH_2CH_3$

c. butanal: $CH_3CH_2CH_2\overset{O}{\overset{\|}{C}}H$, formula = C_4H_8O

$CH_3CH_2\overset{O}{\overset{\|}{C}}CH_3$

d. butylamine: $CH_3CH_2CH_2CH_2NH_2$, formula = $C_4H_{11}N$:

A secondary amine has two R groups bonded to N.

$CH_3{-}N(CH_2CH_2CH_3){-}H$ or $CH_3{-}N(CH_3CHCH_3){-}H$ or $CH_3CH_2{-}N(CH_2CH_3){-}H$

e. A tertiary amine has three R groups bonded to N. (See answer d for structure of butylamine.)

$CH_3{-}N(CH_2CH_3){-}CH_3$

f. 2-methyl-2-propanol: $CH_3C(CH_3)(OH)CH_3$, formula = $C_4H_{10}O$

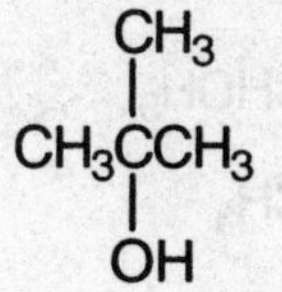

$CH_3—O—CH_2CH_2CH_3$ or $CH_3—O—CH(CH_3)_2$ or $CH_3CH_2—O—CH_2CH_3$

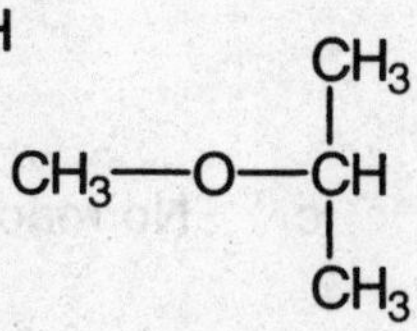

g. A secondary alcohol has two R groups attached to the carbon bonded to the OH group. (See answer f for the structure of 2-methyl-2-propanol.)

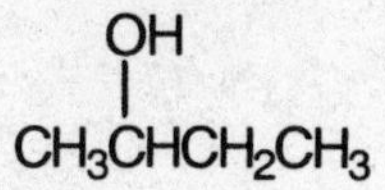

$CH_3CH(OH)CH_2CH_3$

Reactions of Organic Compounds

41. **substitution: An atom or group is replaced by another atom or group.**

e.g., H in benzene is replaced by Cl. $C_6H_6 + Cl_2 \xrightarrow{\text{catalyst}} C_6H_5Cl + HCl$

addition: Atoms or groups are added to a molecule.

e.g., Cl_2 adds to ethene. $CH_2{=}CH_2 + Cl_2 \rightarrow CH_2Cl-CH_2Cl$

43.

a. 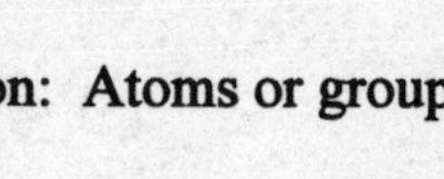 $CH_3CH(H)—CH(H)CH_3$

b. 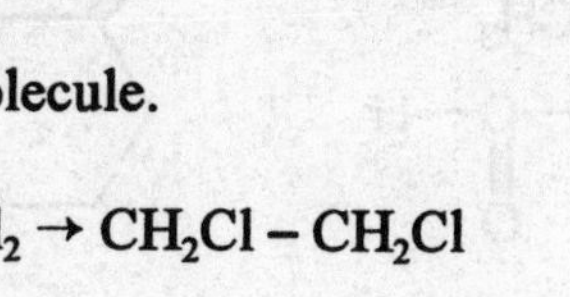 $CH_2(Cl)—CH(Cl)(CH_3)CH(Cl)—CH(Cl)(CH_3)$

c. 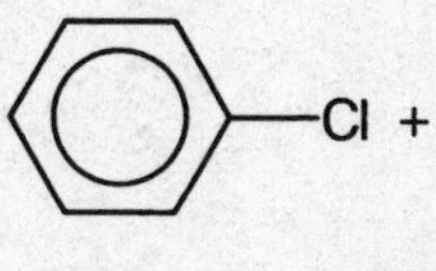 C_6H_5—Cl + HCl

d. $C_4H_8(g) + 6\ O_2(g) \rightarrow 4\ CO_2(g) + 4\ H_2O(g)$

45. **Primary alcohols (a, d and f) are oxidized to aldehydes which can be oxidized further to carboxylic acids. Secondary alcohols (b, e and f) are oxidized to ketones, and tertiary alcohols (c and f) do not undergo this type of oxidation reaction. Note that compound f contains a primary, secondary and tertiary alcohol. For the primary alcohols (a, d and f), we listed both the aldehyde and the carboxylic acid as possible products.**

a. 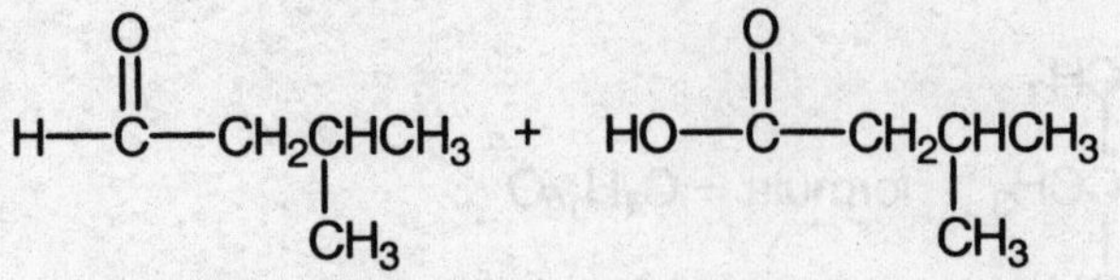 $H-C(=O)-CH_2CH(CH_3)CH_3$ + $HO-C(=O)-CH_2CH(CH_3)CH_3$

b. 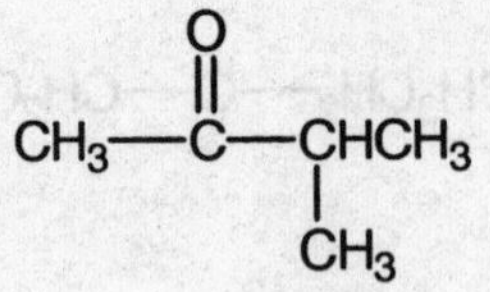 $CH_3-C(=O)-CH(CH_3)CH_3$

c. No reaction

d. 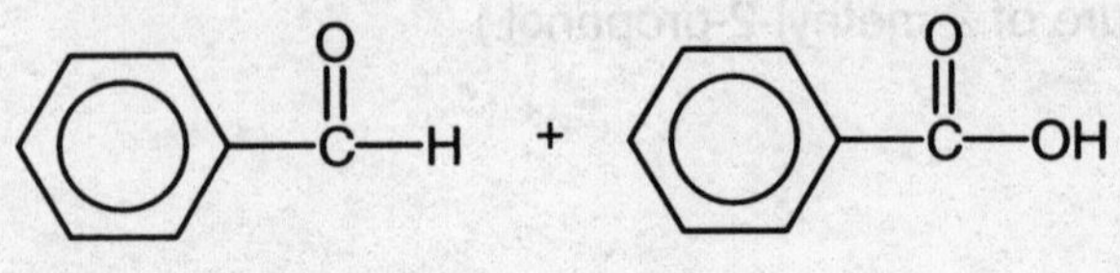 $C_6H_5-C(=O)-H$ + $C_6H_5-C(=O)-OH$

e. 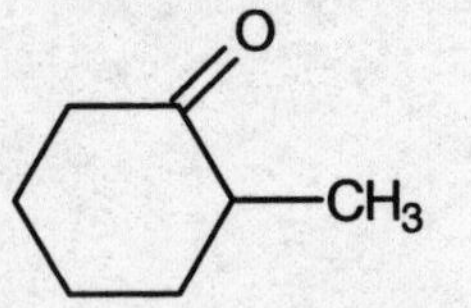 2-methylcyclohexanone (CH_3 on the carbon adjacent to C=O)

f. 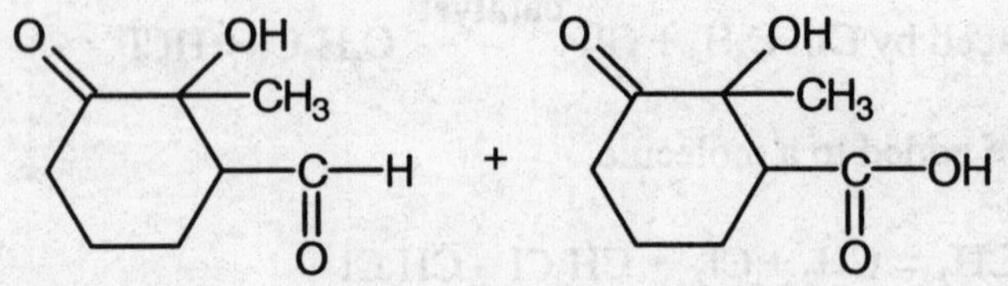 2-hydroxy-2-methyl-3-oxocyclohexane-1-carbaldehyde ($-C(=O)-H$) + corresponding carboxylic acid ($-C(=O)-OH$)

47. a. $CH_3CH{=}CH_2 + Br_2 \rightarrow CH_3CHBrCH_2Br$ **(Addition reaction of Br_2 with propene)**

b.  $CH_3-CH(OH)-CH_3 \xrightarrow{\text{oxidation}} CH_3-C(=O)-CH_3$

Oxidation of 2-propanol yields acetone (2-propanone).

c. 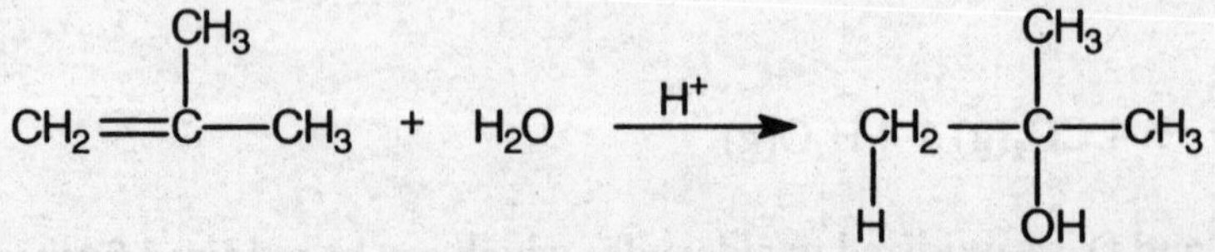 $CH_2{=}C(CH_3)-CH_3 + H_2O \xrightarrow{H^+} CH_2(H)-C(CH_3)(OH)-CH_3$

Addition of H_2O to 2-methylpropene would yield tert-butyl alcohol (2-methyl-2-propanol) as the major product.

d. $CH_3CH_2CH_2OH \xrightarrow{KMnO_4} CH_3CH_2C(=O){-}OH$

Oxidation of 1-propanol would eventually yield propanoic acid. Propanal is produced first in this reaction and is then oxidized to propanoic acid.

49.

(salicylic acid, COOH, O–H) + $HO{-}C(=O)CH_3$ ⟶ (COOH, $O{-}C(=O)CH_3$) + H_2O

acetylsalicylic acid (aspirin)

$CH_3{-}O{-}H$ + $HO{-}C(=O)$–(ring, OH) ⟶ $CH_3{-}O{-}C(=O)$–(ring, OH) + H_2O

methyl salicylate

Polymers

51. a. **addition polymer: Polymer that forms by adding monomer units together (usually by reacting double bonds). Teflon, polyvinyl chloride and polyethylene are examples of addition polymers.**

b. **condensation polymer: Polymer that forms when two monomers combine by eliminating a small molecule (usually H_2O or HCl). Nylon and Dacron are examples of condensation polymers.**

c. **copolymer: Polymer formed from more than one type of monomer. Nylon and Dacron are also copolymers.**

53. **The backbone of the polymer contains only carbon atoms, which indicates that Kel-F is an addition polymer. The smallest repeating unit of the polymer and the monomer used to produce this polymer are:**

$$\left(-\underset{Cl}{\overset{F}{C}}-\underset{F}{\overset{F}{C}}- \right)_n \qquad \underset{Cl}{\overset{F}{C}}=\underset{F}{\overset{F}{C}}$$

55. a.

$H_2N-C_6H_4-NH_2$ and $HO_2C-C_6H_4-CO_2H$

b. **Repeating unit:**

$$\left(-\overset{H}{\underset{|}{N}}-C_6H_4-\overset{H}{\underset{|}{N}}-\overset{O}{\overset{\|}{C}}-C_6H_4-\overset{O}{\overset{\|}{C}}- \right)_n$$

The two polymers differ in the substitution pattern on the benzene rings. The Kevlar chain is straighter and there is more efficient hydrogen bonding between Kevlar chains than between Nomex chains.

57.

$$\left(-\underset{\underset{\underset{O}{\|}}{C-OCH_3}}{\overset{CN}{\overset{|}{C}}}-CH_2-\underset{\underset{\underset{O}{\|}}{C-OCH_3}}{\overset{CN}{\overset{|}{C}}}-CH_2- \right)_n$$

Super glue is an addition polymer formed by reaction of the C=C bond in methyl cyanoacrylate.

59. **Divinylbenzene is a crosslinking agent. Divinylbenzene has two reactive double bonds which are both used when divinylbenzene inserts itself into two adjacent polymer chains. The chains cannot move past each other because the crosslinks bond adjacent polymer chains together making the polymer more rigid.**

61. a.

$$\left(-O-C_6H_4-\underset{CH_3}{\overset{CH_3}{C}}-C_6H_4-O-\overset{O}{\overset{\|}{C}}-O-C_6H_4-\underset{CH_3}{\overset{CH_3}{C}}-C_6H_4-O-\overset{O}{\overset{\|}{C}}- \right)_n$$

b. Condensation; HCl is eliminated when the polymer bonds form.

63. Polyacrylonitrile: $\left(CH_2 - CH(C{\equiv}N) \right)_n$

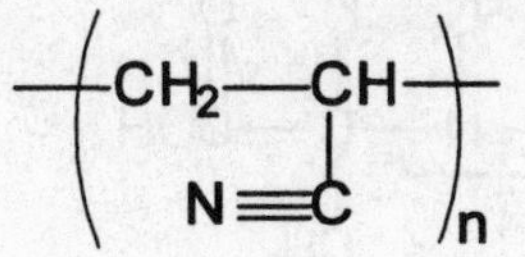

The CN triple bond is very strong and will not easily break in the combustion process. A likely combustion product is the toxic gas hydrogen cyanide, HCN(g).

65.

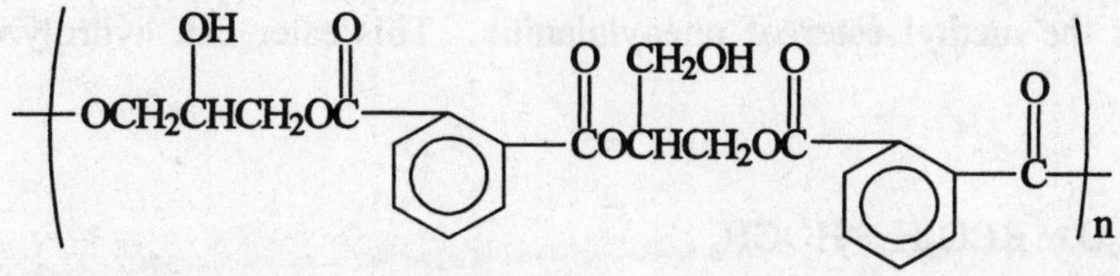

$$\left(-OCH_2CH(OH)CH_2OC(=O)-C_6H_4-C(=O)OCH(CH_2OH)CH_2OC(=O)-C_6H_4-C(=O)- \right)_n$$

Two linkages are possible with glycerol. A possible repeating unit with both types of linkages is shown above. With either linkage, there are free OH groups on the polymer chains. These can react with the acid groups of phthalic acid to form crosslinks between various polymer chains.

Natural Polymers

67. Denaturation changes the three-dimensional structure of a protein. Once the structure is affected, the function of the protein will also be affected.

69. a. Serine, tyrosine and threonine contain the -OH functional group in the R group.

b. Aspartic acid and glutamic acid contain the -COOH functional group in the R group.

c. An amine group has a nitrogen bonded to other carbon and/or hydrogen atoms. Histidine, lysine, arginine and tryptophan contain the amine functional group in the R group.

d. The amide functional group is:

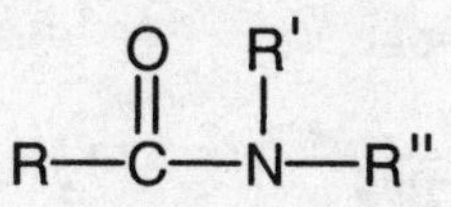

$$R-C(=O)-N(R')-R''$$

This functional group is formed when individual amino acids bond together to form the peptide linkage. Glutamine and asparagine have the amide functional group in the R group.

71 a. **Aspartic acid and phenylalanine make up aspartame.**

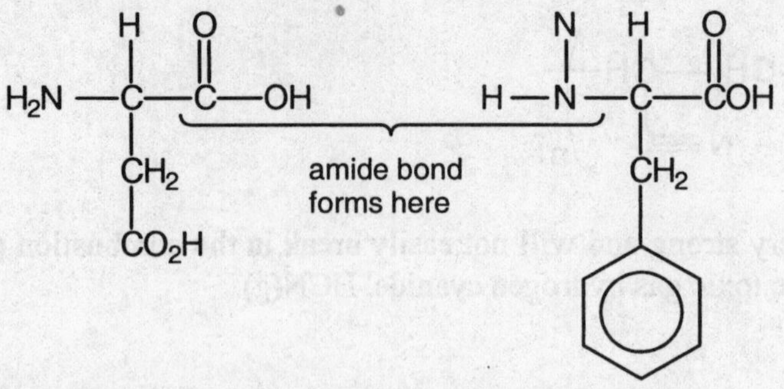

$H_2N-\overset{H}{\underset{CH_2CO_2H}{C}}-\overset{O}{\overset{\|}{C}}-OH \qquad H-\overset{N}{N}-\overset{H}{\underset{CH_2C_6H_5}{C}}-\overset{O}{\overset{\|}{C}}OH$

amide bond forms here

b. **Aspartame contains the methyl ester of phenylalanine. This ester can hydrolyze to form methanol:**

$$R{-}CO_2CH_3 + H_2O \rightleftharpoons RCO_2H + HOCH_3$$

73.

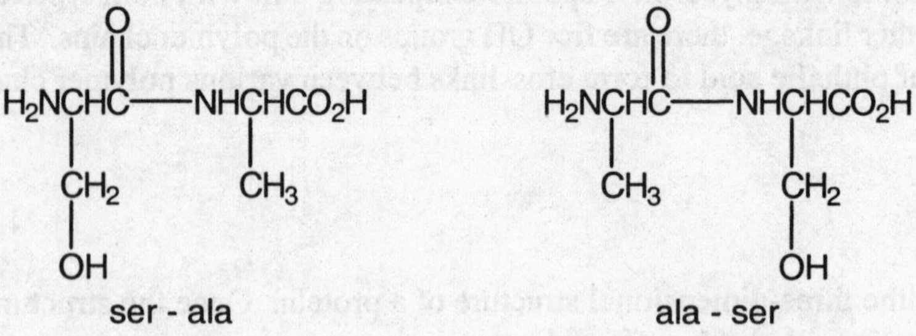

$H_2NCH(CH_2OH)C(=O){-}NHCH(CH_3)CO_2H$ ser - ala

$H_2NCH(CH_3)C(=O){-}NHCH(CH_2OH)CO_2H$ ala - ser

75. a. **Six tetrapeptides are possible. From NH_2 to CO_2H end:**

phe-phe-gly-gly, gly-gly-phe-phe, gly-phe-phe-gly,

phe-gly-gly-phe, phe-gly-phe-gly, gly-phe-gly-phe

b. **Twelve tetrapeptides are possible. From NH_2 to CO_2H end:**

phe-phe-gly-ala, phe-phe-ala-gly, phe-gly-phe-ala,

phe-gly-ala-phe, phe-ala-phe-gly, phe-ala-gly-phe,

gly-phe-phe-ala, gly-phe-ala-phe, gly-ala-phe-phe

ala-phe-phe-gly, ala-phe-gly-phe, ala-gly-phe-phe

77. a. **Covalent (forms a disulfide linkage)**

b. **Hydrogen bonding (need N-H or O-H bond in side chain)**

c. Ionic (need NH_2 group on side chain of one amino acid with CO_2H group on side chain of the other amino acid)

d. London dispersion (need amino acids with nonpolar R groups)

79. Glutamic acid: R = $-CH_2CH_2CO_2H$; Valine: R = $-CH(CH_3)_2$; A polar side chain is replaced by a nonpolar side chain. This could affect the tertiary structure of hemoglobin and the ability of hemoglobin to bind oxygen.

81. Glutamic acid:

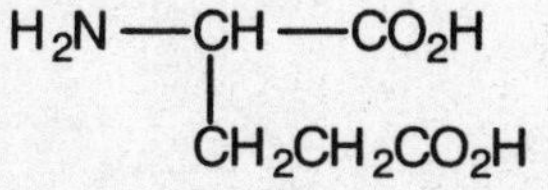

$H_2N-CH-CO_2H$

$CH_2CH_2CO_2H$

One of the two acidic protons in the carboxylic acid groups is lost to form MSG. Which proton is lost is impossible for you to predict.

Monosodium glutamate:

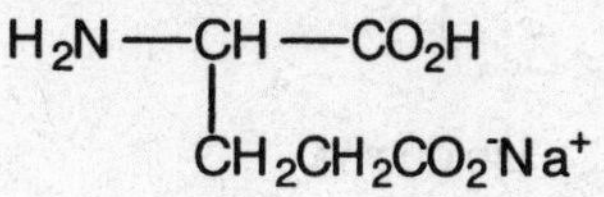

$H_2N-CH-CO_2H$

$CH_2CH_2CO_2^-Na^+$

In MSG, the acidic proton from the carboxylic acid in the R group is lost, allowing formation of the ionic compound.

83. See Figures 22.30 and 22.31 of the text for examples of the cyclization process.

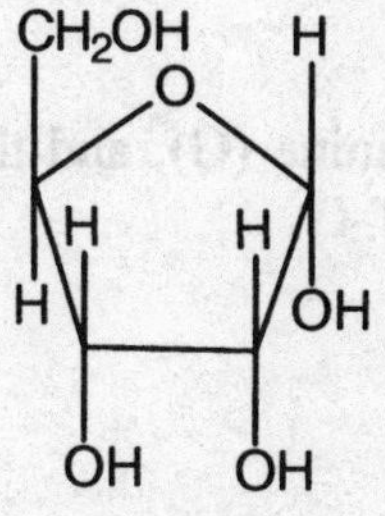

D-Ribose

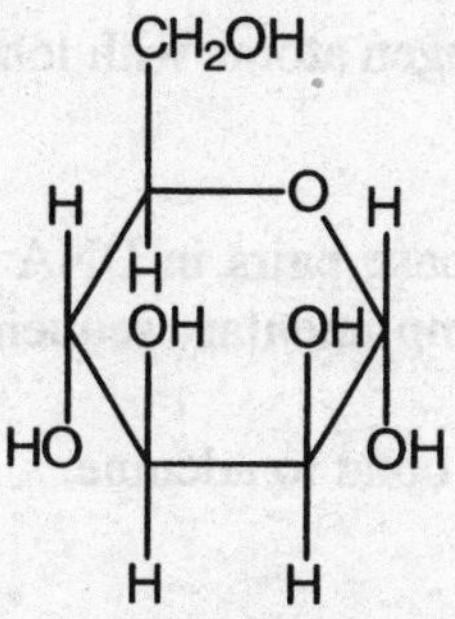

D-Mannose

85. The aldohexoses contain 6 carbons and the aldehyde functional group. Glucose, mannose and galactose are aldohexoses. Ribose and arabinose are aldopentoses since they contain 5 carbons with the aldehyde functional group. The ketohexose (6 carbons + ketone functional group) is fructose and the ketopentose (5 carbons + ketone functional group) is ribulose.

87. A disaccharide is a carbohydrate formed by bonding two monosaccharides (simple sugars) together. In sucrose, the simple sugars are glucose and fructose, and the bond formed between these two monosaccharides is called a glycoside linkage.

89. **The α and β forms of glucose differ in the orientation of a hydroxy group on one specific carbon in the cyclic forms (see Figure 22.31 of the text). Starch is a polymer composed of only α-D-glucose and cellulose is a polymer composed of only β-D-glucose.**

91. **A chiral carbon has four different groups attached to it. A compound with a chiral carbon is optically active. Isoleucine and threonine contain more than the one chiral carbon atom (see asterisks).**

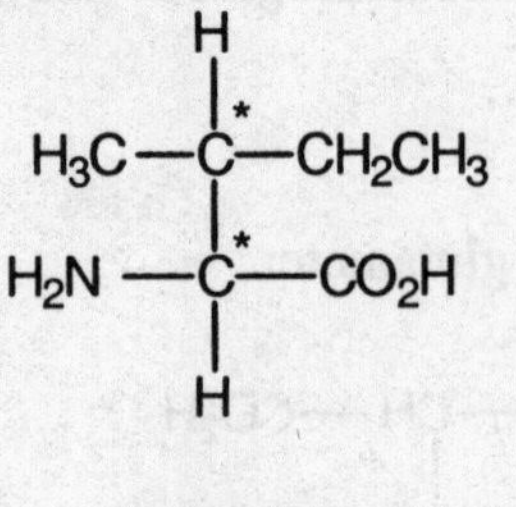

H
$H_3C—C^*—CH_2CH_3$
$H_2N—C^*—CO_2H$
H

isoleucine

H
$H_3C—C^*—OH$
$H_2N—C^*—CO_2H$
H

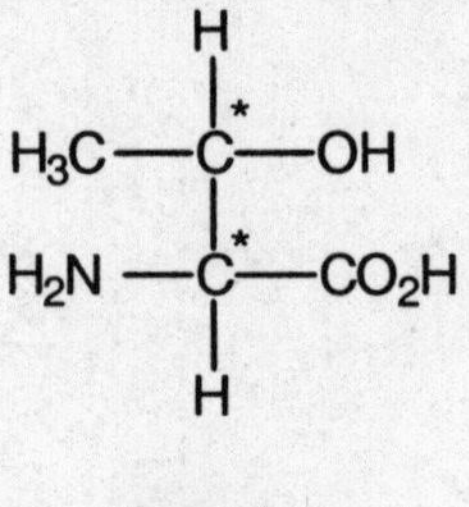

threonine

93. **Only one of the isomers is optically active. The chiral carbon in this optically active isomer is marked with an asterisk.**

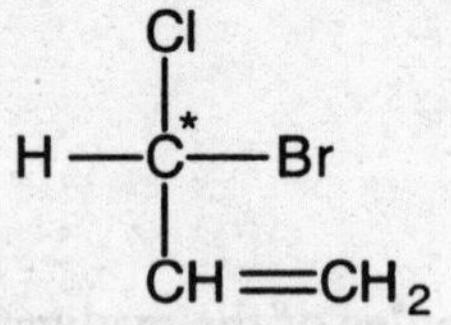

Cl
$H—C^*—Br$
$CH{=}CH_2$

95. **They all contain nitrogen atoms with lone pairs of electrons.**

97. **The complimentary base pairs in DNA are cytosine (C) and guanine (G), and thymine (T) and adenine (A). The complimentary sequence is: C-C-A-G-A-T-A-T-G**

99. **Uracil will hydrogen bond to adenine.**

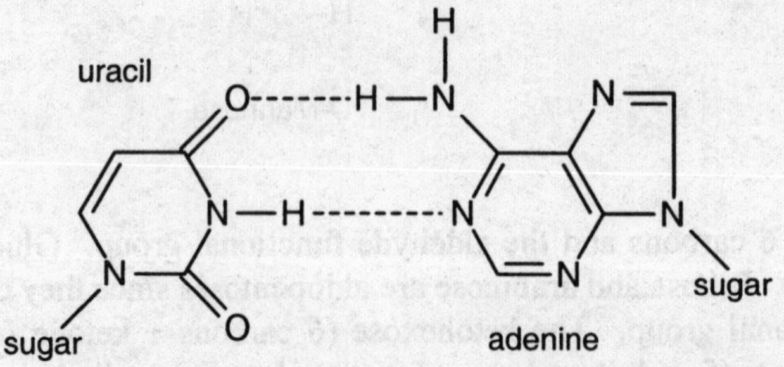

uracil
O·····H—N (H)
N—H·····N
N
O
sugar

adenine
N
N
N
N
sugar

101. **Base pair:**

RNA	DNA
A	T
G	C
C	G
U	A

a. **Glu:** CTT, CTC **Val:** CAA, CAG, CAT, CAC

Met: TAC **Trp:** ACC

Phe: AAA, AAG **Asp:** CTA, CTG

b. **DNA sequence for trp-glu-phe-met:**

ACC - CTT - AAA - TAC
or or
CTC AAG

c. **Due to glu and phe, there is a possibility of four different DNA sequences. They are:**

ACC - CTT - AAA - TAC or ACC - CTC - AAA - TAC or

ACC - CTT - AAG - TAC or ACC - CTC - AAG - TAC

d. T—A—C—C—T—G—A—A—G (met: T—A—C; asp: C—T—G; phe: A—A—G)

e. TAC - CTA - AAG; TAC - CTA - AAA; TAC - CTG - AAA

103. **A deletion may change the entire code for a protein, thus giving an entirely different sequence of amino acids. A substitution will change only one single amino acid in a protein.**

Additional Exercises

105. $CH_3CH_2CH_2CH_2CH_2CH_2CH_2COOH + OH^- \rightarrow CH_3\text{–}(CH_2)_6\text{–}COO^- + H_2O$; **Octanoic acid is more soluble in 1 *M* NaOH. Added OH^- will remove the acidic proton from octanoic acid, creating a charged species. As is the case with any substance with an overall charge, solubility in water increases. When morphine is reacted with H^+, the amine group is protonated creating a positive charge on morphine ($R_3N + H^+ \rightarrow R_3\overset{+}{N}H$). By treating morphine with HCl, an ionic compound results that is more soluble in water and in the bloodstream than is the neutral covalent form of morphine.**

107.

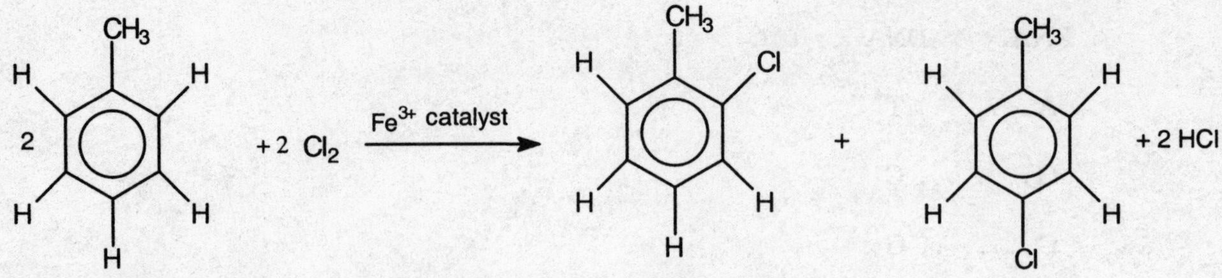

$$2\ C_6H_5CH_3 + 2\ Cl_2 \xrightarrow{Fe^{3+}\text{ catalyst}} \text{ortho-}ClC_6H_4CH_3 + \text{para-}ClC_6H_4CH_3 + 2\ HCl$$

ortho para

$$C_6H_5CH_3 + Cl_2 \xrightarrow{\text{light}} C_6H_5CH_2Cl + HCl$$

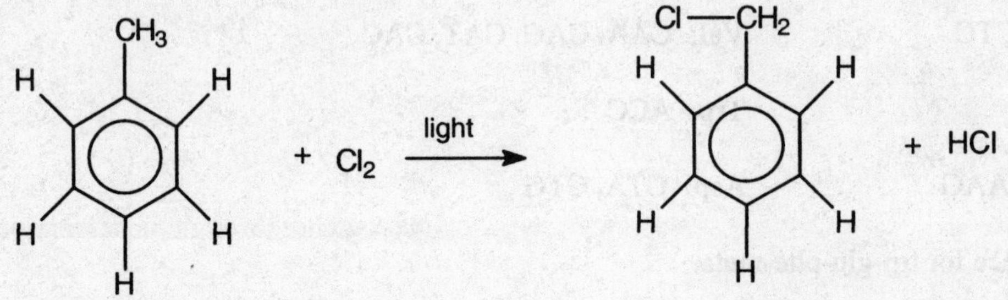

To substitute for the benzene ring hydrogens, an iron(III) catalyst must be present. Without this special iron catalyst, the benzene ring hydrogens are unreactive. To substitute for an alkane hydrogen, specific wavelengths of light must be present. For toluene, the light catalyzed reaction substitutes a chlorine for a hydrogen in the methyl group attached to the benzene ring.

109. $85.63\text{ g C} \times \frac{1\text{ mol C}}{12.01\text{ g C}} = 7.130\text{ mol C};\ 14.37\text{ g H} \times \frac{1\text{ mol H}}{1.008\text{ g H}} = 14.26\text{ mol H}$

Since the mol H to mol C ratio is 2:1 (14.26/7.130 = 2.000), the empirical formula is CH_2. The empirical formula mass ≈ 12 + 2(1) = 14. Since 4 × 14 = 56 puts the molar mass between 50 and 60, the molecular formula is C_4H_8.

The isomers of C_4H_8 are:

$CH_2{=}CHCH_2CH_3$ 1-butene

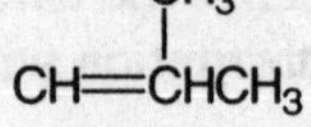

$CH_3CH{=}CHCH_3$ 2-butene

$CH{=}C(CH_3)CH_3$ 2-methyl-1-propene

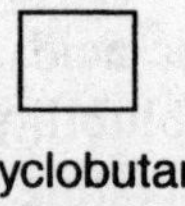

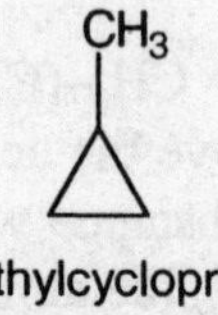

cyclobutane

methylcyclopropane

Only the alkenes will react with H_2O to produce alcohols, and only 1-butene will produce a secondary alcohol for the major product and a primary alcohol for the minor product.

$$CH_2{=}CHCH_2CH_3 + H_2O \longrightarrow \underset{\text{H}}{CH_2}{-}\underset{\text{OH}}{CHCH_2CH_3}$$

2° alcohol, major product

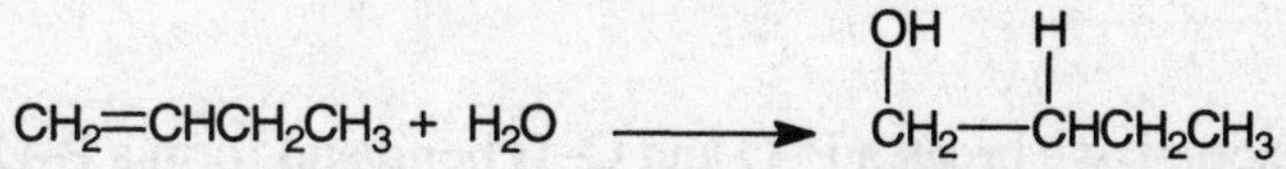

$$CH_2{=}CHCH_2CH_3 + H_2O \longrightarrow \underset{\text{OH}}{CH_2}{-}\underset{\text{H}}{CHCH_2CH_3}$$

1° alcohol, minor product

2-butene will produce only a secondary alcohol when reacted with H_2O, and 2-methyl-1-propene will produce a tertiary alcohol as the major product and a primary alcohol as the minor product.

111. a. **The bond angles in the ring are about 60°. VSEPR predicts bond angles close to 109°. The bonding electrons are closer together than they prefer resulting is strong electron-electron repulsions. Thus, ethylene oxide is unstable (reactive).**

b. **The ring opens up during polymerization; the monomers link together through the formation of O–C bonds.**

$$\left(O{-}CH_2CH_2{-}O{-}CH_2CH_2{-}O{-}CH_2CH_2 \right)_n$$

113.

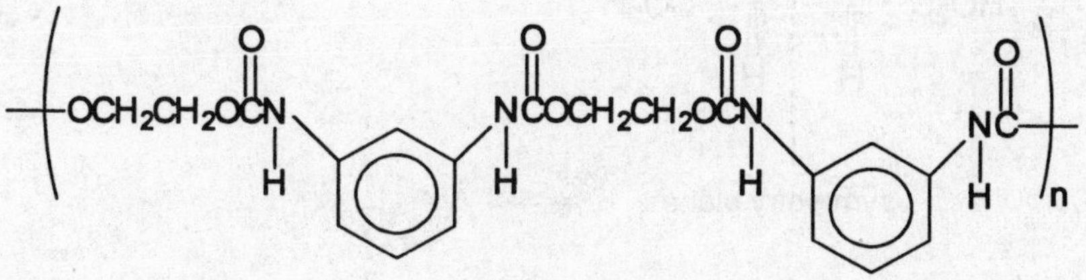

$$\left(-OCH_2CH_2O\overset{O}{\overset{\|}{C}}\underset{H}{N}-C_6H_4-\underset{H}{N}\overset{O}{\overset{\|}{C}}OCH_2CH_2O\overset{O}{\overset{\|}{C}}\underset{H}{N}-C_6H_4-\underset{H}{N}\overset{O}{\overset{\|}{C}}- \right)_n$$

115. **The structures, the types of intermolecular forces exerted and the boiling points for the compounds are:**

$CH_3CH_2CH_2\overset{O}{\overset{\|}{C}}OH$

butanoic acid, 164°C
LD + dipole + H-bonding

$CH_3CH_2CH_2CH_2CH_2OH$

1-pentanol, 137°C
LD + H-bonding

$CH_3CH_2CH_2CH_2\overset{O}{\overset{\|}{C}}H$

pentanal, 103°C
LD + dipole

$CH_3CH_2CH_2CH_2CH_2CH_3$

n-hexane, 69°C
LD only

All these compounds have about the same molar mass. Therefore, the London dispersion (LD) forces in each are about the same. The other types of forces determine the boiling point order. Since butanoic acid and 1-pentanol both exhibit hydrogen bonding (H-bonding) interactions, then these two compounds will have the two highest boiling points. Butanoic acid has the highest boiling point since it exhibits H-bonding along with dipole-dipole forces due to the polar C=O bond.

117. $\Delta G = \Delta H - T\Delta S$; For the reaction, we break a P–O and O–H bond and form a P–O and O–H bond. Thus, $\Delta H \approx 0$. $\Delta S < 0$, since 2 molecules are going to form one molecule (order increases). Thus, $\Delta G > 0$ and the reaction is not spontaneous.

119. a. Even though this form of tartaric acid contains 2 chiral carbon atoms (see asterisks in the following structure), the mirror image of this form of tartaric acid is superimposable. Therefore, it is not optically active. One way to identify optical activity in molecules with two or more chiral carbon atoms is to look for a plane of symmetry in the molecule. If a molecule has a plane of symmetry, then it is never optically active. A plane of symmetry is a plane that bisects the molecule where one side exactly reflects on the other side.

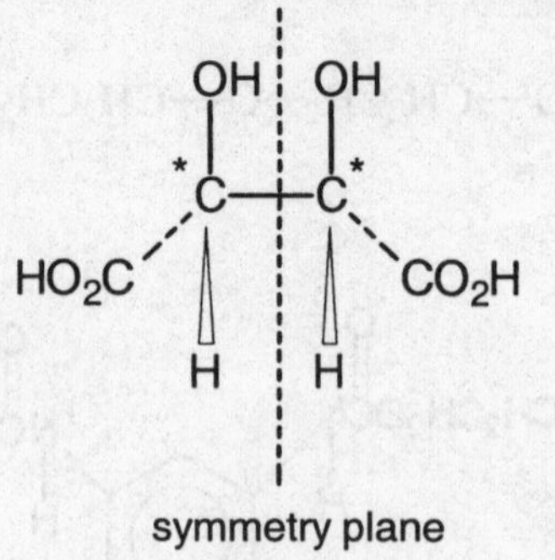

OH OH
C* C*
HO_2C CO_2H
H H
symmetry plane

b. The optically active forms of tartaric acid have no plane of symmetry. The structures of the optically active forms of tartaric acid are:

OH OH | OH OH
C—C | C—C
H CO_2H | HO_2C H
CO_2H H | H CO_2H
mirror

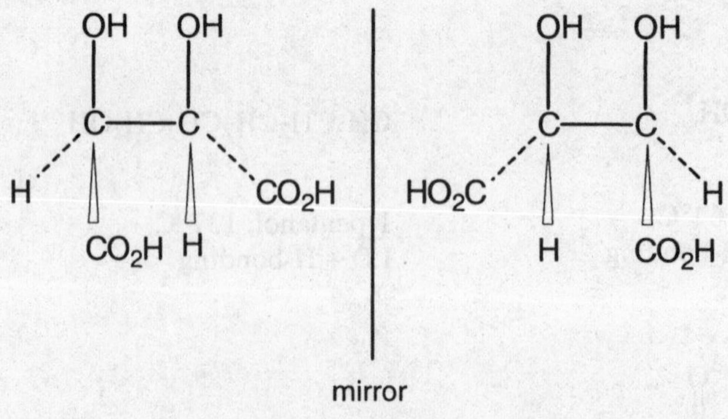

These two forms of tartaric acid are nonsuperimposable.

121. **For the reaction:**

$$^{+}H_3NCH_2CO_2H \rightleftharpoons 2\,H^+ + H_2NCH_2CO_2^-\quad K_{eq} = 7.3 \times 10^{-13} = K_a(\text{-}CO_2H) \times K_a(\text{-}NH_3^+)$$

$$7.3 \times 10^{-13} = \frac{[H^+]^2[H_2NCH_2CO_2^-]}{[^{+}H_3NCH_2CO_2H]} = [H^+]^2,\; [H^+] = (7.3 \times 10^{-13})^{1/2}$$

$[H^+] = 8.5 \times 10^{-7}$; pH = -log $[H^+]$ = 6.07 = isoelectric point

123. a.

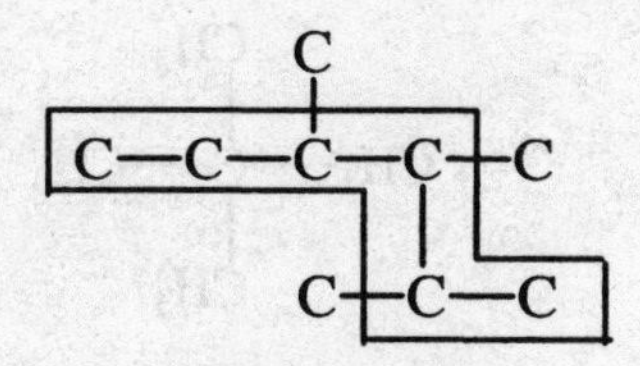

2,3,5,6-tetramethyloctane

b.

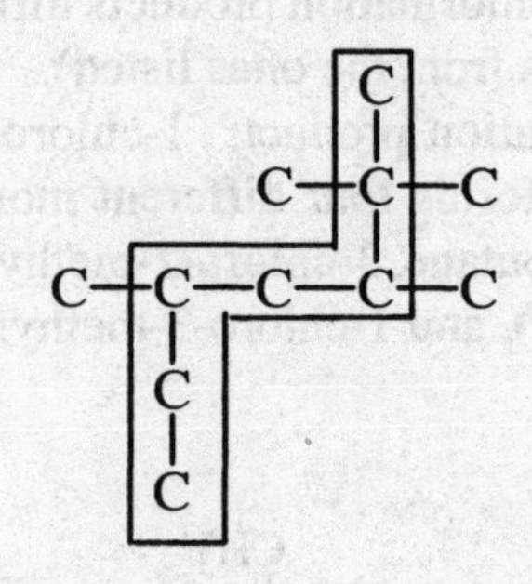

2,2,3,5-tetramethylheptane

c.

2,3,4-trimethylhexane

d.

3-methyl-1-pentyne

125. In nylon, hydrogen bonding interactions occur due to the presence of N–H bonds in the polymer. For a given polymer chain length, there are more N–H groups in Nylon-46 as compared to Nylon-6. Hence, Nylon-46 forms a stronger polymer compared to Nylon-6 due to the increased hydrogen bonding interactions.

Challenge Problems

127. a. The three structural isomers of C_5H_{12} are:

$CH_3CH_2CH_2CH_2CH_3$ — n-pentane

$CH_3CHCH_2CH_3$ with CH_3 on C-2 — 2-methylbutane

CH_3–C(CH_3)(CH_3)–CH_3 — 2,2-dimethylpropane

n-pentane will form three different monochlorination products: 1-chloropentane, 2-chloropentane and 3-chloropentane (the other possible monochlorination products differ by a simple rotation of the molecule; they are not different products from the ones listed). 2-2,dimethylpropane will only form one monochlorination product: 1-chloro-2,2-dimethylpropane. 2-methylbutane is the isomer of C_5H_{12} that forms four different monochlorination products: 1-chloro-2-methylbutane, 2-chloro-2-methylbutane, 3-chloro-2-methylbutane (or we could name this compound as 2-chloro-3-methylbutane), and 1-chloro-3-methylbutane.

b. The isomers of C_4H_8 are:

$CH_2=CHCH_2CH_3$ — 1-butene

$CH_3CH=CHCH_3$ — 2-butene

$CH_2=CCH_3$ with CH_3 on C-2 — 2-methyl-1-propene or 2-methylpropene

cyclobutane

methylcyclopropane

The cyclic structures will not react with H_2O; only the alkenes will add H_2O to the double bond. From Exercise 22.44, the major product of the reaction of 1-butene and H_2O is 2-butanol (a 2° alcohol). 2-butanol is also the major (and only) product when 2-butene and H_2O react. 2-methylpropene forms 2-methyl-2-propanol as the major product when reacted with H_2O; this product is a tertiary alcohol. Therefore, the C_4H_8 isomer is 2-methylpropene.

$$CH_2{=}\underset{}{\overset{CH_3}{\overset{|}{C}}}{-}CH_3 + HOH \longrightarrow CH_3{-}\underset{\underset{OH}{|}}{\overset{\overset{CH_3}{|}}{C}}{-}CH_3$$

2-methyl-2-propanol (a 3° alcohol, 3 R groups)

c. The structure of 1-chloro-1-methylcyclohexane is:

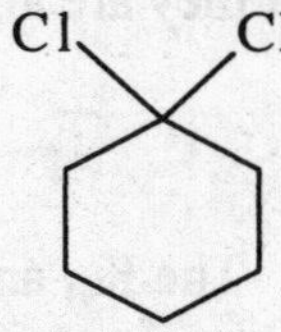

Cl CH$_3$

The addition reaction of HCl with an alkene is a likely choice for this reaction (see Exercise 22.44). The two isomers of C_7H_{12} that produce 1-chloro-1-methylcyclohexane as the major product are:

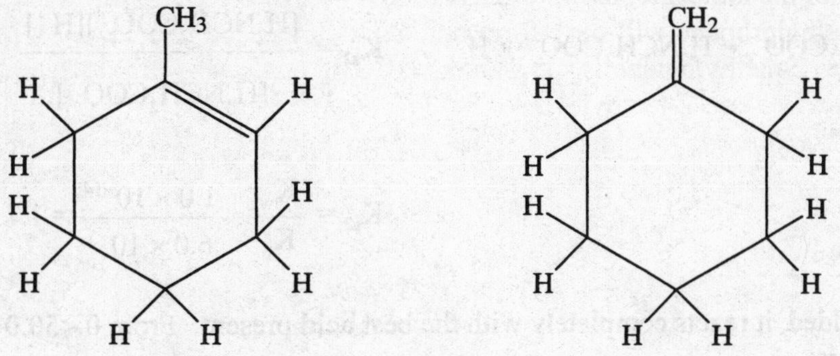

CH$_3$ H H H H H H H H H

CH$_2$ H H H H H H H H H H

d. Working backwards, 2° alcohols produce ketones when they are oxidized (1° alcohols produce aldehydes, then carboxylic acids). The easiest way to produce the 2° alcohol from a hydrocarbon is to add H_2O to an alkene. The alkene reacted is 1-propene (or propene).

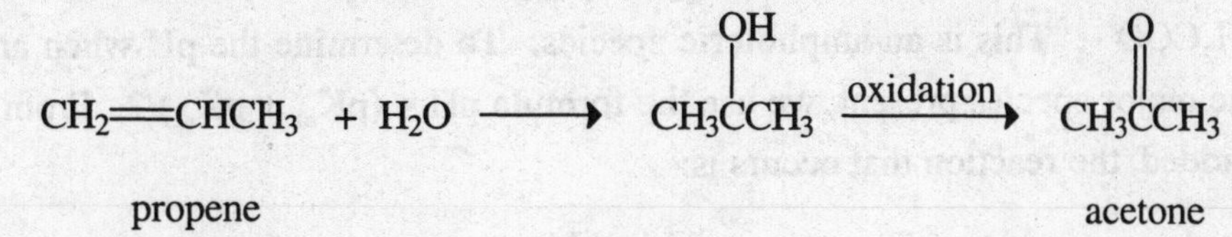

$$\underset{\text{propene}}{CH_2{=}CHCH_3} + H_2O \longrightarrow CH_3\overset{\overset{OH}{|}}{C}CH_3 \xrightarrow{\text{oxidation}} \underset{\text{acetone}}{CH_3\overset{\overset{O}{\|}}{C}CH_3}$$

e. The $C_5H_{12}O$ formula has too many hydrogens to be anything other than an alcohol (or an unreactive ether). 1° alcohols are first oxidized to aldehydes, then to carboxylic acids. Therefore, we want a 1° alcohol. The 1° alcohols with formula $C_5H_{12}O$ are:

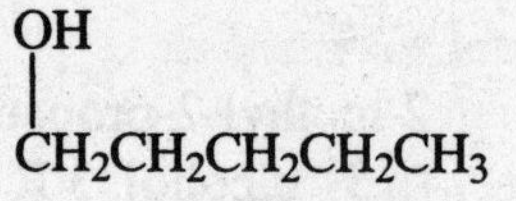

$CH_2(OH)CH_2CH_2CH_2CH_3$	$CH_2(OH)CH(CH_3)CH_2CH_3$	$CH_3CH(CH_3)CH_2CH_2(OH)$	$CH_2(OH)C(CH_3)_2CH_3$
1-pentanol	2-methyl-1-butanol	3-methyl-1-butanol	2,2-dimethyl-1-propanol

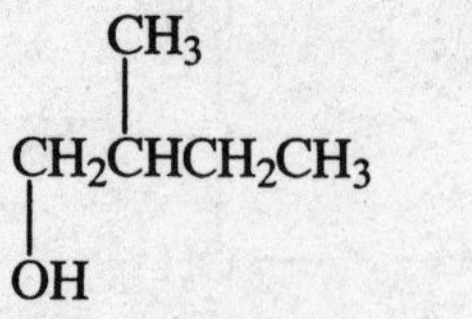

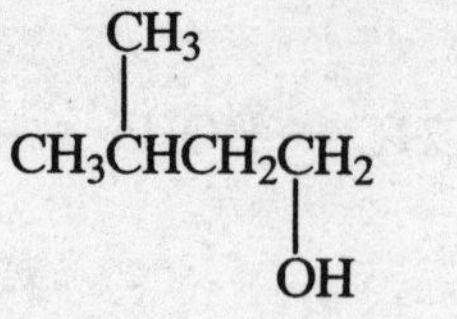

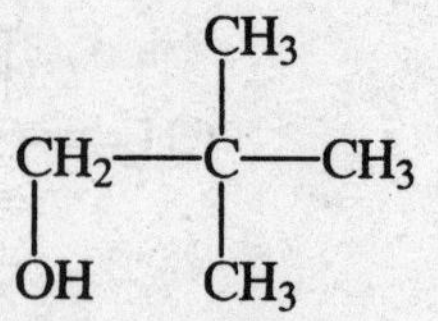

There are other alcohols with formula $C_5H_{12}O$, but they are all 2° or 3° alcohols, which do not produce carboxylic acids when oxidized.

129. **Treat this problem like a diprotic acid (H_2A) titration. The K_{a1} and K_{a2} reactions are:**

$$H_3\overset{+}{N}CH_2COOH \rightleftharpoons H_3\overset{+}{N}CH_2COO^- + H^+ \qquad K_{a1} = \frac{[H_3\overset{+}{N}CH_2COO^-][H^+]}{[H_3\overset{+}{N}CH_2COOH]} = 4.3 \times 10^{-3}$$

$$H_3\overset{+}{N}CH_2COO^- \rightleftharpoons H_2NCH_2COO^- + H^+ \qquad K_{a2} = \frac{[H_2NCH_2COO^-][H^+]}{[H_3\overset{+}{N}CH_2COO^-]}$$

$$K_{a2} = \frac{K_w}{K_b} = \frac{1.0 \times 10^{-14}}{6.0 \times 10^{-5}} = 1.7 \times 10^{-10}$$

As OH^- is added, it reacts completely with the best acid present. From 0 - 50.0 mL of OH^- added, the reaction is:

$$H_3\overset{+}{N}CH_2COOH + OH^- \rightarrow H_3\overset{+}{N}CH_2COO^- + H_2O$$

At 50.0 mL OH^- added (the first equivalence point), all of the $H_3\overset{+}{N}CH_2COOH$ has been converted into $H_3\overset{+}{N}CH_2COO^-$. This is an amphoteric species. To determine the pH when an amphoteric species is the major species present, we use the formula pH = (pK_{a1} + pK_{a2})/2. From 50.1 - 100.0 mL of OH^- added, the reaction that occurs is:

$$H_3\overset{+}{N}CH_2COO^- + OH^- \rightarrow H_2NCH_2COO^- + H_2O$$

100.0 mL of OH^- added represents the second equivalence point where $H_2NCH_2COO^-$ is the major amino acid species present.

a. **25.0 mL of OH^- added represents the first halfway point to equivalence. Here, $[H_3\overset{+}{N}CH_2COOH]$ = $[H_3\overset{+}{N}CH_2COO^-]$. This is a buffer solution where pH = pK_{a1}. At 25.0 mL OH^- added:**

$$pH = pK_{a1} = -\log 4.3 \times 10^{-3} = 2.37$$

50.0 mL of OH^- added represents the first equivalence point. Here, $H_3\overset{+}{N}CH_2COO^-$ is the major amino acid species present. This is amphoteric species. At 50.0 mL OH^- added:

$$pH = \frac{pK_{a1} + pK_{a2}}{2} = \frac{2.37 - \log 1.7 \times 10^{-10}}{2} = \frac{2.37 + 9.77}{2} = 6.07$$

75.0 mL of OH^- added represents the second halfway point to equivalence. Here, $[H_3\overset{+}{N}CH_2COO^-] = [H_2NCH_2COO^-]$ and pH = pK_{a2}. At 75.0 mL OH^- added:

$$pH = pK_{a2} = -\log 1.7 \times 10^{-10} = 9.77$$

b.

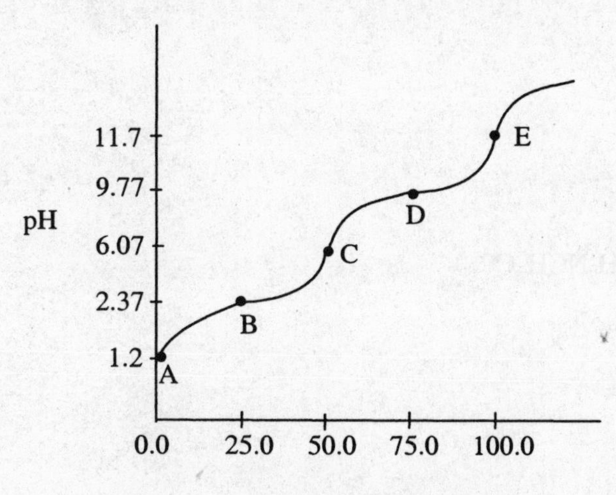

The major amino acid species present are:

point A (0.0 mL OH^-): $H_3\overset{+}{N}CH_2COOH$

point B (25.0 mL OH^-): $H_3\overset{+}{N}CH_2COOH$ **and** $H_3\overset{+}{N}CH_2COO^-$

point C (50.0 mL OH^-): $H_3\overset{+}{N}CH_2COO^-$

point D (75.0 mL OH^-): $H_3\overset{+}{N}CH_2COO^-$ **and** $H_2NCH_2COO^-$

point E (100.0 mL OH^-): $H_2NCH_2COO^-$

c. The various charged amino acid species are:

$H_3\overset{+}{N}CH_2COOH$ **net charge = +1**

$H_3\overset{+}{N}CH_2COO^-$ **net charge = 0**

$H_2NCH_2COO^-$ **net charge = -1**

The net charge is zero at the pH when the major amino acid species present is the $H_3\overset{+}{N}CH_2COO^-$ form; this happens at the first equivalence point in our titration problem. From part a, this occurs at pH = 6.07 (the isoelectric point).

d. The net charge is +1/2 when $[H_3\overset{+}{N}CH_2COOH] = [H_3\overset{+}{N}CH_2COO^-]$; net charge = (+1 + 0)/2 = +1/2. This occurs at the first half-way point to equivalence where $pH = pK_{a1} = 2.37$. The net charge is -1/2 when $[H_3\overset{+}{N}CH_2COO^-] = [H_2NCH_2COO^-]$; net charge = (0 - 1)/2 = -1/2. This occurs at the second halfway point to equivalence where $pH = pK_{a2} = 9.77$.